VOLUME NINETY FIVE

ADVANCES IN BOTANICAL RESEARCH

Seaweeds Around the World: State of Art and Perspectives

ADVANCES IN BOTANICAL RESEARCH

Series Editor

Jean-Pierre Jacquot
Professor, Membre de l'Institut Universitaire de France, Unité Mixte de Recherche INRA, UHP 1136 "Interaction Arbres Microorganismes", Université de Lorraine, Faculté des Sciences, Vandoeuvre, France

VOLUME NINETY FIVE

ADVANCES IN
BOTANICAL RESEARCH

Seaweeds Around the World: State of Art and Perspectives

Edited by

NATHALIE BOURGOUGNON

Laboratoire de Biotechnologie et Chimie Marines
Université Bretagne Sud, France

ELSEVIER

ACADEMIC PRESS
An imprint of Elsevier

Academic Press is an imprint of Elsevier
125 London Wall, London EC2Y 5AS, United Kingdom
525 B Street, Suite 1650, San Diego, CA 92101, United States
50 Hampshire Street, 5th Floor, Cambridge, MA 02139, United States
The Boulevard, Langford Lane, Kidlington, Oxford OX5 1GB, United Kingdom

First edition 2020

ISBN: 978-0-08-102710-3
ISSN: 0065-2296

Cover image: Seaweed textile designed and manufactured by Violaine Buet
Photo credit Laurent Desmaret

Publisher: Zoe Kruze
Acquisitions Editor: Sam Mahfoudh
Editorial Project Manager: Peter J. Llewellyn
Production Project Manager: James Selvam
Cover Designer: Matthew Limbert

Typeset by TNQ Technologies

Contents

Contributors

Shareen A. Abdul Malik
Laboratoire de Biotechnologie et Chimie Marines, EA3884, UBS, IUEM, F-56000 Vannes, France

Serge Andréfouët
French National Research Institute for Sustainable Development, UMR 9220/250 ENTROPIE (IRD-UR-CNRS), Noumea, New Caledonia

Gilles Bedoux
Laboratoire de Biotechnologie et Chimie Marines, EA3884, UBS, IUEM, F-56000 Vannes, France

Achinta Bordoloi
Department of Process Engineering, Stellenbosch University, Stellenbosch, South Africa

Romain Boulho
Gilson Purification, Saint-Avé, France

Nathalie Bourgougnon
Laboratoire de Biotechnologie et Chimie Marines, EA3884, UBS, IUEM, F-56000 Vannes, France

Violaine Buet
Studio in Auray, Bretagne, France

Solène Connan
Univ Brest, CNRS, IRD, Ifremer, LEMAR, Plouzane, France

Alan T. Critchley
Verschuren Centre for Sustainability in Energy and the Environment, Cape Breton University, Sydney, NS, Canada

Gérald Culioli
Université de Toulon, MAPIEM, Toulon, France

Laurence Dartevelle
Sorbonne Université, CNRS, Integrative Biology of Marine Models (LBI2M), Station Biologique de Roscoff (SBR), Roscoff, France

Sarah de J. Cantarino
Postgraduate Program in Marine Biotechnology, Almirante Paulo Moreira Institute of Sea Studies (Brazilian Navy - IEAPM), Fluminense Federal University (UFF), Arraial do Cabo, Rio de Janeiro, Brazil

Chrystèle Dufau
Laboratoire de Biotechnologie et Chimie Marines, EA3884, UBS, IUEM, F-56000 Vannes, France

Nolwen Le Duff
Sorbonne Université, CNRS, Integrative Biology of Marine Models (LBI2M), Station Biologique de Roscoff (SBR), Roscoff, France

Yolanda Freile-Pelegrín
CINVESTAV Marine Resources Department, Mérida, Yucatan, Mexico

Leslie Gager
Univ Brest, CNRS, IRD, Ifremer, LEMAR, Plouzane, France

Jose Q. Garcia Maldonado
CINVESTAV Marine Resources Department, Mérida, Yucatan, Mexico

Neill Goosen
Department of Process Engineering, Stellenbosch University, Stellenbosch, South Africa

Kevin Hardouin
Laboratoire de Biotechnologie et Chimie Marines, EA3884, UBS, IUEM, F-56000 Vannes, France

Leila Hayashi
Federal University of Santa Catarina, Aquaculture Department, Florianópolis, Santa Catarina, Brazil

Anicia Q. Hurtado
Integrated Services for the Development of Aquaculture and Fisheries (ISDA) Inc., Iloilo City, Philippines

Fanny Lalegerie
Univ Brest, CNRS, IRD, Ifremer, LEMAR, Plouzane, France

Marie Lang
Laboratoire de Biotechnologie et Chimie Marines, EA3884, UBS, IUEM, F-56000 Vannes, France

Cédric Leroux
Sorbonne Université, CNRS, FR2424, Station Biologique de Roscoff, Roscoff, France

Rexie Magdugo
Laboratoire de Biotechnologie et Chimie Marines, EA3884, UBS, IUEM, F-56000 Vannes, France

Lydiane Mattio
blue[c]weed, Locmaria-Plouzané, France; Department of Biological Sciences, University of Cape Town, Rondebosch, South Africa

Izabela Michalak
Department of Advanced Material Technologies, Faculty of Chemistry, Wrocław University of Science and Technology, Wrocław, Poland

Antoine De Ramon N'Yeurt
Pacific Centre for Environment and Sustainable Development, The University of the South Pacific, Suva, Fiji

Claude E. Payri
French National Research Institute for Sustainable Development, UMR 9220/250 ENTROPIE (IRD-UR-CNRS), Noumea, New Caledonia

Hugo Pliego-Cortés
Laboratoire de Biotechnologie et Chimie Marines, EA3884, UBS, IUEM, F-56000 Vannes, France

Maya Puspita
Laboratoire de Biotechnologie et Chimie Marines, EA3884, UBS, IUEM, F-56000 Vannes, France; Doctoral Program of Coastal Resources Management, Diponegoro University, Semarang, Jawa Tengah, Indonesia

Ocky Karna Radjasa
Doctoral Program of Coastal Resources Management, Diponegoro University, Semarang, Jawa Tengah, Indonesia

Pascal Riera
Sorbonne Université, CNRS, UMR7144, Station Biologique de Roscoff (SBR), Roscoff, France

Daniel Robledo
CINVESTAV Marine Resources Department, Mérida, Yucatan, Mexico

Nur Azmi Ratna Setyawidati
Agency of Research and Development for Marine and Fisheries, Ministry of Marine and Fisheries, Jakarta Utara, Jakarta, Indonesia; Laboratoire des Sciences de l'environnement Marin (LEMAR), UMR 6539 CNRS UBO Ifremer IRD, Institut Universitaire Européen de la Mer (IUEM), Université de Bretagne Occidentale (UBO), Plouzané, France

Valérie Stiger-Pouvreau
Univ Brest, CNRS, IRD, Ifremer, LEMAR, Plouzane, France

Tohei Theophilus
University of French Polynesia, UMR-EIO, LabEx CORAIL, BP6570, 98702 Faa'a, Tahiti, French Polynesia

François Thomas
Sorbonne Université, CNRS, Integrative Biology of Marine Models (LBI2M), Station Biologique de Roscoff (SBR), Roscoff, France

Olivier P. Thomas
Marine Biodiscovery, School of Chemistry and Ryan Institute, National University of Ireland Galway, University Road, Galway, Ireland

Laurent Vandanjon
Laboratoire de Biotechnologie et Chimie Marines, EA3884, UBS, IUEM, F-56000 Vannes, France

Christophe Vieira
Phycology Research Group and Centre for Molecular Phylogenetics and Evolution, Ghent University, Krijgslaan 281 (S8), Ghent, Belgium

Ita Widowati
Doctoral Program of Coastal Resources Management, Diponegoro University, Semarang, Jawa Tengah, Indonesia

Isuru Wijesekara
Department of Food Science & Technology, Faculty of Applied Sciences, University of Sri Jayewardenepura, Gangodawila, Nugegoda, Sri Lanka

Mayalen Zubia
University of French Polynesia, UMR-EIO, LabEx CORAIL, BP6570, 98702 Faa'a, Tahiti, French Polynesia

Foreword

The sea is the source of life and its resources still serve as a food reservoir for humanity. Very early, the island and coastal populations drew from this rich and varied environment the conditions of their life and sometimes even their survival. Algae, marine plants par excellence, have been consumed since prehistoric times by the inhabitants of the Korean peninsula and the Japanese archipelago. Kombu, Hiziki Wakame or Nori are consumed in a traditional way by these populations and this since time immemorial. The Chinese, the people of the Philippines, Malaysia, Vietnam and the inhabitants of the Marquesas Islands consume many species of red or green algae such as *Caulerpa lentillifera* nicknamed "sea caviar". The West, which is lagging behind in the use of seaweed as sea vegetables, is developing food algae production lines for human consumption. In Europe, Norway and France are pioneers in this area. The use of seaweed as a source of protein for aquaculture-reared animals is a very promising area for research and development. Many species consumed today or exploited for the production of phycocolloids (alginates, agar, carrageenans) are mainly produced by seaweed farming (90% of world production) Knowledge of the life cycle of algae and their ecological characteristics is a decisive condition for the development and valorization of this marine resource, apart from the valorization of seaweed as sea vegetables (direct food recovery) or as a source of food additives (indirect food recovery), these marine plants have many molecules of interest in the therapeutic or biotechnological applications. Ecological sentinels of the quality of the environment, the algae are also good indicators of the disorders generated by the human activity.

The papers in this new book Advanced Botanical Research take stock of recent advances in research in the field of biology, ecology and algae recovery. New ways of valorization are discussed, such as the use of certain algal species in bioremediation or as a source of bioactive compounds. This book is written by recognized experts in the fields treated and appears as a new reference for all those interested in this marine plant resource.

Professor Joel Fleurence
MMS, Nantes, University, France

Preface

Marine macroalgae (seaweeds) are the most common organisms in aquatic environment and a very diverse group in terms of ecological, taxonomic, morphological, and biochemical aspects. The main representatives are marine red algae (Rhodophyta), brown algae (Phaeophyceae), and green algae (Chlorophyta), whose names are derived from the characteristic pigments phycoerythrin, fucoxanthin, and chlorophyll, respectively. Macroalgae are the foundation of algal forests of the most productive and diverse coastal marine ecosystems on the planet acting as a global carbon sink. These ecosystems provide nursery grounds and food for fish and invertebrates, coastline protection from erosion, carbon sequestration, and nutrient fixation. Today, due to their ecological role, the protection of kelp forests is a central subject of environmental concern.

In coastal zones, macroalgae spread over according to their resistance to several abiotic (light, temperature, salinity...) and biotic factors (grazing, fouling, pathogens...). To survive to these important variations, intertidal macroalgae develop physiological and biochemical resistance mechanisms such as the synthesis of special compounds, e.g. photoprotectors, osmolytes or anti-fouling molecules. The Chapters of Fanny Lalegerie et al. and Shareen A Abdul Malik et al. explore some adaptation/protection mechanisms of these algae. Tohei Theophilus et al. focuse particularly on the diversity, distribution and the ecological dynamics of Dictyotaceae species in the coral reef ecosystem and their response to various environmental factors in French Polynesia.

Because of their biodiversity and expansion in various living environments, macroalgae contain many interesting metabolites. They contain significant quantities of vitamins, minerals, dietary fibers, proteins, polysaccharides, and various functional polyphenols while nutrient contents can vary with species and, geographical location, season and temperature. Seaweeds are rich in nutrition elements of interest, which can represent up to 40% of their dry weight: wide diversity of minerals and trace elements: sodium, calcium, chlorine, magnesium, zinc, copper; high protein content, especially red algae with 40% of the dry matter; diversity of storage polysaccharides (starch, laminarin), low cellulose and hemicellulose contents and cell wall phycocolloids (alginates, carrageenan, agars, fucose-containing sulfated polysaccharides, ulvans); Lipids and more particularly polyunsaturated fatty acids and sterols with antioxidant properties and vitamins such as vitamins C or B12.

The usage of macroalgae as traditional food and complementary medicine was recorded in early Neolithic data of ten thou-sand years back. Being naturally enriched in key nutrients and in various health-promoting compounds, they are traditionally consumed in many Asian countries like China, Indonesia, Philippines, South Korea, North Korea, Japan and Malaysia for centuries. Recently, they have attained more reputation in western countries. Seaweeds promising candidates for the design of functional foods have become key foods in current nutritional practices (vegetarians, vegans, health-foods, etc.) and are increasingly widely consumed in the USA, South American and European countries. Global demand for seaweeds has been growing together with increases in usage beyond former traditional applications. According to FAO statistics of the most cultivated seaweed taxa, three were used mainly for hydrocolloid extraction: *Eucheuma* spp. and *Kappaphycus alvarezii* for carrageenans, and *Gracilaria* spp. for agar; *Saccharina japonica, Undaria pinnatifida, Pyropia* spp. and *Sargassum fusiforme* were most important in human food usage. The main producing countries were China, Indonesia and the Philippines, which were also those that cultivated the greatest diversity of seaweed species. In contrast, the European, Canadian and Latin American seaweed industries still rely on harvesting natural resources. Today, seaweed aquaculture techniques are diversifying. The chapters of Leila Hayashi et al. and AQ Hurtado et al. propose respectively how judicious application of knowledge based on the eco-physiological processes of main seaweed species from tropical and cold-water can assist the future developments and scale-up of the global seaweed industry cultivation advances being made in the Philippines. In South Africa, Mexico, Portugal, sustainable production of seaweed and seaweed-based products uses the Integrated Multi-Trophic Aquaculture (IMTA) concept, farming seaweeds in proximity to several species at different trophic levels, allowing the reduction of aquaculture wastes. These original aquaculture aspects to bioactive compounds production will be explained by Hugo Cortés-Pliego et al. and Izabela Michalak.

At present, the main directions in macroalgal biotechnology are biofuels, agricultural biostimulants for crop plants, probiotics for aquaculture, biomaterial, wastewater treatment, and biomedical applications of extracted compounds (polyphenols, mycosporins, polysaccharides). Production of molecules or extracts for direct human consumption or processed into food additives and nutraceuticals, feeds, fertilizers, biofuels, cosmetics and medicines involve innovative eco-friendly biotechnological processes. The main challenge, which determines the industrial utilization of algae potential, is the development of efficient extraction methods while preserving most of the intrinsic qualities of raw material. Several procedures exist for extracting and separate biologically active compounds from algae biomass. The primary challenge when extracting algal bioactive compounds will be to find a compromise between the cost of production of sufficient quantities and quality of compounds in the shortest timeframe, finding the optimum processing condition and meeting the principles of green chemistry and green technology. The chapters of Gilles Bedoux et al., Izabela Michalak, Achinta Bordoloi & Neill Goosen and Hugo Cortés-Pliego et al. will describe you to some bioactive compounds with biotechnological approaches. Maya Puspita et al. focuse possible challenges for sustainable development and bioprospecting of *Sargassum* species in Indonesia, Tohei Theophilus et al. review the active metabolites known from *Dictyotaceae* species and their potential for the development of biotechnological applications in French Polynesia. Violaine Buet approaches the seaweeds with a designer eye and propose a philosophical, mesological and aminist cosmology dialogue with this natural and living material.

Particularly interesting is the integral utilization of the raw material, following a biorefinery approach. In order to make use of all seaweed components requires a rational processing of the whole material, and the algal processing by-products according to the biorefinery concept to allow a complete utilization of biomass. These ideas will be developed in Izabela Michalak and Achinta Bordoloi & Neill Goosen chapters.

In recent years, researchers have also focused on the characterization of endophyte and epiphytic microbiomes of various macroalgal species in order to elucidate host-microbe interactions and understand the function of microbial communities. Although environmental and host-related factors crucially determine microbial composition, mutual microbial agents and symbionts often play a fundamental role in regulating many aspects of the host's physical condition, including ecophysiological and metabolic aspects. In particular, epiphytic bacterial communities of macroalgae play an important role in the process of protecting surfaces from biological fouling, pathogens and herbivores through the production of bioactive metabolites (Shareen A Abdul Malik et al.). To complete the chapter of S. A Abdul Malik et al, François Thomas et al propose a labeling protocol to understand and combine ecosystems trophic interactions and microbial activities towards macroalgal biomass degradation.

Increased physical disturbances and stress, resulting from changing environmental conditions, are known to affect the seaweed abundance, distribution and quality, the macroalgae-associated microbial composition and cause its structural, functional or behavioral changes. For all these reasons, it is indeed imperative to continue research in the fields of metabolism, physiology, alternative production strategies, relations between seaweed and environment, to see and understand the conditions ensuring a culture of quality, to highlight compounds of interest for food, cosmetics, agriculture, bioactives and to explore them as raw material for chemistry and bioenergy.

The authors are all recognized as experts in their respective areas. From Brazil to Poland, from South Africa to Mexico or from Polynesia to Indonesia, they enrich our knowledge about the incredible world of algae. I would like to thank all the contributors for their patience, enthusiasm, hard work and commitment and also to acknowledge the valuable contributions of Elsevier team (specials thanks to Peter Llewellyn), Dr. Erwan Ar Gall (University of Brest, France), Pr. Joël Boustie (Rennes 1 University), Dr. Gérald Culioli (Toulon University, France), Dr. Justine Dumay (Nantes University, France), Pr. Joël Fleurence (Nantes Universiy, France), Pr. Anicia Hurtada (ISDA, Philippines), Pr. Gilles Lepoint (Liège University, Belgium), Dr. Izabela Michalak (WURST University, Poland), Pr. Daniel Robledo (CINVESTAV, Mexico), Dr. Valérie Stiger (University of Brest, France), Dr. Mayalen Zubia (French Polynesia University) and some own authors who provided valuable inputs and support in reviewing this special edition of ABR.

I hope you will enjoy this textbook as much as we have enjoyed putting it together.

Professor Nathalie Bourgougnon
Laboratoire de Biotechnologie et Chimie Marines
Université Bretagne Sud, France

CHAPTER ONE

Isotopic labeling of cultured macroalgae and isolation of ^{13}C-labeled cell wall polysaccharides for trophic investigations

François Thomas[a,*], Nolwen Le Duff[a], Cédric Leroux[b], Laurence Dartevelle[a] and Pascal Riera[c]

[a]Sorbonne Université, CNRS, Integrative Biology of Marine Models (LBI2M), Station Biologique de Roscoff (SBR), Roscoff, France
[b]Sorbonne Université, CNRS, FR2424, Station Biologique de Roscoff, Roscoff, France
[c]Sorbonne Université, CNRS, UMR7144, Station Biologique de Roscoff (SBR), Roscoff, France
*Corresponding author: E mail: fthomas@sb-roscoff.fr

Contents

Abstract

Macroalgae are considered as major primary producers in coastal environments, acting as a global carbon sink. This abundant biomass contains up to 50% of storage or cell wall polysaccharides, which therefore represent a reservoir of organic matter for potential algal consumers. Tracking of natural isotopic abundance ($\delta^{13}C$ vs. $\delta^{15}N$) in macroalgae-colonized habitats such as kelp forests and rocky shores previously evidenced the importance of the algal resource to support local and adjacent trophic webs mostly *via* the microbial detrital pathway. However, such bulk isotopic measures of natural abundances cannot inform precisely on the microbial actors and processes at play for the degradation of selected algal compounds. To overcome these limitations, we developed a stable isotope labeling procedure for cultures of the brown alga *Laminaria*

Advances in Botanical Research, Volume 95
ISSN 0065-2296
https://doi.org/10.1016/bs.abr.2019.11.005

digitata and tested its ability to yield labeled polysaccharides. Sporophytes of *L. digitata* were grown in controlled conditions for three months in seawater regularly amended with ^{13}C-labeled sodium bicarbonate. Elemental analysis – isotope ratio mass spectrometry of algal specimens showed a significant enrichment after 10 days of treatment, reaching a maximum of At%^{13}C = 3.5321% after 80 days. Sequential polysaccharide extraction from this labeled algal biomass allowed retrieving both alginate and fucose-containing sulfated polysaccharide fractions with high isotopic enrichment (At%^{13}C = 3.6279% and 3.5868%, respectively). This labeling protocol opens the way for future studies combining coastal ecosystems trophic interactions and microbial activities toward macroalgal biomass degradation.

1. Introduction

The total biomass of macroalgae on Earth is estimated at 280 million tons (Carpenter & Liss, 2000), covering an area of ca. 3.5 million km^2 for a net primary production of up to 1500 TgC.year^{-1} (Krause-Jensen & Duarte, 2016). Macroalgae are thus considered as major primary producers in coastal areas, acting as a carbon sink and playing a crucial role in biodiversity and in global carbon flows (Dayton, 1985). The algal chemical composition differs drastically from that of plants and animals. It consists primarily of complex cell wall and storage polysaccharides (more than 50% of the dry weight), with additional lipids, proteins, pigments and phlorotannins (Deniaud-Bouët et al., 2014). In particular, macroalgae synthesize unique polysaccharides that are absent from other organisms and specific to each phylum, e.g sulfated agars and carrageenans in red algae, alginate and sulfated fucans in brown algae, and ulvan in green algae (Popper et al., 2011). This specific composition of the algal biomass can influence the processes mediating its trophic use.

In marine environments, the availability of organic matter derived from macrophytes for consumers depends on trophic mediation by associated microorganisms (Adin & Riera, 2003; Crosby, Newell, & Langdon, 1990; Langdon & Newell, 1990). Several studies highlighted that bacteria growing on marine phanerogam detritus can facilitate the transfer of carbon to consumers (Benner, Lay, K'nees, & Hodson, 1988; Tenore, 1977). Laboratory experiments showed that 6–60% of oyster C requirements could be met by utilization of detrital complexes comprising senescing tissues and associated bacteria (Crosby et al., 1990; Langdon & Newell, 1990). Previous studies have pointed out that only about 10% of the net macroalgal production is consumed directly by grazers (Mann, 1982; Pomeroy, 1980). This low direct grazing pressure on living macroalgal tissues is mostly due to the presence of recalcitrant components including lignocellulosic compounds, and polyphenols like phlorotannins in brown algae (Buchsbaum, Valiela, Swain, Dzierzeski, & Allen, 1991; Duggins & Eckman, 1997). On the other hand, about 90% of the net macroalgal production enter detritus food chains mostly *via* microbial degradation (Mann, 1982; Pomeroy, 1980). The pool of detritus derived from macroalgae can be predominant within the total suspended particulate organic matter and represents a readily available food source for many consumers inhabiting kelp forests or nearby areas (Duggins & Eckman, 1997; Kaehler, Pakhomov, Kalin, & Davis, 2006). This microbial reinjection of macroalgal biomass into marine food webs plays a quantitative role of prime importance, especially during winter when the phytoplankton production can be very low (Bustamante & Branch, 1996; Duggins, Simenstad, & Estes, 1989).

Stable isotopes have been successfully used to trace transformations of organic matter during trophic flows and discriminate among various primary producers as food sources in consumers' diets (Bustamante & Branch, 1996). Indeed, the isotopic composition of a consumer depends on the diet isotopic composition, and the isotopic fractionation during food processing. Many natural isotopic signatures have been determined for macroalgae during the last decades (Mercado, de los Santos, Lucas Pérez-Lloréns, & Vergara, 2009; Pinnegar & Polunin, 2000; Schaal, Riera, & Leroux, 2010) indicating a wide range of $\delta^{13}C$ values. In general, average $\delta^{13}C$ are more ^{13}C-enriched in Chlorophyta, intermediate in Phaeophyta and lowest in Rhodophyta (Maberly, Raven, & Johnston, 1992; Marconi, Giordano, & Raven, 2011; Mercado et al., 2009). In a kelp forest in western Brittany (France), $\delta^{13}C$ were more negative for red algae (−38.0 to −15.5‰) compared to brown algae (−24.9 to −13.5‰) (Leclerc et al., 2013). Significant differences among species in $\delta^{13}C$ and/or $\delta^{15}N$ were also observed for living macroalgae or strandline algal species (Adin & Riera, 2003; Dauby, 1989). The $\delta^{13}C$ of marine macroalgae depends largely on their ability to use HCO_3^- rather than dissolved CO_2 as carbon source (reviewed in Raven et al., 2002). Macroalgae displaying very low $\delta^{13}C$ likely use dissolved CO_2 as their main source of inorganic carbon, while $\delta^{13}C$ close to −10‰ corresponds to a preferential fixation of HCO_3^- during primary production. Other factors influencing $\delta^{13}C$ of marine macroalgae include the photosynthesis rates and the molecular composition of algal tissues. These distinct natural isotopic values can be used to assess the contribution of different macroalgal species as food sources for associated consumers. Analyses of stable isotope signatures in kelp forest trophic webs have evidenced the relevance of detrital material as a major food source for the associated benthic communities (Fredriksen, 2003). For example, in a kelp forest dominated by *Laminaria digitata*, filter-feeders derived 35%–50% of their diet from macroalgae-derived detritus (Schaal et al., 2010). Overall, approaches based on natural stable isotopes confirmed the importance of the microbial detrital pathway to re-inject macroalgal primary production into marine food webs. However, such bulk isotopic measurements of natural abundances cannot inform precisely on the specific contribution of selected algal molecules and on the microbial processes involved in their degradation. A thorough knowledge of these bacterial processes is essential because they directly condition the fate of macroalgal organic matter in marine ecosystems.

A powerful approach to overcome these limitations is to use isotopic enrichment experiments. Experiments based on isotopically labeled biomass have allowed to test specific hypotheses on the fate of distinct food sources and to quantify the trophic fluxes in a variety of marine ecosystems. Stable isotopes enrichments were carried out to selectively label various food sources such as benthic diatoms or bacteria and to determine their relative contribution to the diet of macroinvertebrates (Herman, Middelburg, Widdows, Lucas, & Heip, 2000; Middelburg et al., 2000). Similarly, the specific uptake of microalgae-derived carbon by the Polychaeta *Lanice conchilega* or the Echinodermata *Echinocardium cordatum* was quantified through *in situ* experiments using ^{13}C-labeled diatoms (Kamp & Witte, 2005). In addition, differential labeling experiments allowed to assess the preferential uptake of ^{15}N-labeled bacteria or ^{13}C-labeled diatoms by pelagic larvae and foraminiferans (Leroy et al., 2012; Moodley et al., 2000; Pascal, Dupuy, Richard, & Niquil, 2008). So far, stable isotope enrichment studies in marine food webs almost exclusively focused on trophic transfers within planktonic communities or between microorganisms (e.g. microalgae, bacteria) and benthic grazers. By contrast, such approaches are currently lagging behind for macroalgae. Few studies have implemented stable isotope labeling techniques on adult specimens of the red algae *Gracilaria* spp. and *Solieria chordalis* to assess their fate in benthic systems (Hardison, Canuel, Anderson, & Veuger, 2010; Legrand, Martin, Leroux, & Riera, 2018). However, they used the labeled tissues as bulk biomass and did not investigate the allocation of ^{13}C to specific algal compounds. Some high molecular weight compounds such as cell wall structural polysaccharides can have slower turnover time compared to soluble molecules (e.g. osmolytes, carbon storage compounds) (Brinkhuis, 1977; Macler, 1986) and thus require active growth and longer incubation times to incorporate significant amounts of stable isotopes. Given their high proportion in macroalgal tissues, it appears essential to ensure sufficient labeling of cell wall polysaccharides to fully decipher the fate of macroalgal biomass in marine food webs. In the present study, we developed an adapted protocol to enrich actively growing macroalgae in ^{13}C and obtain isotopically labeled cell wall polysaccharide fractions.

2. Material and methods

2.1 Materials

Natural seawater for algal cultures was collected in opaque containers from the Astan site (GPS coordinates 48°46′40″, −2°56′15″) offshore Roscoff (France), filtered on a 100 μm cartridge and autoclaved before use. ^{13}C-enriched sodium bicarbonate $NaH^{13}CO_3$ (99% ^{13}C) was purchased from Eurisotop (Saarbrücken, Germany). Unless otherwise stated, all other chemicals were from Sigma-Aldrich.

2.2 Algal culture procedures

L. digitata life cycle consists of microscopic haploid gametophyte phase, alternating with macroscopic diploid sporophytes. In this study, all experiments were done on the macroscopic diploid individuals. To obtain young laboratory-grown sporophytes, mature *L. digitata* sporophytes showing sorus reproductive tissues were harvested at the "Roches Duon" site (GPS coordinates 48°43′32″, −3°55′20″). Spores were released from sorus pieces in seawater for 3 h at 18°C with an irradiance of 20 μmol s^{-1} m^{-2} and further kept at 13°C at an irradiance of 5 μmol s^{-1} m^{-2} to allow the development of gametophytes and random crosses. The developing young *L. digitata* sporophytes (ca. 5 mm) were cultured in flasks containing 10 L of autoclaved seawater enriched with 1.5 mL of modified Provasoli solution (92 mM H_3BO_3, 1.5 mM $FeCl_3$, 4.9 mM $MnSO_4$, 0.4 mM $ZnSO_4$, 0.09 mM $CoSO_4$, 27 mM EDTA, 9 mM $(NH_4)_2Fe(SO_4)_2$, 824.6 mM $NaNO_3$, 46.3 mM $C_3H_7Na_{206}P$, 0.001 mM cyanocobalamine, 0.3 mM thiamine, 0.004 mM biotin, 413 mM Tris) (Starr & Zeikus, 1993), with continuous bubbling of 0.2 μm-filtered air in a culture cabinet at 13°C illuminated with daylight-type fluorescent lamps at an irradiance of 20 μmol s^{-1} m^{-2} for 12 h per day. Two separate batches were prepared (^{12}C-control and ^{13}C-enriched) and were regularly amended with 1 g L^{-1} solutions of either $NaHCO_3$ with natural isotopic composition (98.9 ^{12}C%) or labeled

$NaHCO_3$ (99 ^{13}C%) previously sterilized on 0.2 μm filters. To favor the growth of larger sporophytes, each batch was subdivided in separate 10L-flasks after 42 and 66 days of cultures. Culture media were renewed biweekly during the first month and weekly afterward. At each medium renewal, sporophytes were visually examined to eliminate individuals showing signs of necrosis or microbial contamination. All manipulations were performed under a sterile laminar flow hood with ethanol-sterilized instruments. At each sampling time (0, 10, 18, 28, 35, 52, 66, 95 days), three sporophytes were sampled from each batch and frozen at −20°C for isotopic analysis. After 95 days of cultures, the total algal biomass was collected on a strainer and excess water was removed by spinning for 1 min in a salad tosser. Algae were weighed to estimate the fresh mass, dried at 40°C in a ventilated oven for 2 days and stored at room temperature until polysaccharide extraction.

2.3 Polysaccharide extraction

Cell wall polysaccharides were retrieved sequentially using a modified chemical extraction (Deniaud-Bouët et al., 2014; Hogsett & Quatrano, 1975; Vauchel, Kaas, Arhaliass, Baron, & Legrand, 2008). Dried algae were pulverized in a MM200 mixer mill (Retsch) at 30 Hz for 90 s. Seven to ten grams of algal powder were homogenized in 300 mL of 70% ethanol and filtered on a 100 μm nylon mesh placed on a G3 glass filter funnel. The powder was repeatedly rinsed with sequential baths (250 mL, 10 min each) in 70% ethanol (3 L total), 80% ethanol (3 L total), 96% ethanol (3 L total), methanol/chloroform 50:50 (v/v, 2 L total) and acetone (0.5 L total) until colourless and finally dried overnight in a ventilated oven at 40°C to yield alcohol-insoluble residues (AIRs). AIRs were resuspended in 0.5 L of 2% (w/v) aqueous $CaCl_2$ and mixed for 4 h at 80°C, followed by centrifugation for 15 min at 10,000 rpm to separate the calcium-soluble supernatant that contains fucose-containing sulfated polysaccharides (FCSPs) from the calcium-insoluble pellet containing alginate. The calcium-insoluble fraction was resuspended in 0.5 L of 4% (w/v) aqueous Na_2CO_3 for 2 h at 80°C followed by centrifugation for 15 min at 10,000 rpm. The supernatant was adjusted to pH 2 with slow addition of 10 mL H_2SO_4 to eliminate carbonates and precipitate alginic acid, which was retrieved by pressing in a 100 μm nylon mesh. The precipitate was dissolved in 200 mL distilled water and neutralized with NaOH. FCSPs and alginate were retrieved from their respective fraction by precipitating with 6 vol of 96% ethanol. For recovery, alginate was pressed on a 100 μm nylon mesh and FCSPs were centrifuged for 20 min at 9000 rpm. Both fractions were lyophilized for long-term storage.

2.4 Isotopic analysis

Triplicate algal samples were dried at 50°C for two days in a ventilated oven, pulverized using a mortar and pestle and inserted in tin capsules (1–2 mg per sample). Carbon isotopic ratios were measured on an elemental analyser (Flash EA1112, ThermoScientific) coupled to an isotope ratio mass spectrometer (IRMS Delta plus, ThermoScientific) *via* a gas interface (Conflo III, ThermoScientific). To measure carbon quantities, expressed in micrograms, a calibration was performed for each run with a certified standard compound (acetanilide, Sylab), a correction for linearity using casein (Sylab) and a two-point linear normalization (Coplen et al., 2006; Paul, Skrzypek, & Fórizs, 2007) using the international standards IAEA-600 and IAEA-CH6. Data were expressed in the standard δ notation in ‰ according to the following equation:

$$\partial^{13}C = \left(\frac{R_{sample} - R_{reference}}{R_{reference}}\right) \times 1000$$

where $R = {}^{13}C/{}^{12}C$

These abundances were calculated in relation to the certified reference materials Vienna Pee Dee Belemnite-limestone (V-PDB), with $R_{reference} = 0.0112372$. The V-PDB was achieved using in-house casein standards, calibrated against IAEA-600 and IAEA-CH6 reference materials. The standard deviation of repeated measurements of $\delta^{13}C$ of nBS-19 was 0.10 ‰ versus V-PDB.

^{13}C atom percent were calculated as follows:

$$At\%^{13}C = \frac{R_{sample}}{R_{sample} + 1} \times 100$$

^{13}C uptake was calculated in two different ways. Cumulative ^{13}C uptakes ($cUp^{13}C$) were calculated for each time point (t) by comparing algae from ^{13}C-enriched batch to algae from control batch at the same time point, as follows:

$$cUp^{13}C_t = \frac{At\%^{13}C_{labeled,t} - At\%^{13}C_{control,t}}{100} \times C_{conc}$$

The first term of the equation represents the ^{13}C atom excess, where At$\%^{13}C_{labeled,t}$ and At$\%^{13}C_{control,t}$ are the ^{13}C atom percent values at time t calculated for algae from batches amended with labeled or natural $NaHCO_3$, respectively. C_{conc} is the carbon concentration in dry algal biomass measured by the elemental analyser.

In addition, time-resolved ^{13}C uptakes (tUp^{13}C) were calculated for each time point (t) by comparing algae from ^{13}C-enriched batch at time t to algae from the same batch at time t−1 and expressed as daily uptake, as follows:

$$tUp^{13}C_t = \frac{At\%^{13}C_{labeled,t} - At\%^{13}C_{labeled,t-1}}{100} \times C_{conc} \times \frac{1}{n_{days}}$$

The first term of the equation represents the ^{13}C atom excess compared to previous time, where At$\%^{13}C_{labeled,t}$ and At$\%^{13}C_{labeled,t-1}$ are the ^{13}C atom percent values at time t and t−1, respectively. C_{conc} is the carbon concentration in dry algal biomass measured by the elemental analyser, and n_{days} is the number of days between sampling times t and t−1.

The percentage of ^{13}C from sodium bicarbonate incorporated into algal biomass was calculated for the final time point as follows:

$^{13}C_{incorp\%} = \frac{M_{tot} \times cUp^{13}C_{tfinal}}{0.99 \times {}^{13}C_{bicarb}}$, where M_{tot} is the total dry algal biomass harvested at the end of the cultivation period, cUp$^{13}C_{tfinal}$ is the cumulative ^{13}C uptake after 95 days of culture, and $^{13}C_{bicarb}$ is the total amount of carbon from sodium bicarbonate added during the cultivation period.

3. Results

The protocol developed to obtain stable isotope-labeled brown algal biomass is summarized in Fig. 1. Labeling of cultured *L. digitata* sporophytes was tested using periodic additions of ^{13}C-labeled sodium bicarbonate into seawater, over a 3-months period. In parallel, a control batch of sporophytes from the same fecundation event was prepared with addition of natural abundance sodium bicarbonate.

In total, each batch was amended with 7.2 g of sodium bicarbonate, corresponding to 1.1 g of carbon. Sporophytes grew from 5 mm to ca. 5–10 cm in length in 3 months, yielding 123 g and 95 g wet biomass in the ^{12}C-control and ^{13}C-enriched batch, respectively. The water content

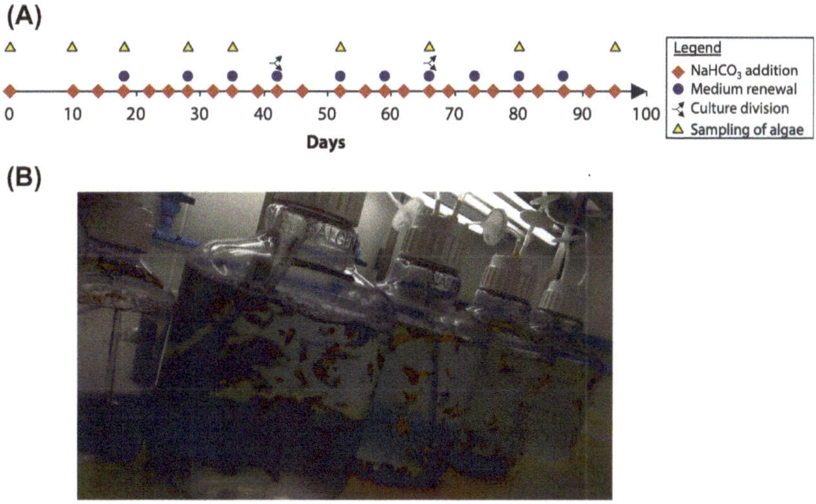

Fig. 1 (A) Schematic overview of the algal cultivation protocol for stable isotope labeling. (B) Photograph of the experimental setup for cultivation of *L. digitata* sporophytes.

estimated from the dry biomass was 85%. The ^{13}C isotopic composition and ^{13}C uptake of algal specimens collected at different time points are reported in Fig. 2.

At the beginning of the culture, juvenile sporophytes showed a $\delta^{13}C$ of -0.50 ± 2.96 ‰. Typical values obtained from wild adult populations of *L. digitata* in the vicinity of Roscoff range from $\delta^{13}C = -17$‰ to -12‰ (Schaal et al., 2010; Schaal, Riera, & Leroux, 2009). The higher ^{13}C-enrichment shown here might reflect differences due to the age of the algal tissues or the specific conditions of cultivation in the laboratory (e.g. amendments with Provasoli solution, air bubbling, etc.). In the control batch, stable isotope compositions remained stable over the 3 months cultivation period (Fig. 2A), ranging from At%$^{13}C = 1.0855$ to 1.1416, with an overall average of At%$^{13}C = 1.1116 \pm 0.0201$ ($n = 27$). By contrast, we observed a rapid enrichment of algal biomass in the ^{13}C-enriched batch (Fig. 2), already showing At%$^{13}C = 1.9735 \pm 0.0987$ after 10 days of culture. The ^{13}C content kept increasing until 80 days of culture, when it reached a plateau. The final isotopic composition of labeled algae after 95 days was At %$^{13}C = 3.3381 \pm 0.3490$% (Fig. 2A). Cumulative ^{13}C uptakes were calculated at each time point (Fig. 2B), and showed a final value of 5.59 mg $^{13}C.g^{-1}$ dry algal biomass after 95 days of culture. When extrapolated to the total dry mass of ^{13}C-enriched sporophytes harvested at the end of the

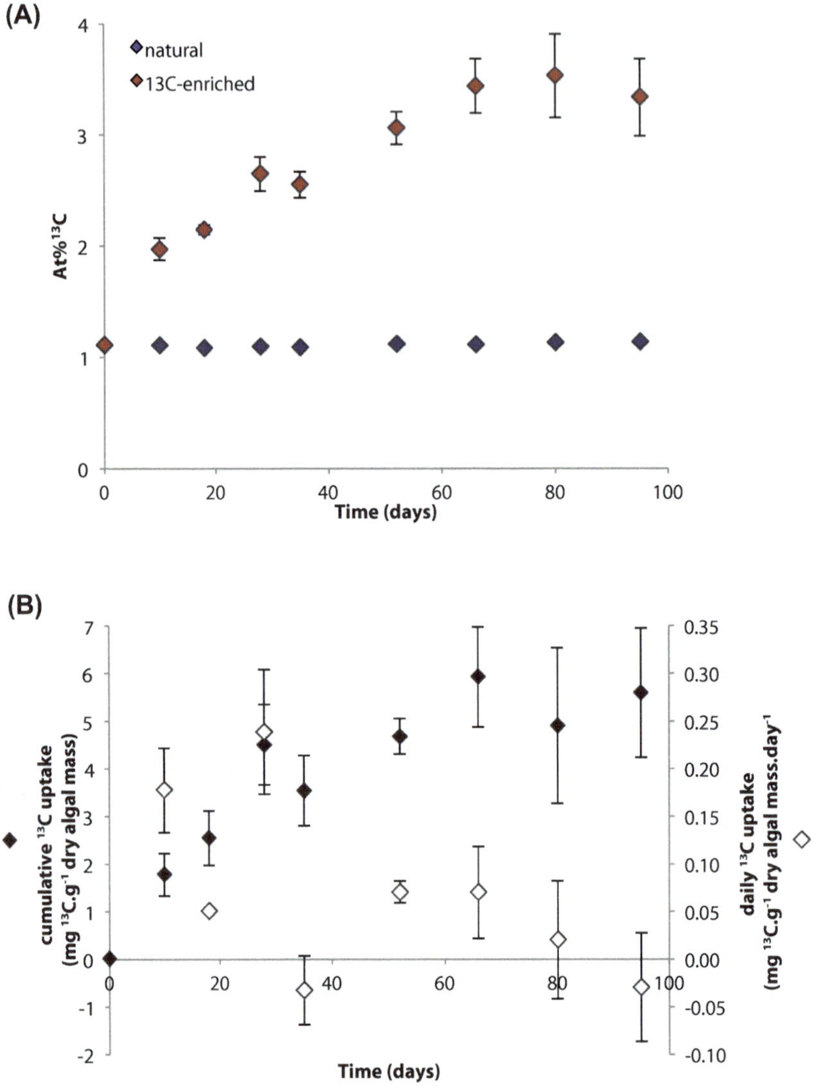

Fig. 2 (A) ^{13}C atom percent of algal specimens collected at different time points in the control (blue) and ^{13}C-enriched culture batches. Values are mean ± standard deviation (n = 3). (B) Uptake in ^{13}C calculated for algae in the ^{13}C-enriched culture batch, calculated either in a cumulative way compared to algae from control batch (black, left axis) or as a daily uptake calculated for each time point compared to previous time point (white, right axis).

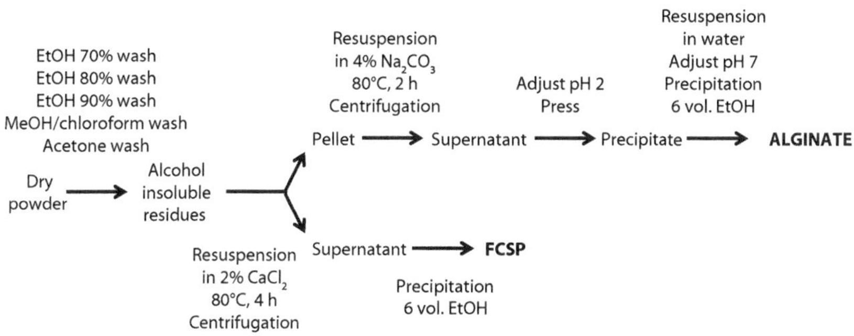

Fig. 3 Schematic overview of the polysaccharide extraction process. *EtOH*, ethanol; *MeOH*, methanol; *FCSP*, fucose-containing sulfated polysaccharides.

experiment (14.14 g), this corresponds to an incorporation of 7.26 ± 1.76% of ^{13}C atoms from sodium bicarbonate into algal biomass. Furthermore, we calculated daily ^{13}C uptakes between each sampling time point (Fig. 2B). This showed that most of the enrichment occurred during the first 40 days, before the plateau. Daily uptakes became close to zero after 80 days of culture.

A sequential chemical extraction was used to purify cell wall polysaccharides from control and ^{13}C-enriched sporophytes (Fig. 3) obtained after 95 days.

Alginate was retrieved as the calcium-insoluble fraction of alcohol-insoluble residues whereas the calcium-soluble fraction contained FCSPs, with final yields of ca. 20% and 7.5% of the dry weight, respectively (Table 1).

Isotopic analysis revealed strongly significant enrichment of polysaccharide fractions extracted from ^{13}C-enriched algal biomass compared to controls (Table 2). Alginate and FCSPs showed stable isotope content of At

Table 1 Yields of cell wall polysaccharide extraction from *L. digitata* cultured sporophytes from the ^{12}C-control and ^{13}C-enriched batch.

	^{12}C-control	^{13}C-enriched
Starting dry biomass	10 g	7.5 g
Alcohol-insoluble residues[a]	5.73 g (57.3%)	4.41 g (58.8%)
Alginate[a]	1.92 g (19.2%)	1.57 g (20.9%)
FCSP[a]	0.83 g (8.3%)	0.53 g (7.1%)

FCSP, fucose-containing sulfated polysaccharides.
[a]Values in brackets are yields based on starting dry biomass.

Table 2 Stable isotope ratios of polysaccharide fractions extracted from ^{12}C-control and ^{13}C-enriched *L. digitata* sporophytes. Values are mean ± standard deviation of technical triplicates.

	At%^{13}C	
Sample	**Alginate**	**FCSP**
^{12}C-control	1.1337 ± 0.0004	1.1454 ± 0.0008
^{13}C-enriched	3.6279 ± 0.0009	3.5868 ± 0.0064

%^{13}C = 3.6279 ± 0.0009% and 3.5868 ± 0.0064%, respectively, corresponding to ^{13}C uptakes of 7.02 ± 0.07 and 7.07 ± 0.03 mg $^{13}C.g^{-1}$ dry polysaccharide, respectively. This shows that carbon from sodium bicarbonate was allocated to cell wall synthesis. In particular, 26% and 9% of ^{13}C incorporated into algal biomass was allocated to alginate and FCSPs, respectively.

4. Discussion

Several studies have shown that it is possible to selectively label different food sources and follow their fate within food chains. The ^{13}C-enriched labeling can be focused on a source of organic matter considered in its entirety (e.g., whole macroalgae). Experiments based on the ^{13}C-labeled red macroalga *Gracilaria* spp (9 At%^{13}C) under decomposition in sediments revealed transfers of isotopic label to bacterial biomass through the ^{13}C enrichment of specific bacterial amino acids or fatty acids (Hardison et al., 2010). Recently, a dual stable isotope labeling experiment was performed on adult specimens of the red alga *Solieria chordalis* using both $H^{13}CO_3^-$ and $^{15}NH_4$ additions to investigate the uptake of whole algal tissues by the sea urchin *Psammechinus miliaris* under different temperature regimes (Legrand et al., 2018). However, stable isotope labeling studies considering macroalgae as a bulk biomass do not inform on (1) the labeling of different molecular fractions of the algae, (2) quantitative transfers of specific algal compounds to different consumers and, (3) the specialization of consumers toward the degradation of specific compounds. When considering the trophic fate of algal constituents, cell wall polysaccharides are of particular interest since they can represent up to 45% of the dry weight (Mabeau & Kloareg, 1987), are chemically distinct from those found in other primary producers (Popper et al., 2011), constitute one of the first physical barrier to be attacked by consumers and are generally recognized as more recalcitrant than other small (e.g. simple sugars) or high molecular weight compounds (e.g. proteins). A deeper understanding of the trophic impact of cell wall polysaccharides and on the microorganisms specialized in their turnover is therefore crucial to fully understand macroalgal biomass recycling.

In the present study, stable isotope labeling was performed on actively growing juvenile brown algae for 3 months, contrasting with the shorter incubation periods (1–2 weeks) previously used on red algae (Hardison et al., 2010; Legrand et al., 2018). This was required to ensure incorporation of ^{13}C-label into cell wall polysaccharides, which can have a long turnover time compared to smaller compounds (Macler, 1986). Indeed, seminal incubation studies of brown algae with $^{14}CO_2$ or $H^{14}CO_3^-$ showed that among sugar constituents, alginate has a slow rate of radioactive carbon incorporation, whereas mannitol is the most rapidly and strongly labeled (Bidwell, 1958; Hellebust & Haug, 1972; Yamaguchi, Ikawa, & Nisizawa, 1966). Isotopic ^{13}C enrichments obtained on brown algae in the present study are consistent with levels previously observed on sources characterized by high polysaccharide contents. Recent labeling experiments were performed on fresh leaves of terrestrial source (i.e. alders) with an isotopic ^{13}C enrichment of about 2.5 At%^{13}C to assess their assimilation by isopods in groundwater (Francois et al., 2016). A mean ^{15}N labeling of the seagrass *Posidonia oceanica* around 4.5 At%^{15}N was also obtained for a quantitative experimental study (Lepoint, Millet, Dauby, Gobert, & Bouquegneau, 2002). Our detailed time course of ^{13}C-incorporation showed that the extent of labeling reached a plateau at the end of the 3-months period and that most of the ^{13}C uptake occurred within 40 days. This suggests that longer incubations in the present experimental setup may not yield a greater incorporation. If even higher ^{13}C enrichments are needed for specific applications, additional ventilation with $^{13}CO_2$ could be tested.

The present results appear very promising for future experimental studies related to the fate and trophic transfers of macroalgal biomass in general and of cell wall polysaccharides in particular. The protocol developed here for the kelp *L. digitata* can be applied to other species of brown, red or green algae for which laboratory cultivation is available. Therefore, the present method complements the toolbox for quantitative trophic ecology approaches to trace ^{13}C-labeled macroalgal biomass within complex pelagic or benthic communities of marine ecosystems. Furthermore, this study is to our knowledge the first to report the purification of ^{13}C-labeled cell wall polysaccharide fractions containing alginate and FCSPs. Macroalgae are generally considered as a unique pure source in trophic models, although they in fact comprise diverse molecular compounds potentially at the base of different food chains due to their specific utilization by micro- and macro-organisms in aquatic ecosystems. By giving access to specific ^{13}C-labeled polysaccharide fractions, the present protocol will help improving the definition of these distinct trophic niches and paves the way for detailed investigations of the diversity and functional processes of uncultured specialized microorganisms responsible for the reinjection of algal polysaccharides into trophic chains *via* the detrital pathway. To this end, various stable-isotope-enabled microbial ecology techniques are available, such as Stable Isotope Probing (SIP) combined to high-throughput sequencing (Dumont & Murrell, 2005), Raman single-cell microspectroscopy (Wagner, 2009) or nano-scale chemical imaging by secondary ion mass spectrometry (nano-SIMS) (Musat, Foster, Vagner, Adam, & Kuypers, 2012). Such studies using ^{13}C-labeled polysaccharides as substrates would allow linking specific members of natural microbial communities to their function as degraders of the macroalgal biomass *in situ*, greatly enhancing our current knowledge on the carbon cycle in marine regions.

Acknowledgments

F.T. acknowledges support from the Center National de la Recherche Scientifique (CNRS) with regards to the "ALGOSIP" project and from the French Agence National de la Recherche (ANR) with regard to the "ALGAVOR" project (grant agreement ANR-18-CE02- 0001-01). All authors are also grateful to ANR for its support with regards to the investment expenditure program IDEALG (grant agreement No. ANR-10-BTBR-04). This work has benefited from the facilities of the Metabomer platform supported by the Biogenouest-Corsaire core facility.

References

Adin, R., & Riera, P. (2003). Preferential food source utilization among stranded macroalgae by *Talitrus saltator* (amphipod, *Talitridae*): A stable isotopes study in the northern coast of Brittany (France). *Estuarine, Coastal and Shelf Science, 56*(1), 91–98. https://doi.org/10.1016/S0272-7714(02)00124-5.

Benner, R., Lay, J., K'nees, E., & Hodson, R. E. (1988). Carbon conversion efficiency for bacterial growth on lignocellulose: Implications for detritus-based food webs. *Limnology & Oceanography, 33*(6, part2), 1514–1526. https://doi.org/10.4319/lo.1988.33.6part2.1514.

Bidwell, R. G. S. (1958). Photosynthesis and metabolism of marine algae: II. A survey of rates and products of photosynthesis in $C^{14}O_2$. *Canadian Journal of Botany, 36*(3), 337–349. https://doi.org/10.1139/b58-029.

Brinkhuis, B. H. (1977). Seasonal variations in salt-marsh macroalgae photosynthesis. I. *Ascophyllum nodosum* ecad *scorpioides*. *Marine Biology, 44*(2), 165–175. https://doi.org/10.1007/BF00386956.

Buchsbaum, R., Valiela, I., Swain, T., Dzierzeski, M., & Allen, S. (1991). Available and refractory nitrogen in detritus of coastal vascular plants and macroalgae. *Marine Ecology Progress Series, 72*(1/2), 131–143. https://doi.org/10.2307/24825436.

Bustamante, R. H., & Branch, G. M. (1996). The dependence of intertidal consumers on kelp-derived organic matter on the west coast of South Africa. *Journal of Experimental Marine Biology and Ecology, 196*, 1–28. https://doi.org/10.1016/0022-0981(95)00093-3.

Carpenter, L. J., & Liss, P. S. (2000). On temperate sources of bromoform and other reactive organic bromine gases. *Journal of Geophysical Research, 105*(D16), 20539–20547. https://doi.org/10.1029/2000JD900242.

Coplen, T. B., Brand, W. A., Gehre, M., Gröning, M., Meijer, H. A. J., Toman, B., et al. (2006). New guidelines for $\delta^{13}C$ measurements. *Analytical Chemistry, 78*(7), 2439–2441. https://doi.org/10.1021/ac052027c.

Crosby, M. P., Newell, R. I. E., & Langdon, C. J. (1990). Bacterial mediation in the utilization of carbon and nitrogen from detrital complexes by *Crassostrea virginica*. *Limnology & Oceanography, 35*(3), 625–639. https://doi.org/10.4319/lo.1990.35.3.0625.

Dauby, P. (1989). The stable carbon isotope ratios in benthic food webs of the Gulf of Calvi, Corsica. *Continental Shelf Research, 9*(2), 181–195.

Dayton, P. K. (1985). Ecology of kelp communities. *Annual Review of Ecology and Systematics, 16*(1), 215–245. https://doi.org/10.1146/annurev.es.16.110185.001243.

Deniaud-Bouët, E., Kervarec, N., Michel, G., Tonon, T., Kloareg, B., & Hervé, C. (2014). Chemical and enzymatic fractionation of cell walls from Fucales: Insights into the structure of the extracellular matrix of brown algae. *Annals of Botany, 114*(6), 1203–1216. https://doi.org/10.1093/aob/mcu096.

Duggins, D. O., & Eckman, J. E. (1997). Is kelp detritus a good food for suspension feeders? Effects of kelp species, age and secondary metabolites. *Marine Biology, 128*(3), 489–495. https://doi.org/10.1007/s002270050115.

Duggins, D. O., Simenstad, C. A., & Estes, J. A. (1989). Magnification of secondary production by kelp detritus in coastal marine ecosystems. *Science, 245*(4914), 170–173.

Dumont, M. G., & Murrell, J. C. (2005). Stable isotope probing — linking microbial identity to function. *Nature Reviews Microbiology, 3*(6), 499–504.

Francois, C. M., Mermillod-Blondin, F., Malard, F., Fourel, F., Lécuyer, C., Douady, C. J., et al. (2016). Trophic ecology of groundwater species reveals specialization in a low-productivity environment. *Functional Ecology, 30*(2), 262–273. https://doi.org/10.1111/1365-2435.12484.

Fredriksen, S. (2003). Food web studies in a Norwegian kelp forest based on stable isotope ($\delta^{13}C$ and $\delta^{15}N$) analysis. *Marine Ecology Progress Series, 260*, 71–81. https://doi.org/10.2307/24867089.

Hardison, A. K., Canuel, E. A., Anderson, I. C., & Veuger, B. (2010). Fate of macroalgae in benthic systems: Carbon and nitrogen cycling within the microbial community. *Marine Ecology Progress Series, 414*, 41–55. https://doi.org/10.3354/meps08720.

Hellebust, J. A., & Haug, A. (1972). Photosynthesis, translocation, and alginic acid synthesis in *Laminaria digitata* and *Laminaria hyperborea*. *Canadian Journal of Botany, 50*(1), 169–176. https://doi.org/10.1139/b72-022.

Herman, P. M. J., Middelburg, J. J., Widdows, J., Lucas, C. H., & Heip, C. H. R. (2000). Stable isotopes as trophic tracers: Combining field sampling and manipulative labelling of food resources for macrobenthos. *Marine Ecology Progress Series, 204*, 79–92. https://doi.org/10.2307/24863624.

Hogsett, W. E., & Quatrano, R. S. (1975). Isolation of polysaccharides sulfated during early embryogenesis in *Fucus*. *Plant Physiology, 55*, 25–29. https://doi.org/10.1104/pp.55.1.25.

Kaehler, S., Pakhomov, E. A., Kalin, R. M., & Davis, S. (2006). Trophic importance of kelp-derived suspended particulate matter in a through-flow sub-Antarctic system. *Marine Ecology Progress Series, 316*, 17–22. https://doi.org/10.3354/meps316017.

Kamp, A., & Witte, U. (2005). Processing of ^{13}C-labelled phytoplankton in a fine-grained sandy-shelf sediment (North Sea): Relative importance of different macrofauna species. *Marine Ecology Progress Series, 297*, 61–70. https://doi.org/10.3354/meps297061.

Krause-Jensen, D., & Duarte, C. M. (2016). Substantial role of macroalgae in marine carbon sequestration. *Nature Geoscience, 9*(10), 737–742. https://doi.org/10.1038/ngeo2790.

Langdon, C. J., & Newell, R. I. E. (1990). *Utilization of detritus and bacteria as food sources by two bivalve suspension-feeders, the oyster* Crassostrea virginica *and the mussel* Geukensia demissa. Marine Ecology Progress Series. Inter-Research Science Center. https://doi.org/10.2307/24842204.

Leclerc, J.-C., Riera, P., Leroux, C., Lévêque, L., Laurans, M., Schaal, G., et al. (2013). Trophic significance of kelps in kelp communities in Brittany (France) inferred from isotopic comparisons. *Marine Biology, 160*(12), 3249–3258. https://doi.org/10.1007/s00227-013-2306-5.

Legrand, E., Martin, S., Leroux, C., & Riera, P. (2018). Effect of temperature on an alga-grazer trophic transfer: A dual stable isotope (^{13}C, ^{15}N) labeling experiment. *Marine Ecology, 39*(2), 1–6. https://doi.org/10.1111/maec.12495.

Lepoint, G., Millet, S., Dauby, P., Gobert, S., & Bouquegneau, J. M. (2002). Annual nitrogen budget of the seagrass *Posidonia oceanica* as determined by in situ uptake experiments. *Marine Ecology Progress Series, 237*, 87–96. https://doi.org/10.3354/meps237087.

Leroy, F., Riera, P., Jeanthon, C., Edmond, F., Leroux, C., & Comtet, T. (2012). Importance of bacterivory and preferential selection toward diatoms in larvae of *Crepidula fornicata* (L.) assessed by a dual stable isotope (^{13}C,^{15}N) labeling approach. *Journal of Sea Research, 70*, 23–31. https://doi.org/10.1016/j.seares.2012.02.006.

Mabeau, S., & Kloareg, B. (1987). Isolation and analysis of the cell walls of brown algae : *Fucus spiralis, F. ceranoides, F. vesiculosus, F. serratus, Bifurcaria bifurcata* and *Laminaria digitata*. *Journal of Experimental Botany, 38*(194), 1573–1580.

Maberly, S. C., Raven, J. A., & Johnston, A. M. (1992). Discrimination between ^{12}C and ^{13}C by marine plants. *Oecologia, 91*(4), 481–492.

Macler, B. A. (1986). Regulation of carbon flow by nitrogen and light in the red alga, *Gelidium coulteri*. *Plant Physiology, 82*(1), 136–141. https://doi.org/10.1104/pp.82.1.136.

Mann, K. H., & Kenneth, H. (1982). *Ecology of coastal waters : A systems approach*. University of California Press.

Marconi, M., Giordano, M., & Raven, J. A. (2011). Impact of taxonomy, geography, and depth on $\delta^{13}C$ and $\delta^{15}N$ variation in a large collection of macroalgae. *Journal of Phycology, 47*(5), 1023–1035. https://doi.org/10.1111/j.1529-8817.2011.01045.x.

Mercado, J. M., de los Santos, C. B., Lucas Pérez-Lloréns, J., & Vergara, J. J. (2009). Carbon isotopic fractionation in macroalgae from Cádiz Bay (Southern Spain): Comparison with other bio-geographic regions. *Estuarine, Coastal and Shelf Science, 85*(3), 449–458. https://doi.org/10.1016/j.ecss.2009.09.005.

Middelburg, J. J., Barranguet, C., Boschker, H. T. S., Herman, P. M. J., Moens, T., & Heip, C. H. R. (2000). The fate of intertidal microphytobenthos carbon: An in situ ^{13}C-labeling study. *Limnology & Oceanography, 45*(6), 1224–1234. https://doi.org/10.4319/lo.2000.45.6.1224.

Moodley, L., Boschker, H. T. S., Middelburg, J. J., Pel, R., Herman, P. M. J., De Deckere, E., et al. (2000). Ecological significance of benthic foraminifera: ^{13}C labelling experiments. *Marine Ecology Progress Series, 202*, 289–295. https://doi.org/10.3354/meps202289.

Musat, N., Foster, R., Vagner, T., Adam, B., & Kuypers, M. M. M. (2012). Detecting metabolic activities in single cells, with emphasis on nanoSIMS. *FEMS Microbiology Reviews, 36*(2), 486–511. https://doi.org/10.1111/j.1574-6976.2011.00303.x.
Pascal, P.-Y., Dupuy, C., Richard, P., & Niquil, N. (2008). Bacterivory in the common foraminifer *Ammonia tepida*: Isotope tracer experiment and the controlling factors. *Journal of Experimental Marine Biology and Ecology, 359*(1), 55–61. https://doi.org/10.1016/J.JEMBE.2008.02.018.
Paul, D., Skrzypek, G., & Fórizs, I. (2007). Normalization of measured stable isotopic compositions to isotope reference scales – a review. *Rapid Communications in Mass Spectrometry, 21*, 3006–3014. https://doi.org/10.1002/rcm.
Pinnegar, J. K., & Polunin, N. V. C. (2000). Contributions of stable-isotope data to elucidating food webs of mediterranean rocky littoral fishes. *Oecologia, 122*(3), 399–409. https://doi.org/10.1007/s004420050046.
Pomeroy, L. R. (1980). Detritus and its role as a food source. In R. S. K. Barnes, & K. H. Mann (Eds.), *Fundamentals of aquatic ecosystems* (pp. 84–102). Oxford, UK: Blackwell Scientific Publications.
Popper, Z. A., Michel, G., Hervé, C., Domozych, D. S., Willats, W. G. T., Tuohy, M. G., et al. (2011). Evolution and diversity of plant cell walls: From algae to flowering plants. *Annual Review of Plant Biology, 62*, 567–590. https://doi.org/10.1146/annurev-arplant-042110-103809.
Raven, J. A., Johnston, A. M., Kübler, J. E., Korb, R., McInroy, S. G., Handley, L. L., et al. (2002). Mechanistic interpretation of carbon isotope discrimination by marine macroalgae and seagrasses. *Functional Plant Biology, 29*(3), 355–378. https://doi.org/10.1071/PP01201.
Schaal, G., Riera, P., & Leroux, C. (2009). Trophic significance of the kelp *Laminaria digitata* (Lamour.) for the associated food web: A between-sites comparison. *Estuarine, Coastal and Shelf Science, 85*(4), 565–572. https://doi.org/10.1016/J.ECSS.2009.09.027.
Schaal, G., Riera, P., & Leroux, C. (2010). Trophic ecology in a Northern Brittany (Batz Island, France) kelp (*Laminaria digitata*) forest, as investigated through stable isotopes and chemical assays. *Journal of Sea Research, 63*, 24–35.
Starr, R. C., & Zeikus, J. A. (1993). UTEX – the culture collection of algae at the university of Texas at Austin. *Journal of Phycology, 29*(s2), 1–106.
Tenore, K. R. (1977). Growth of *Capitella capitata* cultured on various levels of detritus derived from different sources. *Limnology & Oceanography, 22*(5), 936–941.
Vauchel, P., Kaas, R., Arhaliass, A., Baron, R., & Legrand, J. (2008). A new process for extracting alginates from *Laminaria digitata*: Reactive extrusion. *Food and Bioprocess Technology, 1*(3), 297–300. https://doi.org/10.1007/s11947-008-0082-x.
Wagner, M. (2009). Single-cell ecophysiology of microbes as revealed by Raman microspectroscopy or secondary ion mass spectrometry imaging. *Annual Review of Microbiology, 63*, 411–429. https://doi.org/10.1146/annurev.micro.091208.073233.
Yamaguchi, T., Ikawa, T., & Nisizawa, K. (1966). Incorporation of radioactive carbon from $H^{14}CO_3^-$ into sugar constituents by a brown alga, *Eisenia bicyclis*, during photosynthesis and its fate in the dark. *Plant and Cell Physiology, 7*(2), 217–229. https://doi.org/10.1093/oxfordjournals.pcp.a079175.

CHAPTER TWO

Harvesting and potential uses of selected red seaweeds in the Philippines with emerging high-value applications

Anicia Q. Hurtado[a,*], Rexie Magdugo[b] and Alan T. Critchley[c]
[a]Integrated Services for the Development of Aquaculture and Fisheries (ISDA) Inc., Iloilo City, Philippines
[b]Laboratoire de Biotechnologie et Chimie Marines, EA3884, UBS, IUEM, F-56000 Vannes, France
[c]Verschuren Centre for Sustainability in Energy and the Environment, Cape Breton University, Sydney, NS, Canada
*Corresponding author: E-mail: anicia.hurtado@gmail.com

Contents

Abstract

While seaweed cultivation is indeed practiced globally, it is by no means a mature industry. In part, this is because that in spite of the tremendous biodiversity available within this "polyphyletic rag bag" of distantly-related photosynthetic organisms, very

Advances in Botanical Research, Volume 95
ISSN 0065-2296
https://doi.org/10.1016/bs.abr.2019.12.004

few species have actually been studied sufficiently to bring them into domestication for production. This is not unlike the situation with terrestrial agronomy.

The history of extensive cultivation of seaweeds is several centuries old. One of the major, global centers for innovation and production of industrial-scale quantities of biomass for processing has been the Philippines and in turn the Coral Triangle.

However, the species and strains of seaweeds currently in cultivation, often clonally propagated on account of their relative ease of producing propagules, have been victim to their own success. Farmers and industry have simply used the same, limited pool of biomass and techniques for too long. This has led to declines in yield and quality of constituents due to loss of vigor and increased incidences of pests and diseases. Innovation is urgently required both in the choice candidate species for cultivation and also the necessary steps and techniques required for the reliable and sustainable production of industrial quantities of biomass.

Once again, the Philippines is at the forefront of leading technologies for the selection of novel candidate species for responsible, phyconomic activities. These species are economically valuable candidates, not only for their colloidal constituents, but also the more innovative applications of biologically active compounds from seaweeds used for the benefit of human, animal, plant and microbial products. Selected examples of these biologically active constituents and their applications are briefly reviewed.

This chapter summarizes advances being made in the Philippines and outlines developments in several candidate species as future cultivated, marine crops.

1. Introduction

The marine environment has much to offer mankind in terms of the benefits derived from the goods and services supplied by its varied flora and fauna. Seaweeds are one of the major components thriving abundantly, under suitable conditions, distributed from the intertidal to sub-tidal regions of the sea, with some being fully pelagic. Man has explored the sea in search for food, bioactive compounds with medicinal properties, as well as feedstocks for fuel. In spite of a tremendous amount of published work to date, there remains much to discover and develop for both societal and economic benefits to the world.

For more than four decades there has been a great deal of published works presented on economically important red seaweeds including the carrageenophytes — especially the genera *Kappaphycus* and *Eucheuma* (Hayashi et al., 2017). In spite of the fact that there are many success stories related to the socio-economic benefits of eucheumatoid farming in the Philippines and elsewhere, their extensive, open-water farming, for commerce (recently termed phyconomy by Hurtado et al. in press) has been seriously challenged with the deteriorating quality of propagules (seedlings) and lack of strain vigor (Ali, Yashir, Critchley, & Hurtado, 2018; Hurtado et al., 2012).

A number of reasons for this have been put forward including repeated vegetative propagation from a limited genetic stock. Although entirely predictable (with hindsight), the consequences of this practice have contributed to extensive and devastating pest and disease infestations (Critchley et al., 2004; Hurtado & Critchley, 2006; Largo, 2002; Vairappan et al., 2008). Reasons given for this included: increasing abiotic stresses, weak propagules and limited pools from which to select new strains and cultivars, poor crop management, and economic factors such as unstable market prices (Hurtado, 2018).

The rapid geographic expansion of new cultivation sites was the main reason for apparent successes and continued, overall increasing global production of eucheumatoids (FAO, 2017). This was in spite of declining productivity (as in kg per unit area) in some of the already established farming sites affected by pests and disease (Hayashi et al., 2017). Both *Kappaphycus* and *Eucheuma* are commercially cultivated in Southeast Asia and their combined production has come to dominate the rankings of global seaweed production; after 2010 to the present, even surpassing the extensive cultivation of brown seaweeds such as *Laminaria/Saccharina* and *Undaria* (FAO, 2014).

Global seaweed resources are increasingly coming under scrutiny. There are a number scientific reports on the collective biomasses of carrageenophytes which demonstrate their potential as sources of structurally diverse, biologically active compounds with potentially multiple pharmaceutical and biomedical applications (Pere Ribeiro-Claro, 2014; Pereira, 2018b). It must be remembered that seaweeds and their extracts are very complex and, as such, there is no one compound which conveys the many beneficial properties, rather that various seaweeds and their extracts contain many compounds which may work synergistically and/or antagonistically with biological processes. Some of the bioactivity of extracts is as a direct result of the chemical interactions between raw materials and the extraction processes. This is why, while there have been many reports citing beneficial applications of extracts, as demonstrated by model-organism responses and bioassays to date, there is a paucity of available literature on various modes of actions and identification of the constituent bioactive compounds. As an exception to this, some of the more recently published reports have provided an insight into biological activities and neuro-protective effects of marine algal extracts including: antioxidant, anti-neuro-inflammatory conditions, cholinesterase inhibitory activity and the inhibition of neuronal death. Taken together, these studies suggest that specific and selected seaweeds, and the cultivar/strains thereof, have considerable potential to be used as neuro-protective agents which can be assimilated in new products in pharmaceuticals, nutraceuticals, cosmetics and functional foods (Jin et al., 2006; Pangestuti & Kim, 2011).

2. The current status of seaweed resources in the Philippines

The Coral Triangle (CT) which includes: Indonesia, Malaysia, Papua New Guinea, the Philippines, the Solomon Islands and Timor-Leste is one of the most important reef systems in the world. It covers 132,636 km, extending across six countries and occupies just 1.5% of the total ocean area, yet represents 30% of the world's coral reefs. The CT has the highest coral diversity in the world, i.e. 76% of all coral species are found there. Fifteen species are endemic to the region (www.weforum.org). It is a biodiversity-rich region where there is a diversified marine flora and where tropical seaweed farming has become firmly established (Hayashi et al., 2017; Hurtado, Gerung, Yasir, & Critchley, 2014).

The Philippines is an archipelagic country consisting of more than 7000 islands, the shore-lines and reefs of which are endowed with a considerable diversity of seaweeds. Ganzon-Fortes (2012) provided a comprehensive, historical account of studies in to the seaweed biodiversity in the Philippines. The article covered the period from 1750 to the end of the 20th century (1999). The review revealed that the earliest literature contained not only taxonomic studies of commercial species, but also included aspects of their population biology, seasonality, and the yield and quality of their phycocolloids. However, during the last decade of the 20th century, seaweed research trends shifted to more applied aspects of utilization (Buschmann et al., 2017).

The latest report on the seaweed resources of the Philippines was prepared by Trono and Largo (in progress) which provided information on the biodiversity of economically important species of the Philippines, their utilization as food and raw materials in processing and also use of their industrially extracted products such as: agar, carrageenans, alginates, fucoidans, etc. Trono (2010) discussed potential cultivation technologies for the red seaweed *Halymenia durvillei* utilizing both vegetative propagules and spores.

In spite of the tremendous marine biodiversity of the Philippines very few are actually grown commercially and these are really limited to just two genera, viz. *Kappaphycus* and *Eucheuma* (Hayashi et al., 2017). There are still several seaweeds which have a seemingly high potential as candidates for phyconomic development such as: *Asparagopsis taxiformis* (Cordero, 1977; Silva, Menez, & Moe, 1987), *Betaphycus philippinensis* (Doty, 1995; Dumilag, 2018; Dumilag, Liao, & Lluisma, 2014), *Callophyllis adnata* (Hurtado-Ponce, 1983), *Grateloupia filicina* (Hurtado-Ponce, 1983), *H. durvillei* (Trono, 2010) and *Pyropia acanthophora and Pyropia tanegashimensis* (Ame, Ayson, Okuda, & Andres, 2010; Dumilag et al., 2016, 2017).

This chapter presents the case for alternative red seaweeds to be brought on board as candidates to extend phyconomic activities to a greater range of "domesticated" species. Herewith, we discuss seven red seaweeds from the Philippines which are emerging with a high potential for further exploration leading to the development of the required phyconomic techniques for their production. This will ultimately feed a sustainable and reliable supply-chain of raw materials for industrial processing and value-chain-addition and novel applications leading to new products. Such developments have beneficial local, regional and global social and economic values.

3. Brief descriptions, local distribution and uses of red seaweeds with emerging potential applications

Fig. 1 and Table 1 show the distribution of Philippine red seaweeds which are suggested to have considerable potential. Of the seven species, four are restricted in their distribution (i.e. endemic) to northern-most Luzon, e.g. *C. adnata, G. filicina, Pyropia acanthophora and P. tanegashimensis.* The remaining three, i.e. *A. taxiformis, B. philippinensis* (= *Betaphycus gelatinus*) and *Halymenia durvillaei* are distributed around the country.

The marked seasonality of *C. adnata, G. filicina, P. acanthophora* and *P. tanegashimensis* has dictated their market value in the community (Table 2). These seaweeds are predominantly eaten as fresh, sea vegetable salad, especially by Filipinos in the northern-most part of Luzon. *C. adnata* and *G. filicina* cost around \$0.39 per 100 g or \$3.90/kg fresh weight in public markets, while *P. acanthophora* and *P. tanegashimensis* command the higher price \$15.68/ft^2 dry wt. The high cost of these seaweeds is primarily due to their limited supplies during the northeast monsoon (November - early March) in the northern part of the Philippines (Table 3).

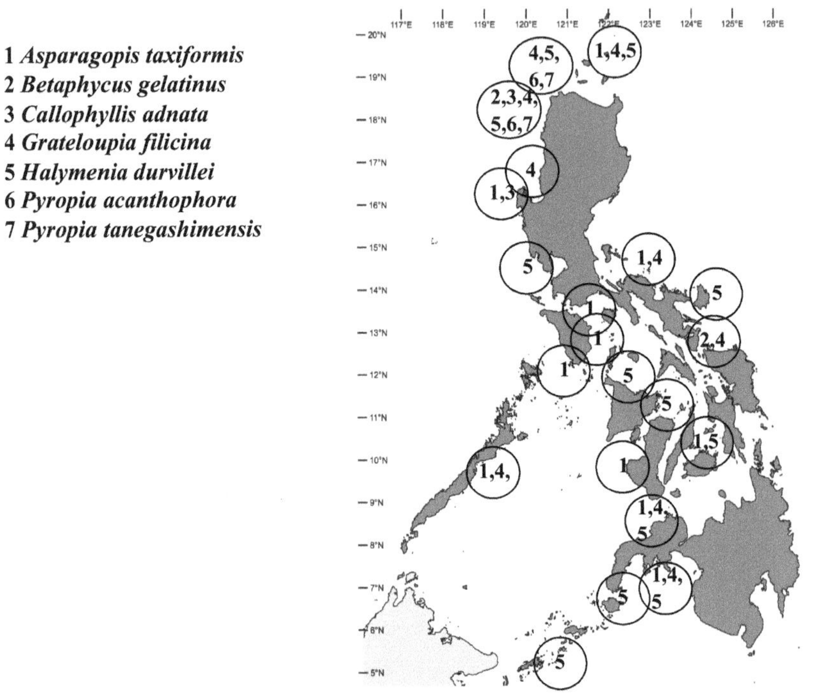

Fig. 1 Distribution of red seaweeds of emerging interest in the Philippines.

Table 1 Distribution of red seaweeds of emerging interest in the Philippines.

Species	Locations	References
Asparagopsis taxiformis (Delile) Trevisan	LUZON: Batanes, Pangasinan, Quezon, Occidental Mindoro, Oriental Mindoro, Palawan	Silva et al., 1987
	ViSAYAS: Aklan, Cebu, Negros Oriental	Cordero, 1977
	MINDANAO: Zamboanga del Sur	Trono, 1977
Betaphycus philippinensis (=*B. gelatinus*) Doty	LUZON: Sorsogon	Doty, 1995; Dumilag et al., 2014
Callophyllis adnata Okamura	LUZON: Ilocos Norte	Hurtado-Ponce, 1983, 1984; Agngarayngay, 1984a,b
Grateloupia filiciina (Lamouroux) C. Agardh	LUZON: Batanes, Cagayan, Ilocos Norte, Pangasinan, Sorsogon	Hurtado-Ponce, 1983; Silva et al., 1987

Table 1 Distribution of red seaweeds of emerging interest in the Philippines.—cont'd

Species	Locations	References
	Palawan, La Union,	Trono, 1977
	VISAYAS: Negros Oriental	Trono, 1977
	MINDANAO: Zamboanga del Sur	Trono, 1977
Halymenia durvillei Bory de Saint-Vincent	LUZON: Batanes, Ilocos Norte, Cagayan, Bataan, Catanduanes, Quezon,	Cordero, 1977; Hurtado-Ponce, 1983
	Batangas,	
	VISAYAS: Iloilo, Aklan, Guimaras, Negros Oriental, Negros Occidental;	Cordero, 1977; Trono, 1977; Hurtado, Luhan, & Guanzon, 2006
	Cebu, Siquijor,	
	MINDANAO: Zamboanga City, Zamboanga del Sur, Sulu, Tawi-Tawi	Cordero, 1977; Trono, 1977
Pyropia acanthophora (E.C.Oliveira & Coll)	LUZON: Cagayan, Ilocos Norte	Dumilag et al., 2017
M.C.Oliveira, D.Milstein & E.C.Oliveira		Dumilag & Monotilla, 2018
Pyropia tanegashimensis (Shinmura)	LUZON: Cagayan, Ilocos Norte	Dumilag et al., 2016; Dumilag & Monotilla, 2018
N.Kikuchi & E. Fujiyoshi		

Table 2 Seasonality of red seaweeds of emerging interest in the Philippines.

Species	Seasonality	References
Asparagopsis taxiformis	Oct–Nov	Cordero, 1977
	Apr–May	Trono, 1977
Betaphycus philippinensis	Dec–Apr	Doty, 1995
Callophyllis adnata	Nov–Apr	Hurtado-Ponce, 1983
		Marcos-Agngarayngay, 1984
Grateloupia filicina	Nov–Apr	Hurtado-Ponce, 1983
	Mar–Jul	Trono, 1977
Halymenia durvillaei	Jan–Dec	Hurtado-Ponce, 1983
		Trono, 1977
Pyropia acanthophora	Oct–Mar	Ame et al., 2010
Pyropia tanegashimensis	Oct–Mar	Ame et al., 2010

Table 3 Summary of the utilization of red seaweeds of emerging interest in the Philippines.

Species	Applications						
	Carrageenan			Pharmaceuticals	Human food	Animal feed	Pigment
	Kappa	Lambda	Beta				
Asparagopsis taxiformis				■		■	
Betaphycus philippinensis (=B. gelatinus)			■		■		
Callophyllis adnata					■		
Grateloupia filiciina		■			■		
Halymenia durvillei		■			■		■
Pyropia acanthophora	■				■		
Pyropia tanegashimensis	■				■		

4. Sustainable harvesting and utilization of biomass removed from natural populations

Among those red seaweeds described in this paper with a high potential for further exploration and development, only *H. durvillei* has, so far, been scaled-up for phyconomic purposes. It is still in its infancy at the pilot-scale (Trono, 2010) in the Visayas. The under-lying feature has been the necessary research required for the sustainable and reliable production of healthy sporelings for land-based culture.

Unfortunately, all other raw materials of red seaweeds mentioned in this chapter are harvested manually and not necessarily sustainably. There are reports of the total removal of the whole, adult thalli (plant bodies), leaving no basal tissue for future regeneration. Likewise, large waves can bring a considerable amount of sand to areas where ephemeral seaweeds grow, thus covering and hampering the (re-)growth of the seaweed (Pers. Observation). Thus, the total biomass harvests have been rather unstable, unreliable and certainly not practiced in a sustainable manner. Furthermore, indiscriminate gathering of these seaweed resources will ultimately result in severe depletion, or even worse, local extinctions as personally observed.

Each of the emerging red seaweed candidates is dealt with individually regarding their descriptions, methods of production and value proposition in the following sections.

A. taxiformis (Fig. 2) is a tropical species widely distributed throughout the Philippines, as reported by Cordero (1977) and Trono (1977). It has a tri-phasic, diplo-haplontic, heteromorphic life cycle. The alga can be found almost year-round, from the upper inter-tidal to sub-tidal regions.

Fig. 2 Photo of *Asparagopsis taxiformis*. (Photo courtesy of N Paul).

Surprisingly there are several reports on the multiple uses of *A. taxiformis*, yet there is no information related to its harvesting, nor commercial cultivation in the Philippines. Recently, *A. taxiformis* became infamous (with considerable global, media attention) in relation to being a potential contributor to the mitigation climate change. A. taxiformis, as a dietary supplement, was found to reduce methane production in cows by more than 99% in laboratory evaluation (Machado, 2015). The mode of action was suggested to be due to the almost total disruption of methanogenic enzymes in the treated animal's gut. This also preserved feed energy for the treated cows which could then be allocated for health and production and not lost as energy/ methane gas in to the atmosphere. Ruminant emissions of methane account for approximately 10% of all greenhouse gases (GHGs) each year in Australia (Kinley, de Nys, Vucko, Machado, & Tomkins, 2016). Reports on the intensive culture of *A. taxiformis* in tanks in Australia are known (Paul, Pers. Comm), however, its commercial, extensive open-water cultivation has yet to be piloted. Furthermore, *in situ* studies have to be made on the biology and eco-physiology of A. taxiformis, especially its heteromorphic alternation of generations, in the Philippines. This information is required in order to understand the needs for up-scaling, as required to provide high quality, consistent and regular, large, industrial volumes of material to service the livestock feed and supplements industry. This seaweed has huge potential value to the Philippines in particular since it is endemic and application of best-practice, phyconomic principles (Hurtado, Neish, & Critchley, 2019) for cultivated production could lead to the reliable supply of raw materials to the livestock industry, which in turn could have a significant impact on reduction of methane GHG in the atmosphere.

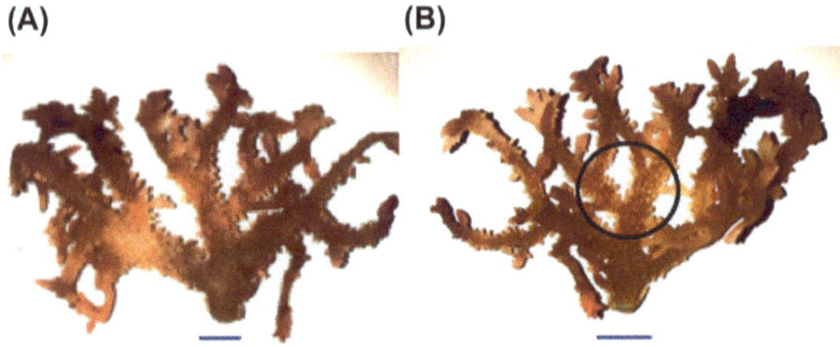

Fig. 3 Photo of *Betaphycus philippinensis*, (A) dorsal, (B) ventral (bar = 1 cm).

B. philippinensis (=*B. gelatinus*) (Fig. 3) thrives on hard coral substrata by means of well-developed haptera; branches are irregular with flattened segments provided with numerous spinous processes at the margin (Cordero, 1977; Doty, 1995; Trono, 1977). Gathering from wild populations of this seaweed is limited to utilization as a source of sea vegetable, especially in northern Luzon (Agngarayngay, 1984b).

C. adnata (Fig. 4) has been reported in Ilocos Norte (Cordero, 1980; Hurtado-Ponce, 1983, 1984; Agngarayngay, 1984a,b). This species grows mostly at the lower edge of the intertidal region where there is moderate wave action. The thallus of *C. adanata* is erect; a flabellate blade arises from a small disc; cystocarps are borne at the margin. This species occurs during the north-east monsoon from December–May. Gathering starts in January reaching a peak in March and April. This seaweed is sold locally in public markets in its fresh form and eaten as a sea vegetable salad. It is a

Fig. 4 Photo of *Callophyllis adnata* bearing cystocarps (bar = 1 cm).

special dish at important parties and occasions. The presence of *C. adnata* in the northernmost part of Ilocos Norte from March—April results in abundant harvests by local people which are preserved by sun-drying (Garcia, pers. comm).

The presence of cystocarps during March—April indicates a potential method of generating new plants from spores. However, there are no known studies in the Philippines on its biology and eco-physiology which would be required to develop methods for its commercial cultivation. This seaweed has much promise due to its relatively high market.

G. filicina (Fig. 5) is a red seaweed which can be found in the northernmost part of Luzon (Agngarayngay, 1984a; Hurtado-Ponce, 1983) and other areas such as Sorsogon, Palawan and Zamboanga del Sur (Trono, 1977) during the colder months of the year (i.e. November—March). This red seaweed is purplish to greenish in color, thin, gelatinous and slippery, erect, caespitose, growing on intertidal rocks or in shallow tide pools. Gatherers start collecting in December reaching a peak in February and the material is normally sold in local markets as a sea vegetable salad.

H. durvillaei (Fig. 6) is a red seaweed, indigenous to the Philippines, which has gained increasing attention because of its high demand in the market as a result of its source of phycobiliproteins (i.e. red-phycoerythrin; R-PE) which has important applications in value-added products such as natural coloring agents in cosmetics, pharmaceuticals and food. Successful

Fig. 5 Photo of *Grateloupia filicina* (bar = 1 cm).

Fig. 6 Photo *Halymenia durvillei* (bar = 1 cm).

attempts to generate new plants from spores was reported by Trono (2010), which led to pilot-scale field cultivation in Luzon and the Visayas. It is expected that this seaweed will be of increasing value and importance in the near future and is one to watch!

There are 10 published species of *Porphyra* in the Philippines (Ame et al., 2010; Codero, 1974, 1976; Tungpalan, 1984). However, out those listed and described species, eight were omitted from the list and only two are confirmed, namely: *Pyropia acanthophora* and *P. tanegashimensis.* These confirmations were based on the comparative analyses of plastid *rbc*L and mitochondrial COI-5P gene sequences (Dumilag et al., 2016, 2017). Furthermore, phylogenetic analysis revealed that Philippine populations of *P. tanegashimensis* were indeed endemic and distinct from the other known endemic populations from Japan and also an introduced strain found in Brazil (Milstein, Madeiros, Oliveira, & Oliveira, 2015).

Pyropia spp. (Fig. 7) are found at the upper inter-tidal on rocks and barnacles in the northernmost part of the Philippines, notably: Claveria, Sta. Prexedes, Sta. Ana and Calayan in Cagayan (Ame et al., 2010) and Burgos and Pagudpud, in Ilocos Norte (Cordero, 1977). These thalli are abundant during the northeast monsoon (December—February) when seawater temperatures drop to 17—18°C. This relatively low, local seawater temperature

Fig. 7 Photo of *Pyropia sp.* (bar = 1 cm).

is believed to be influenced by the cold Kuroshio Current which originates from southeastern Japan and passes by the northernmost part of the Philippines during the northeast monsoon. Gathering of *Pyropia* spp. is restricted to December—February since at other times of the year this seaweed only exists as its microscopic, conchocelis stage of growth and development, which happens to be endolithic, or endo-conchocelic (i.e. living within the protection of porous rocks or shell matrices). The only post-harvest processing of *Pyropia* spp. in the Philippines is sun-drying and bundling into crude sheets for local markets (Agngarayngay, Llaguno, Aquino, Taclan, & Galacgac, 2005). Though some studies have been reported on the biology and ecophysiology of *Pyropia* in the Philippines (Monotilla & Notoya, 2010, 2004), its rearing as a crop within commercial hatcheries and out-door cultivation, has never been introduced in the Philippines. This is in marked contrast to the billion dollar nori industries of China, Japan and Korea (Kim, Yarish, Hwang, Park, & Kim, 2017). In mitigation, one must understand the prevailing characteristics of the sea bordering the Cagayan and Ilocos Norte provinces on the west of the Philippines which during southwest monsoon, and the onset of northeast monsoon, are subject to large, rough waves derived from the Trade Winds. Thus, current methods and technologies for extensive nori cultivation would be too fragile for such conditions.

5. Biochemical compositions

5.1 Protein and amino acids

Generally, red seaweeds are found to be higher in protein content and amino acid composition, as compared to the green and brown algae (Hamed, Özogul, Özogul, & Regenstein, 2015). Sources of variability in algal composition can broadly be divided into taxonomic, ecological and ambient environmental factors (Stengel, Connan, & Popper, 2011). The protein content of various macroalgae differs significantly by: phylum, species, geographic location, season, and environmental parameters (Hamed et al., 2015). However, they have all the essential amino acids, in reasonable proportions. Generally, green or red seaweeds are found to have 10%–47% protein by dry weight, while brown seaweeds contain protein levels as low as 3%–15% on a dry weight basis (Fleurence, 2004). Among red algae, according to McDermid and Stuercke (2003), the highest protein content was found to be in *Halymenia formosa* (21.2%), followed by *Porphyra vietnamensis* (16.5%), and a not insubstantial value for *A. taxiformis* (9.4%). The protein contents of *Porphyra tenera* (up to 47% dry weight) and *Palmaria palmata* (around 35% dry weight) were found to be highest (Fujiwara-Arasakil, Minol, & Kuroda, 1984) and are both comparable to the protein content of soybean; *Grateloupia turuturu* has a reported protein content of up to 27.5% dry weight (Fleurence, 1999). A recent study of three *Halymenia* species from the Philippines found that *Halymenia durvellaei* had the highest protein content (13.02%), followed by *Halymenia maculata* (11.07%), and *Halymenia dilatata* (9.99%) (Kho, Orbita, & Manting, 2016).

The amounts of total protein and profiles of amino acids in red seaweeds of the Philippines makes them interesting sources for use in the industrial food and even aquaculture feed sectors (Baweja, Kumar, Sahoo, & Levine, 2016; Hamed et al., 2015). The percentage proportion of essential amino acids (EAAs) available from red seaweeds (40.4%), was found to be the highest, as compared to greens (38%), and brown (38.9%) seaweeds (Dumay & Morançais, 2016). *Porphyra acanthophora* has a reported EAA of 48% (Lourenço, Barbarino, De-Paula, Pereira, & Marquez, 2002) suggesting that red seaweeds had the highest EAA content per 1 mg of protein in relation to dry algal matter. The dominant forms of algal amino acids are aspartic and glutamic, while other acids (e.g., valine, leucine, and lysine) are in sufficient concentrations (Baweja et al., 2016). In red edible Japanese algae such as *Gloiopeltis tenax*, *G. furcata*, *Gracilaria textorii*, and *Gr. vermiculophylla* have relatively high concentrations of taurine that may be used to create functional foods that are rich in such naturally occurring compounds. In contrast, the taurine content was low or undetectable in brown and green algae (Kawasaki et al., 2017).

5.2 Carbohydrates

Various seaweeds can be a most abundant source of: alginic acid, agar, agarose, and carrageenans — collectively called structural polysaccharides. There are also storage polysaccharides, such as laminarin (*Laminaria* sp., brown seaweeds), ulvan (members of the Ulvales, green seaweeds), and floridean starch (*Porphyra* sp., found in members of the Florideophycidae, red seaweeds) are an accumulation of metabolites with no obvious immediate role and assumed to be useful in some later event (Stiger-Pouvreau, Bourgougnon, & Deslandes, 2016). These polysaccharides are regarded as dietary fibers and cannot be digested by humans due to the general lack of required enzymes (Baweja et al., 2016; Hamed et al.,2015; Mendis & Kim, 2011; Mišurcová, Kráčmar, Klejdus, & Vacek, 2010). There are two groups of dietary fiber, soluble and insoluble. Soluble fibers can form viscous gels in the intestinal tract, which are reported to lower blood cholesterol and glucose levels; whereas insoluble fibers are reported to regulate ingested materials in the digestive tract, thereby increasing faecal-bulking capacity, which can relieve constipation and irregular stools (Baweja et al., 2016; Hamed et al., 2015). Soluble carbohydrates of *A. taxiformis* were determined to comprise 13.2% of dry weight (McDermid & Stuercke, 2003). Consumption of selected seaweeds has been demonstrated to have beneficial effects with anti-tumor, anti-coagulant, and anti-viral properties (Hamed et al., 2015), and through eliminating unnecessary substances by dietary (insoluble) fibers binding to *toxic* compounds from the consuming organism. The total dietary fiber content of some seaweeds was found to be relatively high (25%—75% dry weight) as compared to most fruits and vegetables (MacArtain, Gill, Brooks, Campbell, & Rowland, 2007). Utilization of non-digestible soluble fiber (polysaccharides/oligosaccharides), in the large intestine, can produce short chain fatty acids (SCFAs) (e.g., butyrate, propionate and acetate) which are important, host health-promoting products of beneficial gut bacteria (Cornish, Mouritsen, & Critchley, 2019; Liu et al., 2015). Members of the red algal genus *Porphyra/Pyropia* contain the polysaccharide porphyran, which can make up to 48% of thallus dry weight, and therefore have useful properties for food applications (MacArtain et al., 2007). *P. tenera* had a carbohydrate value (44%—56%), total dietary fiber (34.7%), soluble fiber

(17.9%), and insoluble fiber (16.8% of seaweed dry weight) (Fleurence, 2004). The carbohydrate contents of three *Halymenia* species from the Philippines were reported as: *H. durvillaei* (53.65%), *H. maculata* (51.18%) and *H. dilatata* (40.53%). While its carrageenan yield content comprised about 74% of the carrageenan content of *Kappaphycus alvarezii*, of these *H. durvillaei* had the highest (28.4%), which differed significantly with *H. maculata* (11.5%), and *H. dilatata* (9.7%) (Kho et al., 2016). These data suggested that *H. durvillei* could be a potential source of future carrageenan production.

5.3 Lipid and fatty acid content

In general, total lipids (TLs) and fatty acids (FAs) in seaweeds are characterized by total lipid content which ranges in 1–6% dry weight of TL (Rodrigues et al., 2015), a relatively low content of saturated fatty acids (SFAs), and a substantial amount of polyunsaturated fatty acids (PUFAs) such as omega 3 and omega 6, as compared to land vegetables (Mendis & Kim, 2011).

TL and FA composition may vary during the life cycles of various seaweeds which may be seasonal, geographical, or growth conditions (Melo et al., 2015). For instance, the TL of *Ulva pertusa*, *Laminaria religiosa*, *Undaria pinnatifida*, and *Hetrochordaria obietina* appeared to decrease as depth increased (Ito, Tsuchiya, & Tohoku, 1977). These observations could be linked to light intensity whereby the TL content and FA composition of *G. turuturu* (Hotimchenko, 2002) and *Chondrus crispus* (Koch, Hagen, Graeve, & Bischof, 2017) increased when light intensity decreased, or changes in other abiotic factors (e.g. nutrients, salinity, temperature) and its interactions can affect or influenced lipid metabolism (Schmid, Guihéneuf, & Stengel, 2014). The mean TL content of those Rhodophyceae tested, SFA (42%), PUFA (36%), MUFA (13%), and several red seaweeds (e.g. *A. taxiformis*, *B. philippinensis*, *G. filicina*, *Pyropia acanthophora*, *P. tanegashimensis*, *H. durvillaei*) species varied from less than 0.1%–2.0% (Wielgosz-Collin, Kendel, & Couzinet-Mossion, 2016). The mean TL content of *A. taxiformis* was reported as being 4% (McDermid & Stuercke, 2003), and 12% (Ragonese et al. 2014), and for *G. filicina* as 22% dry weight (Tsai & Sun Pan, 2012). Three *Halymenia* species from the Philippines had a TL content of 1.3% as the highest for *H. durvillaei*, which differed significantly with *H. maculata* (1.1%), and *H. dilatata* (0.3%) (Kho et al., 2016). These are considered beneficial for human health as their consumption reportedly reduced the risk of cardiovascular diseases and osteo-arthritis (Hamed et al., 2015; Poudyal, Panchal, Ward, & Brown, 2013), decreased blood pressure and also improved heart and liver function in animal trials (Poudyal et al., 2013), as well as reducing the risks of diabetes, as long as they are not oxidized or converted into saturated lipids (Hamed et al.,2015). Various seaweeds can also be major sources of long-chain PUFAs (Lordan, Ross, & Stanton, 2011). Since humans are incapable of synthesizing PUFAs with more than 18 carbons, e.g., long-chain fatty acids such as: eicosapentaenoic acid (EPA, ω3 C20:5) and arachidonic acid (AA, ω6 C20:4). EPA and AA are present in red seaweeds, some notable examples are: *Porphyra dioica* (17.0 ± 6.1, 19.7 ± 3.7), *Ceramium virgatum* (17.4 ± 3.7, 3.7 ± 0.6), *C. crispus* (14.8 ± 0.9, 22.5 ± 1.5), *Gracilaria gracilis* (2.6 ± 1.5, 30.8 ± 6.6), *and P. palmata* (36.8 ± 4.3, 1.4 ± 1.2) (Schmid et al., 2014), and docosahexaenoic acid (DHA, C22:6) from *U. pinnatifida* and *U. pertusa* as anti-allergic (Hamed et al., 2015). Certain red seaweed species (*P. palmata*) may be rich sources of PUFAs (up to 20 carbons), with four or five double bonds, e.g. 20:5 n-3 (EPA) with levels ranging from 0.4% to 0.6% of dry weight (Schmid et al., 2014), palmitic (16:0) and oleic (18:1) acids. In *G. turuturu* two unusual non-methylene interrupted (NMI) dienoic FA were identified: (1) the 5,11-octadecadienoic (18:2*n*-7,13) had an equivalent chain length (ECL) value of 17.75 as FAME, firstly characterized in the green seaweed *Cladophora rupestris* and identified through oxidative ozonolysis. It is also known to occur in numerous mollusks; and (2) the 5,9-nonadecadienoic (19:2*n*-10,14) had an ECL value of 19.08 as FAME, was firstly identified in lipids from terrestrial flowers (Kendel, Barnathan, Fleurence, Rabesaotra, & Wielgosz-Collin, 2013).

Table 4 shows the percentages of saturated and unsaturated fatty acids of some selected examples of red seaweeds (Tables 5 and 6).

The galactolipids, monogalactosyl diacylglyceride (MGDG), digalactosyl diacylglyceride (DGDG), sulfoquinovosyl monoacylglyceride (SQMG), and sulfoquinovosyl diacylglyceride (SQDG) were recorded to be present in *G. filicina* (Arao & Yamada, 1989), *Porphyra crispata* (Tsai & Sun Pan, 2012), and *Gracilaria* sp. which was reported with anti-proliferative and anti-inflammatory properties, as determined by monitoring cell viability in human cancer lines. Also, a lipid extract (sulfoglycolipid) from *P. crispata* was demonstrated to inhibit the growth of the human hepato-cellular carcinoma cell-line (HepG2) (Tsai & Sun Pan, 2012), and also decreased cell viability of human T-47D breast cancer cells and of 5637 human bladder cancer cells (da Costa et al., 2017). Furthermore, SQDG is a potent telomerase inhibitor that may act as anti-cancer agent (Eitsuka, Nakagawa, Igarashi, & Miyazawa, 2004; Tsai & Sun Pan, 2012).

Table 4 Biochemical composition (% dw) of selected red seaweeds.

Group/Species	Proteins	Essential AA	TL	Carbohy-drates	Carrageenan	References
Asparagopsis taxiformis	9.4		0.1−12	13.2		(McDermid & Stuercke, 2003; Ragonese et al., 2014; Wielgosz-Collin et al., 2016)
Halymenia formosa	21.2		2.9	16.9		
Porphyra vietnamensis	16.5		4.4	30.5		
Porphyra acanthophora		48				(Lourenço et al., 2002)
Porphyra tenera	47			44−56		(Fujiwara-Arasakil et al., 1984)
Palmaria palmata	35	26.50				(Fleurence, 2004; Okolie, Mason, & Critchley, 2018)
Grateloupia turuturu	27.5					(Joël Fleurence, 1999; Okolie et al., 2018)
Grateloupia filicina			2.2	22		(Tsai & Sun Pan, 2012)
*Halymenia durvellaei**	13.02		1.29	53.65	28.41	(Kho et al., 2016)
*Halymenia maculata**	11.07		1.09	51.18	11.50	
*Halymenia dilatata**	9.99		0.29	40.53	9.7	

TL, total lipid; *AA*, amino acid; *, data obtained in Mindanao, Philippines.

Table 5 The variations in the percentage content of saturated and unsaturated FA of the following class, orders and species of red algae.

Group	No. of studied species	Fatty acids	SFA	MUFA	PUFA	Mean TL (% dw)	References
Class Rhodophyceae			42%	13%	36%		(Wielgosz-Collin et al., 2016)
		C16			1.8		
		C18			4%		
		C20			3.4%		
		C22			1.9%		
Order Bonnemaisoniales	2	from C6 to C28	73.80%			0.1 and 4.0	(Wielgosz-Collin et al., 2016)
A. taxiformis	1		52.47%		1.80%	12.22	(Ragonese et al., 2014)
		C14:0	18.17%				
		C16:0	48.9%				
		EPA			0.96%		
		Linoleic			0.84%		
Order Rhodymeniales	2	from C14 to C24		30.1%		0.1 and 1.4	(Wielgosz-Collin et al., 2016)
Order Halymeniales	11					1.9 (0.1−3.9)	(Wielgosz-Collin et al., 2016)
G. filicina	1	C14:0	13.8%				(Arao & Yamada, 1989)
		C16:0	50.8%				
		C18:1		11.2%			
		AA			9.3%		
G. filicina	1	EPA			43.7%		(Trono, 1977)
H. sinensis		EPA			42.2%		(Trono, 1977)

SFA, saturated fatty acid; *MUFA*, monounsaturated fatty acid; *PUFA*, polyunsaturated fatty acid; *TL*, total lipid.

Table 6 Biological activity of selected red seaweeds.

Species	Fraction	Activity against	IC50 (μg mL^{-1})	References
Asparagopsis taxiformis	Sulfated polysaccharides	Anticoagulant		(Pereira, 2018a)
Gigartina skottsbergii	Carrageenans, ($1C_3$, and $1T_1$, λ)	Anticoagulant, Antiviral (HSV-1&2)	(0.7,0.5 and 0.6, 0.4)	(Carlucci et al., 1997)
Gymnogongrus torulosus	DL-galactan	Anticoagulant, HSV-2 and DENV-2	0.6–16.0, & 0.19–1.7	(Pujol et al., 2002)
Porphyra yezoensis	Carrageenans, Glycoprotein	Anticoagulant, Antiradical Antiangiogenic, Anti-inflammatory		(Liu et al., 2019; Pereira, 2018a, 2018b; Torres, Flórez-Fernández, & Domínguez, 2019)
Gracilaria birdiae	Carrageenans Sulfated polysaccharides	Anticoagulant, Antiradical		
Grateloupia turuturu	carrageenans	Anticoagulant		(Pereira, 2016, 2018a)
Halymenia floresii	Sulfated polysaccharides	Anticoagulant		(Pereira, 2018a)
Kappaphycus alvarezii	*k*-carrageenan Water extract	Apoptotic induction Anti-inflammatory Superoxide, hyroxyl, and lipid peroxidation	0.00011, 0.00033, & 0.00032	(Ariffin et al., 2014; Pereira, 2018b, 2018a)
Gymnogongrus griffithsiae	Kappa/iota/nu carrageenan *G3d* and DL-galactan hybrid *C2S-3*	DENV-2, DENV-3,	1.0, 13.9–14.2	(Talarico et al., 2005)
Porphyra crispata	SQDGs	*HepG2*	126.0	(Tsai & Sun Pan, 2012)
Gigartina tenella	SQDG KM043	*DNA* polymerase α, *DNA* polymerase β, and HIV-reverse transcriptase type 1	0.21, 3.02, & 9.42	(Plouguerné et al., 2014)
Callophycus serratus	Polysaccharides, lipids	anti-thrombotic		(Lin et al., 2010)

$1C_3$, cystocarpic fraction; *$1T_1$*, tetrasporic fraction; *SQDG*, sulfoquinovosyldiacylglycerols; *SQDG KM043*, a new sulfolipid extract belonged to sulfoquinovosyldiacylglycerols; *HSV*, herpes simplex virus; *DENV*, dengue virus; *HepG2*, human hepatocellular carcinoma cell line; *HIV*, human immunodeficiency virus.

6. Biological activity and potential applications

6.1 Anti-coagulant properties

Anti-coagulants or blood thinners are chemical substances which prevent or reduce the coagulation of blood or which prolongs clotting time. The anti-coagulant activity of various seaweed extracts depends on there being, preferably two sulfate groups and a glycosidic linkage on the pyranose ring of its sulfated polysaccharides (Ciancia, Quintana, & Cerezo, 2010), in order to interact with basic groups of proteins. It was suggested that in this interaction, one or more proteases were targeted and that the major components of these activities were thrombin inhibition, as mediated by anti-thrombin and/or heparin co-factor II (Mohamed, Hashim, & Rahman, 2012). An example of anti-coagulant activity by carrageenans extracted from the red seaweeds *Gigartina skottsbergii* (Carlucci et al., 1997), and DL-galactan hybrids isolated from *Gymnogongrus torulosus* (Pujol et al., 2002) were found to have less anti-coagulant action. Another is of agaran fractions from *Acanthophora spicifera* which reportedly largely depended on the interaction affinity of protein-polysaccharide, which varies according to the seaweed-derived, sulfated polysaccharide. Several modes of action(s) were reported for anti-coagulant activities of various seaweed extracts, i.e. direct inhibition of thrombin, catalysis of thrombin inhibition via anti-thrombin and/or HCII, inhibition of protease generation and active complexes of blood coagulation, and interference by fibrin polylmerization (Duarte et al., 2004).

Sulfated galactans from the green alga *Codium fragile* were shown to have no anti-coagulant effects, while its highly sulfated arabinans were reported to have important anti-coagulant activity, with a different mode of action from that of heparin (Pengzhan et al., 2003).

The total sulfate content of the brown seaweed *Dictyopteris delicatula* was not correlated to the anti-coagulant activity of its sulfated polysaccharides (i.e. the heterofucans) which were reported to inhibit the coagulation pathways of both common and intrinsic, but not the extrinsic coagulation pathways (Magalhaes et al., 2011). Anti-coagulant activities were determined using activated partial thromboplastin time (aPTT) and prothrombin time (PT) coagulation assays. Anti-coagulant activity was reported for four fucans from the brown seaweed *Canistrocarpus cervicornis* (with only 0.1 mg/mL of plasma a result similar to that of Clexane, a commercial, low molecular weight heparin) at 1.25-fold less potency than Clexane, but with no clotting inhibition observed in six fucans tested for prothrombin time (PT) (Camara et al., 2011). Brown seaweed fucoidans were as effective as heparin and dermatan sulfate in the potentiation of heparin co-factor II. These anti-coagulant fucoidans might therefore be an alternative source of a cheaper and sustainable resource of alternative drugs, as compared to synthetic anticoagulants (Colliec et al.,1991). Fucoxanthin, a carotenoid from various brown algae was reported to have anti-inflammatory effects (Wijesinghe & Jeon, 2011). The biological effects of these compounds vary from species to species and due to their chemical structure (Cumashi et al., 2007). Philippine seaweeds could be a potential, rich natural resource and used to produce a beneficial and novel new products in the field of biomedical industry.

6.2 Anti-angiogenic properties

Anti-angiogenic compounds are chemical substances that prevent the growth of new blood vessels that provide cells with an adequate supply of oxygen and nutrients, as well as removing waste products. These compounds find applications in preventing cancerous tumors from growing their own blood vessels, or by blocking the vascular, endothelial growth factor (VEGF), a cancer-cell protein, from attaching to receptors lining the blood vessels.

A fucoxanthin extract of the brown seaweed *U. pinnatifida* was found to suppress the development of endothelial progenitor cells and the formation of new blood vessels in cultured human, umbilical vein endothelial cells. It was also observed to reduce the tube length of endothelial cells of a rat aortic ring (Sugawara, Matsubara, Akagi, Mori, & Hirata, 2006). Similar activity was reported from a fucoxanthin extract of *Sargassum* sp. which suppressed expression of VEGF on xeno-grafted, sarcoma 180 (S180) cells in mice (Wang et al., 2012). Another anti-angiogenic effect of a fucoxanthin extracts was proposed to be a molecular mechanism which suppressed fibroblast growth factor 2 (FGF-2) and its receptor (FGFR-1) (Ganesan, Matsubara, Sugawara, & Hirata, 2013). Anti-angiogenic effects of an enzyme-digested fucoidan extracted from *Cladosiphon novae-caledoniae* was reported for human uterine carcinoma HeLa cells by suppressing VEGF. Similar results were also reported with HT1080 human fibrosarcoma cells where the fucoidan extract (commercially known as "Pöwer fucoidan") promoted the VEGF suppression of the matrix metallo-proteinases (MMP-2/9) (Ye et al., 2005).

Carrageenans, from selected red seaweeds are linear, hydrophilic, sulfated polysaccharides (Campo, Kawano, Braz da Silva, & Ivone, 2009), and some have been reported to have anti-angiogenic effects in mice (Dias et al., 2004).

Only a few studies have suggested that carrageenans may have anti-tumoral activities. A study of human intestine (Caco-2) and hepatic (HepG2) cell lines showed apoptotic induction using *k*-carrageenan [from *K. alvarezii* (cottonii)], while no toxicity was observed in corresponding normal and cancerous intestinal and liver cell-lines, using non-degraded carrageenan (Ariffin et al., 2014). According to (Murad, Ghannam, Al-Ktaifani, Abbas, & Hawat, 2015), a pro-apoptotic activity of ι-carrageenan [from *Eucheuma denticulatum* (spinosum)] was shown to be mediated by caspase-3, caspase-9, p53, Bax, and Bcl-2 genes within an MDA-MB-231 metastatic, human, breast cancer cell-line. A review of the harmful, gastro-intestinal effects of carrageenan in animal experiments revealed potentially carcinogenic properties (i.e. colonic ulcerations and gastro-intestinal neoplasms) and cancer-promoting effects of degraded and un-degraded carrageenans in animal and experimental models, respectively (Tobacman, 2001). Therefore, a benefit/risk ratio must be considered in cancer prevention and therapeutic treatments using various carrageenans. A recent study by McKim et al. (2016) on the effects of the three common forms of carrageenan [i.e. lambda (λ)-, kappa (k)-, and iota (i)] on cell permeability, cyto-toxicity, and cytokine gene expression in human intestinal and hepatic cell-lines showed that a standard Caco-2 absorption model had no carrageenan permeability and did not cross the membranes of intestinal epithelial cells at different concentrations (100, 500, 1000 μg/mL) and is was also not cyto-toxic to these cells. In addition, carrageenan did not induce IL-8, IL-6, or MCP-1 (CCL2) or cellular toxicity after 24 h in two human intestinal cell-lines (HT-29 and HCT-8). No cellular oxidative stress was observed in HT-29 cells using k-carrageenan at 0.1, 1.0 and 10.0 μg/mL exposure concentrations. λ -carrageenan (0.1, 1.0, 10.0 and 100.0 μg/mL) had no effect on the expression of IL-8, IL-6, or MCP-1 (CCL2) in the human (HepG2) liver cell line. FMC Biopolymer (Philadelphia, PA) provided a food-blend carrageenan used for the above study consisting of k- and λ-carrageenan, and ι-carrageenan samples were purchased from Sigma-Aldrich (St. Louis, Missouri).

6.3 Anti-viral properties

Seaweeds are known to contain many bioactive compounds, some of which have anti-viral properties which could be used in developing new pharmaceutical agents (Pereira, 2018b; van Hal, Huijgen, & López-Contreras, 2014).

Many preliminary and advanced studies have been made on the anti-viral properties of various macroalgal sources of sulfated polysaccharides (i.e. in particular sulfated galactans and carrageenans) (Pengzhan et al., 2003).

The reported anti-viral compounds from seaweeds include:

(a) carrageenans (such as λ-carrageenan and partially cyclicized μ/ι-carrageenan) against HSV-1 and 2 from *G. skottsbergii* (Carlucci et al., 1997), from *Gymnogongrus griffithsiae* and *Cryptonemia crenulata* (Talarico et al., 2005)
(b) DL-galactan hybrids, against HSV-1 and 2 from *G. griffithsiae* and *C. crenulata* (Talarico et al., 2005), and against HSV-2 and Dengue virus type 2 (DEN-2) from *Gy. torulosus* (Pujol et al., 2002).
(c) fucoidans, against numerous enveloped virii including HSV-1 from *U. pinnatifida* (Hayashi, Nakano, Hashimoto, Kanekiyo, & Hayashi, 2008),
(d) kainic and domoic acids, as anthelmintic agents from *Digenea simplex* and *Chondria armata* (Freile-Pelegrín & Tasdemir, 2019)
(e) major glycolipids: MGDG (from Hydrolithon reinboldii), DGDG (from Sargassum horneri), and SQDG (from Gigartina tenella, Palmaria palmata) as anti-bacterial, anti-tumor, anti-viral and anti-protozoal agents (Plouguerné, da Gama, Pereira, & Barreto-Bergter, 2014),
(f) xylomannan sulfate, against HSV-1 from *Sebdenia polydactyla* (Ghosh, Pujol, Damonte, Sinha, & Ray, 2009), and
(g) galactan sulfate, and sulfated glucurono-galactan (from *Schizymenia binderi*), against HSV-1 & HSV-2, HIV, RSV, HCMV, Dengue virus, Pseudo-rabies virus, and Influenza A and B virii (Pujol et al., 2002; Talarico et al., 2005).

Studies revealed that the above compounds (with particular emphasis on carrageenans) inhibited the cytopathic effect of HIV, and prevented HIV-induced syncytium (giant cell) formation, as well as other sexually transmitted viruses, such as genital warts, human papilloma virus (HPV) and different strains of *Herpes simplex* virus Type 1 and 2 (HSV-1 and 2) (Buck et al., 2006; Mendis & Kim, 2011; Talarico et al., 2005). Reportedly k, ʎ, and ι-carrageenan inhibited the human papilloma virus (HPV) infection process (Buck et al., 2006), and are potential inhibitors of the multiplication of the Dengue virii, Type-2 and Type-3 (DENV-2 & 3) (Talarico et al., 2005). Fucoidans from *Adenocystis utricularis* were also demonstrated to have anti-viral activities against HIV and human cyto-megalovirus (HCMV) (Gupta & Abu-Ghannam, 2011).

Fucoidans extracted from several seaweed species: (i.e. galactofuran and urono-fucoidan) from the alga *A. utricularis* showed that galactofuran had the highest potency against HSV 1 and 2 with no cyto-toxicity, while urono-fucoidan had no anti-viral activity (Ponce, Pujol, Damonte, Flores, & Stortz, 2003); Also fucoidans, inhibited multiple influenza strains such as Mekabu fucoidan from *U. pinnatifida* against (swine flu) H1N1 (Hayashi et al., 2007) and avian influenza A (bird flu) strains H5N1, H5N3, and H7N2 (Synytsya et al., 2014), and fucoidan and L-fucose from *Cladosiphon okamuranus* against para-influenza (Taoda et al., 2008). Similar properties have been observed from polysaccharide fractions of *Sargassum patens* against HSV 1 and 2 (EC_{50} value of 25 mg/mL and 12.5 mg/mL) respectively (Zhu, Ooi, Chan, & Ang, 2003). It is also important to note the presence of the sulfate group and its degree of sulfation, increased potency of the anti-HIV activity. However, the effectiveness of sulfated polysaccharides in the therapy and/or prophylaxis of retro-viral and opportunistic infections must be tested both in animal and human models (Pengzhan et al., 2003). A recent study from the Marinova company showed that their Maritech® fucoidan was a potent inhibitor of HSV-1 and 2 clinical strains and HCMV on human fibroblast cells by strongly inhibiting their activity or preventing entry into cells (Fitton, 2019).

Secondary metabolites (Table 2) from *Ecklonia cava* and *A. utricularis* seaweeds have been reported to have important anti-viral activities, e.g. tannins from *E. cava* showed potency against HIV-1 by inhibiting polymerase and ribo-nuclease activities of reverse transcriptase (RT). Phlorotannin derivatives from *E. cava* (Table 2) were reported to have a high potency against HIV-1 RT, with some inhibition of protease 8,8′-bieckol with (IC50, 0.51 μM) and 8,4‴dieckol (IC50, 5.3 μM) (Ahn et al., 2004). The phloroglucinol derivative 6,6′- bieckol from *E. cava* showed potency against HIV-1 by induced syncytial formation (EC50 1.72 μM), lytic effects (EC50 1.23 μM), and viral p24 antigen production (EC50 1.26 μM). Of these 6,6′- bieckol selectively inhibited HIV-1 RT with an EC50 of 1.07 μM, as well as HIV-1 infection of cells (Artan et al., 2008).

6.4 Anti-thrombotic agents

Thrombosis is a process when a blood clot or thrombus, blocks the normal and smooth flow in the arteries and veins, which culminates in such prevalent diseases including heart attack and strokes. Fibrin and platelets are the most important components. Fibrin is a protein that forms a mesh and traps red blood cells (in veins, causing heart attack). Platelets, are cells within the bloodstream, which form clumps and add to the thrombus mass (causing arterial strokes). Certain marine algal-derived bioactive compounds from *Callophycus serratus* have been reported to have anti-thrombotic effects including: polysaccharides and lipids; small secondary metabolites; such as polyphenols, halogenated phenols, and meroterpenes (Lin et al., 2010). Cardiovascular disease (CVD) is a leading cause of death globally (Cornish, Critchley, & Mouritsen, 2015). An estimated mortality of 17.9 million people or (31%) of all global deaths in 2016 was reportedly due to CVD. Of all of these deaths, 85% were due to heart attack and stroke (Cardiovascular diseases, 2017). Recent studies on human gut microbes or host/microbe inter-dependence (dysbiosis) were reported to be associated with various pathologies, e.g. inflammation, obesity, hypertension, endothelial dysfunction, diabetes, CVD, colitis, and neurological disorders (Cornish, Critchley & Mouritsen, 2019; Mohamed et al., 2012). Since CVD is highly influenced by human food digestion, carbohydrate fermentation, energy harvesting and storage, gene expression, the secondary generation of beneficial compounds (cross-feeding), and the facilitation of metabolic functions, makes applications and utilization of therapeutic dietary impacts of seaweeds to be viable and important for intervention strategies related to food manufacturing and consumption (Cornish et al., 2015, 2019). Judicious consumption of selected seaweeds has been implicated to lower incidence factors involved in CVD occurrence (Mohamed et al., 2012).

A strong-gelling sodium alginate extracted from a particular brown seaweed described in Patent Number WO2007039294 (containing a suggested 1.5 g daily dose) had an indirect, cardiovascular beneficial effect. Sodium alginate was found to limit glucose and cholesterol uptake by reversing the symptoms of hyper-cholesterolemia and diabetes mellitus, thereby reducing CVD risks. Alginates generally are used for tissue engineering and regeneration due to their bio-compatibility and non-thrombogenic nature, also have applications supporting cardiac implants after acute myocardial infarction (Mohamed et al., 2012). Sodium alginate was suggested as a vehicle for the delivery of bioactive molecules, regenerative factors, or even stem-cells into the heart (Paxman, Richardson, Dettmar, & Corfe, 2008).

Another seaweed-derived molecule, i.e. a (400 mg/d) fucoidan fraction from *Laminaria japonica* was administered to nine healthy volunteers for five weeks. This treatment showed an anti-thrombotic effect, as monitored by an *ex vivo* global thrombosis test and measurement of the time required for thrombus lysis (Ren et al., 2013). Dietary fucoidan (derived from

C. okamuranus) stimulated H_2O_2-producing enzymes in intestinal epithelial cells and released H_2O_2 into the blood. This is interpreted as a signaling function which increases prostacyclin (PGI2) production, which in turn shortens the time for thrombolysis. This effect was only found as a result of fucoidan ingestion. No anti-thrombotic effects were observed in the fucoxanthin-treated volunteers. Fucoidan from *C. okamuranus* reportedly provided cardio-protection against isoproterenol-induced myocardial infarction in rats (Thomes, Rajendran, Pasanban, & Rengasamy, 2010) which reduced myocardial damage, oxidative stress, re-equilibrated cholesterol balance (both triglyceride, high-density lipoprotein (HDL), low-density lipoprotein (LDL), and total cholesterol levels) (Huang, Wen, Gao, & Liu, 2010; Thomes et al., 2010). Fucoidan from *L. japonica*, as tested on hyper-lipidemic rats, showed decreased serum levels of (triglyceride, LDL, and total cholesterol), while increasing HDL cholesterol levels and the activities of enzymes linked to the metabolism of lipo-proteins (e.g. lipo-protein lipase, hepatic triglyceride lipase, and lecithin cholesterol acyltransferase) (Huang et al., 2010).

Alpha-linolenic acid (ALA), the precursor to EPA and DHA, inhibits the production of eicosanoids, lowering blood pressure, and blood triglyceride levels (Oomah & Mazza, 1999) while docosapentaenoic acid (DPA) was shown to inhibit platelet aggregation and has strong endothelial cell migration abilities (Kaur, Cameron-Smith, Garg, & Sinclair, 2011). Commercial ALA or hybrid liposomes of ALA ethyl ester (HL-ALAE) seemed to provide benefits against several health problems e.g., decrease in blood cholesterol levels, reduction in the risk of heart attacks, and a growth reduction in breast, colon, and prostate cancer (Tanaka, Goto, Matsumoto, & Ueoka, 2008). Researchers have suggested that elevated blood pressure in adulthood may be associated with peri-natal omega-3 fatty acid deficiency and that early exposure to dietary, long-chain omega-3 played a critical role in supporting heart health and reducing CVD risk in later life (Papanikolaou, Brooks, Reider, & Fulgoni, 2014). Given the situation above, then it is safe to assume that consumption of recommended amounts of omega-3s as ALA in food, or in a supplement, would be generally beneficial to consumers.

6.5 Immune stimulating activities

Beneficial applications of selected seaweed-derived biomolecules is clearly aimed to target nutrition and well-being of humans, whether for nutraceutical, pharmaceutical, or biotechnological applications. Vitamin C is present in all seaweed groups and in relatively large amounts in (green and brown) with substantial levels reportedly in red seaweeds (i.e. from 500 to 3000 mg/kg) (Rajapakse & Kim, 2011). Some seaweeds have larger amounts of vitamin C, i.e. in *Porphyra umbilicalis* (12.88 mg/100 g wet weight) and *P. palmata* (5.5 mg/100 g wet weight) than in some terrestrial foods such as bovine milk (1.7 mg/100 g edible portion) or orange (5 mg/100 g edible portion) (MacArtain et al., 2007; Swarnalatha, 2018).

Consumption of selected seaweeds, as part of the diet and as a source of vitamin C, can strengthen the immune system (Burtin, 2003). Several species of *Eucheuma*, *Ulva*, and *Gracilaria* are reported to have bioactive lectins that interact with specific glycan structures and is also useful in the detection of disease related alterations of glycan synthesis, infectious agents (e.g. virii, bacteria, fungi, parasites), antibiotic, mitogenic, cytotoxic effects, and many others (Holdt & Kraan, 2011; Michalak & Chojnacka, 2015). Peptides from *Porphyra columbina* were reported as immuno-suppressive and anti-hypertensive (Cian, Martinez-Augustin, & Drago, 2012).

Metabolism of polyunsaturated fatty acids (PUFAs) causes the production of different eicosanoids, which are known to regulate cellular processes. The roles of eicosapentaenoic acid (EPA), arachidonic acid (AA), and docosahexaenoic acid (DHA) in human health, are to have an effect on immune-regulation (Gillies, Harris, & Kris-Etherton, 2011; Poudyal et al., 2013), cancer prevention (Slagsvold, Pettersen, Follestad, Krokan, & Schønberg, 2009), and pre- and post-natal development of the brain and retina (Guesnet & Alessandri, 2011). Prostaglandins, thromboxane, oxylipins, and other eicosanoids are produced with the aid of its precursors AA and EPA (Qi et al., 2002). These help to regulate inflammation processes and immune reactions (Poudyal et al., 2013) and were evident in the red seaweed *P. umbilicalis*, and also kelp (*Laminaria*) pertaining to important functions in skin growth and protection (Pereira, 2018a). Inhibition of the pro-lipid mediator, leukotriene B4 synthesis (a pro-inflammatory) by stearidonic acid (STA) was tested for use in cystic fibrosis treatments (Primdahl, Tungen, Aursnes, Hansen, & Vik, 2015; Schubert et al., 2007). These studies suggested that STAs found in *Undaria* and *Ulva* spp. affected the immune system in humans (Ishihara et al., 1998). A palmitic acid extract from the calcareous red seaweed *Amphiroa zonata* showed anti-tumor activity by inducing apoptosis in the human leukemic cell line MOLT-4 at 50 μg/mL, and also in an *in vivo* anti-tumor test in mice (Harada et al., 2002).

7. Conclusions

All the red seaweeds described in this chapter, except for A. taxiformis, have a long history of use as a source of sea vegetables, especially in the northern part of Luzon. These are mainly gathered by the locals during the peak months of growth, from November to February, i.e., the colder months of the year.

There is no known cultivation of these seaweeds except for *H. durvillei* which has just been initiated (Trono In Prep). Basic research on the biology and eco-ecophysiology of wild plants leading to their domestication for cultivation is imperative to ensure meeting the anticipated market demands of these seaweeds.

Only a few studies have been reported on the target species presented here for their different biological activities. It will be noteworthy then to explore and pursue further the potential of these emerging red seaweeds with several applications, considering their wide distribution in the country and potential economic value if produced sustainably.

References

Agngarayngay, Z. M. (1984a). Marine macro-algae of Ilocos Norte. II. Phaeophyta and *Rhodopyta. Ilocos Fisheries Journal, 1,* 1—66.

Agngarayngay, Z. M. (1984b). Species of edible seaweeds of Ilocos Norte. *Ilocos Fisheries Journal, 1,* 118—133.

Agngarayngay, Z. M., Llaguno, A. F. C., Aquino, S. G., Taclan, L. B., & Galacgac, E. S. (2005). *Edible seaweeds of Ilocos Norte: Food reparations other local uses and market potentials.*

Ahn, M.-J., Yoon, K.-D., Min, S.-Y., Lee, J. S., Kim, J. H., Kim, T. G., et al. (2004). Inhibition of HIV-1 reverse transcriptase and protease by phlorotannins from the brown alga *Ecklonia cava. Biological & Pharmaceutical Bulletin, 27*(4), 544—547.

Ali, M. M., Yashir, S., Critchley, A. T., & Hurtado, A. Q. (2018). Impacts of *Ascophyllum* marine plant extract powder (AMPEP) on the growth, incidence of the endophyte *Neosiphonia apiculata* and associated carrageenan quality of three, commercial cultivars of *Kappaphycus. Journal of Applied Phycology, 30,* 1185—1196.

Ame, E. C., Ayson, J. P., Okuda, K., & Andres, R. (2010). Porphyra fisheries in the northern Philippines:
Some environmental issues and the socio- economic impact on the ilocano fisherfolk. *Kuroshio Science, 4,* 53—58.

Arao, T., & Yamada, M. (1989). Positional distribution of fatty acids in galactolipids of algae. *Phytochemistry, 28*(3), 805—810. https://doi.org/10.1016/0031-9422(89)80119-0.

Ariffin, S. H. Z., Yeen, W. W., Abidin, I. Z. Z., Abdul Wahab, R. M., Ariffin, Z. Z., & Senafi, S. (2014). Cytotoxicity effect of degraded and undegraded kappa and iota carrageenan in human intestine and liver cell lines. *BMC Complementary and Alternative Medicine, 14,* 508. https://doi.org/10.1186/1472-6882-14-508.

Artan, M., Li, Y., Karadeniz, F., Lee, S.-H., Kim, M.-M., & Kim, S.-K. (2008). Anti-HIV-1 activity of phloroglucinol derivative, 6,6'-bieckol, from *Ecklonia cava. Bioorganic & Medicinal Chemistry, 16*(17), 7921—7926. https://doi.org/10.1016/j.bmc.2008.07.078.

Baweja, P., Kumar, S., Sahoo, D., & Levine, I. (2016). Biology of seaweeds. In J. Fleurence, & I. Levine (Eds.), *Seaweed in health and disease prevention* (pp. 41–106). Amsterdam Boston Heidelberg London: Elsevier, Academic Press.
Buck, C. B., Thompson, C. D., Roberts, J. N., Müller, M., Lowy, D. R., & Schiller, J. T. (2006). Carrageenan is a potent inhibitor of papillomavirus infection. *PLoS Pathogens, 2*(7), e69. https://doi.org/10.1371/journal.ppat.0020069.
Burtin, P. (2003). *Nutritional value of seaweeds*. Retrieved December 6, 2018, from ResearchGate website: https://www.researchgate.net/publication/228554296_Nutritional_value_of_seaweeds.
Buschmann, A. H., Camus, C., Infante, J., Neori, A., Israel, A., Hernández-González, M. C., Pereda, S. V., Pinchetti, J. L., Golberg, A., Tadmor-Shalev, N., & Critchley, A. T. (2017). Seaweed production: Overview of the global state of exploitation, farming and emerging research activity. *European Journal of Phycology, 52*(4), 391–406.
Camara, R. B. G., Costa, L. S., Fidelis, G. P., Nobre, L. T. D. B., Dantas-Santos, N., Cordeiro, S. L., et al. (2011). Heterofucans from the brown seaweed *Canistrocarpus cervicornis* with anticoagulant and antioxidant activities. *Marine Drugs, 9*(1), 124–138. https://doi.org/10.3390/md9010124.
Campo, V., Kawano, D., Braz da Silva, D., & Ivone, C. (2009). Carrageenans: Biological properties, chemical modifications and structural analysis – A review. *Carbohydrate Polymers, 77*, 167–180. https://doi.org/10.1016/j.carbpol.2009.01.020.
Cardiovascular diseases (CVDs), 2017. Retrieved December 6, 2018, from: https://www.who.int/news-room/fact-sheets/detail/cardiovascular-diseases-(cvds).
Carlucci, M. J., Pujol, C. A., Ciancia, M., Noseda, M. D., Matulewicz, M. C., Damonte, E. B., et al. (1997). Antiherpetic and anticoagulant properties of carrageenans from the red seaweed Gigartina skottsbergii and their cyclized derivatives: Correlation between structure and biological activity. *International Journal of Biological Macromolecules, 20*(2), 97–105. https://doi.org/10.1016/S0141-8130(96)01145-2.
Ciancia, M., Quintana, I., & Cerezo, A. S. (2010). Overview of anticoagulant activity of sulfated polysaccharides from seaweeds in relation to their structures, focusing on those of green seaweeds. *Current Medicinal Chemistry, 17*(23), 2503–2529.
Cian, R. E., Martinez-Augustin, O., & Drago, S. R. (2012). Bioactive properties of peptides obtained by enzymatic hydrolysis from protein byproducts of *Porphyra columbina. Food Research International, 49*(1), 364–372.
Codero, P. A., Jr. (1974). Phycological observations I: Genus *Porphyra* of the Philippines, its species and their occurrences. *Bulletin of the Japanese Society of Phycology, 22*, 134–142.
Colliec, S., Fischer, A. M., Tapon-Bretaudiere, J., Boisson, C., Durand, P., & Jozefonvicz, J. (1991). Anticoagulant properties of a fucoïdan fraction. *Thrombosis Research, 64*(2), 143–154.
Cordero, P. A., Jr. (1976). Phycological observations – II: *Porphyra marcosii*, a new species from the Philippines. *Acta Manilana Research Center, University of Sto Tomas, Series. A, 24*(15), 14–24.
Cordero, P. A., Jr. (1977). Studies of Philippine marine red algae. *Special Publications from the Seto Marine Biological Laboratory, 4*, 1–258.
Cordero, P. A., Jr. (1980). Taxonomy and distribution of Philippine useful seaweeds. *National Research Council of the Philippines Bulletin No, 81*, 36–41.
Cornish, M. L., Critchley, A. T., & Mouritsen, O. G. (2015). A role for dietary macroalgae in the amelioration of certain risk factors associated with cardiovascular disease. *Phycologia, 54*(6), 649–666. https://doi.org/10.2216/15-77.1.
Cornish, M. L., Mouritsen, O. G., & Critchley, A. T. (2019). A mini-review on the microbial continuum: Consideration of a link between judicious consumption of a varied diet of macroalgae and human health and nutrition. *Journal of Oceanology and Limnology*. https://doi.org/10.1007/s00343-019-8104-2.

da Costa, E., Melo, T., Moreira, A. S. P., Bernardo, C., Helguero, L., Ferreira, I., et al. (2017). Valorization of lipids from *Gracilaria* sp. through lipidomics and decoding of antiproliferative and anti-inflammatory activity. *Marine Drugs, 15*(3). https://doi.org/10.3390/md15030062.

Critchley, A. T., Largo, D., Wee, W., Bleicher L'honneur, G., Hurtado, A. Q., & Schubert, J. (2004). A preliminary summary on *Kappaphycus* farming and the impact of epiphytes. *Japanese Journal of Phycology, 52*(Supplement), 231–232.

Cumashi, A., Ushakova, N. A., Preobrazhenskaya, M. E., D'Incecco, A., Piccoli, A., Totani, L., et al. (2007). A comparative study of the anti-inflammatory, anticoagulant, antiangiogenic, and antiadhesive activities of nine different fucoidans from brown seaweeds. *Glycobiology, 17*(5), 541–552. https://doi.org/10.1093/glycob/cwm014.

Dias, P. F., Siqueira, J. M., Vendruscolo, L. F., de Neiva, T. J. C., Gagliardi, A. R., Maraschin, M., et al. (2004). Antiangiogenic and antitumoral properties of a polysaccharide isolated from the seaweed *Sargassum* stenophyllum. *Cancer Chemotherapy and Pharmacology, 56*, 436–446. https://doi.org/10.1007/s00280-004-0995-7.

Doty, M. S. (1995). *Betaphycus philippinensis* gen. et. sp. nov. and related species (Solieriaceae, Gigartinales). In I. A. Abbott (Ed.), *Taxonomy of economic seaweeds with reference to some Pacific species* (Vol. 5, pp. 237–245).

Duarte, M. E. R., Cauduro, J. P., Noseda, D. G., Noseda, M. D., Gonçalves, A. G., Pujol, C. A., et al. (2004). The structure of the agaran sulfate from *Acanthophora* spicifera (Rhodomelaceae, Ceramiales) and its antiviral activity. Relation between structure and antiviral activity in agarans. *Carbohydrate Research, 339*(2), 335–347. https://doi.org/10.1016/j.carres.2003.09.028.

Dumay, J., & Morançais, M. (2016). Proteins and pigments. In J. Fleurence, & I. Levine (Eds.), *Algae: A way of life and health* (pp. 275–318). Amsterdam Boston Heidelberg London: Elsevier, Academic Press.

Dumilag, R. V. (2018). *Betaphycus gelatinus* and *B. philippinensis* (Gigartinales, Rhodophyta) are conspecific. *Phycotaxa, 372*, 022–034.

Dumilag, R. V., Aguinaldo, Z. A., Mintu, C. B., Quinto, M. P., Ame, E. C., Andres, R. C., et al. (2016). Morphological and molecular confirmation of the occurrence of *Pyropia tanegashimensis* (Bangiales, Rhodophyta) from Palaui is., Sta. Ana, Cagayan, Philippines. *Phytotaxa, 255*, 083–090.

Dumilag, R. V., Aguinaldo, Z. A., Mintu, C. B., Quinto, M. P., Ame, E. C., Andres, R. C., et al. (2017). A review of the current taxonomic status of foliose *Bangiales* (Rhodophyta) in the Philippines. *Phytotaxa, 312*, 047–059.

Dumilag, R. V., Liao, L. M., & Lluisma, A. O. (2014). Phylogeny of *Betaphycus* (Gigartinales, Rhodophyta) as inferred from COI sequences and morphological observations on *B. philippinensis*. *Journal of Applied Phycology, 26*, 587–595.

Dumilag, R. V., & Monotilla, W. D. (2018). Molecular diversity and biogeography of Philippine foliose *Bangiales* (Rhodophyta). *Journal of Applied Phycology, 30*, 173–186.

Eitsuka, T., Nakagawa, K., Igarashi, M., & Miyazawa, T. (2004). Telomerase inhibition by sulfoquinovosyldiacylglycerol from edible purple laver (Porphyra yezoensis). *Cancer Letters, 212*(1), 15–20. https://doi.org/10.1016/j.canlet.2004.03.019.

Fitton, H. (2019). *Maritech organic fucoidans: The science* (Vol. 28).

Fleurence, J. (1999). Seaweed proteins: Biochemical, nutritional aspects and potential uses. *Trends in Food Science & Technology, 10*(1), 25–28. https://doi.org/10.1016/S0924-2244(99)00015-1.

Fleurence, J. (2004). Seaweed proteins. In R. Y. Yada (Ed.), *Proteins in food processing* (pp. 197–213). Woodhead Publishing in Food Science and Technology. Google Search. Retrieved January 9, 2019, from https://www.google.com/search?source=hp&ei=r401XPqaDJaBoATA0Ygo&q=Fleurence%2C+J.%2C+2004.+Seaweed+proteins.+In%3A+Yada%2C+R.Y.+%28Ed.%29%2C+Proteins+in+Food+Processing.

+Woodhead+Publishing+in+Food+Science+and+Technology%2C+pp.+197%E2%80%93213.&btnK=Google+Search&oq=Fleurence%2C+J.%2C+2004.+Seaweed+proteins.+In%3A+Yada%2C+R.Y.+%28Ed.%29%2C+Proteins+in+Food+Processing.+Woodhead+Publishing+in+Food+Science+and+Technology%2C+pp.+197%E2%80%93213.&gs_l=psy-ab.3...6472.6472..7373...0.0..0.717.1335.5-1j1......0....1j2..gws-wiz.....0.RYMGm03V_7w.

Food and Agriculture Organization of the United Nations. (2014). *The state of world fisheries and aquaculture* (Rome, Italy).

Food and Agriculture Organization of the United Nations. (2017). *The state of world fisheries and aquaculture* (Rome, Italy).

Freile-Pelegrín, Y., & Tasdemir, D. (2019). Seaweeds to the rescue of forgotten diseases: A review. *Botanica Marina*. https://doi.org/10.1515/bot-2018-0071.

Fujiwara-Arasakil, T., Minol, N., & Kuroda, M. (1984). The protein value in human nutrition of edible marine algae in Japan. *Hydrobiologia, 116*, 513–516.

Ganesan, P., Matsubara, K., Sugawara, T., & Hirata, T. (2013). Marine algal carotenoids inhibit angiogenesis by down-regulating FGF-2-mediated intracellular signals in vascular endothelial cells. *Molecular and Cellular Biochemistry, 380*, 1–9. https://doi.org/10.1007/s11010-013-1651-5.

Ganzon-Fortrs, E. T. (2012). A historical account of biodiversity studies on Philippine seaweeds (1800–1999). *Coastal Marine Science, 35*(1), 182–201.

Ghosh, T., Pujol, C. A., Damonte, E. B., Sinha, S., & Ray, B. (2009). Sulfated xylomannans from the red seaweed *Sebdenia polydactyla* : Structural features, chemical modification and antiviral activity. *Antiviral Chemistry and Chemotherapy, 19*(6), 235–242. https://doi.org/10.1177/095632020901900603.

Gillies, P. J., Harris, W. S., & Kris-Etherton, P. M. (2011). Omega-3 fatty acids in food and pharma: The enabling role of biotechnology. *Current Atherosclerosis Reports, 13*(6), 467–473. https://doi.org/10.1007/s11883-011-0206-z.

Guesnet, P., & Alessandri, J. M. (2011). Docosahexaenoic acid (DHA) and the developing central nervous system (CNS) – implications for dietary recommendations. *Biochimie, 93*(1), 7–12. https://doi.org/10.1016/j.biochi.2010.05.005.

Gupta, S., & Abu-Ghannam, N. (2011). Bioactive potential and possible health effects of edible brown seaweeds. *Trends in Food Science & Technology, 22*(6), 315–326. https://doi.org/10.1016/j.tifs.2011.03.011.

van Hal, J. W., Huijgen, W. J. J., & López-Contreras, A. M. (2014). Opportunities and challenges for seaweed in the biobased economy. *Trends in Biotechnology, 32*(5), 231–233. https://doi.org/10.1016/j.tibtech.2014.02.007.

Hamed, I., Özogul, F., Özogul, Y., & Regenstein, J. M. (2015). Marine bioactive compounds and their health benefits: A review: Importance of functional seafood. *Comprehensive Reviews in Food Science and Food Safety, 14*(4), 446–465. https://doi.org/10.1111/1541-4337.12136.

Harada, H., Yamashita, U., Kurihara, H., Fukushi, E., Kawabata, J., & Kamei, Y. (2002). Antitumor activity of palmitic acid found as a selective cytotoxic substance in a marine red alga. *Anticancer Research, 22*(5), 2587–2590.

Hayashi, T., Hayashi, K., Kanekiyo, K., Ohta, Y., Lee, J.-B., Hashimoto, M., et al. (2007). *Promising antiviral glyco-molecules from an edible alga*. Hoboken, NJ, USA: John Wiley & Sons.

Hayashi, K., Nakano, T., Hashimoto, M., Kanekiyo, K., & Hayashi, T. (2008). Defensive effects of a fucoidan from brown alga *Undaria pinnatifida* against herpes simplex virus infection. *International Immunopharmacology, 8*(1), 109–116. https://doi.org/10.1016/j.intimp.2007.10.017.

Hayashi, L., Reis, R. P., Alves dos Santos, A. A., Castelar, B., Robledo, D., de Vega, G. B., et al. (2017). The cultivation of *Kappaphycus* and *Eucheuma* in tropical and sub-tropical waters. In A. Q. Hurtado, A. T. Critchley, & I. C. Neish (Eds.), *Tropical seaweed farming trends, problems and opportunities: Focus on spinosum and cottonii of commerce* (pp. 55–90). Netherlands: Springer.

Holdt, S. L., & Kraan, S. (2011). Bioactive compounds in seaweed: Functional food applications and legislation. *Journal of Applied Phycology, 23*(3), 543–597. https://doi.org/10.1007/s10811-010-9632-5.

Hotimchenko, S. V. (2002). Fatty acid composition of algae from habitats with varying amounts of illumination. *Russian Journal of Marine Biology, 28*(3), 218–220. https://doi.org/10.1023/A:1016861807103.

Huang, L., Wen, K., Gao, X., & Liu, Y. (2010). Hypolipidemic effect of fucoidan from *Laminaria* japonica in hyperlipidemic rats. *Pharmaceutical Biology, 48*(4), 422–426. https://doi.org/10.3109/13880200903150435.

Hurtado, A. Q. (2018). Seaweed-carrageenan industry of the Philippines: Quo vadis?. In *Paper presented at the 2018 national seaweed congress, Cebu city, Philippines, 10–12 October 2018.*

Hurtado-Ponce, A. Q. (1983). Marine macrobenthic algae of Currimao, Ilocos Norte. *Ilocos Fisheries Journal, 1*, 104–149.

Hurtado-Ponce, A. Q. (1984). List of edible seaweeds of Currimao, Ilocos Norte. *Ilocos Fisheries Journal, 1*, 179–181.

Hurtado, A. Q., & Critchley, A. T. (2006). Seaweed industry of the Philippines and the problem of epiphytism in *Kappaphycus* farming. In S. M. Phang, A. T. Critchley, & P. Ang, Jr. (Eds.), *Advances in seaweed cultivation and utilization in Asia* (pp. 21–28). Kuala Lumpur: University of Malaya Maritime Research Centre, University of Malaya.

Hurtado, A. Q., Gerung, G. S., Yasir, S., & Critchley, A. T. (2014). Cultivation of tropical red seaweeds in the BIMP-EAGA region. *Journal of Applied Phycology, 26*, 702–718.

Hurtado, A. Q., Joe, M., Sanares, R. C., Fan, D., Prithiviraj, B., & Critchley, A. T. (2012). Investigation of the application of Acadian Marine plant extract powder (AMPEP) to enhance the growth, phenolic content, free radical scavenging, and iron chelating activities of *Kappaphycus* Doty (Solieriaceae, Gigartinales, Rhodophyta). *Journal of Applied Phycology, 24*, 601–611.

Hurtado, A. Q., Luhan, M. R. J., & Guanzon, N. G., Jr. (2006). *Seaweeds of panay*. Tigbauan, Iloilo: SEAFDEC Aquaculture Department (Monograph). 50pp.

Hurtado, A. Q., Neish, I. C., & Critchley, A. T. (2019). Phyconomy: The extensive cultivation of seaweeds, sustainability and economic value – with reference from experiences in eucheumatoid farming. *Phycologia*. https://doi.org/10.1080/00318884.2019.1625632.

Ishihara, K., Murata, M., Kaneniwa, M., Saito, H., Shinohara, K., & Maeda-Yamamoto, M. (1998). Inhibition of icosanoid production in MC/9 mouse mast cells by n-3 polyunsaturated fatty acids isolated from edible marine algae. *Bioscience, Biotechnology, and Biochemistry, 62*(7), 1412–1415. https://doi.org/10.1271/bbb.62.1412.

Ito, K., Tsuchiya, Y., & Tohoku, U. (1977). Differential fatty acid composition of some marine algae associated with their habitat depths. *Tohoku Journal of Agricultural Research*. Retrieved from http://agris.fao.org/agris-search/search.do?recordID=JP7806824.

Jin, D.-Q., Lim, C. S., Sung, J.-Y., Choi, H. G., Ha, I., & Han, J.-S. (2006). *Ulva conglobata*, a marine algae, has neuroprotective and anti-inflammatory effects in murine hippocampal and microglial cells. *Neuroscience Letters, 402*, 154–158.

Kaur, G., Cameron-Smith, D., Garg, M., & Sinclair, A. J. (2011). Docosapentaenoic acid (22:5n-3): A review of its biological effects. *Progress in Lipid Research, 50*(1), 28–34. https://doi.org/10.1016/j.plipres.2010.07.004.

Kawasaki, A., Ono, A., Mizuta, S., Kamiya, M., Takenaga, T., & Murakami, S. (2017). The taurine content of Japanese seaweed. In *Advances in experimental medicine and biology* (Vol. 975, pp. 1105–1112). https://doi.org/10.1007/978-94-024-1079-2_88.

Kendel, M., Barnathan, G., Fleurence, J., Rabesaotra, V., & Wielgosz-Collin, G. (2013). Non-methylene interrupted and hydroxy fatty acids in polar lipids of the alga Grateloupia turuturu over the four seasons. *Lipids, 48*(5), 535–545. https://doi.org/10.1007/s11745-013-3783-5.

Kho, F. B., Orbita, M. L. S., & Manting, M. M. E. (2016). Mindanao, Philippines. *Carrageenan and biochemical composition of three species of Halymenia (Halymenia durvillaea bory de Saint-Vincent, Halymenia maculata J. Agardh and Halymenia dilatata zanardini) in initao, misamis oriental* (Vol. 8).

Kim, J. K., Yarish, C., Hwang, E. K., Park, M., & Kim, Y. (2017). Seaweed aquaculture: Cultivation technologies, challenges and its ecosystem services. *Algae, 32*, 1–13.

Kinley, R. D., de Nys, R., Vucko, M. J., Machado, L., & Tomkins, N. W. (2016). The red macroalgae *Asparagopsis* taxiformis is a potent natural anti-methanogenic that reduces methane production during in vitro fermentation with rumen fluid. *Animal Production Science, 256*, 282–289.

Koch, K., Hagen, W., Graeve, M., & Bischof, K. (2017). Fatty acid compositions associated with high-light tolerance in the intertidal rhodophytes *Mastocarpus stellatus* and *Chondrus crispus. Helgoland Marine Research, 71*(1), 15. https://doi.org/10.1186/s10152-017-0495-x.

Largo, D. B. (2002). Recent development in seaweed diseases. In A. Q. Hurtado, N. G. Guanzon, Jr., T. R. de Castro-Mallare, & M. R. J. Luhan (Eds.), *Proc nat seaweed planning workshop. Tigbauan, Iloilo* (pp. 35–42).

Lin, A.-S., Stout, E. P., Prudhomme, J., Roch, K. L., Fairchild, C. R., Franzblau, S. G., et al. (2010). Bioactive bromophycolides R–U from the Fijian red alga Callophycus serratus. *Journal of Natural Products, 73*(2), 275–278. https://doi.org/10.1021/np900686w.

Liu, J., Kandasamy, S., Zhang, J., Kirby, C. W., Karakach, T., Hafting, J., et al. (2015). Prebiotic effects of diet supplemented with the cultivated red seaweed *Chondrus crispus* or with fructo-oligo-saccharide on host immunity, colonic microbiota and gut microbial metabolites. *BMC Complementary and Alternative Medicine, 15*, 279. https://doi.org/10.1186/s12906-015-0802-5.

Liu, Z., Gao, T., Yang, Y., Meng, F., Zhan, F., Jiang, Q., & Sun, X. (2019). Anti-cancer activity of porphyran and carrageenan from red seaweeds. *Molecules, 24*, 4286. https://doi.org/10.3390/molecules24234286.

Lordan, S., Ross, R. P., & Stanton, C. (2011). Marine bioactives as functional food ingredients: Potential to reduce the incidence of chronic diseases. *Marine Drugs, 9*(6), 1056–1100. https://doi.org/10.3390/md9061056.

Lourenço, S. O., Barbarino, E., De-Paula, J. C., Pereira, L. O. da S., & Marquez, U. M. L. (2002). Amino acid composition, protein content and calculation of nitrogen-to-protein conversion factors for 19 tropical seaweeds. *Phycological Research, 50*(3), 233–241. https://doi.org/10.1046/j.1440-1835.2002.00278.x.

MacArtain, P., Gill, C. I. R., Brooks, M., Campbell, R., & Rowland, I. R. (2007). Nutritional value of edible seaweeds. *Nutrition Reviews, 65*(12), 535–543. https://doi.org/10.1111/j.1753-4887.2007.tb00278.x.

Machado, L. (2015). *Tropical macroalgae as a natural alternative for the mitigation of methane emissions in ruminant livestock systems*. PhD thesis. Australia: James Cook University http://researchonline.jcu.edu.au/46419/.

Magalhaes, K. D., Costa, L. S., Fidelis, G. P., Oliveira, R. M., Nobre, L. T. D. B., Dantas-Santos, N., et al. (2011). Anticoagulant, antioxidant and antitumor activities of heterofucans from the seaweed Dictyopteris delicatula. *International Journal of Molecular Sciences, 12*(5), 3352–3365. https://doi.org/10.3390/ijms12053352.

Marcos-Agngarayngay, Z. D. (1984). Marine macro-algae of Ilocos Norte, II: Phaeophyta and Rhodophyta. *Ilocos Fisheries Journal, 1*(2), 1–66.

McDermid, K. J., & Stuercke, B. (2003). Nutritional composition of edible Hawaiian seaweeds. *Journal of Applied Phycology, 15*(6), 513–524. https://doi.org/10.1023/B:JAPH.0000004345.31686.7f.

McKim, J. M., Baas, H., Rice, G. P., Willoughby, J. A., Weiner, M. L., & Blakemore, W. (2016). Effects of carrageenan on cell permeability, cytotoxicity, and cytokine gene expression in human intestinal and hepatic cell lines. *Food and Chemical Toxicology, 96*, 1–10. https://doi.org/10.1016/j.fct.2016.07.006.

Melo, T., Alves, E., Azevedo, V., Martins, A. S., Neves, B., Domingues, P., et al. (2015). Lipidomics as a new approach for the bioprospecting of marine macroalgae — unraveling the polar lipid and fatty acid composition of *Chondrus crispus. Algal Research, 8*, 181–191. https://doi.org/10.1016/j.algal.2015.02.016.

Mendis, E., & Kim, S.-K. (2011). Present and future prospects of seaweeds in developing functional foods. *Advances in Food & Nutrition Research, 64*, 1–15. https://doi.org/10.1016/B978-0-12-387669-0.00001-6.

Michalak, I., & Chojnacka, K. (2015). Algae as production systems of bioactive compounds. *Engineering in Life Science, 15*(2), 160–176. https://doi.org/10.1002/elsc.201400191.

Milstein, D., Madeiros, A. S., Oliveira, E. C., & Oliveira, M. C. (2015). Native or introduced? A re-evaluation of *pyropia* species (*Bangiales*, Rhodophyta) from Brazil based on molecular analysis. *European Journal of Phycology, 50*, 37–45. https://doi.org/10.1080/09670262.2014.982202.

Mišurcová, L., Kráčmar, S., Klejdus, B., & Vacek, J. (2010). Nitrogen content, dietary fiber, and digestibility in algal food products. *Czech Journal of Food Sciences*. https://doi.org/10.17221/111/2009-CJFS.

Mohamed, S., Hashim, S. N., & Rahman, H. A. (2012). Seaweeds: A sustainable functional food for complementary and alternative therapy. *Trends in Food Science & Technology, 23*(2), 83–96. https://doi.org/10.1016/j.tifs.2011.09.001.

Monotilla, W. D., & Notoya, M. (2004). Morphological and physiological responses of *Porphyra suborbiculata* Kjellman (*Bangiales*, Rhodophyta) blades from five localities. *Botanica Marina, 47*, 323–334.

Monotilla, W. D., & Notoya, M. (2010). Growth and development of *Porphyra marcosii* (*Bangiales*, Rhodophyta) under different temperatures and photoperiod. *Philippine Journal of Science, 139*, 197–206.

Murad, H., Ghannam, A., Al-Ktaifani, M., Abbas, A., & Hawat, M. (2015). Algal sulfated carrageenan inhibits proliferation of MDA-MB-231 cells via apoptosis regulatory genes. *Molecular Medicine Reports, 11*(3), 2153–2158. https://doi.org/10.3892/mmr.2014.2915.

Okolie, C., Mason, B., & Critchley, A. T. (2018). Seaweeds as a source of proteins for use in pharmaceuticals and high-value applications. In M. Hayes (Ed.), *Novel Proteins for Food, Pharmaceuticals and Agriculture: Sources, Applications and Advances*. Wiley. Chapter: 11 (Edition: 1). https://onlinelibrary.wiley.com/doi/abs/10.1002/9781119385332.ch11.

Oomah, B., & Mazza, G. (1999). Health benefits of phytochemicals from selected Canadian crops. *Trends in Food Science & Technology, 10*(6–7), 193–198. https://doi.org/10.1016/S0924-2244(99)00055-2.

Pangestuti, R., & Kim, S. K. (2011). Neuroprotective effects of marine algae. *Marine Drugs, 9*, 803–818.

Papanikolaou, Y., Brooks, J., Reider, C., & Fulgoni, V. L. (2014). U.S. Adults are not meeting recommended levels for fish and omega-3 fatty acid intake: Results of an analysis using observational data from NHANES 2003–2008. *Nutrition Journal, 13*, 31. https://doi.org/10.1186/1475-2891-13-31.

Paxman, J. R., Richardson, J. C., Dettmar, P. W., & Corfe, B. M. (2008). Daily ingestion of alginate reduces energy intake in free-living subjects. *Appetite, 51*(3), 713–719. https://doi.org/10.1016/j.appet.2008.06.013.

Pengzhan, Y., Quanbin, Z., Ning, L., Zuhong, X., Yanmei, W., & Zhi'en, L. (2003). Polysaccharides from Ulva pertusa (*Chlorophyta*) and preliminary studies on their antihyperlipidemia activity. *Journal of Applied Phycology, 15*(1), 21–27. https://doi.org/10.1023/A:1022997622334.

Pereira, L. (2016). *Edible Seaweeds of the World*. CRC Press. https://doi.org/10.1201/b19970.

Pereira, L. (2018a). Seaweeds as source of bioactive substances and skin care therapy—cosmeceuticals, algotheraphy, and thalassotherapy. *Cosmetics, 5*(4), 68. https://doi.org/10.3390/cosmetics5040068.

Pereira, L. (2018b). Biological and therapeutic properties of seaweed polysaccharides. *International Biology Review, 2*, 1–50.

Pereira, L., & Ribeiro-Claro, P. J. A. (2014). Analysis by vibrational spectroscopy of seaweed polysaccharides with potential use in food, pharmaceutical and cosmetic industries. In L. Pereira, & M. N. Joao (Eds.), *Marine algae: Biodiversity, taxonomy, environmental assessment and biotechnology* (pp. 225–247). CRC Press. Taylor & Francis Group.

Plouguerné, E., da Gama, B. A. P., Pereira, R. C., & Barreto-Bergter, E. (2014). Glycolipids from seaweeds and their potential biotechnological applications. *Frontiers in Cellular and Infection Microbiology, 4*. https://doi.org/10.3389/fcimb.2014.00174.

Ponce, N. M. A., Pujol, C. A., Damonte, E. B., Flores, M. L., & Stortz, C. A. (2003). Fucoidans from the brown seaweed *Adenocystis utricularis*: Extraction methods, antiviral activity and structural studies. *Carbohydrate Research, 338*(2), 153–165.

Poudyal, H., Panchal, S. K., Ward, L. C., & Brown, L. (2013). Effects of ALA, EPA and DHA in high-carbohydrate, high-fat diet-induced metabolic syndrome in rats. *The Journal of Nutritional Biochemistry, 24*(6), 1041–1052. https://doi.org/10.1016/j.jnutbio.2012.07.014.

Primdahl, K. G., Tungen, J. E., Aursnes, M., Hansen, T. V., & Vik, A. (2015). An efficient total synthesis of leukotriene B4. *Organic and Biomolecular Chemistry, 13*(19), 5412–5417. https://doi.org/10.1039/C5OB00473J.

Pujol, C. A., Estevez, J. M., Carlucci, M. J., Ciancia, M., Cerezo, A. S., & Damonte, E. B. (2002). Novel DL-galactan hybrids from the red seaweed Gymnogongrus torulosus are potent inhibitors of herpes simplex virus and dengue virus. *Antiviral Chemistry & Chemotherapy, 13*(2), 83–89. https://doi.org/10.1177/095632020201300202.

Qi, B., Beaudoin, F., Fraser, T., Stobart, A. K., Napier, J. A., & Lazarus, C. M. (2002). Identification of a cDNA encoding a novel C18-Delta(9) polyunsaturated fatty acid-specific elongating activity from the docosahexaenoic acid (DHA)-producing microalga, Isochrysis galbana. *FEBS Letters, 510*(3), 159–165.

Ragonese, C., Tedone, L., Beccaria, M., Torre, G., Cichello, F., Cacciola, F., et al. (2014). Characterisation of lipid fraction of marine macroalgae by means of chromatography techniques coupled to mass spectrometry. *Food Chemistry, 145*, 932–940. https://doi.org/10.1016/j.foodchem.2013.08.130.

Rajapakse, N., & Kim, S.-K. (2011). Nutritional and digestive health benefits of seaweed. In *Advances in food and nutrition research* (Vol. 64, pp. 17–28). https://doi.org/10.1016/B978-0-12-387669-0.00002-8.

Ren, R., Azuma, Y., Ojima, T., Hashimoto, T., Mizuno, M., Nishitani, Y., et al. (2013). Modulation of platelet aggregation-related eicosanoid production by dietary F-fucoidan from brown alga *Laminaria* japonica in human subjects. *British Journal of Nutrition, 110*(5), 880–890. https://doi.org/10.1017/S000711451200606X.

Rodrigues, D., Freitas, A. C., Pereira, L., Rocha-Santos, T. A. P., Vasconcelos, M. W., Roriz, M., et al. (2015). Chemical composition of red, brown and green macroalgae from Buarcos bay in central west coast of Portugal. *Food Chemistry, 183*, 197–207. https://doi.org/10.1016/j.foodchem.2015.03.057.

Schmid, M., Guihéneuf, F., & Stengel, D. B. (2014). Fatty acid contents and profiles of 16 macroalgae collected from the Irish coast at two seasons. *Journal of Applied Phycology, 26*(1), 451–463. https://doi.org/10.1007/s10811-013-0132-2.

Schubert, R., Kitz, R., Beermann, C., Rose, M. A., Baer, P. C., Zielen, S., et al. (2007). Influence of low-dose polyunsaturated fatty acids supplementation on the inflammatory response of healthy adults. *Nutrition, 23*(10), 724–730. https://doi.org/10.1016/j.nut.2007.06.012.

Silva, P. C., Menez, E. G., & Moe, R. L. (1987). Catalog of the benthic marine algae of the Philippines. *Smithsonian Contributions to the Marine Sciences,* 27, 190pp.

Slagsvold, J. E., Pettersen, C. H. H., Follestad, T., Krokan, H. E., & Schønberg, S. A. (2009). The antiproliferative effect of EPA in HL60 cells is mediated by alterations in calcium homeostasis. *Lipids, 44*(2), 103–113. https://doi.org/10.1007/s11745-008-3263-5.

Stengel, D. B., Connan, S., & Popper, Z. A. (2011). Algal chemodiversity and bioactivity: Sources of natural variability and implications for commercial application. *Biotechnology Advances, 29*(5), 483–501. https://doi.org/10.1016/j.biotechadv.2011.05.016.

Stiger-Pouvreau, V., Bourgougnon, N., & Deslandes, E. (2016). Carbohydrates from seaweeds. In J. Fleurence, & I. Levine (Eds.), *Algae: A way of life and health* (pp. 223–274). Amsterdam Boston Heidelberg London: Elsevier, Academic Press.

Sugawara, T., Matsubara, K., Akagi, R., Mori, M., & Hirata, T. (2006). Antiangiogenic activity of brown algae fucoxanthin and its deacetylated product, fucoxanthinol. *Journal of Agricultural and Food Chemistry, 54*(26), 9805–9810. https://doi.org/10.1021/jf062204q.

Swarnalatha, G. (2018). *Comparison of nutritive values of seaweed with other terrestrial foods — A review*. Retrieved April 11, 2019, from https://www.ijfans.com/ijfansadmin/upload/ijfans_5a7935678f6ed.pdf.

Synytsya, A., Bleha, R., Synytsya, A., Pohl, R., Hayashi, K., Yoshinaga, K., et al. (2014). Mekabu fucoidan: Structural complexity and defensive effects against avian influenza a viruses. *Carbohydrate Polymers, 111,* 633–644. https://doi.org/10.1016/j.carbpol.2014.05.032.

Talarico, L. B., Pujol, C. A., Zibetti, R. G. M., Faría, P. C. S., Noseda, M. D., Duarte, M. E. R., et al. (2005). The antiviral activity of sulfated polysaccharides against dengue virus is dependent on virus serotype and host cell. *Antiviral Research, 66*(2–3), 103–110. https://doi.org/10.1016/j.antiviral.2005.02.001.

Tanaka, Y., Goto, K., Matsumoto, Y., & Ueoka, R. (2008). Remarkably high inhibitory effects of docosahexaenoic acid incorporated into hybrid liposomes on the growth of tumor cells along with apoptosis. *International Journal of Pharmaceutics, 359*(1–2), 264–271. https://doi.org/10.1016/j.ijpharm.2008.03.045.

Taoda, N., Shinji, E., Nishii, K., Nishioka, S., Yonezawa, Y., Uematsu, J., et al. (2008). Fucoidan inhibits parainfluenza virus type 2 infection to LLCMK2 cells. *Biomedical Research, 29*(6), 331–334.

Thomes, P., Rajendran, M., Pasanban, B., & Rengasamy, R. (2010). Cardioprotective activity of *Cladosiphon* okamuranus fucoidan against isoproterenol induced myocardial infarction in rats. *Phytomedicine, 18*(1), 52–57. https://doi.org/10.1016/j.phymed.2010.06.006.

Tobacman, J. K. (2001). Review of harmful gastrointestinal effects of carrageenan in animal experiments. *Environmental Health Perspectives, 109*(10), 983–994. https://doi.org/10.1289/ehp.01109983.

Torres, M. D., Flórez-Fernández, N., & Domínguez, H. (2019). Integral Utilization of Red Seaweed for Bioactive Production. *Marine Drugs, 17*(6), 314. https://doi.org/10.3390/md17060314.

Trono, G. C., Jr. (1977). *Field guide and atlas of the seaweed resources of the Philippines*. Bookmark, Makati City, 306 pp.

Trono, G. C., Jr. (2010). *A primer on the land-based culture of* Halymenia *durvillaei bory de Saint-Vincent (Rhodophyta).* Quezon City: Marine Science Institute, University of the Philippines Diliman.

Tsai, C.-J., & Sun Pan, B. (2012). Identification of sulfoglycolipid bioactivities and characteristic fatty acids of marine macroalgae. *Journal of Agricultural and Food Chemistry, 60*(34), 8404–8410. https://doi.org/10.1021/jf302241d.

Tungpalan, A. Y. (1984). Ethnobotanical study of the seaweeds of Ilocos Norte. *Ilocos Fisheries Journal, 1*, 134–146.

Vairappan, C. S., Chung, C. S., Hurtado, A. Q., Msuya, F. E., Bleicher L'honneur, G., & Critchley, A. T. (2008). Distribution and symptoms of epiphyte infection in major carrageenophyte-producing farms. *Journal of Applied Phycology, 20*, 477–483.

Wang, J., Chen, S., Xu, S., Yu, X., Ma, D., Hu, X., et al. (2012). In vivo induction of apoptosis by fucoxanthin, a marine carotenoid, associated with down-regulating STAT3/EGFR signaling in sarcoma 180 (S180) xenografts-bearing mice. *Marine Drugs, 10*(9), 2055–2068. https://doi.org/10.3390/md10092055.

Wielgosz-Collin, G., Kendel, M., & Couzinet-Mossion, A. (2016). Lipids, fatty acids, glycolipids, and phospholipids. In J. Fleurence, & I. Levine (Eds.), *Algae: A way of life and health* (pp. 185–222). Amsterdam Boston Heidelberg London: Elsevier, Academic Press.

Wijesinghe, W. A. J. P., & Jeon, Y. J. (2011). Biological activities and potential cosmeceutical applications of bioactive components from brown seaweeds: A review. *Phytochemistry Reviews, 10*(3), 431–443. https://doi.org/10.1007/s11101-011-9214-4. Request PDF.

Ye, J., Li, Y., Teruya, K., Katakura, Y., Ichikawa, A., Eto, H., et al. (2005). Enzyme-digested fucoidan extracts derived from seaweed mozuku of *Cladosiphon* novae-caledoniae kylin inhibit invasion and angiogenesis of tumor cells. *Cytotechnology, 47*(1–3), 117–126. https://doi.org/10.1007/s10616-005-3761-8.

Zhu, W., Ooi, V. E., Chan, P. K. S., & Ang, P. O., Jr. (2003). *Inhibitory effect of extracts of marine algae from Hong Kong against herpes simplex viruses*. Request PDF. Retrieved December 6, 2018, from ResearchGate website: https://www.researchgate.net/publication/291158077_Inhibitory_effect_of_extracts_of_marine_algae_from_Hong_Kong_against_Herpes_simplex_viruses.

CHAPTER THREE

Challenges to the future domestication of seaweeds as cultivated species: understanding their physiological processes for large-scale production

Leila Hayashi[a,*,1], Sarah de J. Cantarino[b] and Alan T. Critchley[c]

[a]Federal University of Santa Catarina, Aquaculture Department, Florianópolis, Santa Catarina, Brazil
[b]Postgraduate Program in Marine Biotechnology, Almirante Paulo Moreira Institute of Sea Studies (Brazilian Navy - IEAPM), Fluminense Federal University (UFF), Arraial do Cabo, Rio de Janeiro, Brazil
[c]Verschuren Centre for Sustainability in Energy and the Environment, Cape Breton University, Sydney, NS, Canada
*Corresponding author: E-mail: leila.hayashi@ufsc.br

Contents

Abstract

A detailed understanding of physiological and reproductive processes in seaweeds has repeatedly proven to be an essential pre-requisite in the successful development of a sustainable industry. The prime example of this was the classical discovery of the "conchocelis"-phase of *Pyropia* (*Porphyra*) by Kathleen Mary Drew-Baker in 1949. Such elegant research proved to be pivotal to the development of a globally important "nori" industry which transitioned from the simple provision of the enhanced surface area of the substrata for spore settlement to the sophisticated, mechanized and computerized operations in modern hatcheries supplied by seedling banks of selected

[1] Senior contributor

Advances in Botanical Research, Volume 95
ISSN 0065-2296
https://doi.org/10.1016/bs.abr.2019.11.010

species and their cultivars. All of the pre-requisite knowledge was acquired through intensive applied research. However, not all solutions need to be high-tech; problems caused by epiphytes and contaminants have been achieved by exposing *Pyropia* nets periodically and was found to be effective. This protocol was achieved based on fundamental observations by farmers which were then complemented and refined by laboratory trials. Techniques must be adapted for site-specific differences, as abiotic factors such as water current and movement, surface seawater temperature, light regime and photoperiod, nutrients dispersion and water quality are interrelated, either positively or negatively, influencing seaweed productivity and the end-use of the biomass. Unfortunately, positive techniques that have been shown *in vitro/silica* can prove to be impractical once attempted at large-scale cultivation and/or the return on investment is not justified by the commercial value of the resultant seaweed biomass. This chapter presents a summary of how the judicious application of knowledge based on the eco-physiological processes of common seaweed species from tropical and cold-waters can assist the future development and scale-up of the global seaweed industry.

1. The physiological process associated with seaweeds farmed commercially

One of the most famous species around the world thanks to the popularity of sushi, *Pyropia* (*Porphyra*) is also one example of how the knowledge of reproductive and physiological aspects can promote a whole industry. The classic work of Kathleen Drew Baker (Drew, 1949) showed how researchers were struggling to find out what was the relationship between both the leafy and the filamentous thalli, after germination of some "spores." Clearly, there was some misunderstanding about the role of monospores. The knowledge of the life cycle of various species of the genus *Pyropia*, besides the understanding of the optimal conditions for each phase of the life cycle, allowed for domestication and development of a very valuable industry. Now, it is possible to stimulate or inhibit the sporulation of all phases, controlling nutrient availability, water temperature, and irradiance. For example, to increase the release of conchospores, mechanical stimulation can be used, as with water circulation using compressed air, as well as physical stimulation, such as reducing water temperature from 23°C to 18–20°C (Blouin, Brodie, Grossman, Xu, & Brawley, 2011; Oohusa, 1993). However, differences driven by light, temperature and photoperiod on conchospores release have being reported to be species dependent (López-Vivas, Riosmena-Rodríguez, Llave, Pacheco-Ruíz, & Yarish, 2015).

The reproduction of the main cultivated seaweeds may occurs after some stress stimulus, as well explained in Laminariales by Lüning and Dieck

(1989). One day of low temperature or the absence of light is enough to stimulate spore or gamete release when the seaweeds are cultivated under standard conditions. The authors described that the temperature, photoperiod and internal control systems interacts in the seasonal development of organisms. Even the simple water exchange in cultivation tanks may boost the release of spores/gametes (Kawashima, 1993; Ohno & Matsuoka, 1993). The farming of the red seaweed *Pyropia* and browns seaweeds such as *Saccharina*, and *Undaria*, are based on asexual propagation by using meiospores. The kelps (Laminariales) have a high level of specialization and thallus fragments of sporophytes do not regenerate (Khan & Satam, 2003), so zoospores (i.e. meiospores) are parts of the reproductive cycle. These examples basically reflect that some seaweeds have a hatchery phase easier to manage in their production cycle which has allowed commercial "exploitation" and domestication. However, the biomass production of the most commonly cultivated tropical seaweeds does not necessary require a hatchery phase (in the sense that gametes or spores are rarely involved in mass cultivation); most, e.g. *Gracilaria, Kappaphycus*, and *Eucheuma* are cultivated by vegetative propagation of fragments of thalli. In addition, the maintenance of genotypes of commercial interest by controlling the life cycles would be a factor increasing the success of seaweed production in the long term (Charrier et al., 2017). Thorough knowledge of the physiological processes which regulate productivity in marine farms is therefor essential to the industry.

There is also an increasing need to control the quality of seaweed biomass grown out in sea farms, especially in times of global warming. According to Azanza and Ask (2017), irradiance and Surface Seawater Temperature (SST) are the main factors affecting eucheumatoids (i.e. *Kappaphycus* and *Eucheuma*) production. Specifically for *Kappaphycus alvarezii* and *Eucheuma denticulatum*, the most studied carrageenophyte species, earlier on basic and applied studies of Dawes and Koch (1991) and Dawes, Trono and Lluisma (1993) on the effects of physiological aspects were driven by the necessity to increase the productivity of marine farms. This is also the case for *Gracilaria* cultivation, which has been extensively studied to understand the relationship between abiotic factors and the growth and agar yield and quality. Buschmann, Hernández-González and Varela (2008) reported variations in the productivity of *Gracilaria chilensis*, depending on those area of the coastal shore which were cultivated (i.e. subtidal or intertidal), the region of Chile (i.e. southern or northern) or season (highest growth in spring, declining in the summer and lowest in winter). The authors concluded that several (abiotic) environmental factors were also inter-related to this variation.

2. Optimization of large-scale cultivation: challenges and potential

In order to enhance the production of commercially important seaweeds, it is important to understand and modify the main parameters which influence their cultivation, such as water current and movement, seawater temperature, irradiance and photoperiod, nutrients dispersion and water quality, and understand the relationships of these factors and the intrinsic physiological response.

An understanding of the physiological process made it possible to improve the mass production of biomass in cultivation, for example, of the brown seaweed *Saccharina latissima.* The use of stress conditions, such as thallus dissection and thermal shock, can stimulate the production of meiospores. Also, if the meiospores are maintained under red light it can prevent gametophyte fertilization, while the fertilization and sporophyte development are promoted when cultivated in white light (Kerrison, Stanley, & Hughes, 2018). The hatchery phase is still a challenge if mass cultivation is considered since seeded ropes need to be kept until the young sporophytes achieve 1−2 mm prior to being transferred to open water. This phase can take up to eight months, thereby increasing the costs of operation. To solve that, Kerrison et al. (2018) tested textile sheets as a physical substratum for meiospores, gametophytes, and sporophytes, and observed that these sheets produced twice the biomass, after four months, as compared to the traditional method. In addition, sporophytes grew faster than meiospores and gametophytes, when attached to the sheets with a binder (AT ~ SEA Technologies, BE). In this new sporophyte seeding technique, the selection of a suitable twine was essential for the sucess of juveniles development (Kerrison et al., 2019). For *Palmaria palmata* cultivation, new hatchery strategies to improve seedling have being develop based on Germination, Maceration and Agitation (GMA) method - from release and handling of tetraspores to produce seedlings (propagules) on cultivation substrate (Schmedes, Nielsen, & Petersen, 2019).

For the production of those seaweeds of mass cultivation by propagules, such as *Gracilaria*, for agar and, *Kappaphycus* and *Eucheuma,* for carrageenan extraction, several methods have been traditionally employed for rapid biomass production. Examples are off-bottom, rafts, and long-line ropes (Hayashi et al., 2017). Basically, the propagules are attached by various tieing methods on to ropes which are then fixed to stakes hammered into soft substrata (off-bottom) or floating, suspended by buoys (rafts and long-lines)

(Buschmann et al., 2008). The use of tubular nets, similar to those used in mollusk farms, replace the ropes, thereby allowing the improvement of productivity of carrageenophyte farms and also increased the possibility of mechanizing the process (Hayashi et al., 2017). Tubular nets works well in cultivations with muddy bottoms and increased sea water movement. This technique could retain the seedlings on the raft, avoiding invasiveness on the environment, and improve the manage of the cultivar by reducing the time spent in comparison with the tie-tie technique (Góes & Reis, 2011; Reis, Pereira, & Goes, 2015). For this, however, it may be extremely efficient in fast-growing species because the terete, side branches grow out through the mesh, before the apertures can be filled with sediments or epiphytic over-growth (depending on the area).

Although many commercial seaweed species are cultivated in slow open water, they are not entirely free from damage caused by natural hazards such as typhoons and storms, which can occur with variable frequency and severity, mainly in tropical regions. This problem is not easy to solve, and according to Kim, Yarish, Hwang, Park and Kim (2017), the only solution is to remove the seaweeds from the sea (as in growing them on land). However, the costs to keep the seedlings in tanks are expensive and laced with difficulties, since they need to acclimate and be further selected for the prevailing conditions in these new conditions which can add further layers of stress.

The first example of intensively managed, on-land seaweed cultivation is the Irish moss (*Chondrus crispus)* aquaculture in Charlesville, Nova Scotia, Canada (Craigie, Cornish, & Deveau, 2019). The growth and photosynthetic rates, zinc contamination, disease control, and clone selection are severe challenges to overcome. Technologies applied included drone monitoring, real-time adjustments of air delivery systems, compressed air-agitation in each tank, regulation of pH by carbon dioxide with carbon recovery up to 90% in the large-scale cultivation tanks. Water issues pipelines were solved by accessing deeper, cooler seawater and providing low incoming particulate matter. The large tanks management use subdivided to facilitate management and harvesting. Prophylactic measures were required in addition to mechanical sorting and handling techniques. This industry is focused on biomass for food production with all requisites to meet the stringent demands of the export market (Japan). The success of this operation was due to the continuous P & D work undertaken a strain selection, biology, engineering and marketing of a high quality product.

There are also other approaches to economically feasible and ecologically sustainable land-based cultivation as in the association of seaweed cultivation with other aquatic organisms in the Integrated Multi-trophic Aquaculture System (IMTA) (Sebök, Herppichb, & Hanelta, 2017). Seaweed-based integrated farms on full scale achieved commercial sucess in some countries, like China, South Africa and Canada, and have being widely studied and reported in Israel (Neori, Guttman, Israel, & Shpigel, 2019; Neori, Sphigel, Guttman, & Israel, 2017). The selection of species depends on economical and logistical factors as well as expertize and equipments required. Suitable models were establish with species of *Gracilaria* and *Ulva*. The seaweed growth works as a function of light intensity, temperature, water motion, plant density and nutrient enrichment, which can be sourced from fertilizers and effluents (Neori, Shpigel, Guttman, & Israel, 2017). For example, Abreu, Pereira, Buschmann, Sousa-Pinto, and Yarish (2011) integrated *Gracilaria vermiculophylla* as a biofilter for fish-effluents rich in nitrogen. The requirements of nutrients for the growth marine seaweed also changes and interact with the season and other factors such as the prevailing light and temperature regimes (Usandizaga et al., 2018). Nobre, Valente, and Neori (2017) elaborated a model to assess water renewal rates and nitrogen limitation (uptake and growth) in commercial seaweed farms based on *Ulva spp.* parameter for the management of the nutrient limitation that occurs when expanding seaweed production and for planning the seaweed integration with other aquaculture products in an IMTA system. According to them, this model helps to support the calculation of the renewals required to expand from a planning stage to a production scale. Shpigel et al. (2018) employed *Ulva lactuca* to reduce nitrogen loads from the effluents of sea urchin tank and fishponds in a semi-commercial land-based IMTA system, and observed that the seaweed assimilated 74% of the dissolved nitrogen and became itself a valuable crop for sea urchin.

Effluents of bioflocs from the farming of marine shrimp were observed by Pedra, Ramlov, Maraschin and Hayashi (2017) to improved not only the growth of *K. alvarezii* in tanks, but also the bioactive content in the biomass, offering the possibility to utilize the raw materials for value-added products, other than carrageenan. Nevertheless, the economic value and long term viability of these systems still needs to be demonstrated before wider adoption will take place. It is claimed that some of the production costs can be off-set, however, monetizing the treatment of (marine) effluents is a very recent area, and it remains to be seen when and where cost-effective operations will be viable.

High productivity of outdoor tank cultivation of several nori species in relation to seeded nets in open sea in Israel, for an entire growth season, were reported by Israel, Levy and Friedlander (2006), suggesting the tank and pond cultivation of the foliose phase as a controlled high yield alternative (50–300% higher than yields indicated for crops grown on seeded nets). However, the known operational costs needed to be highlighted, where the water agitation in the cultivation of seaweeds may be significant. The physical movement and low stocking density are required so as to regulate the temperature and ensure adequate distribution of carbon and nutrients inside the tank, which in turn promote movement facilitating absorption of nutrients by *Pyropia haitanensis*. In high stocking density cultivation, extra carbon supply may be necessary to avoid negative effects on the photosynthesis due to lowered carbon (Jiang, Zou, Lou, & Ye, 2017). Agitation is required as there is a need to break down the laminar barriers on the surface of the seaweed thalli—which affect gaseous exchange (i.e. morphology plays a role here). Energy required for the pumping/agitation is an important consideration: simple movement can be achieved with a paddle wheel. Using compressed air with spargers creates movement (there are physical constraints which determine the shape of tank and where to place spargers), and CO_2 can be added which is both a pH adjustment and a source of carbon. Seaweeds in tanks are stressed, they need to circulate so as to be in the photic zone as much as possible. The amount of agitation determines if they move in a circular cellular motion, or move with laminar flow. In terms of the plants, at the surface of a pond they may well be photo-inhibited. Circulating them from the surface to the bottom means they go through periods of light/dark very rapidly, thereby improving their photosynthetic efficiency.

Development of technologies are required. One example is the closed ring-shaped cultivation model by Sebök et al. (2017). They evaluated growth rates from *P. palmata, C. crispus,* and *U. lactuca* over 7 days in these experimental models. According to these authors, their system reduced the volume of cultivation media required and increased control by separating the supply of CO_2 and nutrients, as well as the temperature control from agitation. To design a new system for optimal growth conditions and efficiency requires an in-depth analysis of every parameter involved in the cultivation process which in many cases has to be established and tested experimentally under laboratory-scale conditions.

Another physiological aspect of seaweeds is that there can be temporal changes in chemical composition of commercial interest, which has an impact on when harvesting should be conducted (Taelman, Champenois, Edwards, De Meester, & Dewulf, 2015). One example is the work of Hayashi et al. (2007), which demonstrated that, depending on the cultivation period, in Brazil, the end-product of carrageenan from *K. alvarezii* could have significant variations in the relative composition of *kappa* and *iota* fractions. The biomass increased according to the duration of the cultivation, giving the highest gel strength, viscosity, and molecular weight values.

One of major challenges to the seaweed industry is the outbreak of diseases, which affect most of the major commercially important species and can cause significant economic losses. Production of the kelp *Saccharina* was far of be affected by many diseases, endo- and epibionts, causing significant economic losses (Wang et al., 2019). *Pyropia* is mainly affected by oomycetes *Olpidiopis* spp. and *Pythium* spp., as well as red rot and chytrid diseases (Gachon, Sime-Ngando, Strittmatter, Chambouvet, & Kim, 2010). Eucheumatoid genera (i.e. *Kappaphycus* and *Eucheuma*) are frequently affected by "ice-ice" disease although there is evidence that bleached thalli are primarily a consequence of physiological stress and not necessarily provoked by pathogens (i.e. they are secondary in nature, Loureiro, Hurtado, & Critchley, 2017). A common point to all of these diseases is that their spread seems to be directly related to intensification of large-scale production (Gachon et al., 2010). There are some short-term solutions to minimize the impacts of these diseases, such as treating the blades of *Pyropia* with regular acid-washing (Gachon et al., 2010), but this is not viable for large-scale production. Hafting et al. (2015) warn of the necessity to develop strains resistant to diseases, which can be supported by the emerging field of phycopathology.

Epiphytes, fouling, and herbivores can be responsible for significant economic losses. Desiccation and pH control can be used in *Pyropia*, and briefly washing the thalli with freshwater works relatively well for tropical species such as *Kappaphycus, Eucheuma* and *Gracilaria* (Kim et al., 2017). Growing seaweeds in high densities can also help minimize the presence of epiphytes since some of these, i.e. *Ulva* and *Ectocarpus* require high irradiances and high densities remove the light. One strategy used in *Saccharina* cultivation is the growth in the laboratory under conditions of short photoperiod, this helps to prevent the growth of epiphytes, activate the basal blade meristem, stimulates continuous growth throughout the year and interrupted the seasonal cycle of growth (Lüning & Pang, 2003). On the other hand, with the increase of the density, space can be limited even for the hosts, so the balance of the epiphytes control and productivity is fundamental. Depending on the situation, the only solution is to change the cultivation site or grow a limited amount of "carry over" biomass, temporarily in tanks (Kim et al., 2017).

The use of biostimulants—extracts of brown seaweeds, i.e., *Ascophyllum nodosum* (AMPEP), or *Ecklonia maxima* (commercially know as Kelpak) could be used in *Kappaphycus* for land-sea-based nursery cultivation operations (Tibubos, Hurtado, & Critchley, 2017) and also to increase carrageenan yield (Loureiro, Reis, Berrogain, & Critchley, 2014). These extracts can also improve the growth of this species and mitigate the damage of epiphytic *Neosiphonia* infestation (Ali, Yasir, Critchley, & Hurtado, 2018; Borlongan, Tibubos, Yunque, Hurtado, & Critchley, 2011; Marroig, Loureiro, & Reis, 2016), and the physiological stress (Borlongan, Gerung, Kawaguchi, Nishihara, & Terada, 2017; Borlongan, Gerung, Nishihara, & Terada, 2017; Borlongan, Luhan, Padilla, & Hurtado, 2016). This concept is interesting if one considers that induction of haloperoxidase seems to be a generic mechanism of brown and red algae against biotic and abiotic stresses (Gachon et al., 2010), and the use of both can help to minimize damage. The treatment with AMPEP induced an oxidative burst response, caused by the interaction between the β-1,3 glucan present in the extract and the β-1,3 glucanase enzyme present in *K. alvarezii,* which stimulated the production of volatile halogenated compounds which were elicited through NADPH oxidase in the cellular membrane. Thus, the β-1,3 glucan, together with betaines and cytokinin-elicitors, also reported to be present in the AMPEP, shielded the treated seaweed seedlings against the effects of constant H_2O_2 exudation and exposure, acting as "vaccine" and growth promoter (Loureiro et al., 2017; Loureiro, Reis, Berrogain, & Critchley, 2012).

Wang et al. (2019) also suggested the use of a specific polysaccharides, the oligoalginate, to prevented and controlled diseases of kelp juveniles in open sea-cultivation. Oligoalginate elicitation was found to decrease the density of endobionts and the number of bacterial cells on the juvenile sporophytes of *S. latissima* and *S. japonica.* However, the same was not true for the adult sporophytes where elicitation increased their susceptibility to epibionts settlement of both species and accelerated the aging of *S. japonica* adults.

3. From the laboratory to the sea: product development

The production of seaweed biomass is predominantly directed to markets and processing for human food, e.g., edible seaweeds—*Laminaria* (kombu), *Porphyra* (nori) and *Undaria* (wakame), or the phycocolloids extracted from *Kappaphycus, Eucheuma,* and *Gracilaria.* However, there are other seaweed-derived products with commercial potential that are emerging and being developed by a diversifying seaweed industry.

New extraction technologies, applications of the biorefinery principle and the discovery of novel bioactive compounds, are stimulating value-added products with applications in several markets such as functional foods, nutraceuticals, cosmeceuticals, pharmaceuticals, dietary feed supplements, fertilizers, biostimulants, bioregulators, bioremediation and nano-particle technologies (e.g., Duan, Critchley, Fu, & Pereira, 2019; Hafting et al., 2015; Martins, Vieira, Gaspar, & Santos, 2014; Michalak & Chojnacka, 2015; Vijayan et al., 2016). There are several studies in the literature about the bioactive compounds, but here we give some examples of the physiology related to bioactive compounds.

Bioactive compounds are chemical products resulting from metabolic processes, such as: proteins, minerals, vitamins, polysaccharides, oils, polyunsaturated fatty acids, plant growth-promoting substances, pigments, and secondary metabolites, which also express biological activity in other organisms, i.e. anti-bacterial, -viral, -fungal, -oxidative, -inflammatory, -diabetic, -hypertensive, -allergic, -tumor properties and -fouling (Admassu, Gasmalla, Yang, & Zhao, 2018; Dahms & Dobretsov, 2017; Hindu, Chandrasekaran, Mukherjee, & Thomas, 2019; Jaswir & Monsur, 2011; Michalak & Chojnacka, 2015; Seca & Pinto, 2018; Vijayan et al., 2016). A wide range of bioactive compounds is reportedly found in marine organisms, as a response to the effect of an exigent, competitive and aggressive aquatic environment (Kolanjinathan, Ganesh, & Saranraj, 2014).

The marine natural products are already a consolidate research field (e.g., *Marine Drugs*) with more than 16,000 products isolated from diverse marine organisms, but screenings are still finding new compounds of interest (Bhakuni & Rawat, 2005). There are at least 3000 bioactive from seaweeds reported (Freile-Pelegrín & Tasdemir, 2019). In a review of the literature for 2017, 89 new compounds from seaweed were reported, in a total of 1490 new compounds from diverse origens (Carroll, Copp, Davis, Keyzers, & Prinsep, 2019).

The popularization of nutraceuticals or functional foods stimulated the commercialization of seaweeds as important ingredients because of the appeal as antihypertensive (mainly polysaccharides, peptides, and phlorotannins) and anti-obesity agents (phlorotannins, carotenoids, fucoxanthin, indole-derivatives, and sterols) found in many edible seaweed (*eg. Pyropia yezoensis, Py. columbina, P. palmata, Undaria pinnatifida, C. crispus, Ulva clathrata, Ulva rigida, U. pertusa, Ecklonia cava, E. stolonifera* and *Pelvetia canaliculata*) (Seca & Pinto, 2018). Mouritsen, Rhatigan and Pérez-Lloréns (2019) discuss these resources as human food and highlights the sensory perception and the important role of cookbooks, media-communications, and recipes to ordinary consumer. There are at least 665 species of edible macroalgae with bioactive compounds described (see Pereira, 2016). In this way, bioactive compounds from various seaweeds can occupy new markets. For example, some red seaweeds seems to be a good source of protein for new markets and also a source of ingredients (Cerná, 2011). *Pyropia* is a good source of vitamin B12, and besides their antioxidant activity, they can also have activities against hypertension, preventing cardiovascular disease, and reducing the risk of cancer (Skrovánková, 2011). Besides the diversity of vitamins, all essential minerals are available in dietary seaweeds, and they are a healthy alternative to sodium consumption. The salt made from seaweed has nearly 9%–12% sodium, while normal salt has 98% sodium (Drum, 2013).

Seaweeds can also be used as nutraceuticals in ingredients or as feed additives for other marine organisms. The inclusion of *U. pinnatifida* and *Sargassum filipendula* in the diet of the Pacific white shrimp *Litopenaeus vannamei* was reported to enhance resistance against White Spot Syndrome Virus (WSSV), besides improving digestive physiology, acting in the nutrient absorption and morphology of digestive tract. The addition of only 4% *U. pinnatifida* dry biomass to the feed reduced the cumulative mortality of shrimp at 72 h after WSSV infection (Schleder et al., 2018). *U. pinnatifida* was also reported to improve hemato-immunological parameters and reduce the quantity of *Vibrio* sp., an important group of pathogenic bacteria for shrimp farming. A feed with 0.5% and 2% of *S. filipendula* helped with thermal shock resistance, as indicated higher survival rates in treated animals. The higher content of total phenolic compounds, flavonoids, fucoxanthin, and lutein of the brown seaweeds (*S. filipendula* and *U. pinnatifida*) were reported (Schleder et al., 2017).

New ranges of products with active seaweed-derived ingredients are emerging in multiple markets. In the cosmeceutical industry, the potential applications were well-reviewed by Wijesinghe and Jeon (2011) (*eg. Ecklonia spp., Fucus spp., Ishige okamurae, Hizikia fusiformis, Laminaria japonica, Sargassum spp., Stocheospermum marginatum, Turbinaria spp.*, and *U. pinnatifida*) and Thomas and Kim (2013) (*eg. Dermatophagoides farina, Ecklonia* spp., *Eisenia* spp., *L. japonica, Laurencia* spp.). They are highlighted as promising as brown seaweed polyphenols and phloroglucinol derivatives (e.g. phlorotannins), have emerging applications as cancer chemo-preventive agents against photo-carcinogenesis and other adverse effects of UV-B exposure, photo-aging, inflammatory skin disorder, anti-oxidants, antiallergic and skin whitening, or protective agents. Sulfated polysaccharides (e.g., fucoidans and carrageenan) can also act against degradation and skin photo-aging, helping elasticity by increasing hydration, while carotenoids (e.g. fucoxanthin) can protect against oxidative stress-induced by UV-B radiation. Seaweed-based cosmeceuticals have been produced on a large-scale in the form of bioactive algae-based ingredients (Martins et al., 2014).

As for pharmaceutical applicability, there are several investigations on new drug delivery systems or in direct effect against diseases. An example of these new mechanisms are the food bioactive peptides. These peptide-based therapeutics demonstrated a similar mechanism of drug action, but when incorporated into foods it transports the substances through cell membranes for applications of medical diagnostics, nutraceutical, antimicrobial or cosmetic uses. The development of large-scale production (e.g. from *P. palmata, Porphyra yezoensis, Porphyra columbina, Pyropia columbina, U. pinnatifida, and A. nodosum*), optimization of protein extraction methods, and peptide isolation processes are important to be invested (Admassu et al., 2018).

As effective drugs, at the moment, there is only one seaweed-based product being commercialized, the innovative antiviral nasal spray made of iota-carrageenan (Carragelose®) by the Marinomed Biotech AG. It was demonstrated to be clinically effective against early symptoms of common cold due a protective physical antiviral barrier in the nasal cavity (Martins et al., 2014). This sulfated polysaccharide is cited for having biological effect preventing virus due to the structure in the surface that inhibit target cell binding (Chen et al., 1997). Other seaweed-based products have been sold but without proceed in all clinical test phases. *Laminaria* and *Sargassum* species have been used for the treatment of cancer in China, capsule forms of *Undaria* are used as anti-viral to inhibit the *Herpes simplex* virus, *Asparagopsis taxiformis* and *Sarconema* sp. are used to control and cure goiter (Khan & Satam, 2003). Natural products are preferred over non-natural because of the biochemical specificity, binding efficiency, high chemical diversity, and propensity to interact with biological targets (Martins et al., 2014).

Different medical applications are being studied, it ranges from bioactive against human immunodeficiency virus (HIV) (Besednova et al., 2019) until neglected tropical diseases (Freile-Pelegrín & Tasdemir, 2019). There are many metabolites such as sulfated polysaccharides (e.g. fucoidans and carrageenans), lectins and phenolic compounds with anti-HIV activity. This interest concerns the relatively low cost, high antiviral activity, low or absence of toxicity, the formation of pathogens resistance, good solubility and the possibility of cultivation of seaweed. However, currently there are no seaweed-based drugs on the pharmaceutical market for treating and preventing HIV infection (Besednova et al., 2019). Against several neglected tropical diseases, one of the substances of interest is elatol, the major component of the Brazilian red seaweed *Laurencia dendroidea* (Freile-Pelegrín & Tasdemir, 2019). The natural products can be stimulated depending on the seaweed; this was tested to produce the compound elatol as a rapid response to inducible chemical defenses to herbivory with *L. dendroidea* (Sudatti, Fujii, Rodrigues, Turra, & Pereira, 2018). Special vesicles called *corps en cerise* were described to be responsible for stored halogenated compounds such aselatol, which is transported to the seaweed thallus surface by exocytosis, in higher or lower concentrations depending on the cellular activities (Salgado et al., 2008). Although this species has been extensively studied, the ecological roles and production dynamics of such bioactive are still under investigation (Sudatti et al., 2018).

Another good example is the interactions of various components of seaweeds and bacteria (not just as a prebiotic), such has been highlighted in the recent years—not only by pathogenicity but also the promotion of beneficial or mutualistic interactions, which reflected on the production of primary and secondary metabolites. However the understanding of these relations is still in the beginning and needs further studies (Charrier et al., 2017).

The production of secondary metabolites in *K. alvarezii* and its growth by land-based aquaculture can be modified by treatments with Ascophyllum Marine Extract Powder (AMPEP), a commercial, biostimulant based on the brown seaweed *A. nodosum*. An extensive review of the AMPEP effects on cultivation of *K. alvarezii* and its selected cultivars was made by Hurtado and Critchley (2018). The AMPEP can be applied in the seaweed cultivation fertilized with effluents from shrimp cultivation in a biofloc system (BFT). The combined treatment of extracts of the BFT and AMPEP enhanced by a factor of two the concentration of some carotenoids (e.g. trans-β-carotene, cis-β-carotene, α-carotene, zeaxanthin, lutein, and others). Significantly higher content of total phenolics and flavonoids were found in treatments using the BFT effluents, thereby suggesting that the re-use of aquaculture effluents, in an IMTA for the benefits seaweed (Pedra et al., 2017).

In the context of IMTA, certain seaweeds may play fundamental roles, once the by-products from one species become inputs for another—in a sequential, bio-refinery approach, and the constituents of the algal biomass become feed resources for various extractive processes or for another organism. Such an approach not only mitigates effluents but can also assist economic viability due to product diversification and better management practices (Chopin & Sawhney, 2009). Seaweed aquaculture along affected coasts may also work as a biofilter and used as one treatment to mitigate the effects of marine eutrophication (Lüning & Pang, 2003).

The intensive cultivation of seaweeds also provides opportunities to remediate nitrogen and carbon wastes from wastewater producing biochar once the biomass can sequester carbon, nitrogen, and phosphorous. Biochar offers an interesting commercial solution because there is an increasing necessity to mitigate carbon production in industries. In addition, there is a search for the use of various biomass sources of seaweeds as natural animal feed/supplements, bio-fertilizers, soil conditioners, and feedstock - instead of (or in synergistic combinations) with synthetic chemicals and mineral fertilizers, so minimizing the environmental impacts of these products (Michalak & Chojnacka, 2015). Changes in the pre-treatment and temperature parameters of biochar production can guide its use for enhanced crop productivity and health, amelioration of soils and their microbial communities (i.e. as prebiotics), application to acidic soils (pH control) and carbon sequestration, and also maximize the benefits this seaweed feedstock (Bird, Wurster, Silva, Bass, & De Nys, 2011; Roberts & De Nys, 2016).

The bioremediation properties of selected seaweeds and their biomass has many applications, including the removal of certain metals from solutions. In a biomedical view, sodium alginate has reported the ability to remove strontium 85 and strontium 87 from living beings without seriously affecting their metabolism (Bhakuni & Rawat, 2005). The effective metal-reducing capacity of phytochemicals such as hydroxyl, carboxyl, and amino-functional groups raises the potential applications of certain seaweeds as an efficient, cost-effective and eco-friendly mediated synthesis of metal nanoparticles in the biotechnology industry. In this industry, the use is diverse, for example for application of polysaccharides greatly used as thickening agent in microbiology (agar) and in development of bioplastic (agar and alginate) or for development of coating materials on the surfaces of aquaculture installations and other marine systems to prevent fouling (Vijayan et al., 2016).

Some of the barriers, bottlenecks and limitations facing the global seaweed industry to further innovation and rapid development are suggested here to include the necessary product registrations due to their seasonal and geographic variation of chemical composition, product viability, generation of new markets, design of production systems, selection of efficient cultivars, the difference in the scale studies and technologies (Michalak & Chojnacka, 2015). The variability in chemical composition can be individual, caused by environment variations, and populational, due to genetic differentiation into locally adapted strains or phenotypically acclimated individuals (Martin et al., 2014). Innovative solutions need to be built from the academic knowledge along with the awareness and business expertise, and to bring it to the market, standardization, efficacy, traceability and clinical tests are needed (Hafting et al., 2015; Martin et al., 2014).

Industrial-scale, extensive biomass production of value-added products depends on net growth and the seaweed capacity to grow in specific structures and compositions. Since the hatchery phase needs parameters adjusted for the directional demands of the juveniles, investments in impacts of the surrounding physical and chemical environment (light, nutrients, salinity, and water movement) are required (Charrier et al., 2017). The search is to maximize the benefits of the sustainable seaweed raw materials and the potential for even larger volumes of cultivated biomass (Duan et al., 2019). Intensive, land-based cultivation enables crop manipulation, and management focused on genetic selection and control of parameters that directly act in the seaweed physiology and bioactive production. However, viability depends on the value of the product within various markets, in part as production and capital infrastructure costs must be covered by the value of the end product. A major challenge for savvy producers will be to activate and maintain the profiles of bioactive compounds in cultivated seaweed biomass (Hafting et al., 2015). The manufacturing process, technology, and cost in relation to yields of the product and market cost per kilogram are crucial to define the viability of the product (Martin et al., 2014). Technological advances in extraction process of bioactive compounds from various seaweeds can be driven by innovation and market demand (Michalak & Chojnacka, 2015).

Another challenge would be to bring the extension and application of the results from the laboratory to industrial scales since the new technologies and techniques require re-sizing up to a given scale of production; this is sometimes where enterprises fail (Michalak & Chojnacka, 2015). Besides, laws and regulations can have limited support in cases of new pharmaceuticals and food derived from seaweed (Hafting et al., 2015; Michalak & Chojnacka, 2015).

4. Climate change challenges

As with other sustainable, commercial activities developed in coastal areas, seaweed farming is also seen to be negatively impacted by the effects of climate change. Storm surge/raising sea levels/tsunamis, warming oceans, changes in CO_2/pH, earthquakes, prolonged rainy periods, typhoons and strong currents are being noted with greater frequencies, causing significant production losses (Largo, Chung, Phang, Gerung, & Sondak, 2017).

The higher frequency and intensity of storms in Canada may cause a large loss of rockweed (*A. nodosum*) biomass during the winter each year and with permanent loss in exposed areas that are not able to recover during the summer (Ugarte, Craigie, & Critchley, 2010). The North Atlantic relic kelps forests (such as *Alaria esculenta, Laminaria digitata, L. hyperborea, L. ochroleuca, L. solidungula, S. latissima, Saccorhiza polyschides* and *Saccorhiza dermatodea*) are estimated by niche modeling to be in threat of disappearing due to ongoing climate changes (Assis, Araújo, & Serrão, 2018). Migration of marine species has been noted, modifying ecosystems and seasonal activities, and affecting the seaweed crops. The increasing grazing and herbivorous fish and competition with other species can maximize the sedimentation in coastal, damaging the farms (Kronen, 2013; Largo et al., 2017).

In the coastal environments, seaweed aquaculture beds provide many ecosystem services associated with natural seaweed communities like CO_2 sequestration, food provision, and the supply of chemicals (Chung, Sondak, & Beardall, 2017). Thus, seaweeds acts directly in CO_2 sequestration by photosynthesis, accumulating biomass through nitrogen assimilation (Chen, Xia, Zou, & Zhang, 2019). Even though according to Largo et al. (2017) the idea of seaweed as a carbon sink may seem unrealistic because their life history is short and the accumulation of carbon in their thalli may not last long periods, the farms can contribute to minimizing the effects of ocean acidification.

Kim et al. (2017) made an interesting quantification of nitrogen and carbon removed from seawater in marine, macroalgal farms, considering the most cultivated genera: *Pyropia, Gracilaria, Kappaphycus/Eucheuma,* kelps, and *Sargassum*. According to these authors, these groups can remove approximately 65,000 tons of nitrogen and 760,000 tons of carbon *per* year.

Bjerregaard et al. (2016) calculated that the assimilation of nitrogen by marine farms, considering the production of 500,000 tons of seaweed and assuming nitrogen content of 3% for dry seaweed, could be 10 million tons from seawater or 30% of nitrogen inputs in to the ocean, in wastewater from agriculture, for example. These last authors also calculated that the same 500,000 tons of the seaweed could absorb 135 million tons of carbon or 3.2% of all carbon added in seawater from greenhouse gas emission.

Jiang, Zou, Lou, Deng and Zeng (2018) evaluated the effects of intensive production of *P. haitanensis* in large-scale and high density under conditions of seawater acidification. They verified that the increase in density caused by the growth of biomass and high photosynthetic range helped to reduce acidification of seawater. However, as density increased, the photosynthetic performance reduced. In another work, Jiang et al. (2017) observed that the maintenance of a high stocking density demonstrated a negative effect on the farms once the carbon supply was reduced and compromise photosynthesis of the same species. Intense photosynthesis can deplete dissolved inorganic carbon (DIC) in the water, which makes seaweed cultivation beds subject to low carbon.

Chen et al. (2019) tried to simulate the effects of aspects of climate change, incubating *P. haitanensis* for 10 days at elevated CO_2 levels, and changing diurnal temperature conditions. They observed that in an acid condition and suitable diurnal temperature, the carbon fixation and nitrogen assimilation were enhanced and consequently, the growth of the seaweed thalli. Several authors observed that high concentrations of dissolved inorganic carbon (DIC) in acidified seawater could change the content of soluble proteins, soluble carbohydrates, photosynthetic pigments and other biochemicals in *P. haitanensis* (Chen et al., 2019; Zou & Gao, 2004). This happens because commercial seaweed species can utilize increased DIC efficiently in photosynthesis, in an environment rich in CO_2, since the enhancement of photosynthesis was attributed to the increased activity of intra- and extra-cellular carbonic anhydrase (CA), thereby helping assimilation of carbon and nitrogen, when nutrients were in a sufficient supply (Chen et al., 2019; Zou & Gao, 2004). However, increased carbon and nitrogen metabolic rates were not in parallel and can also be influenced by temperature. Temperature fluctuations in the sea surface would be conducive to carbon fixation and nitrogen assimilation by the enhancement of photosynthesis and nitrate reductase activity in acidified seawater.

In excessively high seawater temperatures, photosynthesis, and energy consumption can negatively affect the cultivated species (Chen et al.,

2019). Although, under high-temperature conditions rather than low-temperature, *Gracilaria lemaneiformis* exhibited higher growth rates, regardless of CO_2 levels. In high temperatures, enhanced nitrogen accumulation resulted in an increased content of nitrogen and amino acids as the ratios of C:N was reduced, while at lower temperatures no significant effects on the carbon and nitrogen acquisition were observed. Seasonal temperature changes during the culture period associated with an increase in CO_2 levels might be effective for the growth and nutritional accumulation of *G. lemaneiformis*—if the nutrients in seawater were sufficient (Chen, Zou, Du, & Ji, 2018).

An understanding of the responses to the temperature inside the context of ocean acidification of commercially important seaweeds is an important issue for the management of seaweed aquaculture into the future. Productivity may suffer intensively from the elevation of surface seawater temperature and CO_2 as also their interactions with other environmental factors (Chung et al., 2017). Therefore, the industry may be prepared for the effects of climate change through the development of thermo-tolerant strains (Kim et al., 2017).

Further efforts to identify species for use in seaweed aquaculture in the future will need to take into account their physiological tolerance to environmental changes and to ocean acidification and warming (Chung et al., 2017). This is particularly true for cold temperate species, such as kelps. Efforts to develop fast-growing, disease-resistant, and higher temperature tolerant strains are needed—urgently.

Carrageenophytes, in general, are well-adapted to the tropical seawater temperatures variation in their cultivation site. Temperature experiments revealed that while they exhibit maximum photosynthesis at 17–33°C, they are likely close to their threshold temperatures for thermal inhibition. This knowledge is essential for the improvement of seaweed cultivation practices to ensure sustainable production and supply, specially in the midst of increasing seawater temperatures brought about by global climate change (Borlongan, Gerung, Nishihara, & Terada, 2017). In relation to the most cultivated genera of sub-tropical red seaweeds *Eucheuma*/*Kappaphycus*, a generally warming ocean facilitated its successful introduction to sub-tropical regions (e.g., Brazil and Japan), and it seems that the increase in atmospheric CO_2 was favorable for their growth with no direct evidence of the impact of the associated lower pH in the field (not the case for coralline algae); Although experiments under laboratory conditions showed that lower pH could result in lower growth rates and a change in protein profile

(Largo et al., 2017). For coralline algae, water chemistry parameters with positive calcification regulatory, control of surface pH and light-independent carbon fixation are important to the maintenance in future ocean conditions (Schoenrock et al., 2018).

One of the main concerns is the production using vegetative propagation, resulting in a reduction of productivity and weak strains, susceptible to disease and epiphyte/pathogen attacks, besides variation in environmental conditions. Attempts to solve this problem have been made for several researchers around the world, raising new crops from spores or improving techniques of micropropagation and tissue culture (e.g., Jong, Thien, Yong, Rodrigues, & Yong, 2015; Luhan & Sollesta, 2010; Tibubos et al., 2017).

Risk assessment of expansion of seaweed farm to larges scale that brings drivers of environmental change (light, nutrients, carbon, kinetic energy, artificial material, noise, dissolved and particulate matter, etc), such as Campbell et al. (2019) for operations in Europe, needs to be made worldwide.

For mitigation, the climate changes impacts on seaweeds modeling approaches with physiological and ecological insights, integrated with genetic, epigenetic, and microbiome levels of intra-specific variation can help to predict a more the scenario (Duarte et al., 2018).

5. Future investigations

The necessary phyconomical developments required for the large-scale production of seaweeds for economic and industrial purposes, needs the interaction of academia, private sector, and government agencies for sharing the learning lessons and to strengthening the field (Hurtado, Neish, & Critchley, 2019). One goal for the development of sustainable cultivation of multiple species of seaweeds is to ensure that commercial aquaculture has minimal adverse effects on the environment and vice-versa. One way to achieve this goal is through the development of improved methods of waste management for land-based and coastal aquaculture (Kim et al., 2017). Carrageenophyte farming can have positive effects on the environment because it demonstrably improves the benthic ecosystem and sequesters carbon, thereby offering the potential for carbon credits (Valderrama, Cai, Hishamunda, & Ridler, 2013). Cultivation of various seaweeds can provide a low footprint for the environment so as to reduce pressure on land, freshwater and other resources, since *S. latissima* production was estimated to have an equivalent footprint as the microalgal production and farmed food (such as sugar beets, maize and potatoes) (Taelman et al., 2015).

Intensification of activities in seaweed cultivation can be achieved in certain locations, particularly in areas where production is seasonal. However, intensification carries the risk of bringing imbalances for farming areas that have reached the limits of their carrying capacity. In particular, the integrated multi-trophic aquaculture (IMTA), provides ecological benefits on overall production and enable the production of two or more cultivated species (Neori et al., 2019).

The aquaculture of some species still uses propagules/plantlets originally derived from natural populations and needs legislation and regulations so as to avoid their overexploitation (Kim et al., 2017), although the seaweed farming already contributed against the overexploitation of fishery resources (Hurtado, 2013). New technologies such as remote sensing data from high-resolution drones and satellites can make estimative of natural stocks of seaweed and identify potential areas for seaweed collection and cultivation (Setyawidati et al., 2018). Ensuring the biodiversity is a necessity for further value of seaweeds in the environment and their applications (Critchley et al., 2019).

A thorough understanding of the commercial species physiology is increasingly important to all forms of seaweed exploitation. The end-user industries and environmental concerns demand more control, quality, and greater utilization of the by-products, something even more difficult if marine farms due to effects of climate changes are considered. New technologies and processes need to be developed to address market needs and promote environmental sustainability.

In spite of the foregoing, there remains a tremendous need for continues research, development, innovation and refinement so as to find and improve on "new" applications for seaweeds and their bioactives, this includes "disruptive technologies" which provide things the industry "does not yet know it needs". By no means is everything we need to know about seaweeds and their properties fully known. However, we "need to know what we do not know". We need to recognize "seaweeds" as a polyphyletic "rag bag" of some very desperately associated genera of marine photosynthetic organisms. There is no "one size fits all" category for seaweeds and their bioactivities. We will only get the most benefits from their goods and services if we understand more about their fundamental biology - but then also have the skills and knowledge to take that information and apply it to beneficial services and products which provide the "greatest bang for the buck" for human society and the sustained existence of this planet.

A serious analysis must be undertaken related to biomass production and the viability of the production and value chain. Land-based, intensive cultivation has high costs, and biomass selection and maintenance is not easy. Back to basics, fundamental studies must be made to the current commercial species and the potential new (novel) ones, so as to provide specific insights as was made to cultivation protocols for *Pyropia* and *Saccharina* production based in their physiology.

Acknowledgments

Leila Hayashi thanks Conselho Nacional de Desenvolvimento Científico e Tecnológico (CNPq) for the Productivity Fellowship (Process number 308631/2017-0).

References

Abreu, M. H., Pereira, R., Buschmann, A. H., Sousa-Pinto, I., & Yarish, C. (2011). Nitrogen uptake responses of *Gracilaria vermiculophylla* (Ohmi) Papenfuss under combined and single addition of nitrate and ammonium. *Journal of Experimental Marine Biology and Ecology, 407*, 190–199.

Admassu, H., Gasmalla, M. A. A., Yang, R., & Zhao, W. (2018). Bioactive peptides derived from seaweed protein and their health benefits: Antihypertensive, antioxidant, and antidiabetic properties. *Journal of Food Science, 83*, 6–16.

Ali, M. K. M., Yasir, S. M., Critchley, A. T., & Hurtado, A. Q. (2018). Impacts of *Ascophyllum* marine plant extract powder (AMPEP) on the growth, incidence of the endophyte Neosiphonia apiculata and associated carrageenan quality of three commercial cultivars of *Kappaphycus*. *Journal of Applied Phycology, 30*, 1185–1195.

Assis, J., Araújo, M. B., & Serrão, E. A. (2018). Projected climate changes threaten ancient refugia of kelp forests in the North Atlantic. *Global Change Biology, 24*, 55–66.

Azanza, R. V., & Ask, E. (2017). Reproductive biology and eco-physiology of farmed *Kappaphycus* and *Eucheuma*. In A. Q. Hurtado, A. T. Critchley, & I. C. Neish (Eds.), *Tropical seaweed farming: Trends, problems and opportunities – focus on* Kappaphycus *and* Eucheuma *of commerce* (pp. 45–54). Cham, Switzerland: Springer.

Besednova, N. N., Zvyagintseva, T. N., Kuznetsova, T. A., Makarenkova, I. D., Smolina, T. P., Fedyanina, L. N., et al. (2019). Marine algae metabolites as promising therapeutics for the prevention and treatment of HIV/AIDS. *Metabolites, 9*, 87.

Bhakuni, D. S., & Rawat, D. S. (2005). *Bioactive marine natural products, department of chemistry*. New Delhi, India: Anamaya Publishers.

Bird, M. I., Wurster, C. M., Silva, P. H. P., Bass, A. M., & De Nys, R. (2011). Algal biochar–production and properties. *Bioresource Technology, 102*, 1886–1891.

Bjerregaard, R., Valderrama, D., Radulovich, R., Diana, J., Capron, M., Forster, J., et al. (2016). *Seaweed aquaculture for food security, income generation and environmental health in tropical developing countries*. World Bank Group. Report #107147. Retrieved from: http://documents.worldbank.org/curated/en/947831469090666344/pdf/107147-WP-REVISED-Seaweed-Aquaculture-Web.pdf.

Blouin, N. A., Brodie, J. A., Grossman, A. C., Xu, P., & Brawley, S. H. (2011). *Porphyra*: A marine crop shaped by stress. *Trends in Plant Science, 16*, 29–37.
Borlongan, I. A., Gerung, G. S., Kawaguchi, S., Nishihara, G. N., & Terada, R. (2017). Thermal and PAR effects on the photosynthesis of *Eucheuma denticulatum* and *Kappaphycus striatus* (so-called sacol strain) cultivated in shallow bottom of Bali, Indonesia. *Journal of Applied Phycology, 29*, 395–404.
Borlongan, I. A. G., Gerung, G. S., Nishihara, G. N., & Terada, R. (2017). Light and temperature effects on photosynthetic activity of *Eucheuma denticulatum* and *Kappaphycus alvarezii* (brown and green color morphotypes) from Sulawesi Utara, Indonesia. *Phycological Research, 65*, 69–79.
Borlongan, I. A. G., Luhan, M. R. J., Padilla, P. I. P., & Hurtado, A. Q. (2016). Photosynthetic responses of '*Neosiphonia sp.* epiphyte-infected'and healthy *Kappaphycus alvarezii* (Rhodophyta) to irradiance, salinity and pH variations. *Journal of Applied Phycology, 28*, 2891–2902.
Borlongan, I. A. G., Tibubos, K. R., Yunque, D. A. T., Hurtado, A. Q., & Critchley, A. T. (2011). Impact of AMPEP on the growth and occurrence of epiphytic *Neosiphonia* infestation on two varieties of commercially cultivated *Kappaphycus alvarezii* grown at different depths in the Philippines. *Journal of Applied Phycology, 23*, 615–621.
Buschmann, A. H., Hernández-González, M. D., & Varela, D. (2008). Seaweed future cultivation in Chile: Perspectives. *International Journal of Environment and Pollution, 33*, 432–456.
Campbell, I., Macleod, A., Sahlmann, C., Neves, L., Funderud, J., Øverland, M., et al. (2019). The environmental risks associated with the development of seaweed farming in Europe – prioritizing key knowledge gaps. *Frontiers in Marine Science, 6*, 107.
Carroll, A. R., Copp, B. R., Davis, R. A., Keyzers, R. A., & Prinsep, M. R. (2019). Marine natural products. *Natural Product Reports, 36*, 122–173.
Cerná, M. (2011). Seaweed proteins and amino acids as nutraceuticals. In S. Kim, & S. Taylor (Eds.), *Advances in food and nutrition research* (pp. 297–312). St Louis: Elsevier.
Charrier, B., Abreu, M. H., Araujo, R., Bruhn, A., Coates, J. C., De Clerck, O., et al. (2017). Furthering knowledge of seaweed growth and development to facilitate sustainable aquaculture. *New Phytologist, 216*, 967–975.
Chen, Y., Maguire, T., Hileman, R. E., Fromm, J. R., Esko, J. D., Linhardt, R. J., et al. (1997). Dengue virus infectivity depends on envelope protein binding to target cell heparan sulfate. *Nature Medicine, 3*, 866–871.
Chen, B., Xia, J., Zou, D., & Zhang, X. (2019). Responses to ocean acidification and diurnal temperature variation in a commercially farmed seaweed, *Pyropia haitanensis* (Rhodophyta). *European Journal of Phycology, 54*, 184–192.
Chen, B., Zou, D., Du, H., & Ji, Z. (2018). Carbon and nitrogen accumulation in the economic seaweed *Gracilaria lemaneiformis* affected by ocean acidification and increasing temperature. *Aquaculture, 482*, 176–182.
Chopin, T., & Sawhney, M. (2009). Seaweeds and their mariculture. In S. E. Jörgensen, & B. D. Fath (Eds.), *The encyclopedia of ecology: Vol. 3. Ecological engineering* (pp. 4477–4486). Oxford: Elsevier.
Chung, I. K., Sondak, C. F. A., & Beardall, J. (2017). The future of seaweed aquaculture in a rapidly changing world. *European Journal of Phycology, 52*, 495–505.
Craigie, J. S., Cornish, M. L., & Deveau, L. E. (2019). Commercialization of Irish moss aquaculture: The Canadian experience. *Botanica Marina, 62*, 411–432.
Critchley, A., Hurtado, A., Pereira, L., Cornish, M., Largo, D., & Paul, N. (2019). Seaweed resources of the world: A 2020 vision. *Botanica Marina, 62*, 191–193.
Dahms, H., & Dobretsov, S. (2017). Antifouling compounds from marine macroalgae. *Marine Drugs, 15*, 265.

Dawes, C. J., & Koch, E. W. (1991). Branch, micropropagule and tissue culture of the red algae *Eucheuma denticulatum* and *Kappaphycus alvarezii* farmed in the Philippines. *Journal of Applied Phycology, 3*, 247–257.
Dawes, C. J., Trono, G. C., & Lluisma, A. O. (1993). Clonal propagation of *Eucheuma denticulatum* and *Kappaphycus alvarezii* for Philippine seaweed farms. *Hydrobiologia, 260*(1), 379–383.
Drew, K. M. (1949). Conchocelis-phase in the life-history of *Porphyra umbilicalis* (L.) Kütz. *Nature, 164*, 748–749.
Drum, R. (2013). Sea vegetables for food and medicine. *Well Being Journal*. Retrieved from http://www.eidon.com/SeaweedArticle.pdf.
Duan, D., Critchley, A. T., Fu, X., & Pereira, L. (2019). Preface: Bioactive substances of various seaweeds and their applications and utilization. *Journal of Oceanology and Limnology, 37*, 779–782.
Duarte, B., Martins, I., Rosa, R., Matos, A. R., Roleda, M. Y., Reusch, T. B. H., et al. (2018). Climate change impacts on seagrass meadows and macroalgal forests: An integrative perspective on acclimation and adaptation potential. *Frontiers in Marine Science, 5*, 190.
Freile-Pelegrín, Y., & Tasdemir, D. (2019). Seaweeds to the rescue of forgotten diseases: A review. *Botanica Marina, 62*, 211–226.
Gachon, C. M., Sime-Ngando, T., Strittmatter, M., Chambouvet, A., & Kim, G. H. (2010). Algal diseases: Spotlight on a black box. *Trends in Plant Science, 15*, 633–640.
Góes, H. G., & Reis, R. P. (2011). An initial comparison of tubular netting versus tie–tie methods of cultivation for *Kappaphycus alvarezii* (Rhodophyta, Solieriaceae) on the south coast of Rio de Janeiro state, Brazil. *Journal of Applied Phycology, 23*, 607–613.
Hafting, J. T., Craigie, J. S., Stengel, D. B., Loureiro, R. R., Buschmann, A. H., Yarish, C., et al. (2015). Prospects and challenges for industrial production of seaweed bioactives. *Journal of Phycology, 51*, 821–837.
Hayashi, L., Oliveira, E. C., Bleicher-Lhonneur, G., Boulenguer, P., Pereira, R. T. L., Von Seckendorff, R., et al. (2007). Carrageenan analyses of *Kappaphycus alvarezii* (Rhodophyta, Solieriaceae) cultivated in different conditions in Ubatuba Bay, São Paulo, Brazil. *Journal of Applied Phycology, 19*, 505–511.
Hayashi, L., Reis, R. P., dos Santos, A. A., Castelar, B., Robledo, D., de Vega, G. B., et al. (2017). The cultivation of *Kappaphycus* and *Eucheuma* in tropical and sub-tropical waters. In A. Q. Hurtado, A. T. Critchley, & I. C. Neish (Eds.), *Tropical seaweed farming trends, Problems and opportunities: Vol. 9. Developments in applied phycology* (pp. 55–90). Cham: Springer.
Hindu, S. V., Chandrasekaran, N., Mukherjee, A., & Thomas, J. (2019). A review on the impact of seaweed polysaccharide on the growth of probiotic bacteria and its application in aquaculture. *Aquaculture International, 27*, 227–238.
Hurtado, A. Q. (2013). Social and economic dimensions of carrageenan seaweed farming in the Philippines. Fisheries and aquaculture technical paper: N. 580. In D. Valderrama, J. Cai, N. Hishamunda, & N. Ridler (Eds.), *Social and economic dimensions of carrageenan seaweed farming* (pp. 91–113). Rome: FAO.
Hurtado, A. Q., & Critchley, A. T. (2018). A review of multiple biostimulant and bioeffector benefits of AMPEP, an extract of the brown alga *Ascophyllum nodosum*, as applied to the enhanced cultivation and micropropagation of the commercially important red algal carrageenophyte *Kappaphycus alvarezii* and its selected cultivars. *Journal of Applied Phycology, 30*, 2859–2873.
Hurtado, A. Q., Neish, I. C., & Critchley, A. T. (2019). Phyconomy: The extensive cultivation of seaweeds, their sustainability and economic value, with particular reference to important lessons to be learned and transferred from the practice of eucheumatoid farming. *Phycologia, 58*, 472–483.

Israel, A., Levy, I., & Friedlander, M. (2006). Experimental tank cultivation of *Porphyra* in Israel. In R. Anderson, J. Brodie, E. Onsøyen, & A. T. Critchley (Eds.), *Developments in applied phycology: Vol. 1. Eighteenth international seaweed symposium*. Dordrecht, Netherlands: Springer.
Jaswir, I., & Monsur, H. A. (2011). Anti-inflammatory compounds of macro algae origin: A review. *Journal of Medicinal Plants Research, 5*, 7146–7154.
Jiang, H., Zou, D., Lou, W., Deng, Y., & Zeng, X. (2018). Effects of seawater acidification and alkalization on the farmed seaweed *Pyropia haitanensis* (Bangiales, Rhodophyta), grown under different irradiance conditions. *Algal Research, 31*, 413–420.
Jiang, H., Zou, D., Lou, W., & Ye, C. (2017). Effects of stocking density and decreased carbon supply on the growth and photosynthesis in the farmed seaweed, *Pyropia haitanensis* (Bangiales, Rhodophyta). *Journal of Applied Phycology, 29*, 3057–3065.
Jong, L. W., Thien, V. Y., Yong, Y. S., Rodrigues, K. F., & Yong, W. T. L. (2015). Micropropagation and protein profile analysis by SDS-PAGE of *Gracilaria changii* (Rhodophyta, Solieriaceae). *Aquaculture Reports, 1*, 10–14.
Kawashima, S. (1993). Cultivation of the brown alga, *Laminaria* "kombu". In M. Ohno, & A. T. Critchley (Eds.), *Seaweed cultivation and marine ranching* (pp. 25–40). Yokosuka, Japan: JICA.
Kerrison, P. D., Stanley, M. S., & Hughes, A. D. (2018). Textile substrate seeding of *Saccharina latissima* sporophytes using a binder: An effective method for the aquaculture of kelp. *Algal Research, 33*, 352–357.
Kerrison, P. D., Twigg, G., Stanley, M., De Smet, D., Buyle, G., Pina, A. M., et al. (2019). Twine selection is essential for successful hatchery cultivation of *Saccharina latissima*, seeded with either meiospores or juvenile sporophytes. *Journal of Applied Phycology, 31*, 3051–3060.
Khan, S. I., & Satam, S. B. (2003). Seaweed mariculture: Scope and potential in India. *Aquaculture Asia, 8*, 26–29.
Kim, J. K., Yarish, C., Hwang, E. K., Park, M., & Kim, Y. (2017). Seaweed aquaculture: Cultivation technologies, challenges and its ecosystem services. *Algae, 32*, 1–13.
Kolanjinathan, K., Ganesh, P., & Saranraj, P. (2014). Pharmacological importance of seaweeds: A review. *World Journal of Fish and Marine Sciences, 6*, 1–15.
Kronen, M. (2013). Social and economic dimensions of carrageenan seaweed farming in Solomon Islands. Fisheries and aquaculture technical paper: N. 580. In D. Valderrama, J. Cai, N. Hishamunda, & N. Ridler (Eds.), *Social and economic dimensions of carrageenan seaweed farming* (pp. 147–161). Rome: FAO.
Largo, D. B., Chung, I. K., Phang, S. M., Gerung, G. S., & Sondak, C. F. A. (2017). Impacts of climate change on *Eucheuma-Kappaphycus* farming. In A. Q. Hurtado, A. T. Critchley, & I. C. Neish (Eds.), *Tropical seaweeds farming trends, problems and opportunities - focus on* Kappaphycus *and* Eucheuma *of commerce* (pp. 111–120). Cham, Switzerland: Springer.
López-Vivas, J. M., Riosmena-Rodríguez, R., Pacheco-Ruíz, I., & Yarish, C. (2015). Growth and reproductive responses of the conchocelis phase of *Pyropia hollenbergii* (Bangiales, Rhodophyta) to light and temperature. *Journal of Applied Phycology, 27*, 1561–1570.
Loureiro, R. R., Hurtado, A. Q., & Critchley, A. T. (2017). Impacts of AMPEP on epiphytes and diseases in *Kappaphycus* and *Eucheuma* cultivation. In A. Q. Hurtado, A. T. Critchley, & I. C. Neish (Eds.), *Tropical seaweeds farming trends, problems and opportunities - focus on* Kappaphycus *and* Eucheuma *of commerce* (pp. 111–120). Cham, Switzerland: Springer.
Loureiro, R. R., Reis, R. P., Berrogain, F. D., & Critchley, A. T. (2012). Extract powder from the brown alga *Ascophyllum nodosum* (Linnaeus) Le Jolis (AMPEP): A "vaccine-like" effect on *Kappaphycus alvarezii* (Doty) Doty ex P.C. Silva. *Journal of Applied Phycology, 24*, 427–432.

Loureiro, R. R., Reis, R. P., Berrogain, F. D., & Critchley, A. T. (2014). Effects of a commercial extract of the brown alga *Ascophyllum nodosum* on the biomass production of *Kappaphycus alvarezii* (Doty) Doty ex P. C. Silva and its carrageenan yield and gel quality cultivated in Brazil. *Journal of Applied Phycology, 26*, 763–768.
Luhan, M. R. J., & Sollesta, H. (2010). Growing the reproductive cells (carpospores) of the seaweed, *Kappaphycus striatum*, in the laboratory until outplanting in the field and maturation to tetrasporophyte. *Journal of Applied Phycology, 22*, 579–585.
Lüning, K., & Dieck, T. I. (1989). Environmental triggers in algal seasonality. *Botanica Marina, 32*, 389–398.
Lüning, K., & Pang, S. (2003). Mass cultivation of seaweeds: Current aspects and approaches. *Journal of Applied Phycology, 15*, 115–119.
Marroig, R. G., Loureiro, R. R., & Reis, R. P. (2016). The effect of *Ascophyllum nodosum* (Ochrophyta) extract powder on the epibiosis of *Kappaphycus alvarezii* (Rhodophyta) commercially cultivated on floating rafts. *Journal of Applied Phycology, 28*, 2471–2477.
Martins, A., Vieira, H., Gaspar, H., & Santos, S. (2014). Marketed marine natural products in the pharmaceutical and cosmeceutical industries: Tips for success. *Marine Drugs, 12*, 1066–1101.
Michalak, I., & Chojnacka, K. (2015). Algae as production systems of bioactive compounds. *Engineering in Life Science, 15*, 160–176.
Mouritsen, O. G., Rhatigan, P., & Pérez-Lloréns, J. L. (2019). The rise of seaweed gastronomy: Phycogastronomy. *Botanica Marina, 62*, 195–209.
Neori, A., Guttman, L., Israel, A., & Shpigel, M. (2019). Israeli-developed models of marine integrated multi trophic aquaculture (IMTA). *Journal of Coastal Research, 86*, 11–20.
Neori, A., Shpige, M., Guttman, L., & Israel, A. (2017). Development of polyculture and integrated multi-trophic aquaculture (IMTA) in Israel: A review. *The Israeli Journal of Aquaculture, 69*, 1–19.
Nobre, A. M., Valente, L. M. P., & Neori, A. (2017). A nitrogen budget model with a user-friendly interface, to assess water renewal rates and nitrogen limitation in commercial seaweed farms. *Journal of Applied Phycology, 29*, 3039–3055.
Ohno, M., & Matsuoka, M. (1993). *Undaria* cultivation "wakame". In M. Ohno, & A. T. Critchley (Eds.), *Seaweed cultivation and marine ranching* (pp. 25–40). Yokosuka, Japan: JICA.
Oohusa, T. (1993). The cultivation of *Porphyra* "nori". In M. Ohno, & A. T. Critchley (Eds.), *Seaweed cultivation and marine ranching* (pp. 57–73). Yokosuka, Japan: JICA.
Pedra, A. G. L. M., Ramlov, F., Maraschin, M., & Hayashi, L. (2017). Cultivation of the red seaweed *Kappaphycus alvarezii* with effluents from shrimp cultivation and brown seaweed extract: Effects on growth and secondary metabolism. *Aquaculture, 479*, 297–303.
Pereira, L. (2016). *Edible seaweeds of the world*. Boca Raton, USA: CRC Press/Taylor & Francis Group.
Reis, R. P., Pereira, R. R. C., & Góes, H. G. (2015). The efficiency of tubular netting method of cultivation for *Kappaphycus alvarezii* (Rhodophyta, Gigartinales) on the south-eastern Brazilian coast. *Journal of Applied Phycology, 27*, 421–426.
Roberts, D. A., & De Nys, R. (2016). The effects of feedstock pre-treatment and pyrolysis temperature on the production of biochar from the green seaweed *Ulva*. *Journal of Environmental Management, 169*, 253–260.
Salgado, L. T., Viana, N. B., Andrade, L. R., Leal, R. N., Da Gama, B. A. P., Attias, M., et al. (2008). Intra-cellular storage, transport and exocytosis of halogenated compounds in marine red alga *Laurencia obtusa*. *Journal of Structural Biology, 162*, 345–355.
Schleder, D. D., Perucha, L. G. B., Polia, M. A., Ferreira, T. H., Silva, C. P., Andreatta, E. R., et al. (2018). Effect of brown seaweeds on Pacific white shrimp growth performance, gut morphology, digestive enzymes activity and resistance to white spot virus. *Aquaculture, 495*, 359–365.

Schleder, D. D., Rosa, J. R., Guimarães, A. M., Ramlov, F., Maraschin, M., Seiffert, W. Q., et al. (2017). Brown seaweeds as feed additive for white-leg shrimp: Effects on thermal stress resistance, midgut microbiology, and immunology. *Journal of Applied Phycolology, 29*, 2471–2477.

Schmedes, P. S., Nielsen, M. M., & Petersen, J. K. (2019). Improved *Palmaria palmata* hatchery methods for tetraspore release, even settlement and high seedling survival using strong water agitation and macerated propagules. *Algal Research, 40*, 101494.

Schoenrock, K., Vad, J., Muth, A., Pearce, D., Rea, B., Schofield, J., et al. (2018). Biodiversity of kelp forests and coralline algae habitats in southwestern Greenland. *Diversity, 10*, 117.

Sebök, S., Herppichb, W. B., & Hanelta, D. (2017). Development of an innovative ring-shaped cultivation system for a land-based cultivation of marine macroalgae. *Aquacultural Engineering, 77*, 33–41.

Seca, A., & Pinto, D. (2018). Overview on the antihypertensive and anti-obesity effects of secondary metabolites from seaweeds. *Marine Drugs, 16*, 237.

Setyawidati, N., Kaimuddin, A., Wati, I., Helmi, M., Widowati, I., Rossi, N., et al. (2018). Percentage cover, biomass, distribution, and potential habitat mapping of natural macroalgae, based on high-resolution satellite data and in situ monitoring, at Libukang island, Malasoro Bay, Indonesia. *Journal of Applied Phycology, 30*, 159.

Shpigel, M., Shauli, L., Odintsov, V., Ben-Ezra, D., Neori, A., & Guttman, L. (2018). The sea urchin, *Paracentrotus lividus*, in an integrated multi-trophic aquaculture (IMTA) system with fish (*Sparus aurata*) and seaweed (*Ulva lactuca*): Nitrogen partitioning and proportional configurations. *Aquaculture, 490*, 260–269.

Skrovánková, S. (2011). Seaweed vitamins as nutraceuticals. In K. Se-Kwon (Ed.), *Advances in food and nutrition research marine medicinal foods implications and applications, macro and microalgae* (pp. 357–369). London: Academic Press.

Sudatti, D. B., Fujii, M. T., Rodrigues, S. V., Turra, A., & Pereira, R. C. (2018). Prompt induction of chemical defenses in the red seaweed *Laurencia dendroidea*: The role of herbivory and epibiosis. *Journal of Sea Research, 138*, 48–55.

Taelman, S. E., Champenois, J., Edwards, M. D., De Meester, S., & Dewulf, J. (2015). Comparative environmental life cycle assessment of two seaweed cultivation systems in north west Europe with a focus on quantifying sea surface occupation. *Algal Research, 11*, 173–183.

Thomas, N. V., & Kim, S. K. (2013). Beneficial effects of marine algal compounds in cosmeceuticals. *Marine Drugs, 11*, 146–164.

Tibubos, K. R., Hurtado, A. Q., & Critchley, A. T. (2017). Direct formation of axes in new plantlets of *Kappaphycus alvarezii* (Doty) Doty, as influenced by the use of AMPEP K^+, spindle inhibitors, and plant growth hormones. *Journal of Applied Phycology, 29*, 2345–2350.

Ugarte, R. A., Craigie, J. S., & Critchley, A. T. (2010). Fucoid flora of the rocky intertidal of the Canadian Maritimes: Implications for the future with rapid climate change. In J. Seckbach, R. Einav, & A. Israel (Eds.), *Seaweeds and their role in globally changing environments: Vol. 15. Cellular origin, life in extreme habitats and astrobiology* (pp. 69–90). Dordrecht: Springer.

Usandizaga, S., Camus, C., Kappes, J. L., Guillemin, M. L., & Buschmann, A. H. (2018). Nutrients, but not genetic diversity, affect Gracilaria chilensis (Rhodophyta) farming productivity and physiological responses. *Journal of Phycology, 54*, 860–869.

Valderrama, D., Cai, J., Hishamunda, N., & Ridler, N. (2013). *Social and economic dimensions of carrageenan seaweed farming*. Fisheries and aquaculture technical paper: No. 580. Rome: FAO.

Vijayan, S. R., Santhiyagu, P., Ramasamy, R., Arivalagan, P., Kumar, G., Ethiraj, K., et al. (2016). Seaweeds: A resource for marine bionanotechnology. *Enzyme and Microbial Technology, 95*, 45–57.

Wang, G., Chang, L., Zhang, R., Wang, S., Wei, X., Rickert, E., et al. (2019). Can targeted defense elicitation improve seaweed aquaculture? *Journal of Applied Phycology, 31*, 1845–1854.

Wijesinghe, W. A. J. P., & Jeon, Y. J. (2011). Biological activities and potential cosmeceutical applications of bioactive components from brown seaweeds: A review. *Phytochemistry Reviews, 10*, 431–443.

Zou, D., & Gao, K. (2004). Exogenous carbon acquisition of photosynthesis in *Porphyra haitanensis* (Bangiales, Rhodophyta) under emersed state. *Progress in Natural Science, 14*, 138–144.

CHAPTER FOUR

The application of seaweeds in environmental biotechnology

Izabela Michalak*
Department of Advanced Material Technologies, Faculty of Chemistry, Wrocław University of Science and Technology, Wrocław, Poland
*Corresponding author: E-mail: izabela.michalak@pwr.edu.pl

Contents

Abstract

Apart from being applied as a source of human food, animal feed, fertilizers, components of cosmetics, pharmaceuticals and nutraceuticals, biofuels, fast-growing seaweeds can also be used to prevent environmental pollution, for bioremediation of polluted environment and in environmental biomonitoring (bioindication of marine ecosystem pollution). They are known to effectively remove toxic metals (e.g., arsenic, cadmium, chromium, cobalt, copper, lead, mercury, nickel, zinc) and organic compounds (e.g., DDT, PCB) from contaminated soil, industrial effluents and wastewater treatment. Biosorption of toxic metal ions by dry seaweeds is a kind of biotreatment of polluted water. Also the bioremediation of nutrients, generated by intensive aquaculture (e.g., fish, shrimps and other aquatic animals) in so-called integrated multitrophic aquaculture makes use of seaweeds. Macroalgae derived from unpolluted water – e.g., eutrophicated natural water reservoirs or remediated wastewater from aquaculture can be used for the production of components of fertilizers, feed additives and biochar which provides a multitude of benefits in terms of environmental management, e.g., climate change mitigation due to carbon sequestration, remediation of

Advances in Botanical Research, Volume 95
ISSN 0065-2296
https://doi.org/10.1016/bs.abr.2019.11.006

contaminated sites (water, soil) due to very good sorption properties, plus the mitigation of contaminants movement in soil due to action as a landfill filter medium. The proposed biotechnological tools based on seaweeds biomass can improve the environment and assure its sustainability.

1. Introduction

According to the European Federation of Biotechnology, "Environmental biotechnology provides effective tools, which are sustainable from economic, environmental and social points of view and can be applied to: (1) monitor and reduce the risk to humans from contaminated sites and from the storage of municipal and industrial (bio)wastes, (2) clean up water, soil, and air to achieve good quality standards and reuse treated wastewater to reduce the demand of natural water for industrial, agricultural and municipal purposes and (3) turn (bio)waste, including wastewater, into bio-based biodegradable/biocompatible and renewable chemicals for material and fuel production" (EFB, 2019; Gavrilescu, 2010; Wang, Ivanov, Tay, & Hung, 2010).

In the present chapter, the possibilities of the seaweeds applications in the bioremediation of polluted water and soil and in the environmental monitoring are evaluated. Industrialization, urbanization, population growth and agriculture largely contribute to the degradation of the natural environment (Bilal et al., 2018; Chekroun & Baghour, 2013). Humankind needs to remove from the environment toxic metals (e.g., arsenic, cadmium, mercury and lead) and organic compounds used in numerous industrial activities, such as: chemical, pigments, textile, plastics, storage batteries, electroplating, mining, smelting, metallurgical processes etc. (Bilal et al., 2018; Bulgariu & Bulgariu, 2017). Heavy metals are especially dangerous because they are toxic, persistent in the environment, cannot be destroyed or degraded and tend to bioaccumulate in the food chain (Arumugam et al., 2018; Bulgariu & Bulgariu, 2017; Ortiz-Calderon, Silva, & Vásquez, 2017; Sati, Verma, Bora, & Rai, 2016; van Ginneken & de Vries, 2018). An important issue is also bioremediation of the contaminated soil with organic (e.g., pesticides, DDT (1,1,1-trichloro-2,2-bis (*p*-chlorophenyl) ethane), PCBs (polychlorinated biphenyls), PAHs (polycyclic aromatic hydrocarbons)) and toxic metals (e.g., As, Cd, Cr, Cu, Hg, Mn, Ni, Pb, Zn) (Chekroun & Baghour, 2013; Cheney, Rajic, Sly, Meric, & Sheahan, 2014; Farias et al., 2017; Kantachote et al., 2004). Also eutrophicated water and effluents generated by intensive aquaculture (farming of fish, shellfish, shrimp, sea urchin, abalone etc.) that are rich in biogens such as nitrogen and phosphorus pose a problem (Chopin, 2013; Neori et al., 2004). The effective removal of pollutants from industrial/aquaculture effluents and soil is one of the tasks of environmental biotechnology and conditions of sustainable development.

In order to reduce the level of contamination in water bodies and in soil, several remediation technologies have been implemented, for example the application of low-cost adsorbents in biosorption, chelating agents and biology-based techniques for example phytoremediation (Singh & Prasad, 2015). Seaweeds, also called marine macroalgae, which occur mainly in three phyla – brown (B; phylum Ochrophyta, class Phaeophyceae), green (G; phylum Chlorophyta) and red macroalgae (R; phylum Rhodophyta) – are promising candidates for the remediation technologies (Dawes, 2016). Macroalgae are of great ecological importance because they supply oxygen to seawater and constitute the most important element in the marine food chain – they are primary producers (Chan, Ho, & Phang, 2006). Seaweeds grow fast and abound in many regions all over the world, especially if the water is warm and acidified. An intensive seaweed proliferation is a symptom of eutrophication. Their biomass captures carbon and some nutrients like nitrogen and phosphorous (Arumugam et al., 2018; Bird, Wurster, Silva, Paul, & de Nys, 2012; Gao, Clare, Rose, & Caldwell, 2018; Seghetta, Torring, Bruhn, & Thomsen, 2016). The remediation technologies aiming at the reduction of water and soil contamination (and the same in the food chain) should render a great service in agriculture (Singh & Prasad, 2015). The examples of seaweeds applications in environmental biotechnology are presented in Fig. 1.

2. Bioremediation of toxic metal ions from wastewater

2.1 Biosorption by seaweeds

Bioremediation of toxic metals from wastewater by algae can be carried out through biosorption (using non-living organisms) or bioaccumulation (using living cells) (Aksu, 1998; Henriques et al., 2017). The former is known to be much faster and effective than the latter where higher concentrations of pollutants in wastewater can cause the death of living organisms (Aksu, 1998; Bilal et al., 2018). Therefore, biosorption, as a subcategory of adsorption is more often used for the removal of

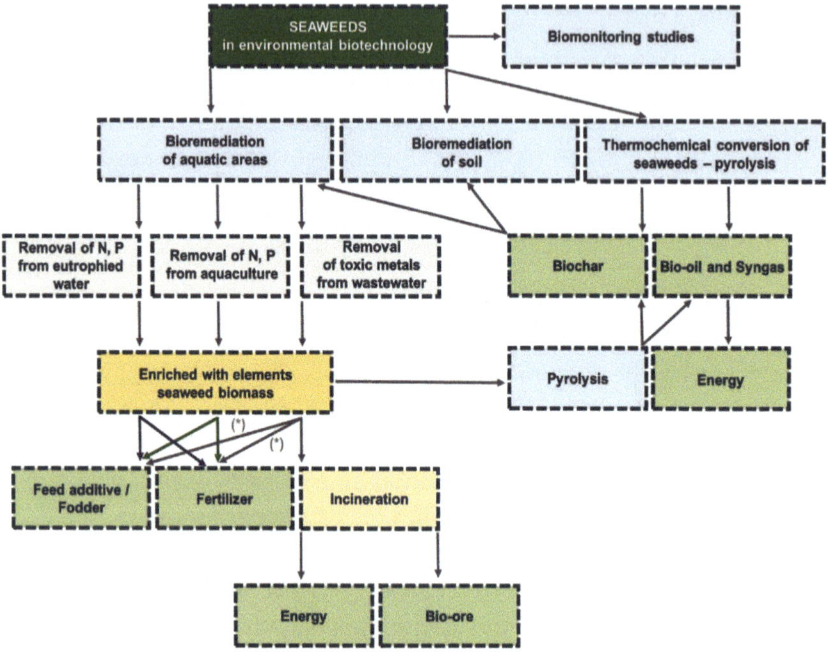

Fig. 1 The examples of seaweeds applications in the environmental biotechnology.

contaminants from wastewater than bioaccumulation. Biosorption is also recognized as one of the tools of environmental biotechnology (Bulgariu & Bulgariu, 2017; Deniz & Ersanli, 2018). This process is used when other technologies such as chemical precipitation, coagulation, membrane separation, electrochemical reduction fail — incomplete metal removal especially when the concentration of metal ions is low and ranges from 1 to 100 mg/L or are too expensive (Bilal et al., 2018; Bordoloi, Goswami, Kumar, & Kataki, 2017; Davis, Volesky, & Mucci, 2003; Gadd, 2009; Ortiz-Calderon et al., 2017; Sati et al., 2016).

Biosorbents of metal ions used in this process should be widely available, renewable, cost-effective and environment friendly (Arumugam et al., 2018; Bilal et al., 2018; Bordoloi et al., 2017; Bulgariu & Bulgariu, 2017; Gadd, 2009; Ortiz-Calderon et al., 2017). Seaweeds have very good biosorption properties, which result from the presence of macromolecules (e.g., polysaccharides, proteins) in the algal cell wall which offer different functional groups such as amino, hydroxyl, carboxyl and sulfate, capable of binding metal ions (Abd-Elhady & El-Zabalawy, 2014; Bilal et al., 2018; Davis et al., 2003; Sati et al., 2016). Macroalgae can efficiently remove different pollutants, such as dyes from paper, textile and printing industry, nitrogen, phosphorous and phenolic compounds (e.g., from aquaculture), and heavy metals from various sources (Arumugam et al., 2018; Vijayaraghavan & Ashokkumar, 2019). Ion exchange is indicated as a principal mechanism for biosorption of metal ions by dry macroalgal biomass — light metal ions from seawater such as calcium, magnesium, sodium, potassium naturally bound to the algal functional groups are exchanged with metal ions from wastewater (Bilal et al., 2018; Davis et al., 2003; Zeraatkar, Ahmadzadeh, Talebi, Moheimani, & McHenry, 2016).

Researchers still take an interest in this technique (Fig. 2): the total number of papers on biosorption by seaweed is 625 (according to the Web of Science database; July 12, 2019) within the timespan of 1900—2019. Keywords "biosorption by seaweed" was searched in the topic of papers.

Among all seaweed species, the most often examined in terms of biosorption are brown algae — 303 papers, then green algae — 131 papers and finally red algae — 61 papers (according to the Web of Science database; July 17, 2019). Search was done for keywords, for example "biosorption by green seaweed" in the topic of a paper. A similar analysis was performed for metal ions, which are the most often selected for biosorption by seaweeds. The total number of papers on biosorption of a given metal ion by seaweeds

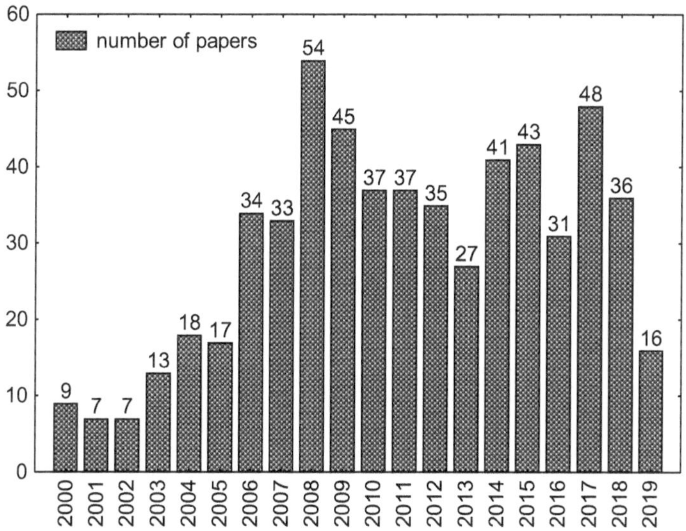

Fig. 2 Number of papers on biosorption by seaweed according to the Web of Science database (July 12, 2019).

is presented in Fig. 3. Keywords for example "biosorption of Cu by seaweed" were searched in the topic of papers (according to the Web of Science; July 12, 2019) for the timespan 1900–2019.

Effective removal of toxic metal ions from wastewater depends on many factors, such as biosorbent type (phylum) and size and the operating parameters of biosorption, such as initial solution pH, biosorbent dosage (C_S), initial metal ion concentration (C_0), temperature, contact time, single- or multi-metal system (the presence of competing ions), etc. These factors have all been described in detail (e.g., Bilal et al., 2018; Bulgariu & Bulgariu, 2017; Gadd, 2009; Sati et al., 2016; Zeraatkar et al., 2016). The examples of the removal of heavy metal ions such as Cd(II), Cu(II), Ni(II), Pb(II), Zn(II) by different seaweed species have been reported in many review papers or book chapters (Arumugam et al., 2018; Bilal et al., 2018; Bulgariu & Bulgariu, 2017; Chekroun & Baghour, 2013; Davis et al., 2003; Gadd, 2009; He & Chen, 2014; Michalak, Witek-Krowiak, Chojnacka, & Bhatnagar, 2015b; Ortiz-Calderon et al., 2017; Sati et al., 2016; Zeraatkar et al., 2016). The detailed biosorption performance of different algal-based biosorbents for the removal of heavy metals is presented by He and Chen (2014). The authors showed that brown algae are able to remove efficiently Pb(II) ions (with a biosorption capacity ranging from 166 to 284 mg/g), Cu(II) ions (from 50.8 to 105 mg/g), Cd(II) ions (from 59.6 to 131 mg/g), Zn(II) ions (from 32.7 to 59.5 mg/g), Ni(II) ions (from 5.3 to 62.8 mg/g), and Cr(III) ions (from 60.8 to 62.9 mg/g). In the case of green macroalgae, the biosorption capacity ranged from 26.9 to 302 mg/g for Pb(II) ions, from 5.1 to 114 mg/g for Cu(II) ions, from 4.5 to 88.8 mg/g

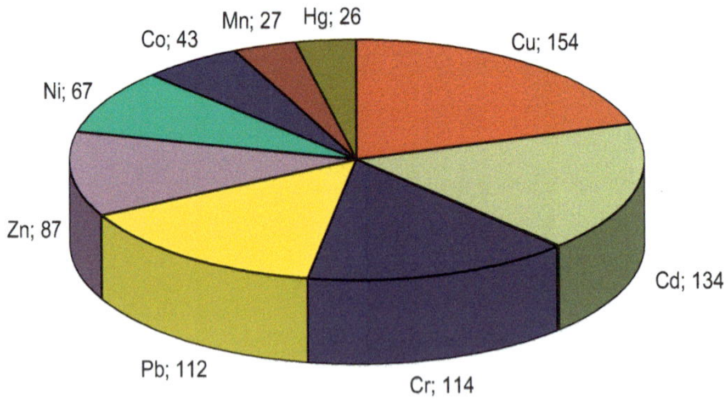

Fig. 3 Number of papers on biosorption of a given metal ion by seaweed according to the Web of Science (July 12, 2019).

for Cd(II) ions, from 2.6 to 35.3 mg/g for Zn(II) ions, from 12.9 to 66.9 mg/g for Ni(II) ions and from 36.9 to 53.0 mg/g for Cr(III) ions. For red seaweeds, these values were as follows: Pb(II) ions 29.0–203 mg/g, Cu(II) ions 21.0–40.0 mg/g, Cd(II) ions 16.9–85.4 mg/g, Zn(II) ions 21.6–45.1 mg/g, Ni(II) ions 16.4–37.0 mg/g, Cr(III) ions 28.1–105 mg/g, Co(II) ions 32.4–76.0 mg/g.

2.2 Biosorption by seaweed biochar

Waste algal biomass can itself be a biosorbent or can be transformed into another sorbent that is biochar. Biochar is known as a carbon-rich "biological charcoal" (Roberts, Cole, Paul, & de Nys, 2015a) and is produced during pyrolysis, a thermochemical conversion of the biomass (Bulgariu & Bulgariu, 2017; Michalak, 2018; Milledge, Smith, Dyer, & Harvey, 2014), which yields also bio-oil and syngas (Bird et al., 2012; Kim et al., 2016; Milledge et al., 2014) successfully used for energy production. Seaweed-derived energy is included in the category of "clean technologies" (Bulgariu & Bulgariu, 2017). Finding the application for biochar will enhance the economic feasibility of pyrolysis (Kim et al., 2016).

Seaweed biochar is used for sequestration of carbon in soil, soil amelioration, remediation of degraded soils, improvement of soil fertility (increase in the retention of nutrients and reduction of N_2O emissions from agricultural soils), as a source of bioenergy, and in the treatment of contaminated wastewater, etc. (Roberts, Paul, Dworjanyn, Bird, & de Nys, 2015b). Biochar produced from seaweeds, cultivated in unpolluted water does not contain excessive amounts of toxic elements that is why it can amend soil (Bird et al., 2012). All types of waste seaweed biomass, generated also by low temperature combustion or algae processing, for example oil or agar extraction can be used for the production of biochar rather than be disposed of in landfills or incinerated (Bulgariu & Bulgariu, 2017; Roberts et al., 2015d). Waste management after extraction of biologically active compounds from seaweeds will improve also the sustainability of macroalgae cultivation (Roberts et al., 2015d). Seaweed biochar can be suitable not only as a biosorbent of pollutants, but also in terms of resource recycling and energy efficiency (Poo et al., 2018). When waste is treated with waste, many contamination problems can find an effective solution (Roberts et al., 2015d).

The abundance of usually green seaweeds resulting from eutrophication (*Cladophora coelothrix*, *Cladophora patentiramea*, *Cladophora vagabunda*, *Chaetomorpha*, *Chaetomorpha linum*, *Cladophoropsis* sp., *Ulva flexuosa*) makes them attractive as a source of biomass for biochar production (Bird, Wurster, de Paula Silva, Bass, & de Nys, 2011; Kim et al., 2016). Very popular are also seaweeds, which are not only abundant but also widely distributed such as *Undaria pinnatifida* (Cho et al., 2013), *Saccharina japonica* (Poo et al., 2018), *Sargassum fusiforme* (Poo et al., 2018) and *Gracilaria* sp. (Johansson, Paul, de Nys, & Roberts, 2016; Roberts et al., 2015d).

Seaweeds biochar is not as popular a biosorbent as seaweeds. Most of the papers describing biosorption properties of macroalgal biochar have been published in the recent few years (Table 1). Macroalgal biochar is rich in alkali and alkaline earth metals (calcium, magnesium, sodium and potassium), which originate from seawater (nutrients for the seaweed growth). These elements have a large exchangeable cation capacity, which enables the removal of heavy metal ions (Davis et al., 2003; Kim et al., 2016). Cho et al. (2013) and Kim et al. (2016) showed that steam activation could increase the surface area and the amount of cation exchange capacity of seaweed biochar. The temperature of pyrolysis can also influence the biosorption properties of biochar. Poo et al. (2018) demonstrated that *S. japonica* biochar obtained at temperatures higher than 400°C (among tested 250, 400, 500, 600 and 700°C), had removal efficiencies for Cd(II), Cu(II) and Zn(II) ions higher than 98%. For *S. fusiforme* biochar, the highest removal efficiencies (higher than 86%) for the same three heavy metal ions were obtained for seaweeds pyrolysis at temperatures over 500 °C.

Seaweed biochar can also remove oxyanionic elements such as As, Mo and Se as shown in the work of Kidgell, de Nys, Paul, and Roberts (2014) and Johansson et al. (2016). The authors used biochar and iron-biochar produced from *Oedogonium* sp. to remediate metals and metalloids from a coal-fired power station effluent. The biosorbent was shown to be more effective when the algal biomass was treated with a ferric solution before conversion to biochar through pyrolysis (Kidgell et al., 2014; Roberts et al., 2015d; Johansson et al., 2016). Biochar can be loaded with iron also after pyrolysis (Roberts et al., 2015d). Biosorption with iron-based sorbents is recognized as the most promising method for metalloid bioremediation (Kidgell et al., 2014; Roberts et al., 2015d; Johansson et al., 2016). Algal biochar may have better sorption properties than the alga itself. For example, Thivya and Vijayaraghavan (2019) demonstrated that biochar produced from *Kappaphycus alvarezii* exhibited dye sorption capacities 3.1 times higher than those of raw biomass. Algal biochar can be applied several times if washed with deionized water and dried (Song, Guo, Li, Zhao, & Yu, 2019). In the work of Vijayaraghavan and Ashokkumar (2019), low pH

Table 1 Removal of toxic metal ions and organic contaminants by seaweed biochar.

Seaweed	Sorbate	Experimental conditions	Maximum biosorption capacity (q_{max}); mg/g	Reference
Inorganic – metal ions				
Ulva compressa (G); coast of Korea	Cu(II)	C_0 5–50 mg/L, C_S 0.1 g/L, 20 °C, 48 h, pH 5.5	[a]33 mg/g (biochar); 137 mg/g – biochar activated with water vapor stream, 20 mg/g – activated with KOH	Kim et al. (2016)
Undaria pinnatifida (B); southern coastal areas in Korea	Cu(II)	C_0 5–50 mg/L, C_S 0.1 g/L, 20 °C, 48 h, pH 5.0	[a]125.9 mg/g – biochar activated with water vapor stream	Cho et al. (2013)
Saccharina japonica and *Sargassum fusiforme* (B); Busan, Korea	Cu(II), Cd(II), Zn(II)	C_0 1500 mg/L, C_S 17 g/L, 20 °C, 24 h	Cu(II): *S. japonica* 98.6 mg/g (pyrolysis temp. 500 °C), *S. fusiforme* 94.1 mg/g (600 °C); Cd(II): both 700 °C – *S. japonica* 60.7 mg/g and *S. fusiforme* 37.2 mg/g; Zn(II): both 700 °C – *S. japonica* 84.3 mg/g and *S. fusiforme* 43.0 mg/g	Poo et al. (2018)
Turbinaria turbinata (B); Caribbean area, Guadeloupe	Cr(VI)	C_0 0–10 mg/L, pH 5.0, 25 °C	[a]12.6 mg/g – biochar activated with stream and 11.0 mg/g – activated with H_3PO_4	Yacou, Altenor, Carene, & Gaspard (2018)
Gracilaria sp. (R); (waste after agar extraction); Bogatama, Indonesia	Se, Mo, As	C_0 0.1 and 200 mg/L, C_S 10 g/L, 15 °C, 1 h	Mo 78.5 mg/g, As 62.5 mg/g and Se 14.9 mg/g	Johansson et al. (2016)

(*Continued*)

Table 1 Removal of toxic metal ions and organic contaminants by seaweed biochar.—cont'd

Seaweed	Sorbate	Experimental conditions	Maximum biosorption capacity (q_{max}); mg/g	Reference
Gracilaria sp. (R); (waste after agar extraction); Indonesia	Se(IV), Se(VI)	C_0 0.5 mg/L, C_S 10 g/L, pH 2.5, 4 and 8, 4 h, 20 °C	no effect of pH on biosorption: pH 2.5 −2.60 mg/g; pH 4.0 −2.72 mg/g; pH 8.0 −2.66 mg/g	Roberts et al. (2015d)
Hizikia fusiformis (B); Jeju Island, Korea	Ni(II), Zn(II), Cu(II), Pb(II), Cd(II)	C_0 5−100 mg/L, C_S 1 g/L, pH 4, 25 ± 0.5 °C, 6 h	[a] in a mixed solute system: Cu(II) 2.24 mg/g, Pb(II) 2.89 mg/g, Ni(II) 12.1 mg/g, Cd(II) 22.0 mg/g, Zn(II) 22.2 mg/g	Shin (2017)
Organic				
Kappaphycus alvarezii (R)	Reactive blue 4 (RB4), Reactive orange 16 (RO16)	C_0 50−500 mg/L, C_S 4 g/L	RB4 − 0.324 mmol/g, RO16−0.140 mmol/g	Thivya & Vijayaraghavan (2019)
Sargassum sp. (B); coast of Singapore	tetracycline; cefradine	C_0 20 mg/L, C_S 0.4 g/L, 24 h	[a] tetracycline 128 mg/g; cefradine 61.7 mg/g	Song et al. (2019)
Turbinaria conoides (B); Mandapam region, a coastal area of Tamilnadu (India)	Remazol brilliant blue R (Reactive blue 19)	C_0 50−500 mg/L, C_S 4 g/L, pH 2, 32 °C, 8 h	[a] 92.5 mg/g	Vijayaraghavan & Ashokkumar (2019)

[a]— determined from Langmuir equation.

conditions were favorable for the biosorption of dye – Remazol brilliant blue R (Reactive blue 19) by biochar produced from brown seaweed *Turbinaria conoides*. Therefore, the elution of dye from loaded biochar was carried out by means of alkaline elutants (0.1 M Na_2CO_3 and 0.1 M NaOH). Table 1 presents examples of the removal of inorganic (toxic metal ions) and organic compounds (dye, antibiotic) by seaweed biochar. The results point to the conclusion that seaweed biochar is a prospective and effective adsorbent for wastewater treatment systems. Biochar produced from waste algal biomass is believed to be able to replace expensive commercial activated carbon (Kim et al., 2016).

3. Bioremediation of nutrients by seaweeds

3.1 Eutrophicated water

Marine eutrophication occurs in many places all over the world. The cultivation of seaweeds in coastal waters can be useful to alleviate eutrophication and to control harmful algae blooms that result in animal death and economic loss (Wu et al., 2015a; Xiao et al., 2017; Xu, Fang, & Wei, 2008; Yang et al., 2015; Zheng, Jin, Zhang, Wang, & Wu, 2019). The efficient cultivation of algae requires suitable environmental factors, such as light, appropriate temperature, content of nutrients, salinity, cultivation depth, water movement and the presence of epiphytes and herbivorous fish (Yang et al., 2015). During bioremediation of eutrophicated water, the nitrogen content in seaweed tissues increases with the increase in the concentrations of dissolved nitrogen in seawater (Chung, Kang, Yarish, Kraemer, & Lee, 2002). The production of seaweeds with a high tissue protein and carbohydrate content has the potential for biorefinery (Ashkenazi, Israel, & Abelson, 2019; Ross et al., 2018). The macroalgal biomass can be also sustainably returned to agricultural land or be used for the manufacture of valuable products, thus meeting both economic and ecological objectives. In this context, seaweeds aquaculture can be considered as "vegetables cultivation in the sea" with high nutritional value. When compared with terrestrial vegetable cultivation, farming of macroalgae in eutrophicated water saves the fertilizer, pesticides, and consequently farmland (Zheng et al., 2019). Other environmental benefits offered by seaweed aquaculture include carbon sequestration, CO_2 absorption, release of oxygen into coastal waters and iodine production (Yang et al., 2015; Zheng et al., 2019). The removal of nutrients from eutrophicated water through seaweeds cultivation is the most popular in China where approximately 9500 tons of phosphorus and 75,000 tons of nitrogen is removed annually (Xiao et al., 2017). China's large-scale seaweed aquaculture accounts for over ⅔ of the global production (Xiao et al., 2017; Zheng et al., 2019).

Green seaweeds *Ulva* sp. (Amosu, Robertson-Andersson, Kean, Maneveldt, & Cyster, 2016; Aníbal et al., 2014; Chung et al., 2002; Zheng et al., 2019), *Ulva clathrata* (Aníbal et al., 2014), *C. coelothrix* and *Cladophora parriaudii* (Ross et al., 2018); red seaweeds *Porphyra* sp. (Carmona, Kraemer, & Yarish, 2006; Chung et al., 2002; Wu et al., 2015a; Zheng et al., 2019) and *Gracilaria* sp. (Marinho-Soriano et al., 2011; Xu et al., 2008; Yang et al., 2015; Zheng et al., 2019; Zhou et al., 2006) are some of the most popular algal species used in the bioremediation of eutrophicated water. Green seaweeds, which abundantly occur in the marine environment and are characterized by a rapid growth rate and a broad spectrum of environmental tolerance, are the best candidates to sequester carbon, nitrogen and phosphorus (Aníbal et al., 2014; Bird et al., 2011; Lawton, Mata, de Nys, & Paul, 2013; Ross et al., 2018). *Ulva* and *Cladophora* are also responsible for seaweed blooms, called green tides around the world (Abd-Elhady, 2015; Abd-Elhady & El-Zabalawy, 2014; Cheney et al., 2014; Farias et al., 2017; Ross et al., 2018). Red algae can better be used in aquaculture than brown algae because they usually have higher average nitrogen content, mainly due to their photosynthetic pigments (N-containing phycobiliproteins) and protein levels (Carmona et al., 2006; Chung et al., 2002). Some red seaweed species (e.g., *Gracilaria* sp.) offer a high yield and additionally produce valuable compounds, for example agar-agar, a type of food for humans and fodder for animals (Zhou et al., 2006). Yang et al. (2015) indicated that *Gracilaria* in Chinese coastal waters remediates contaminants from aquaculture, sequesters carbon dioxide, mitigates eutrophication, and controls harmful algal blooms, thus maintaining healthy mariculture systems.

Wu et al. (2015a) showed that during the large-scale cultivation of *Porphyra yezoensis* in open sea in China (South Yellow Sea), mean nutrient concentrations in the cultivation area were significantly lower than in the non-cultivation one. Reduction efficiency of the nutrients in seawater was NH_4—N 44%, NO_3—N 49%, and PO_4—P 45%. Huo et al. (2011) showed that the nutrient removal by red seaweed *Gracilariopsis longissimi* in Hangzhou Bay, Shanghai (East China Sea) was slightly higher (54% 75% 49%, respectively) than for *P. yezoensis* examined in the work of Wu et al. (2015a). Marinho-Soriano et al. (2011) indicated that in order to remove nutrients efficiently, macroalgae cultivation should be combined with animal farming. The authors performed experiments consisting of three treatments: (1) macroalgae (*Gracilaria caudata*), (2) macroalgae and microcrustacean (*Artemia franciscana*) and (3) *Artemia* alone, with the best results being obtained for the second option (removal was NO_2 100%; NO_3 72.4%; NH_4 29.8% and dissolved inorganic nitrogen 44.5%). Seaweed cultivation acting as a nutrient scrubber system can be one of the best reasons for bioremediation of coastal eutrophicated waters (Chung et al., 2002; Ross et al., 2018).

3.2 Aquaculture

Aquaculture, which covers the farming of both animals (including crustaceans, fish and molluscs) and plants (including seaweeds and freshwater macrophytes) can be practised in inland (freshwater) and coastal (brackish water, seawater) areas (Edwards & Demaine, 1997; Neori et al., 2004). Intensive aquaculture production requires a high input of water, fertilizers, feeds and other chemicals, which produces considerable amounts of wastes with all accompanying environmental problems (Chung et al., 2002; Neori et al., 2004; Zhou et al., 2006). Fish excretion and feces production induce eutrophication, harmful algal blooms, loss of biodiversity, anoxia in bottom waters and coastal-water pollution (Silva, Yáñez, Martín-Díaz, & DelValls, 2012; Yang et al., 2015; Zhou et al., 2006).

Integrated Multitrophic Aquaculture (IMTA) is a system in which effluents from aquaculture production are used as a nutrient for further production of animals and seaweed biomass. Here seaweeds can be used for wastewater bioremediation (as biological filters to remove inorganic nutrients) or/and as feed additives for aquaculture (fish, shrimp, molluscs etc.) (Arumugam et al., 2018; Ashkenazi et al., 2019; Bird et al., 2011; Elizondo-González, Quiroz-Guzmán, Escobedo-Fregoso, Magallón-Servín, & Peña-Rodríguez, 2018; Grote, 2016; Hou et al., 2012; Lamprianidou, Telfer, & Ross, 2015; Largo, Diola, & Marababol, 2016; Lavania-Baloo, Azman, Said, Ahmad, & Mohamad, 2014; Marinho-Soriano et al., 2011; Neori et al., 2004; Silva et al., 2012; Wei et al., 2017; Wu, Huo, Hu, Wei, & He, 2015b; Yang et al., 2015; Zhang, Hansen, Fang, Wang, & Jiang, 2009; Zhou et al., 2006). Oxygen produced by seaweed through photosynthesis can improve the respiration of cultivated animals (Neori et al., 2004; Yang et al., 2015) and provide them with a positive environment (Wei et al., 2017). Some seaweeds, for example *Gracilaria* sp., are able to inhibit the growth of phytoplankton including species such as *Prorocentrum donghaiense*, *Scrippsiella trochoidea*, *Alexandrium tamarense*, *Dunaliella salina*, *Skeletonema costatum*, *Heterosigma akashiwo* that cause harmful microalgal blooms (Yang et al., 2015).

Decomposing animal waste produced by farms releases nutrients in abundance, which pollute the cultivation, and the surrounding waters (Yang et al., 2015). Total ammonia nitrogen (unionized ammonia; NH_3) and ionized ammonia (NH_4^+) are the main excretory products of fish nitrogen metabolism. Concentrations higher than 1.5 mg NH_3—N/L can be toxic to most fish (Dosdat, Servais, Métailler, Huelvan, & Desbruyères, 1996) and shrimp species (Lavania-Baloo et al., 2014). Special attention should also be paid to the selection of appropriate macroalgal candidate for the IMTA system. Suitable seaweeds should be characterized by a high growth rate, productivity, nutrient uptake and insensitivity to high concentrations of ammonium (NH_4^+) (Grote, 2016; Elizondo-González et al., 2018). Among all seaweed species, green alga *Ulva* spp. and red algae *Gracilaria* spp. and *Porphyra* spp. are well-established aquaculture species. Their nutrient uptake ability and growth rates are high when compared to other seaweed species (Grote, 2016; Ashkenazi et al., 2019). Red alga *Palmaria palmata* is described as NH_4^+-sensitive (Grote, 2016). Elizondo-González et al. (2018) showed that green macroalga *Ulva lactuca* removed nitrogen (80%) and phosphorus (64%) from shrimp's wastewater coming from the integrated recirculation system and improved their growth performance and the content of carotenoids. The same conclusion was reached by Lavania-Baloo et al. (2014): in the IMTA system, macroalgae *Gracilaria edulis* and *U. lactuca* removed significant ammonium amounts from shrimp wastewater (70% and 45%, respectively) and improved water quality, which is essential for shrimp growth and survival. Table 2 presents the examples of integrated multitrophic aquaculture in different areas of the world. China, Japan and South Korea dominate. Farms of fish, shellfish and seaweed are placed next to each other in bays and lagoons (Neori et al., 2004).

The IMTA system is more and more popular as evidenced by the number of published scientific papers concerning this topic. According to the Web of Science database (July 19, 2019), the number of papers concerning the bioremediation of aquaculture with seaweed was 70. The rationale, evolution and state of the art of the integrated aquaculture was described in detail by Neori et al. (2004). The IMTA system is the example of sustainable development of aquaculture (Chung et al., 2002; Fleurence et al., 2012; Marinho-Soriano et al., 2011). Seaweed cultivation combined with fed species in the integrated multi-trophic aquaculture is an important tool in environmental management. It improves ecological efficiency, sustainability and

Table 2 The examples of the integrated multitrophic aquaculture.

Seaweed	Aquaculture	Region	Reference
Ulva lactuca (G) *Gracilaria edulis* (R)	tiger shrimp	outdoor experimental setup, Malaysia	Lavania-Baloo et al. (2014)
Ulva lactuca (G)	shrimp	La Paz bay in Baja California Sur, Mexico	Elizondo-González et al. (2018)
Ulva rigida (G) *Gracilaria conferta* (R) *Hypnea musciformis* (R)	gilthead sea bream (*Sparus aurata*)	Haifa, Israel	Ashkenazi et al. (2019)
Gracilaria lemaneiformis (R)	fish (*Sebastodes fuscescens*)	Jiaozhou Bay, north China	Zhou et al. (2006)
Gracilaria lichenoides (R)	shrimp (*Litopenaeus vannamei*), fish (*Epinephelus awoara*)	Xiangshan Bay, Ningbo, China	Xu et al. (2008)
Gracilaria caudata (R)	microcrustacean (*Artemia franciscana*)	Brazil	Marinho-Soriano et al. (2011)
Gracilaria verrucosa (R)	fish – large yellow croaker (*Pseudosciaena crocea*)	Xiangshan harbor, China	Hou et al., 2012
Gracilaria heteroclada (R) *Eucheuma denticulatum* (R)	donkey's ear abalone (*Haliotis asinina*)	open sea Cebu, Philippines	Largo et al. (2016)
Gracilaria lemaneiformis (R)	fish (*Pseudosciaena crocea*)	Yantian Bay, China	Wei et al. (2017)
Gracilaria lemaneiformis (R) *Laminaria japonica* (B)	fish (*Pseudosciaena crocea*), Japanese oyster (*Crassostrea gigas*), sea cucumber (*Apostichopus japonicas*)	Sansha Bay, an enclosed bay in China	Wu et al. (2015b)
Laminaria japonica (B)	shellfish (*Crassostrea gigas, Chlamys farreri*)	Sungo Bay, China	Zhang et al. (2009)

safeguards cost-efficient production (Carmona et al., 2006; Grote, 2016; Wei et al., 2017; Xu et al., 2008). Marine aquaculture is a key future direction for the production of food (Ashkenazi et al., 2019).

4. Utilization of seaweeds from bioremediation

Seaweeds from the remediation of wastewater from industry, aquaculture, agriculture, and eutrophicated water could provide a significant source of biomass for the production of valuable products (Bird et al., 2011, 2012; Roberts, Paul, Bird, & de Nys, 2015c). It is proposed that algae grown in wastewater can be used for the production of biofuels, biosorbents (e.g., biochar), and soil amendments etc., sometimes as feed additives for animals. Combining algal biomass production with wastewater treatment can reduce the cost of algae cultivation through minimization of freshwater consumption. This approach provides also a pathway to combatting industrial pollution and eutrophication (Lyon, Ahmadzadeh, & Murry, 2015). For example, Roberts et al. (2015c) cultivated macroalgae in contaminated ash water from a coal-fired power station. During macroalgal growth, the intracellular metal ions sequestration reduced their concentration in treated water. The authors proposed that enriched biomass be used for the production of biochar via slow pyrolysis and energy. Algal biochar has a low content of leachable metals and therefore can be used as an ameliorant for low-fertility soils. The bioremediation technology of wastewater combined with seaweeds cultivation (water recycling and biomass production) can be recommended for a water-limited industry (e.g., power stations). Deniz and Ersanli (2018) used the mixture of four species of marine seaweeds — green *Chaetomorpha* sp. and *Ulva* sp., red *Vertebrata* sp. and brown *Cystoseira* sp. to remove Cu(II) ions from aqueous medium. The maximum biosorption capacity was 180 mg of Cu(II) ions per gram of the biosorbent (for the optimal experimental conditions: pH 6, C_S: 0.1 g/L, C_0: 10–30 mg/L, 2 h). The authors proposed to integrate this bioremediation system with biofuel production as sustainable waste management and renewable energy production. The loaded algal biomass can be incinerated to produce "bio-ore" and energy (Brooks, Chambers, Nicks, & Robinson, 1998; Roberts et al., 2015c). The fully "metal-loaded" biosorbent may concentrate heavy metals a thousand fold from their concentration in aqueous solutions (Davis et al., 2003).

Special care should be taken when searching for new management methods of the algal biomass after bioremediation. Besides the removal of biogens such as nitrogen and phosphorus, seaweeds can also bioaccumulate other elements that are present in water, for example toxic metal ions such as copper, cadmium, lead, zinc, etc. Therefore, before utilization, seaweed biomass should be examined for multielemental composition. Regular monitoring of minerals in seaweeds would prevent toxic and other undesirable situations (Makkar et al., 2016). The toxic metals content can disqualify algae from various applications, especially as components of human and animal food. For example, Amosu et al. (2016) showed that green *Ulva armoricana* could serve as a biological filter for dissolved nutrient from aquaculture effluents. Algal biomass had a relatively good quality with limited toxic metals content (Cu, Pb, Zn), but due to the presence of small amounts of cadmium, it cannot be used as a feed additive for livestock or the food industry. The authors recommend its application as a plant biostimulant.

Many of toxic elements that are removed by seaweeds from wastewater are typical heavy metals such as As, Cd, Hg, Pb, but some of them are microelements, e.g., Zn, Cu, Mn, Co that are necessary in small quantities for the proper growth and development of plants or animals (Bilal et al., 2018; Ozyigit, Uyanik, Sahin, Yalcin, & Demir, 2017). Therefore, biomass enriched with elements via biosorption/bioaccumulation can be utilized for the production of components of fertilizers or feed additives with microelements (Bădescu, Bulgariu, & Bulgariu, 2017; Michalak et al., 2015b). Bădescu et al. (2017) used marine algal biomass (*Ulva* sp.) as biosorbent of Zn(II) ions. The enriched biomass was then applied as a fertilizer for the improvement of soil quality. In the biosorption experiment, *Ulva* sp. bound 29.6 mg of Zn(II) ions per gram of the biomass (for pH 5.5, C_0: 10–220 mg/L, C_S: 8 g/L and the contact time 2 h). After 8 weeks from the use of Zn-loaded *Ulva*, the content of zinc in soil increased over four times. Tuhy, Samoraj, Michalak, and Chojnacka (2014) applied a similar approach. Baltic seaweeds and post-extraction residues obtained from the supercritical fluid extraction of seaweeds were used as biosorbents of Zn(II) ions in order to produce micronutrient fertilizer. Germination tests on garden cress (*Lepidium sativum*) showed that the application of enriched biomass caused the biofortification of plants in zinc. The weight of plants fertilized with seaweeds loaded with Zn(II) ions was higher than in the control group made up of conventional fertilizers such as inorganic salt and chelate.

Rarely are enriched algae used as feed additives for livestock although seaweeds have a long history of use in livestock feeding (ruminant, pig, poultry e.g., broilers, laying hens, horse, rabbit diet). Seaweeds contribute to the mineral, protein and energy requirements of livestock and due to the content of the prebiotic compound, they enhance livestock production and health (Makkar et al., 2016). Michalak et al. (2011 and 2015a) used the mixture of green marine macroalgae *Ulva prolifera* and *Cladophora* sp., enriched with microelement ions Cu(II), Zn(II), Co(II), Mn(II), Cr(III) via biosorption as feed additives for laying hens and pigs. This biomass replaced inorganic salts traditionally used in animal feeding. It was shown that the bioavailability of microelements from algal feed additives was higher than in the case of inorganic form. Supplementation of enriched macroalgae resulted in higher microelement transfer to eggs of laying hens and meat of pigs. In the case of hens, seaweeds increased egg weight, eggshell thickness and body weight of hens and enhanced yolk color. These results show that seaweeds can be used both in aquaculture and in livestock feeding.

5. Bioremediation of contaminated soil

Soil bioremediation is a process in which living organisms have the ability to accumulate or degrade chemical contaminants into less toxic forms (Juwarkar, Singh, & Mudhoo, 2010). Seaweed ability to bioremediate contaminated soil can be limited by the pollutants content in the soil and the ability of algae to survive under these conditions. Green seaweeds (e.g., *Ulva* sp.) have been used for soil bioremediation mainly due to their wide distribution, fast growth, relatively simple structure and high tolerance to metal pollution (Abd-Elhady, 2015; Abd-Elhady & El-Zabalawy, 2014; Cheney et al., 2014; Farias et al., 2017). Farias et al. (2017) used macroalga *Ulva australis* for the removal of Zn(II) ions from the contaminated soil in the *in situ* assessment. *U. australis* thalli were transplanted to sites with different levels of metal pollution. It was shown that after 45 days, the content of zinc in the biomass was 700 mg/kg. During the first 12 days, zinc accumulation did not affect *U. australis* physiology (photosynthesis performance, photosynthetic pigment content, growth rate). After 45 days, Zn accumulation increased with the accompanying deterioration of the thalli, which led to the conclusion that a time period of 20 days was set for bioremediation. Abd-Elhady and El-Zabalawy (2014) and Abd-Elhady (2015) used dried seaweeds (*Ulva sp.* and *Gelidium* sp.) mixed with soil not only for the remediation of soil contaminated with Cu, Mn, Pb and Zn, but also to assess their effect on the growth (roots and shoots) and the content of the examined elements in the red radish (*Raphanus sativus* L.) and lettuce (*Lactuca sativa*). Pot experiments showed that the addition of seaweeds to soil improved its total porosity, the content of available water and organic matter and increased the content of the examined elements in red radish and lettuce biomass. In the work of Azmat and Askari (2015), dried green seaweed (*Codium iyengrii*) was added to soil to alleviate the lethal effect of Hg on the growth of fenugreek (*Trigonella foenumgraecum*). It was shown that seaweeds successfully controlled the mobility of mercury (binding of Hg by seaweeds) only at lower applied dose of metal (5, 10 and 15 mg/kg). At higher doses (20 and 25 mg/kg) seaweeds were effective to a much lesser extent: plants could not cope with the toxic effects of Hg.

Kantachote et al. (2004) used seaweeds to remove DDT from the environment, which is especially dangerous due to its low solubility, toxicity, bioaccumulation and biomagnification in the higher trophic levels. In the experiments, performed in a pilot-scale research bioremediation facility, it was shown that dried and ground seaweeds of green (*Ulva* sp.) and red algae (*Gelidium* sp.) enhanced the biodegradation of DDT in waterlogged soils. The highest percentages of DDT biodegradation (80%) after 6 weeks was found in the soil with the addition of 0.5% of seaweeds (among tested doses 1, 3, 5, 13 (*w/w*)). Anaerobic biodegradation of DDT occurred, probably due to an increased bacterial population after addition of seaweeds. Cheney et al. (2014) examined the uptake of PCBs contained in marine sediments by dead and live *Ulva rigida*. Macroalga was shown to be capable of removing PCBs in sediments at a rapid rate and under laboratory conditions the total PCBs uptake was significantly greater in live than dead biomass. Cheney et al. (2014) emphasized that the remediation of estuarine and coastal river sediments is much more difficult than that of soil due to the overlying water column and the possibility of contaminants resuspension during clean-up (Gomes, Dias-Ferreira, & Ribeiro, 2013).

Also seaweed biochar can be used for the bioremediation of anthropogenically impacted natural environments (Bird et al., 2011). In many countries, the disturbed land after mining has to be rehabilitated with native vegetation. This process can be difficult and slow due to absence or low content of nutrients and organic matter. Therefore it is proposed to apply algal biochar to enhance the rehabilitation of the stockpiled soil (Roberts et al., 2015a). Roberts et al. (2015a) used macroalgae (*Oedogonium*) cultivated additionally in wastewater at a coal-fired power station as a feedstock for the production of biochar, which reduced the germination time of the grass seeds and enhanced the growth and production of a native plant i.e. the Kangaroo grass (*Themeda australis*). Although laboratory experiments provide useful information concerning the soil bioremediation with seaweeds, there is still a need to understand how macroalgae can perform this process under real field conditions.

6. Seaweeds in biomonitoring studies

As shown in previous sections, seaweeds have the ability to accumulate metal ions from the surrounding water (Ali, Idris, Ebrahim, & Eltayeb, 2017; Burger et al., 2007; García-Seoane, Fernández, Villares, & Aboal, 2018; Henriques et al., 2017; Singh & Prasad, 2015; van Ginneken & de Vries, 2018), which results from their excellent survival strategies that enable them to withstand the environmental stresses (Chan et al., 2006; Chaudhuri, Mitra, Havrilla, Waguespack, & Schwarz, 2007). These biological systems are able to detect any toxic, harmful or deleterious change in the environment and adapt to different environmental conditions (Ali et al., 2017). Additionally, seaweeds are sessile, constitute representatives of the study area and are abundant along most of the world's coasts (Chaudhuri et al., 2007). Therefore, algae are crucial in the biomonitoring of environmental pollution (biological monitoring) and have been used in biomonitoring studies since the 1950s (García-Seoane et al., 2018). Seaweeds constitute also the first level in the trophic chain: they are primary producers that are eaten by organisms on the higher trophic level, including people and livestock (Burger et al., 2007).

The analysis of aquatic components such as water or sediment by seaweeds can provide an overview of the contamination and quality of the environment (Chaudhuri et al., 2007; Ozyigit et al., 2017). García-Seoane et al. (2018) lists seaweed species that are most often used in biomonitoring studies. Some of the more important green seaweeds are *Ulva* sp., *Cladophora* spp., *Codium* spp., *Caulerpa* spp., *Chaetomorpha* spp., some of the most important brown seaweeds are *Fucus* spp., *Sargassum* spp., *Padina* spp., *Cystoseira* spp., *Ecklonia* spp., *Ascophyllum* spp., and some of the most important red seaweeds are *Gracilaria* spp., *Laurencia* spp., *Hypnea* spp., *Corallina* spp., *Porphyra* spp., *Gracilaria* spp., *Ceramium* spp. and *Phyllophora* spp. Green macroalgae e.g., *U. lactuca* are known to be good biomonitors of toxic metal ions, such as Cd, Hg, Pb (for example, Henriques et al., 2017; Ozyigit et al., 2017). Usually there is a close relation between the concentration of toxic metals in the surrounding water and in the living seaweed-biomass (van Ginneken & de Vries, 2018). García-Seoane et al. (2018) found that most biomonitoring studies (97%) involved inorganic contaminants and very few organic contaminants (3%). Biomonitoring studies using macroalgae are conducted with the aim of detecting the presence of elements, mainly Cu, Zn, Cd, Pb, Fe, Mn, Ni, Cr, Co, N, As and Hg in the environment. In order to assure the sustainability and health of the aquatic environment, the concentrations of heavy metals should be regularly monitored (Chaudhuri et al., 2007; Ozyigit et al., 2017). Seaweeds will reflect any change in the marine environment (Chaudhuri et al., 2007). To perform the biomonitoring study correctly, the following factors should be considered: sampling site, number of sampling sites; temporal representativeness, timing of surveys and the selection of the seaweed species used as biomonitors (García-Seoane et al., 2018). García-Seoane et al. (2018) proposed the methodology for the biomonitoring with marine macroalgae. Biomonitoring based on seaweeds can be a promising and a reliable tool of quantifying the negative effect of the environmental contaminant (Ezeonu, Tagbo, Anike, Oje, & Onwurah, 2012).

7. Conclusions

Seaweeds have the capacity to remove heavy metal ions and nutrients from water and therefore can potentially be used in the bioremediation and biomonitoring of such pollutants. Marine macroalgae are an undiscovered or underestimated raw material that can clean industrial effluents, eutrophicated water, effluents from aquaculture including marine integrated multitrophic aquaculture and polluted soils. Seaweeds can serve in environmental biotechnology as a dead (biosorption) or live (bioaccumulation) biomass. Waste streams especially from agriculture and aquaculture can constitute a nutrient source for macroalgae growth. This solution provides reuse options: algal biomass can be utilized for the production of fertilizers, feed additives, biofuels or additional value adding products. Waste can be treated with waste, which can solve many environmental problems. Wastewater treatment with seaweeds is an additional argument for the application of the marine resources in the generation of exploitable biomass.

Acknowledgments

This work was supported by the project entitled: "The effect of bioactive algae enriched by biosorption in the certain minerals such as Cr(III), Mg(II) and Mn(II) on the status of glucose in the course of metabolic syndrome horses. Evaluation *in vitro* and *in vivo*"; No 2015/18/E/NZ9/00607; the National Science Centre in Poland.

References

Abd-Elhady, E. S. E. (2015). Evaluation of algae dry biomass as a biochemical soil remediation for polluted soil. *International Journal of Environment, 4*(4), 309–314.

Abd-Elhady, E. S. E., & El-Zabalawy, K. M. (2014). Remediation of a soil contaminated with heavy metals using some seaweeds. *Journal of Soil Sciences and Agricultural Engineering, 5*(12), 1623–1633.

Aksu, Z. (1998). Biosorption of heavy metals by microalgae in batch and continous systems. In Y. S. Wong, & N. F. Y. Tam (Eds.), *Algae for wastewater treatment* (pp. 37–53). Berlin: Springer.

Ali, A. Y. A., Idris, A. M., Ebrahim, A. M., & Eltayeb, M. A. H. (2017). Brown algae (*Phaeophyta*) for monitoring heavy metals at the Sudanese Red Sea coast. *Applied Water Science, 7*, 3817–3824.

Amosu, A. O., Robertson-Andersson, D. V., Kean, E., Maneveldt, G. W., & Cyster, L. (2016). Biofiltering and uptake of dissolved nutrients by *Ulva armoricana* (*Chlorophyta*) in a land-based aquaculture system. *International Journal of Agriculture and Biology, 18*, 298–304.

Aníbal, J., Madeira, H. T., Carvalho, L. F., Esteves, E., Veiga-Pires, C., & Rocha, C. (2014). Macroalgae mitigation potential for fish aquaculture effluents: An approach coupling nitrogen uptake and metabolic pathways using *Ulva rigida* and *Ulva clathrata*. *Environmental Science and Pollution Research, 21*, 13324–13334.

Arumugam, N., Chelliapan, S., Kamyab, H., Thirugnana, S., Othman, N., & Nasri, N. S. (2018). Treatment of wastewater using seaweed: A review. *International Journal of Environmental Research and Public Health, 15*, 2851. https://doi.org/10.3390/ijerph15122851.

Ashkenazi, D. Y., Israel, A., & Abelson, A. (2019). A novel two-stage seaweed integrated multi-trophic aquaculture. *Reviews in Aquaculture, 11*, 246–262.

Azmat, R., & Askari, S. (2015). Improvement in the bioenergetics system of plants under Hg stress environment via seaweeds. *Pakistan Journal of Botany, 47*(3), 851–858.

Baloo, L.-, Azman, S., Said, M. I. M., Ahmad, F., & Mohamad, M. (2014). Biofiltration potential of macroalgae for ammonium removal in outdoor tank shrimp wastewater recirculation system. *Biomass and Bioenergy, 66*, 103–109.

Bilal, M., Rasheed, T., Sosa-Hernández, J. E., Raza, A., Nabeel, F., & Iqbal, H. M. N. (2018). Biosorption: An interplay between marine algae and potentially toxic elements—a review. *Marine Drugs, 16*, 65. https://doi.org/10.3390/md16020065.

Bird, M. I., Wurster, C. M., de Paula Silva, P. H., Bass, A. M., & de Nys, R. (2011). Algal biochar – production and properties. *Bioresource Technology, 102*, 1886–1891.

Bird, M. I., Wurster, C. M., Silva, P. H. D., Paul, N. A., & de Nys, R. (2012). Algal biochar: Effects and applications. *Global Change Biology Bioenergy, 4*, 61–69.

Bordoloi, N., Goswami, R., Kumar, M., & Kataki, R. (2017). Biosorption of Co(II) from aqueous solution using algal biochar: Kinetics and isotherm studies. *Bioresource Technology, 244*, 1465–1469.

Brooks, R. R., Chambers, M. F., Nicks, L. J., & Robinson, B. H. (1998). Phytomining. *Trends in Plant Science, 3*(9), 359–362.

Bulgariu, L., & Bulgariu, D. (2017). Sustainable utilization of marine algae biomass for environmental bioremediation. In B. N. Tripathi, & D. Kumar (Eds.), *Prospects and challenges in algal biotechnology* (pp. 179–217). Springer Nature Singapore Pte Ltd.

Burger, J., Gochfeld, M., Jeitner, C., Gray, M., Shukla, T., Shukla, S., et al. (2007). Kelp as a bioindicator: Does it matter which part of 5 m long plant is used for metal analysis? *Environmental Monitoring and Assessment, 128*, 311–321.

Bădescu, I. S., Bulgariu, D., & Bulgariu, L. (2017). Alternative utilization of algal biomass (*Ulva* sp.) loaded with Zn(II) ions for improving of soil quality. *Journal of Applied Phycology, 29*, 1069–1079.

Carmona, R., Kraemer, G. P., & Yarish, C. (2006). Exploring Northeast American and Asian species of *Porphyra* for use in an integrated finfish—algal aquaculture system. *Aquaculture, 252*, 54—65.
Chan, C. X., Ho, C. L., & Phang, S. M. (2006). Trends in seaweed research. *Trends in Plant Science, 11*(4), 165—166.
Chaudhuri, A., Mitra, M., Havrilla, C., Waguespack, Y., & Schwarz, J. (2007). Heavy metal biomonitoring by seaweeds on the Delmarva Peninsula, east coast of the USA. *Botanica Marina, 50*, 151—158.
Chekroun, K. B., & Baghour, M. (2013). The role of algae in phytoremediation of heavy metals: A review. *Journal of Materials and Environmental Science, 4*(6), 873—880.
Cheney, D., Rajic, L., Sly, E., Meric, D., & Sheahan, T. (2014). Uptake of PCBs contained in marine sediments by the green macroalga *Ulva rigida*. *Marine Pollution Bulletin, 88*, 207—214.
Cho, H. J., Baek, K., Jeon, J. K., Park, S. H., Suh, D. J., & Park, Y. K. (2013). Removal characteristics of copper by marine macroalgae-derived chars. *Chemical Engineering Journal, 217*, 205—211.
Chopin, T. (2013). Aquaculture, integrated multi-trophic (IMTA). In P. Christou, R. Savin, B. A. Costa-Pierce, I. Misztal, & C. B. A. Whitelaw (Eds.), *Sustainable food production*. New York: Springer.
Chung, I. K., Kang, Y. H., Yarish, C., Kraemer, G. P., & Lee, J. A. (2002). Application of seaweed cultivation to the bioremediation of nutrient-rich effluent. *Algae, 17*(3), 187—194.
Davis, T. A., Volesky, B., & Mucci, A. (2003). A review of the biochemistry of heavy metal biosorption by brown algae. *Water Research, 37*, 4311—4330.
Dawes, C. (2016). Macroalgae systematics. In J. Fleurence, & I. Levine (Eds.), *Seaweed in health and disease prevention* (pp. 107—138). Amsterdam: Elsevier Inc.
Deniz, F., & Ersanli, E. T. (2018). An ecofriendly approach for bioremediation of contaminated water environment: Potential contribution of a coastal seaweed community to environmental improvement. *International Journal of Phytoremediation, 20*(3), 256—263.
Dosdat, A., Servais, F., Métailler, R., Huelvan, C., & Desbruyères, E. (1996). Comparison of nitrogenous losses in five teleost fish species. *Aquaculture, 141*, 107—127.
Edwards, P., & Demaine, H. (1997). *Rural aquaculture: Ooverview and framework for country reviews*. Bangkok (Thailand): FAO.
EFB. (2019). *European federation of biotechnology*. Available from: http://www.efbiotechnology.org/environmental/.
Elizondo-González, R., Quiroz-Guzmán, E., Escobedo-Fregoso, C., Magallón-Servín, P., & Peña-Rodríguez, A. (2018). Use of seaweed *Ulva lactuca* for water bioremediation and as feed additive for white shrimp *Litopenaeus vannamei*. *PeerJ, 6*, e4459. https://doi.org/10.7717/peerj.4459.
Ezeonu, C. S., Tagbo, R., Anike, E. N., Oje, O. A., & Onwurah, I. N. E. (2012). Biotechnological tools for environmental sustainability: Prospects and challenges for environments in Nigeria — a standard review. *Biotechnology Research International*, 450802. https://doi.org/10.1155/2012/450802.
Farias, D. R., Hurd, C. L., Eriksen, R. S., Simioni, C., Schmidt, E., Bouzon, Z. L., et al. (2017). *In situ* assessment of *Ulva australis* as a monitoring and management tool for metal pollution. *Journal of Applied Phycology, 29*, 2489—2502.
Fleurence, J., Morançais, M., Dumay, J., Decottignies, P., Turpin, V., Munier, M., et al. (2012). What are the prospects for using seaweed in human nutrition and for marine animals raised through aquaculture? *Trends in Food Science & Technology, 27*, 57—61.
Gadd, G. M. (2009). Biosorption: Critical review of scientific rationale, environmental importance and significance for pollution treatment. *Journal of Chemical Technology and Biotechnology, 84*(1), 13—28.

Gao, G., Clare, A. S., Rose, C., & Caldwell, G. S. (2018). *Ulva rigida* in the future ocean: Potential for carbon capture, bioremediation and biomethane production. *GCB Bioenergy, 10*, 39–51.
García-Seoane, R., Fernández, J. A., Villares, R., & Aboal, J. R. (2018). Use of macroalgae to biomonitor pollutants in coastal waters: Optimization of the methodology. *Ecological Indicators, 84*, 710–726.
Gavrilescu, M. (2010). Environmental biotechnology: Achievements, opportunities and challenges. *Dynamic Biochemistry, Process Biotechnology and Molecular Biology, 4*(1), 1–36.
van Ginneken, V., & de Vries, E. (2018). Seaweeds as biomonitoring system for heavy metal (HM) accumulation and contamination of our oceans. *American Journal of Plant Sciences, 9*, 1514–1530.
Gomes, H., Dias-Ferreira, C., & Ribeiro, A. (2013). Overview of *in situ* and *ex situ* remediation technologies for PCB-contaminated soils and sediments and obstacles for full-scale application. *The Science of the Total Environment, 445–446*, 237–260.
Grote, B. (2016). Bioremediation of aquaculture wastewater: Evaluating the prospects of the red alga *Palmaria palmata* (Rhodophyta) for nitrogen uptake. *Journal of Applied Phycology, 28*, 3075–3082.
He, J. S., & Chen, J. P. (2014). A comprehensive review on biosorption of heavy metals by algal biomass: Materials, performances, chemistry, and modelling simulation tools. *Bioresource Technology, 160*, 67–78.
Henriques, B., Rocha, L. S., Lopes, C. B., Figueira, P., Duarte, A. C., Vale, C., et al. (2017). A macroalgae-based biotechnology for water remediation: Simultaneous removal of Cd, Pb and Hg by living *Ulva lactuca*. *Journal of Environmental Management, 191*, 275–289.
Huo, Y., Wu, H., Chai, Z., Xu, S., Han, F., Dong, L., et al. (2012). Bioremediation efficiency of *Gracilaria verrucosa* for an integrated multi-trophic aquaculture system with *Pseudosciaena crocea* in Xiangshan harbor, China. *Aquaculture, 326–329*, 99–105.
Huo, Y. Z., Xu, S. N., Wang, Y. Y., Zhang, J. H., Zhang, Y. J., Wu, W. N., et al. (2011). Bioremediation efficiencies of *Gracilaria verrucosa* cultivated in an enclosed sea area of Hangzhou Bay, China. *Journal of Applied Phycology, 23*, 173–182.
Johansson, C. L., Paul, N. A., de Nys, R., & Roberts, D. A. (2016). Simultaneous biosorption of selenium, arsenic and molybdenum with modified algal-based biochars. *Journal of Environmental Management, 165*(1), 117–123.
Juwarkar, A. A., Singh, S. K., & Mudhoo, A. (2010). A comprehensive overview of elements in bioremediation. *Reviews in Environmental Science and Biotechnology, 9*, 215–288.
Kantachote, D., Naidu, R., Williams, B., McClure, N., Megharaj, M., & Singleton, I. (2004). Bioremediation of DDT-contaminated soil: Enhancement by seaweed addition. *Journal of Chemical Technology and Biotechnology, 79*, 632–638.
Kidgell, J. T., de Nys, R., Paul, N. A., & Roberts, D. A. (2014). The sequential application of macroalgal biosorbents for the bioremediation of a complex industrial effluent. *PLoS One, 9*(7), e101309. https://doi.org/10.1371/journal.pone.0101309.
Kim, B. S., Lee, H. W., Park, S. H., Baek, K., Jeon, J. K., Cho, H. J., et al. (2016). Removal of Cu^{2+} by biochars derived from green macroalgae. *Environmental Science and Pollution Research, 23*(2), 985–994.
Lamprianidou, F., Telfer, T., & Ross, L. G. (2015). A model for optimization of the productivity and bioremediation efficiency of marine integrated multitrophic aquaculture. *Estuarine, Coastal and Shelf Science, 164*, 253–264.
Largo, D. B., Diola, A. G., & Marababol, M. S. (2016). Development of an integrated multitrophic aquaculture (IMTA) system for tropical marine species in southern Cebu, Central Philippines. *Aquaculture Reports, 3*, 67–76.
Lawton, R. J., Mata, L., de Nys, R., & Paul, N. A. (2013). Algal bioremediation of waste waters from land-based aquaculture using *Ulva*: Selecting target species and strains. *PLoS One, 8*(10), e77344. https://doi.org/10.1371/journal.pone.0077344.

Lyon, S., Ahmadzadeh, H., & Murry, M. (2015). Algae-based wastewater treatment for biofuel production: Processes, species, and extraction methods. In N. R. Moheimani, M. P. McHenry, K. de Boer, & P. A. Bahri (Eds.), *Biomass and biofuels from microalgae* (pp. 95–115). Springer International Publishing.
Makkar, H. P. S., Tran, G., Heuzé, V., Giger-Reverdin, S., Lessire, M., Lebas, F., et al. (2016). Seaweeds for livestock diets: A review. *Animal Feed Science and Technology, 212*, 1–17.
Marinho-Soriano, E., Azevedo, C. A. A., Trigueiro, T. G., Pereira, D. C., Carneiro, M. A. A., & Camara, M. R. (2011). Bioremediation of aquaculture wastewater using macroalgae and *Artemia*. *International Biodeterioration & Biodegradation, 65*, 253–257.
Michalak, I. (2018). Experimental processing of seaweeds for biofuels. *Wiley Interdisciplinary Reviews, 7*, e288. https://doi.org/10.1002/wene.288.
Michalak, I., Chojnacka, K., Dobrzański, Z., Górecki, H., Zielińska, A., Korczyński, M., et al. (2011). Effect of macroalgae enriched with microelements on egg quality parameters and mineral content of eggs, eggshell, blood, feathers and droppings. *Journal of Animal Physiology and Animal Nutrition, 95*, 374–387.
Michalak, I., Chojnacka, K., & Korniewicz, D. (2015a). New feed supplement from macroalgae as the dietary source of microelements for pigs. *Open Chemistry, 13*, 1341–1352.
Michalak, I., Witek-Krowiak, A., Chojnacka, K., & Bhatnagar, A. (2015b). Advances in biosorption of microelements – the starting point for the production of new agrochemicals. *Reviews in Inorganic Chemistry, 35*(3), 115–133.
Milledge, J. J., Smith, B., Dyer, P. W., & Harvey, P. (2014). Macroalgae-derived biofuel: A review of methods of energy extraction from seaweed biomass. *Energies, 7*, 7194–7222.
Neori, A., Chopin, T., Troell, M., Buschmann, A. H., Kraemer, G. P., Halling, C., et al. (2004). Integrated aquaculture: Rationale, evolution and state of the art emphasizing seaweed biofiltration in modern mariculture. *Aquaculture, 231*(1–4), 361–391.
Ortiz-Calderon, C., Silva, H. C., & Vásquez, D. B. (2017). Metal removal by seaweed biomass. In J. Shankar Tumuluru (Ed.), *Biomass volume Estimation and Valorization for energy* (Vol. 16, pp. 361–380). IntechOpen. Available from: https://www.intechopen.com/books/biomass-volume-estimation-and-valorization-for-energy/metal-removal-by-seaweed-biomass.
Ozyigit, I. I., Uyanik, O. L., Sahin, N. R., Yalcin, I. E., & Demir, G. (2017). Monitoring the pollution level in Istanbul coast of the sea of Marmara using algal species *Ulva lactuca* L. *Polish Journal of Environmental Studies, 26*(2), 773–778.
Poo, K. M., Son, E. B., Chang, J. S., Ren, X., Choi, Y. J., & Chae, K. J. (2018). Biochars derived from wasted marine macro-algae (*Saccharina japonica* and *Sargassum fusiforme*) and their potential for heavy metal removal in aqueous solution. *Journal of Environmental Management, 206*, 364–372.
Roberts, D. A., Cole, A. J., Paul, N. A., & de Nys, R. (2015a). Algal biochar enhances the re-vegetation of stockpiled mine soils with native grass. *Journal of Environmental Management, 161*, 173–180.
Roberts, D. A., Paul, N. A., Dworjanyn, S. A., Bird, M. I., & de Nys, R. (2015b). Biochar from commercially cultivated seaweed for soil amelioration. *Scientific Reports, 5*, 9665. https://doi.org/10.1038/srep09665.
Roberts, D. A., Paul, N. A., Bird, M. I., & de Nys, R. (2015c). Bioremediation for coal-fired power stations using macroalgae. *Journal of Environmental Management, 153*, 25–32.
Roberts, D. A., Paul, N. A., Dworjanyn, S. A., Hu, Y., Bird, M. I., & de Nys, R. (2015d). *Gracilaria* waste biomass (sampah rumput laut) as a bioresource for selenium biosorption. *Journal of Applied Phycology, 27*, 611–620.
Ross, M. E., Davis, K., McColl, R., Stanley, M. S., Day, J. G., & Semião, A. J. C. (2018). Nitrogen uptake by the macro-algae *Cladophora coelothrix* and *Cladophora parriaudii*:

Influence on growth, nitrogen preference and biochemical composition. *Algal Research, 30*, 1–10.
Sati, M., Verma, M., Bora, M., & Rai, J. P. N. (2016). Potential of algae in bioremediation of heavy metals: A review. *Bulletin of Environment, Pharmacology and Life Sciences, 5*(11), 86–97.
Seghetta, M., Torring, D., Bruhn, A., & Thomsen, M. (2016). Bioextraction potential of seaweed in Denmark – an instrument for circular nutrient management. *The Science of the Total Environment, 563–564*, 513–529.
Shin, W. S. (2017). Adsorption characteristics of phenol and heavy metals on biochar from *Hizikia fusiformis*. *Environmental Earth Sciences, 76*, 782. https://doi.org/10.1007/s12665-017-7125-4.
Silva, C., Yáñez, E., Martín-Díaz, M. L., & DelValls, T. A. (2012). Assessing a bioremediation strategy in a shallow coastal system affected by a fish farm culture – application of GIS and shellfish dynamic models in the Rio San Pedro, SW Spain. *Marine Pollution Bulletin, 64*, 751–765.
Singh, A., & Prasad, S. M. (2015). Remediation of heavy metal contaminated ecosystem: An overview on technology advancement. *International Journal of Environmental Science and Technology, 12*, 353–366.
Song, G., Guo, Y., Li, G., Zhao, W., & Yu, Y. (2019). Comparison for adsorption of tetracycline and cefradine using biochar derived from seaweed *Sargassum* sp. *Desalination and Water Treatment, 160*, 316–324.
Thivya, J., & Vijayaraghavan, J. (2019). Single and binary sorption of reactive dyes onto red seaweed-derived biochar: Multi-component isotherm and modelling. *Desalination and Water Treatment, 156*, 87–95.
Tuhy, L., Samoraj, M., Michalak, I., & Chojnacka, K. (2014). The application of biosorption for production of micronutrient fertilizers based on waste biomass. *Applied Biochemistry and Biotechnology, 174*, 1376–1392.
Vijayaraghavan, K., & Ashokkumar, T. (2019). Characterization and evaluation of reactive dye adsorption onto biochar derived from *Turbinaria conoides* biomass. *Environmental Progress & Sustainable Energy, 38*(4). https://doi.org/10.1002/ep.
Wang, L. K., Ivanov, V., Tay, J. H., & Hung, Y. T. (2010). *Environmental biotechnology*. Springer Science+Business Media, LLC.
Wei, Z., You, J., Wu, H., Yang, F., Long, L., Liu, Q., et al. (2017). Bioremediation using *Gracilaria lemaneiformis* to manage the nitrogen and phosphorous balance in an integrated multi-trophic aquaculture system in Yantian Bay, China. *Marine Pollution Bulletin, 121*, 313–319.
Wu, H., Huo, Y., Hu, M., Wei, Z., & He, P. (2015b). Eutrophication assessment and bioremediation strategy using seaweeds co-cultured with aquatic animals in an enclosed bay in China. *Marine Pollution Bulletin, 95*, 342–349.
Wu, H., Huo, Y., Zhang, J., Liu, Y., Zhao, Y., & He, P. (2015a). Bioremediation efficiency of the largest scale artificial *Porphyra yezoensis* cultivation in the open sea in China. *Marine Pollution Bulletin, 95*, 289–296.
Xiao, X., Agusti, S., Lin, F., Li, K., Pan, Y., Yu, Y., et al. (2017). Nutrient removal from Chinese coastal waters by large-scale seaweed aquaculture. *Scientific Reports, 7*, 46613. https://doi.org/10.1038/srep46613.
Xu, Y., Fang, J., & Wei, W. (2008). Application of *Gracilaria lichenoides* (Rhodophyta) for alleviating excess nutrients in aquaculture. *Journal of Applied Phycology, 20*, 199–203.
Yacou, C., Altenor, S., Carene, B., & Gaspard, S. (2018). Chemical structure investigation of tropical *Turbinaria turbinata* seaweeds and its derived carbon sorbents applied for the removal of hexavalent chromium in water. *Algal Research, 34*, 25–36.

Yang, Y., Chai, Z., Wang, Q., Chen, W., He, Z., & Jiang, S. (2015). Cultivation of seaweed *Gracilaria* in Chinese coastal waters and its contribution to environmental improvements. *Algal Research, 9*, 236–244.

Zeraatkar, A. K., Ahmadzadeh, H., Talebi, A. F., Moheimani, N. R., & McHenry, M. P. (2016). Potential use of algae for heavy metal bioremediation, a critical review. *Journal of Environmental Management, 181*, 817–831.

Zhang, J., Hansen, P. K., Fang, J., Wang, W., & Jiang, Z. (2009). Assessment of the local environmental impact of intensive marine shellfish and seaweed farming – application of the MOM system in the Sungo Bay, China. *Aquaculture, 287*, 304–310.

Zheng, Y., Jin, R., Zhang, X., Wang, Q., & Wu, J. (2019). The considerable environmental benefits of seaweed aquaculture in China. *Stochastic Environmental Research and Risk Assessment, 33*, 1203–1221.

Zhou, Y., Yang, H., Hu, H., Liu, Y., Mao, Y., Zhou, H., et al. (2006). Bioremediation potential of the macroalga *Gracilaria lemaneiformis* (Rhodophyta) integrated into fed fish culture in coastal waters of north China. *Aquaculture, 252*, 264–276.

CHAPTER FIVE

Indonesian *Sargassum* species bioprospecting: potential applications of bioactive compounds and challenge for sustainable development

Maya Puspita[a,b,*], Nur Azmi Ratna Setyawidati[c,d], Valérie Stiger-Pouvreau[e], Laurent Vandanjon[a], Ita Widowati[b], Ocky Karna Radjasa[b], Gilles Bedoux[a] and Nathalie Bourgougnon[a]

[a]Laboratoire de Biotechnologie et Chimie Marines, EA3884, UBS, IUEM, F-56000 Vannes, France
[b]Doctoral Program of Coastal Resources Management, Diponegoro University, Semarang, Jawa Tengah, Indonesia
[c]Agency of Research and Development for Marine and Fisheries, Ministry of Marine and Fisheries, Jakarta Utara, Jakarta, Indonesia
[d]Laboratoire des Sciences de l'environnement Marin (LEMAR), UMR 6539 CNRS UBO Ifremer IRD, Institut Universitaire Européen de la Mer (IUEM), Université de Bretagne Occidentale (UBO), Plouzané, France
[e]Univ Brest, CNRS, IRD, Ifremer, LEMAR, Plouzane, France
*Corresponding author: E-mail: maya.puspita@hotmail.co.id

Contents

Advances in Botanical Research, Volume 95
ISSN 0065-2296
https://doi.org/10.1016/bs.abr.2019.12.002

Abstract

Due to the urge of human interest in isolating natural products from marine biodiversity for food or health commercial and industrial purposes, marine bioprospecting has increased rapidly in recent years and holds important promise for the future. Indonesian *Sargassum* species are one of the marine species of brown algae with substantial values. Their wide range of potentiality, especially their bioactive compounds, make them a natural resource for various applications in food, cosmetic and pharmacological sectors. Thus, the bioprospecting of *Sargassum* species is becoming an interesting topic to discuss, especially for Indonesia where only the red macroalgae *Kappaphycus sp* and *Gracilaria sp* are exported. This review illustrates the potential applications of *Sargassum* species concerning their bioactive chemodiversity. Further, possible challenges for sustainable development of Indonesia are also be described.

1. Introduction

The word bioprospecting consisted of a mixture of "biodiversity" and "prospecting". Bioprospecting covers commercial purpose research and development, building on the use of natural occurring compound, all the way from first discovery, over patenting, benchmarking, improvement, development and commercialization (Bhatia & Chugh, 2015; Harvey & Gericke, 2011; Wynberg, 2016). It became a new term emerging as the interest in the search of alternative biochemical compounds or any other useful and valuable biological discoveries in nature are still increasing. The main focus of bioprospecting research is a value from the results of scientific study involving biological samples. Further, this added-value approach has been enhanced by developments in laws on intellectual property rights, new bio-rational approaches in specimen collection and drug and other product-development research as well as evolving trade/business practices (Scott, 2001). This bioprospecting has frequently been mentioned as sustainable use of biodiversity presumably providing a motivation for conservation. However, in Indonesia — well known by its biodiversity-richness — the level of bioprospecting is much below the potential (Kursar et al., 2006).

Being the second longest coastline after Canada with 99.093 km^2 and positioned within the coral triangle, Indonesia is a tropical country, widely known for its high marine biodiversity. Among Indonesian marine organisms, the taxonomical group of algae is notorious for its species-richness. "Sibolga Expedition" reported that 555 species of algae are living around Indonesia (Akrim, 2014). Atmadja and Prud'homme van Rein (2014) recorded the presence of 201 species belonging to 7 orders, 24 families and 48 genera of Chlorophyta (Ulvophyceae/green algae), 138 species in 6 orders, 12 families and 31 genera of Ochrophyta (Phaeophyceae/brown algae). Meanwhile, the total species from Rhodophyta (Florideophyceae and Bangiophyceae/red algae) is 452 species.

For time being, despite its outstanding diversity of species, there are two genera that become the main export commodity in Indonesia, *i.e. Kappaphycus* sp. or *Eucheuma* sp and *Gracilaria* sp. While others genera such as *Caulerpa, Cladophora, Codium, Halimeda, Ulva* (Chlorophyta) *and Dictyota, Padina, Sargassum, Turbinaria* (Phaeophyta) have been under the radar or have not been well explored and exploited yet. As a result, there are still a lot of works to be done on marine Indonesian algae, in particular for bioprospecting purpose.

Marine algae have been historically important sources of phycocolloids and natural products with relevant applications, such as in the cosmetics, nutraceutical, and pharmaceutical industries (Agatonovic-Kustrin & Morton, 2013; Bedoux et al., 2014; Bourgougnon & Stiger-Pouvreau, 2012; Cardozo et al., 2007; Gupta & Abu-ghannam, 2011; Holdt & Kraan, 2011; Le Lann et al., 2016; Mayer, Rodriguez, Taglialatela-Scafati, & Fusetani, 2013; Pangestuti & Kim, 2015; Surget et al., 2017; Vo, Ngo, & Kim, 2012; Wijengsihe & Jeon, 2011).

Sargassum species constitute an abundant biomass in Indonesia, especially in the areas directly adjacent to the Pacific Ocean and the Indian Ocean, i.e. Lombok and Sumba at West Nusa Tenggara: South of Java, Spermonde islands (Kadi, 2005). On this subject, (Setyawidati, Kaimuddin et al., 2017; Setyawidati, Liabot et al., 2017; Setyawidati et al., 2018) mapped populations of brown algae in two Indonesian bays, Ekas and Malasoro Bays, and showed that the potential of both bays is very interesting in term of diversity and potential biomasses to be harvested. The high biomasses of *Sargassum* species and their pseudo-perennial life history, making them available almost the whole year, is an attractive feature for their sustainable management. Furthermore, previous studies have revealed that *Sargassum* species from Indonesia present pharmacological potentialities with strong antioxidant activities, due to the production of interesting phenolic compounds (Hidayati, Yudiati, Pringgenies, Arifin, & OktaviyantI, 2019;

Widowati, Susanto, Puspita, Stiger-Pouvreau, & Bourgougnon, 2014) and alginates (Setyawidati et al., 2018). For that reason, this review covers the potential of Indonesian macroalgae, particularly those belonging to the genus *Sargassum,* related to their bioactive compounds and the possible challenges that might be encountered in the future, in the context of macroalgal sustainable development.

2. Current status and applications of Indonesian algae

Marine algae have attracted many attentions for the past few years and have been commercially harvested for industrial purposes. This is due to their various potential that can be exploited into diverse industrial applications for the sake of human interest. The worldwide seaweed industry has an estimated total value of US$10 billion per year (Valderama et al., 2013). This demand has led to the development of cultivation industries that recently produce more than 96% of the market's demand; with Asiatic countries dominating the algae culture production. Global production of seaweeds has outstandingly grown from 13.5 million tonnes in 1995 to over than 30 million tonnes in 2016 (FAO, 2018). In 2011, 96% of the global production of algae came from aquaculture (99.05% considering the quantity and 99.36% considering the value; FAO, 2014) with five genera, e.g. *Saccharina, Undaria, Porphyra/Pyropia, Eucheuma/Kappaphycus,* and *Gracilaria,* representing 98% of the world's cultivated seaweed production. China and Indonesia are by far considered as the largest seaweed producers of *Saccharina japonica, Undaria pinatifida, Gracilaria* and *Pyropia* for China and *Eucheuma/Kappaphycus* for Indonesia based on the data provided by FAO in 2016 (Buschmann et al., 2017). Indonesia is considered as one of the major contribution to the rapid growth of seaweed farming, particularly for its carragenophyte species *Kappaphycus/Eucheuma* (FAO, 2018; Setyawidati, Kaimuddin et al., 2017; Setyawidati, Liabot et al., 2017).

Food products based on algae for human consumption contribute about US$ 5 billion while the total annual value provided by this activity was estimated around US$ 5.5 billion culture production. The weeds aquaculture outside Asia is scarce triggering a worldwide search of unexploited sources (Rebours et al., 2014). In 2011, 786.466 tonnes of algae were commercially harvested in almost 28 countries ranging from cold to tropical regions with over 55% of the biomass harvested in Latin America and almost 32% in

Europe. In addition, Chile stands out as the top producer followed by Norway respectively accounting for 51.3% and 19.2% of the global catches of natural algae (FAO, 2014).

China might dominate the global seaweed production (48%) due to its kelp production; however, Indonesia produces almost 72% of the world production coming from the genera *Eucheuma, Kappaphycus* and *Gracilaria* (FAO, 2018). Specifically for *Kappaphycus,* Indonesia represents even prominently ~87% of the world even prominently production (FAO, 2018), therefore, Indonesia is known as the biggest producer for this alga.

In total, there are six species that are commercially harvested from Indonesia; four of these species, *i.e. Eucheuma spinosum, Kappaphycus alvarezii* (also known in commercial name as *Eucheuma cottonii*), *Gracilaria* sp. and *Caulerpa* sp., have been cultivated while the other two (*Gelidium* spp. and *Sargassum* spp.) are mostly collected from natural populations (Setyawidati et al., 2017). The largest seaweed production based on the wet weight in tonnes comes from the Sulawesi province (50.63% of total production) followed by the West and East Nusa Tenggara province with 31.85% of total production (FAO, 2018). As for Java, Bali, Kalimantan, Maluku and Papua contribute merely 17% of total production. Fig. 1 shows the national production of Indonesian algae from 2005 to 2017 based on the production of wet weight in tonnes. This production comes from the cultivated

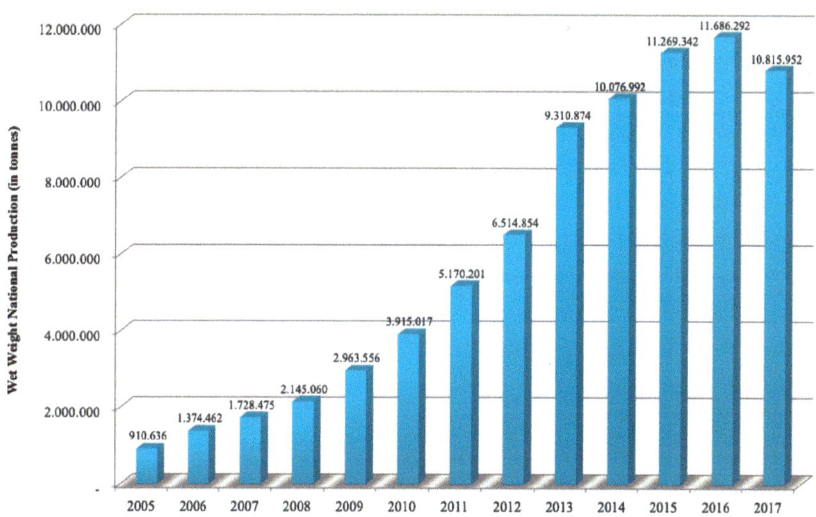

Fig. 1 National algae production of Indonesia from 2005 to 2017 according to the department of aquaculture.

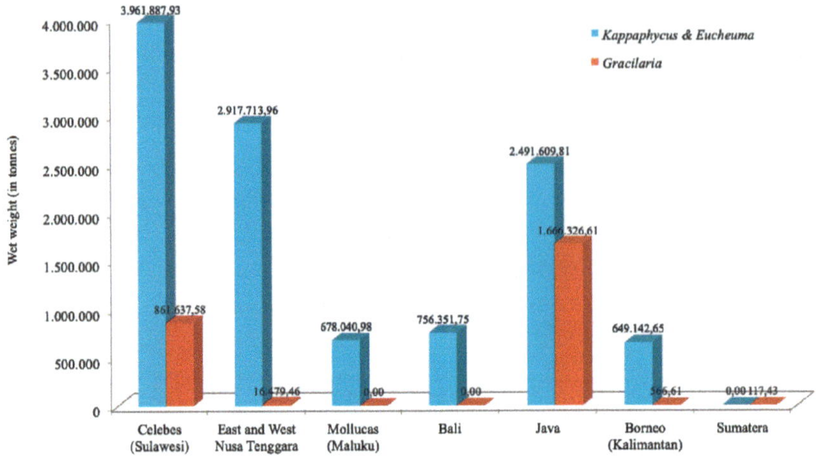

Fig. 2 Algae production in 2015 (weight based on fresh algae in tonnes) in different provinces of Indonesia according to the department of aquaculture.

Gracilaria sp. and *E. cottonii* as presented by the Ministry of Marine and Fisheries in 2015 (source: Statistic of Aquaculture Fisheries). It is important to note that the maximal production came from Sulawesi for *K. alvarezii* for more than 3.9 million tonnes of wet weight and in Java for *Gracilaria* for more than 1.6 million tonnes of wet weight according to the national production in 2015. Fig. 2 illustrates the national production of red algae in 2015 from seven provinces in Indonesia.

Neish (2013) stated that Red Algal Galactans (RAGSs), from the genera *Kappahycus/Eucheuma* and *Gracilaria,* are the most polysaccharides obtained from cultivated algae in Indonesia. In other word, most of Indonesian algae are processed to extract their hydrocolloids, *i.e.* agar and carrageenan. Hydrocolloid productions existing in Indonesia are the Alkali Treated Cottonii (ATC), Alkali Treated *Gracilaria* (ATG), Semi Refined and Refined Carrageenan. In total, there are 51 industries, both local and foreign investments producing carrageenan (31 industries) and agar (20 industries). Table 1 presents the value for Indonesian algae export from 2014 to 2018 according to the Central Agency of Indonesian Statistic. Hydrocolloids production contributed around 3%–5% from the total export, while 94%–97% of export contribution came from the seaweed dry matter. The average export of Indonesian seaweed dry material from 2014 to 2018 was 171.699.298 USD. Fig. 3 presents the classification of exported algal-based products from Indonesia showing that the biggest contribution of export mostly comes from the raw material. The top three of export destination country

Table 1 Indonesian seaweed export from 2014 to 2018 based on different products expressed in value of Kg and in USD according to the Central Agency of Indonesian Statistic.

	2014		2015		2016		2017		2018	
Products	**Value in kg**	**Value in USD**	**Value in kg**	**Value in USD**	**Value in kg**	**Value in USD**	**Value in kg**	**Value in USD**	**Value in kg**	**Value in USD**
Semi-refined carrageenan	4.164.174	34.503.839	3.474.267	24.820.259	4.564.896	26.794.385	4.101.101	25.050.397	4.942.442	36.352.707
Refined carrageenan	0	0	0	0	0	0	528.765	2.671.750	543	2.824
Alkali treated carrageenan	648.430	3.955.431	780.494	5.327.790	416.751	2.093.173	197.375	761.848	360.537	1.614.752
Other carrageenan (other than ATC)	0	0	0	0	0	0	1.301.072	7.240.396	5.007.359	30.555.497
Agar-agar	933.253	14.811.551	606.621	8.041.336	945.874	8.908.223	1.355.118	10.047.571	1.488.871	14.169.036
Raw material[a]	210.705.767	226.228.693	177.957.645	139.902.668	182.371.133	124.007.687	184.592.534	159.215.032	201.161.863	209.142.411

[a]Dry raw material consists of *K. alvarezii*, *E. denticulatum*, *Gracilaria sp.*, *Sargassum sp.* and other species that is not specifically mentioned in the data.

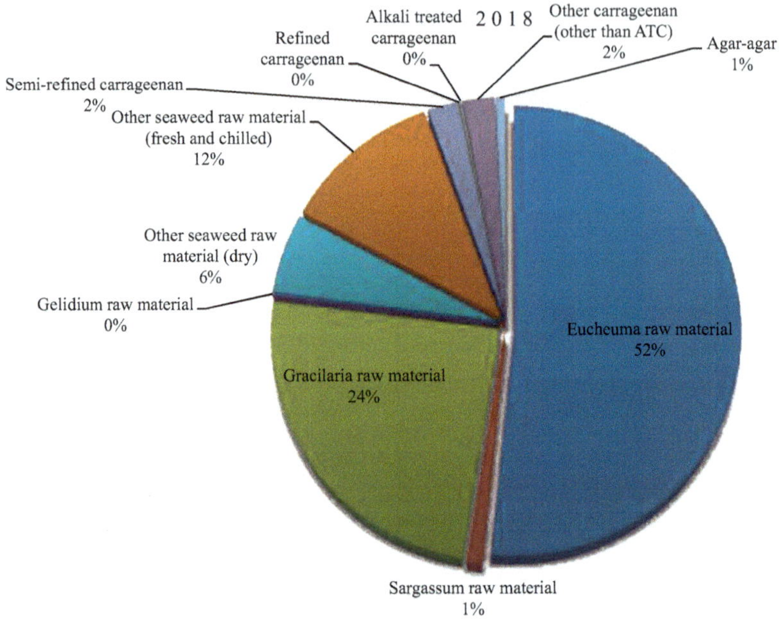

Fig. 3 Classification of exported algal-based products in 2017 according to central agency of Indonesian statistic.

for Indonesian algal-based products is China (75%), United States of America (9%) and Vietnam (6%) (Figs. 4 and 5).

3. Interest for the genus *Sargassum*

Classification of the brown genus *Sargassum* according to Guiry and Guiry (2018) is as follow: phylum Ochrophyta, class Phaeophyceae, order Fucales and family Sargassaceae. There are five genera belonging to the class Phaeophyceae, which are widely distributed in Indonesian water of which members of the genera *Hydroclathrus, Padina, Sargassum,* and *Turbinaria* have economic value as traditional food and vegetables. In addition, *Dictyota, Padina, Sargassum,* and *Turbinaria,* known as the most species-rich group, belong to the Order Dictyotales (for the two first genera) and Fucales (for the two last ones), respectively. As for *Sargassum,* there are about 58 species detected in Indonesian waters. Brown algae are mostly inhabited the intertidal and subtidal zone with muddy, sandy or rocky substrates (Atmadja & Prudhomme van Reine, 2014).

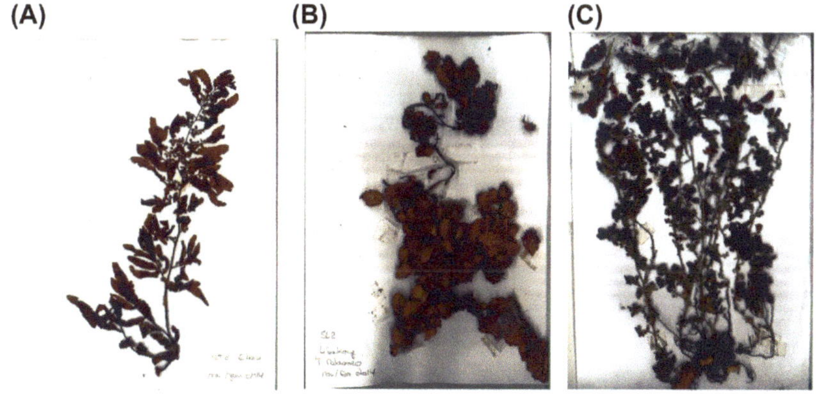

Fig. 4 Three most common species of *Sargassum* in Indonesia. (A) S. aquifolium (Turner) (C) Agardh (previously regarded as S. echinocarpum J. Agardh). (B) S. ilicifolium (Turner) (C) Agardh (previously regarded as S. duplicatum J. Agardh). (C) S. polycystum (C) Agardh.

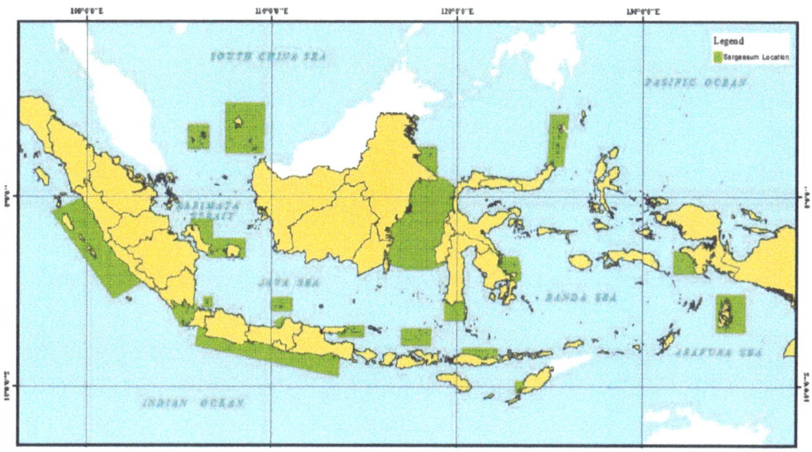

Fig. 5 Distribution of *Sargassum* in Indonesian water based on Atmadja and Prud-homme van Reine (2014).

To date, there have been 353 species and variant of *Sargassum*, which were taxonomically validated from 941 species, and variants listed in www.algabase.org (Guiry & Guiry, 2018). Based on the literature reviewed by Kursar et al. (2011), Indonesia has around 79 species and variants of *Sargassum*. Fig. 2 shows the most common and most abundant *Sargassum* species found in Indonesia: *Sargassum aquifolium*, *Sargassum ilicifolium* and *Sargassum polycystum* can be collected in Indonesian waters. The genus

Sargassum has a large distribution all around Indonesia as shown on Fig. 3, which illustrates the distribution of specimens of *Sargassum* in Indonesian waters.

In Fucales, the reproduction occurs by releasing the haploid gametes and the biological life cycle is a diplophasic monogenetic cycle. Adult individuals, also called meiosporophytes (2n chromosomes), are monecious or diecious depending on the species, and are producing gametes after a chromatic reduction (de Reviers, 2002; Heesch et al., 2019). Gametes are produced in conceptacles. Small, numerous and motile male gametes are produced within antheridia and/while large and not motile female gametes are produced within oogonia. Conceptacles are grouped into receptacles along the fertile laterals in the genus *Sargassum* (Le Lann, 2009). Conceptacles are found inside small finger-like branchlets called receptacles and usually bear one (monecious) or two (diecious) type/s of gametes (Critchley, Peddemors, & Pienaar, 1991; Ogawa, 1976). The thallus of *Sargassum* is a diploid meiosporophyte, reproducing both by vegetative multiplication and sexual ways (production of gametes). Fragmentation is the only known method of vegetative reproduction for this alga. Each fragment grows into new plant, which is genetically identical to the original thallus. The fragmentation commonly occurs accidently or by death and decay of older parts separating the younger parts of the branches.

Both the growth and the distribution of *Sargassum* are influenced by environmental parameters. The most important parameters regulating algal growth are nutrient and substrate, quantity and quality of light, pH, salinity, temperature, and hydro dynamism. The most optimal parameters as well as the tolerated ranges are species-specific and the various factors may be inter-independent and a parameter optimal for one set condition is not necessarily optimal for another (Barsanti & Gualtieri, 2006; Lalegerie, Gager, Stiger-Pouvreau, & Connan, 2019).

Indonesia, as a tropical country, is characterized by two seasons, *i.e.* dry and rainy seasons, which have implication on the phenology of *Sargassum*. In a work done in two Indonesian Bays, it was demonstrated that the higher biomasses of *Sargassum* species in Malasoro Bay (South Sulawesi, Setyawidati, Kaimuddin et al., 2017) and in Ekas Bay (Lombok, Setyawidati et al., 2018) occurred during May/June period while in November, the biomasses in both regions were very low. Apparently, the optimum temperature ranges from 27.25 to 29.30°C (Kadi, 2005). For other species of *Sargassum*, such as *S. polycystum* in China, it exists in both seasons because it has a wider range of survival temperature estimated at 20—30°C (Hwang,

Tsai, & Lee, 2004). The salinity fluctuation initiates an osmotic stress, which has significant implication on the growth and survival of *Sargassum* species. *Sargassum* tolerates a salinity range between 32 and 33.5 PSU (Kadi, 2005). Others environmental parameters, such as temperature, salinity, precipitation, irradiance and pH are known to affect the seasonal growth of *Sargassum* (Ang, 2006; Le Lann, Connan, & Stiger-Pouvreau, 2012; Steen, 2004; Yeong & Wong, 2013), which further affects the yield and constituents of alginate as well as its economic value (Saraswathi, Babu, & Rengasamy, 2003). Zou et al. (2017) reported that *S. polycystum* had broad tolerance to various environmental conditions, therefore it survived under a wide range of salinity (24–36 PSU), temperature (10–30°C) and irradiance (10–120 μmol photons m^{-2} s^{-1}) conditions. The optimal conditions from these conditions could maximize the growth and improve the yields.

The nutrient availability will affect algae's growth rate. Moreover, C, H, O, P, Mg, Cu, Mn, Zn and Mo are considered to be required by all algae (DeBoer & Ryther, 1977; O'Kelley, 1974). Among these possible limiting elements, nitrogen appears to be the most frequently element which limits the growth of algae (Topinka & Robbins, 1976 reviewed by; Lalegerie et al., 2019), followed by phosphorus as the second most limiting nutrient (Manley & North, 1984). Most often employed light intensities range between 100 and 200 μE s^{-1}, which corresponds to about 5%–10% of full daylight (2000 μE m^{-2} sec^{-1}) (Barsanti & Gualtieri, 2006). Kadi (2005) also reported that *Sargassum* needed light intensity ranged between 6000 and 7500 lux equivalents to 100–200 μE m^{-2} sec^{-1}. Nevertheless, the optimum light intensity for maximum growth required by marine algae is species-specific.

4. Bioactive compounds from *Sargassum* species and their promising potential

Algae have been subjected to many researches these past years and some of them have been introduced and used in food, cosmeceutical and pharmaceutical industries. For brown algae, the three most studied families of molecules are polysaccharides, proteins and phlorotannins, especially related to their biochemically active properties (Bedoux et al., 2014; Bourgougnon & Stiger-Pouvreau, 2012; Costa et al., 2010; Samarakoon & Jeon, 2012; Shin, Ahn, Hyun, Kim, & Moon, 2014; Stiger-Pouvreau, Jégou, Cérantola, Guérard, & Le Lann, 2014; Wijengsihe and Jeon, 2012; Wijesekara, Yoon, & Kim, 2010; Yang & Zhang, 2009).

4.1 Polysaccharides

Algae are widely known for their high polysaccharide content, acting as a component of their cell wall and reserve/storage material. Phaeophyceae cell walls are generally composed of two layers with cellulose making up the main structural skeleton (Kloareg & Quatrano, 1988; reviewed by; Stiger-Pouvreau, Bourgougnon, & Deslandes, 2016) and contents range from 2% to 10% (Kraan, 2012). The cell wall system in Fucales is mostly an assembly of two polysaccharide-based networks, the first being composed of fucose-containing sulfated polysaccharides interlocking a -glucan scaffold embedded in the second made up of alginates and cross-linking polyphenols. Proteins, glycoproteins, halogenated and/or sulfated, phlorotannins, iodide and calcium, are additional components (Balboa, Conde, Moure, Falque, & Dominguez, 2013; Deniaud-Bouët, Hardouin, Potin, Kloareg, & Hervé, 2017). Alginate generally makes up to 10%–30% of the seaweed dry weight, as for fucoidan, their content range from 1% to 20% (Park, Jang, & Kim, 1997, pp. 283–336). Concerning laminarins – glucans of lower molecular masses, their content may reach 35% of seaweed dry weight and function as storage polysaccharides. These polysaccharides provide strength and flexibility of brown algae as well as maintaining the ionic equilibrium and preventing desiccation (Balboa et al., 2013).

4.1.1 Alginates

Alginic acid consists of (1–4) linked β-D-mannuronic acid (M) and α-L-guluronic acid (G) residues (Fig. 6 (1) A, B, C). In the mixed salts form – known as alginates – they contain several cations such as Na^+, K^+, Mg^{2+} and Ca^{2+} (Percival & McDowell, 1967, pp. 157–174). To isolate alginic acids, the algal biomass is pretreated with dilute acid to remove polyvalent cations, and the polymer is extracted in the form of readily soluble sodium salt by treatment with sodium carbonate solution at elevated temperature. Alginic acids are precipitated after acidification of the extract, whereas sodium alginate may be precipitated with ethanol; alternatively, insoluble calcium alginate may be obtained by addition of calcium chloride. Alginic acids isolated from various sources may differ in the M/G ratio depending on the species, the part of the thallus, their surrounding conditions and the harvest period (Apoya, Ogawa, & Nanba, 2002; Chandia, Matsuhiro, Mejias, & Moenne, 2004; Skriptsova, Khomenko, & Isakov, 2004). Moreover, the molecules may differ in block composition, since uronic acid residues are

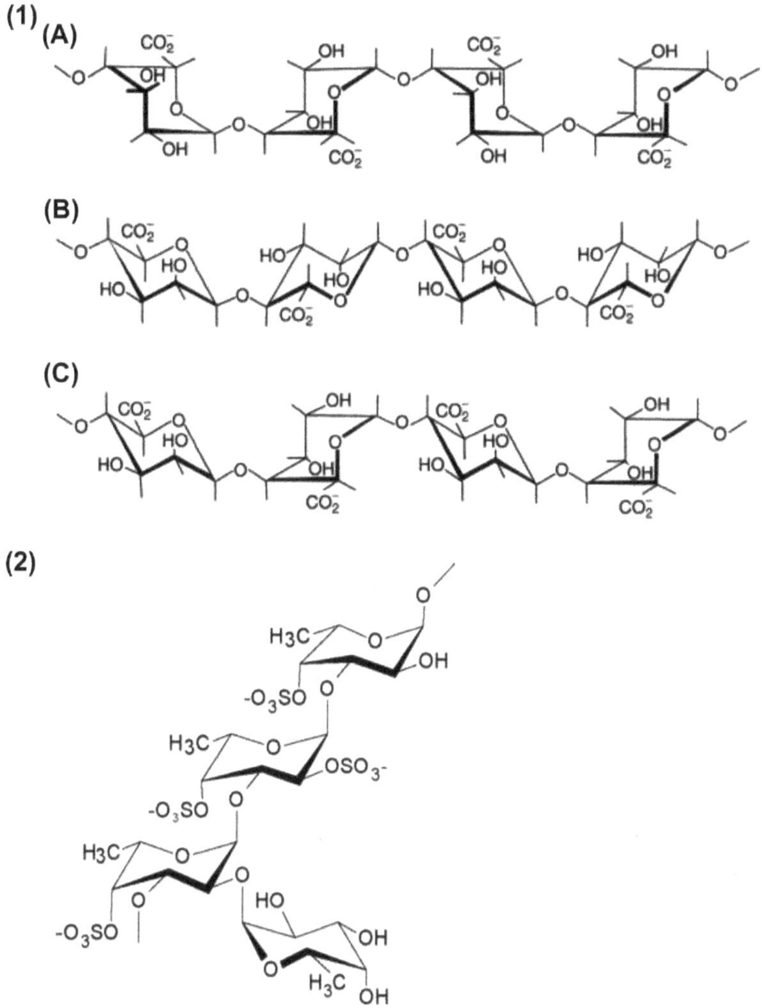

Fig. 6 (1) Structure of alginate in different blocks unit (A) (1,4)-α-L-guluronnic acid blocks (GG) (B) (1,4)-β-D-mannuronic acid blocks (MM) (C) heteropolymeric sequences of M and G blocks (MGs) (Rinaudo, 2014). (2) Structure of fucoidan isolated from Saccharina latissima (previously regarded as Laminaria saccharina) (Ale & Meyer, 2013).

usually arranged in two homopolymeric blocks, poly-M and poly-G, interspersed with alternating structure of MG-blocks. Such blockwise distribution of monomeric units is the result of specific biosynthetic pathway, where the formation of linear polymannuronan precursors is followed by epimerization of some β -D-mannuronic acid residues at C5 under the action of mannuronan-C(5)-epimerase, giving rise to the appearance of

α)-L-guluronic acid residues, single or contiguous, within the polymeric molecule (Draget, Smidsrød, & Skjåk-Bræk, 2005; Ertesvag, Valla, & Skjak-Brak, 1996).

Sargassum alginates are found to have a wide range of molecular weights ranging from 13 to 330 kDa. This depends upon species, season, age, part of plant used for extraction and also upon extractive procedures. As alginates have hydrophilic and viscous properties, they have been used as thickeners, stabilizers and gelling agents in food and pharmaceutical industries (Liu, HeinrichMyers, & Dworjanyn, 2012).

4.1.2 Fucose-containing sulfated polysaccharides

Cell walls in Fucales feature fucose-containing sulfated polysaccharides (FCSPs) including fucoidans (backbone structure based on a variety of glycosyl residues), and fucans (backbone structure based on fucopyranosyl residues) (Balboa et al., 2013; Holdt & Kraan, 2011). FCSPs comprise highly sulfated homofucan molecules together with a variety of highly branched polysaccharides: Galactofucan structures in *Sargassum crassifolium*, sulfated (xylo)fucoglucuronomannans in *Sargassum fusiforme* (Deniaud-Bouët et al., 2017; Pomin & Mourao, 2008). They account for 10%–20% of dry weight and are not found in other divisions of algae or in terrestrial plants. These carbohydrates are very soluble and do not develop highly viscous solutions (Balboa et al., 2013). The chemical composition of FCSPs depends on the algal species, part of the thallus and the harvesting period (Chojnacka, Saeid, Witkowska, & Tuhy, 2012; Rioux, Turgeon, & Beaulieu, 2009; Skriptsova, Shevchenko, Zvyagintseva, & Imbs, 2010). Fig. 6(2) illustrates fucoidan structure isolated from *Saccharina latissima* (known before as *Laminaria saccharina*).

Many previous studies have revealed the wide range potential showed by fucoidans especially related to their bioactivity. They are known to possess anti-tumoral (Alekseyenko, Zhanayeva, Venediktova, Zvyagintseva, & Kuznetsova, 2007; Maruyama, Tamauchi, Iizuka, & Nakano, 2006), anticoagulant (Ananthi et al., 2010; Karaki, Sebaaly, Chahine, Faour, & ZinchenkoRachidKanaan, 2013; Zoysa, Nikapitiya, Jeon, Jee, & Lee, 2008), anti-inflammatory (Dore et al., 2013), antithrombotic (Cumashi et al., 2007; Dore et al., 2013), antioxidant (Chattopadhyay et al., 2010; Dore et al., 2013; Yuan et al., 2005), antiviral (Adhikari et al., 2006; Hemmingson, FalshawFurneaux, & Thompson, 2006; Lee, Takeshita, Hayashi, & Hayashi, 2011) and anticancer properties (Moghadamtousi et al., 2014; Yang, Zeng, Dong, Liu, & Li, 2010; Yamasaki-Miyamoto, 2009).

Fucoidan from *Sargassum* are found to have antiviral against Herpes Simplex Virus Type I (HSV-I), Type 2 (HSV-2), Hepatitis A-Virus (HAV), coxsackie virus (CVB-3), human sytomegalovirus (HCMV) and Immunodeficiency Virus Type-1 (HIV-1) (Harden, Falshaw, Carnachan, Kern, & Mark, 2010; Soares et al., 2012; Zhu, Chiu, Ooi, Chan, & Ang, 2004). Sinha et al. (2010) stated that the fucoidan has the ability to inhibit the initial stages of viral infection, attachment and penetration into host cells. Isolated fucoidan from *Sargassum tenerrimum* have been proven to have low cytotoxicity when it was tested toward the cell lines such as normal Vero cells (African green monkey kidney cells) (Sinha et al., 2010). This result suggests the potential for developing fucoidan as safe antiviral drugs (Mayer et al., 2013). Marinova – a Tasmanian company – has managed to isolate fucoidans from a range of species and provide characterized fractions for either investigational research or as ingredients for nutraceutical and cosmetic applications. This SME has successfully isolated and supplied commercial volumes of fucoidan extracted from different genera as follow: *Undaria, Lessonia, Macrocystis, Cladosiphon, Durvillaea, Laminaria, Ecklonia, Fucus, Sargassum, Ascophyllum* and *Alaria* with Halal- and Kosher-certified (Holdt & Kraan, 2011).

4.2 Proteins

In general, the proteins content of brown algae is low ranging between 6% and 13% of dry material (Hernandez-Carmona et al., 2009; Kim, 2010). For this reason, proteins of marine algae is low ranging between 6% and 13% of dry material of red algae – have attracted less attention and their structure and biological properties are not as widely documented as their polysaccharides (Chojnacka et al., 2012). However, they recently have gained increasing attention due to their functional properties and bioactivity that is promising for further applications (Michalak & Chojnacka, 2015). They are present in a diverse range of forms and cellular locations, i.e. as a component of cell wall, enzymes and bound to pigments and carbohydrates (Stengel, Connan, & Popper, 2011).

Previous studies have revealed the potentiality of brown algae proteins as source of natural products. Proteoglycan hypertensive present in *Sargassum horneri* had shown a potential as anticoagulant agent (Athukorala, Lee, Kim, & Jeon, 2007). Protein enzymatic hydrolyzates of *Sargassum fulvellum* with Protamex, *S. horneri* with Alcalase, and *Sargassum thunbergii* with Flavourzyme and Alcalase generated a significant H_2O_2-scavenging activity (Heo & Jeon, 2008). Moreover, Alcalase digests of *S. horneri* had a strong cytoprotective effects against H_2O_2-induced DNA damage *in vitro* (Heo & Jeon, 2008).

4.3 Phlorotannins

Phlorotannins are structural classes of polyketides found exclusively in brown algae and classified into six groups based upon variations in their assemblage from the polymerization of phloroglucinol units (Arnold & Targett, 2000, 2002). Meslet-Cladière et al. (2013) demonstrated the biosynthetic pathway of phlorotannins in isolating a polyketide synthase (type III) from the brown filamentous alga *Ectocarpus siliculosus,* which catalyzes the synthesis of phloroglucinol from malonyl-CoA. Stored within cells in vessels called physodes, phlorotannins have been reported from almost all brown algal orders. According to Vo et al. (2012), phlorotannins are classified into three primary groups based on their inter-phloroglucinol linkages. They are fucols, phlorethols and fucophlorethols. Species of brown seaweeds are able to produce different kinds of phlorotannins as illustrated on Table 2. Nevertheless, some brown seaweeds are able to produce fuhalols and eckols/carmalols which represented two other classes of phlorotannins, as described in Stiger-Pouvreau et al. (2014).

The total content of phlorotannins in brown algae ranges from $< 1\%$ to 14% of dry weight and has been shown to depend on environmental factors such as salinity, light and nutrient availability (Hemmi, Mäkinen, Jormalainen, & Honkanen, 2005; Holdt & Kraan, 2011; Jormalainen & Honkanen, 2008; Le Lann et al., 2012). Variation of polyphenol content seems to occur among local populations of algae (Hemmi & Jormalainen, 2004; Koivikko, Eranen, Loponen, & Jormalainen, 2008; Pavia, Toth, & Åberg, 1999). Such variation can be explained by genetic differentiation among populations and/or response to local changes of environmental factors (Koivikko et al., 2008).

Spatial variation in levels of brown algal polyphenolic compounds has been examined across very large spatial scales, such as between hemispheres and between temperate and tropical habitats (Stiger, Deslandes, & Payri, 2004; Le Lann et al., 2012; Tanniou, Vandajon, Goncalves et al., 2014; Tanniou, Vandanjon, Incera et al., 2014). Variation in polyphenol levels has also been examined across regional scales (tens to hundreds of km) for several species of Australasian brown algae (Steinberg, 1989), *Lobophora variegata* in Caribbean and Atlantic waters (Arnold, Tanner, & Hatch, 1995), *Fucus vesiculosus* in Atlantic waters (Targett et al., 1992), *Ascophyllum nodosum* in Europe (Pavia & Åberg, 1996), and also in *Sargassum* species, such as in

Table 2 Main groups and sub-groups of phlorotannins in different species of brown algae.

Main group of phlorotannins	Sub-group	Species	Interphloroglucinol linkages and molecular size	Reference
Fucols			With only phenyl linkages	
	Tetrafucol-A and B	*Fucus vesiculosus* L		Glombitza, Rauwald, and Eckhardt (1975); Truus, Vaher, Koel, Mähar, and Taure (2004)
Phlorethol			With only ether bonds	
	Diphlorethol	*Halidrys siliquosa* (L) Lyngbye		Le Lann et al. (2016)
	Triphlorethol C and tetraphlorethols A and B	*Laminaria ochroleuca*	—	Koch, Glombitza, and Eckhardt (1980)
	Tetraphlorethol C	*Ecklonia maxima* (Osbeck) Papenfuss	—	Glombitza and Gerstberger (1985)
	Pentaphlorethol B and Hexaphlorethol A	*Cystophora retroflexa* (Labilladiere) J. Argadh	—	Sailler et al. (1999)
Fuhalols			Phlorethol with additional hydroxyl group(s)	
	Trifuhalol	*Halidrys siliquosa*		Le Lann et al. (2016)
	Isotrifuhalol	*Chorda filum*	—	Grosse-Damhues and Glombitza (1984)

(Continued)

Table 2 Main groups and sub-groups of phlorotannins in different species of brown algae.—cont'd

Main group of phlorotannins	Sub-group	Species	Interphloroglucinol linkages and molecular size	Reference
	Hydroxytrifuhalol B, Hydroxypentafuhalol A, hydroxyheptafuhalol B, Hydroxynonafuhalol A	*Sargassum spinuligerum*	—	Glombitza, Keusgen, and Hauperich (1997); Glombitza and Keusgen (1995); Keusgen et al. (1997); Glombitza and Schmidt (1999)
Eckols			Phlorethols including 1,4-dibenzodioxin bond(s)	
	Trimers eckol and Dioxinodehydroeckol	*Ecklonia maxima* and *Eisenia arborea* Areschoug	—	Glombitza and Vogels (1985); Glombitza and Gerstberger (1985)
	Eckol, Phlorofucoeckol-A, Dieckol, 6,6′-bieckol and 8,8′-bieckol	*Eisenia bicyclis*	—	(Kim et al., 2013); Fuji et al. (2013)
	Tetramers 2-phloroeckol and 7-phloroeckol	*Eisenia bicyclis* (Kjellman) Setchell, *E. kurome* Okamura and *E. stolonifera* Okamura	—	Shibata et al. (2004, 2008); Kim et al. (2011); Kita, Fujimoto, Nakajima, Hayashi, and Shibuya (1990); (Nagayama, Shibata, Fujimoto, Honjo, & Nakamura, 2003); Kang et al. (2012); Kim, An, Yoon, Nam, and Choi (2005); Goo, Choi, and Na (2010)

	Diphlorethohydroxycarmalol	*Carpophyllum maschalocarpum*		Li and Glombitza (1991)
	Triphlorethohydroxycarmalol	*Ishige okamurae* Yendo		Zou et al. (2008); Yoon, Lee and Kim, 2009; Yoon, Eom, Kim, M.M. and Kim, S.K., 2009
	Eckol, Phlorofucoeckol-A, Dieckol, 6,6′-bieckol and 8,8′-bieckol	*Eisenia bicyclis*	–	Kim et al. (2013); Fuji et al. (2013)
	Tetramers 2-phloroeckol and 7-phloroeckol	*Eisenia bicyclis* (Kjellman) Setchell, *E. kurome* Okamura and *E. stolonifera* Okamura	–	Shibata et al. (2004, 2008); Kim et al. (2011); Kita et al. (1990); Nagayama et al. (2003); Kang et al. (2012); Kim et al. (2005); Goo et al. (2010)
	Diphlorethohydroxycarmalol	*Carpophyllum maschalocarpum*		Li and Glombitza (1991)
	Triphlorethohydroxycarmalol	*Ishige okamurae* Yendo		Zou et al. (2008); Yoon, Lee and Kim, 2009; Yoon, Eom, Kim, M.M. and Kim, S.K., 2009
Fucophlorethols			With aryl ether and phenyl linkages	
	Fucophlorethols	*Himanthalia elongata* (Linnaeus)	~994 Da	Glombitza and Grosse (1985)

(Continued)

Table 2 Main groups and sub-groups of phlorotannins in different species of brown algae.—cont'd

Main group of phlorotannins	Sub-group	Species	Interphloroglucinol linkages and molecular size	Reference
	Quaterfucononaphlorethol	*Carpophyllum maschalocarpum* (Turner)	1.738 Da	Li and Glombitza (1991)
	Fucophlorethols	*Sargassum spinuligerum* and *Cystophora torulosa* (R. Brown)	—	Glombitza et al. (1997)
	Fucophlorethols-B	*Laminaria ochroleuca*		Koch et al. (1980);
	Fucodiphlorethol-C and G	*Eisenia bicyclis, Ecklonia cava*	—	Martinez and Castaneda (2013); Ham, Baik, Hyun, and Lee (2007);
	Trifucodiphlorethol A	*Fucus vesiculosus*		Parys et al. (2010)
	Bisfucotriphlorethol A, Bisfucoetraphlorethol A, Bisfucopentahlorethol A, Hydroxyfucodiphlorethol B, Hydroxyfucotriphlorethol B, Bisfucotriphlorethol, Hydroxybisfucopentaphlorethol A and Fucodifucotetraphlorethol	*Cystophora torulosa*	—	Glombitza, Hauperich, and Keusgen (1997)

Sargassum mangarevense from French Polynesia (Stiger et al., 2004), and *Sargassum muticum* along European coasts, from Portugal to Norway (Tanniou, Vandajon, Goncalves et al., 2014; Tanniou, Vandanjon, Incera et al., 2014). Steinberg (1989) reported that polyphenol levels of Australasian kelps and rockweeds often differed significantly among sites, but within a species the highest polyphenol concentration was generally less than twice the lowest concentration. Table 3 presents the seasonal variation of brown algal phlorotannins and illustrated the great variability of phenolic content depending on species, season and location.

Table 3 Polyphenol contents variation in different species of brown algae.

Species	Polyphenols content	Type of variation	References
Cystoseira tamariscifolia	~2% in winter ~8% in summer ~0.1% in autumn ~0.6% in spring	Seasonal	Abdala-Diaz, Cabello-Pasini, Perez-Rodriguez, Conde Alvarez, and Figueroa (2006) Jégou, Kervarec, Cérantola, Bihannic, & Stiger-Pouvreau, 2015
Stypopodium zonale	~0.5% (without herbivore) ~0.7% (with herbivore)	Simulated herbivory	Ank, da Gama Perez, Pereira, da Gama, & Pereira, 2013
	~0.6% (low epibiosis) ~0.6% (medium epibiosis) ~0.4% (high epibiosis)	Epibiosis effects	
Halidrys siliquosa	18.25% in winter	Photoprotection	Le Lann et al., 2016
Fucus serratus *Ascophyllum nodosum* *Bifurcaria bifurcata* *Himanthalia elongata* *Laminaria digitata*	4.27% in lower intertidal zone ~5.8% in lower intertidal zone 3.73% in lower intertidal zone 2.17% in lower intertidal zone	Interspecific	Connan, Goulard, Stiger, Deslandes, & Gall, 2004

(*Continued*)

Table 3 Polyphenol contents variation in different species of brown algae.—cont'd

Species	Polyphenols content	Type of variation	References
F. spiralis *F. vesiculosus* *Pelvetia canaliculata*	0.13% in subtidal zone 3.88% in upper intertidal zone ~5.8% in upper intertidal zone 3.39% in upper intertidal zone		
L. digitata *F. vesiculosus* *H. elongata* *P. canaliculata* *A. nodosum*	0.09% in summer 0.19% in winter 7.84% in spring ~7% in summer 0.49 % in winter 4.84% in summer 3.72% in summer ~3% in winter 7.18% in summer ~6% in winter	Seasonal	Connan et al., 2004
Sargassum cristaefolium *S. aquifolium* *S. polycystum*	0.08 % 0.10–0.71 % 0.09–0.74%	Geographical location with depth and reef geomorphology	Le Lann et al., 2012

The members of brown seaweeds have been extensively studied for their high antioxidant potential as described in many previous studies (Chakraborty, Praveen, Vijayan, & Rao, 2013; Farvin & Jacodsen, 2013; Keyrouz et al., 2011; Kim & Kim, 2010; Le Lann et al., 2012; Mwangi et al., 2013; Nakai, Kageyama, Nakahara, & Miki, 2006; Wang, Wang, & Guan, 2012). One of the bioactive compounds responsible for this activity is the phenolic compound. It was shown that phlorotannins, a group of phenolic compounds, restricted to the polymers of phloroglucinol (1,3,5-trihydroxybenzene) and found only in the member of brown algae, is responsible of a large range of activities (Agatonovic-Kustrin & Morton, 2013; Stiger-Pouvreau et al., 2014).

Strong correlation between the total phenolic content and antioxidant activity in brown algae has been reported by many previous studies (Cox, Gupta, & Abu-Ghannam, 2012; Eom et al., 2011; Harnita et al., 2013; Le Lann, Jégou, & Stiger-Pouvreau, 2008; Namvar et al., 2013; Tanniou et al., 2013; Zubia et al., 2009, 2019). Other studies have revealed that phlorotannins demonstrated a good potential against 1,1-diphenyl 1,2-picryhydrazyl (DPPH), hydroxyl, superoxide and peroxyl radicals (Li et al., 2009; Nakai et al., 2006; Santoso et al., 2013; Siriwardhana, Lee, & Jeon, 2005; Tierney et al., 2012).

Apparently, phlorotannins do not only perform strong antioxidant but also antimicrobial properties (Ahn et al., 2004; Choi, Lee, B Lee, Kim, & Kim, 2014; Eom et al., 2011, 2012; Le Lann et al., 2016; Lee, Lee, Oh, Lee, & Chee, 2010; Lee et al., 2014; Lopes et al., 2012; Plouguerné, Hellio, Deslandes, Véron, & Stiger-Pouvreau, 2008; Tanniou, Vandajon, Goncalves et al., 2014; Tanniou, Vandanjon, Incera et al., 2014). They also perform other biological activities such as anti-proliferative and antitumor (Imbs et al., 2009; Nwosu et al., 2011; Pérez-Larrán, Balboa, Torres, & Domínguez, 2018; Yang, Ham, Lee, Lee, & Hyun, 2013; Zubia et al., 2009); antidiabetic (Azizi, Ekantari, & Husni, 2019; Jung, Yoon, Woo, & Choi, 2008; Nwosu et al., 2011); angiotensin-I-converting enzyme (ACE I) inhibition (Jung, Hyun, Kim, & Choi, 2006), anti-inflammatory (Kim & Kim, 2010); matrix metalloproteinases (MMPs) inhibition (Kim et al., 2006; Thomas & Kim, 2011), hyaluronidase inhibition (Ferreres, Lopes, Gil-izquierdo, & Andrade, 2012), tyrosinase inhibition (Chan, Kim, & Cheah, 2011; Heo et al., 2009; Puspita, 2017; Yoon et al., 2009), photoprotective (Heo et al., 2009; Le Lann et al., 2008; Le Lann et al., 2012; Le Lann et al., 2016) and antiallergic activities (Sugiura, Kunio, & Amano, 2011).

4.4 Other bioactive compounds

Terpenoids are others bioactive compounds that are quite abundant not only in higher plants but also in marine algae (Agatonovic-Kustrin & Morton, 2013). Marine terpenoids show a great promising potential as source for new antioxidant agents in cosmetics preparation (Paduch et al., 2007) due to their low systemic toxicity, low irritation and good-penetration-enhancing abilities. Fucosterol is one of the steroidal terpenoid extracted from several brown algae, such as *Sargassum caryophyllum* (Tang et al., 2002) and *Ecklonia stolonifera* (Jung et al., 2006). This compound has strong antioxidant activity by increasing the concentration of superoxidase dismutase (SOD), catalase and glutathione peroxidase (GSH-px); enzymes responsible in the fine control of cellular H_2O_2 concentration (Lee et al., 2005; Pillai et al., 2005).

Lipid content in marine algae ranges from 0.3% to 6% for macro- and >70% for microalgae. The low content of lipids in macroalgae may not be commercially interesting and valuable target to explore more (Milledge, Nielsen, & Bailey, 2015). Even though macroalgae have low lipid content, their polyunsaturated fatty acids (PUFAs) can be as high or even higher as that of land plants (El-Shoubaky et al., 2008). Palmitic acid (C16:0) was the major fatty acid in *Stoechospermum marginatum* as reported by Narayan, Kazuo, and Masashi (2004) as well as in *S. polycystum* (Matanjun, Mohamed, Mustapha, & Muhammad, 2009). Most of the PUFAs have good antioxidant activities (Henry et al., 2002). A number of fatty acids have a specific function in restoring the permeability barrier, and preventing scaly dermatitis and skin dehydration associated with the lack of PUFAs in the skin.

Sterols of *Sargassum* species have shown to have anti-atherosclerotic. Then, their saringosterol shows potent activity in lowering the cholesterol (Chen et al., 2014). Thunberol, a new sterol isolated from *S. thunbergii* exhibits significant inhibitory effect against protein tyrosinase phosphatase 1B. It indicates that they are potential for the treatment of Type II-diabetes and obesity (He, YaoLiu, & Guo, 2014). There are many other bioactive compounds reported in *Sargassum* species, such as glycolipids (Plouguerné et al., 2010), farnesylacetones, dipeptides, loliodilides. However, compared to *Sargassum* polysaccharides, phlorotannins and meroterpenoids, terpenoids are less studied (Liu et al., 2012).

5. Potential applications of *Sargassum*

Sargasso Sea Alliance has revealed that there are about 90 patents referencing *Sargassum,* for example, WO1991007946 A1 for application of *S. muticum* extract in pharmaceuticals, cosmetic and foods. Various industrial, medical and nutritional uses are proposed including applications focused on inhibiting HIV infection, as an antibiotic, antifungal and antifouling substance, biofuel, fertilizer and animal feeds (Laffoley et al., 2011). Several studies on bioactive compounds as well as their potentiality in several domains have made them a valuable source to explore. The rapid progress of biotechnology research related to seaweed bioactive compounds seems to change the previous status of *Sargassum* in Indonesia, from extracts with high volumes to high added value molecules with small volume.

5.1 Bioenergy

Some countries have started considering exploiting the massive biomass of *Sargassum*. For example, Japan has recognized *Sargassum* as a source of sustainable bioenergy by converting alginate, mannitol and fiber into ethanol, butanol, and others (Aizawa, Asaoka, Atsumi, & Sakou, 2007). Exploitation of *Sargassum* biomass is considered one of the best options for increasing renewable energy due to its massive influx along the coast. Bioethanol production from seaweeds in Japan, entitled the "Ocean Sunrise Project," aims to produce seaweed-based bioethanol by farming and harvesting *S. horneri*, utilizing 4.47 million km^2 (the sixth largest in the world) of unused areas of the Exclusive Economic Zone (EEZ) and maritime belts of Japan (Aizawa et al., 2007).

Potential of *Sargassum* as source of bioenergy continues to be reported in many studies by using different techniques. Despite the massive biomass of *Sargassum* that is available, the selection of appropriate technique is crucial in order to obtain higher yield yet with promising and more efficient technique. Production of ethanol from the crude laminarin of *Sargassum* sp. through saccharification treatments combined with cellulase and laminarinase has been reported by Perez, Pajares, Alcantara, and Simbahan (2018). Bio-oil production from *Sargassum* sp. using hydrothermal liquefaction has also been reported and seems to be promising process since it uses less energy to produce biofuel (Rahbari, Akram, Pazoki, & Aghbashlo, 2019).

5.2 Animal feed and health

As *Sargassum* contains interesting proteins, fiber, macro- and micro-minerals (especially potassium, nitrates, calcium, iron and PUFAs), this genus has been targeted as alternative source of animal feeds. Dry material of *Sargassum* integrated as supplements in animal feeds could improve the growth rate, survival rate and immune system of cultivated animals as reported in shrimp *Litopenaeus vannamei, Fenneropenaeus chinensis,* and rainbow trout *Oncorhynchus mykiss* (Hafezieh, Ajdari, & Hosseini, 2014; Huynh et al., 2011; Sudaryono, Aditomo, & Isnansetyo, 2015; Zamannejad, Emadi, & Hafezieh, 2016).

Mushollaeni and Endang Rusdiana (2012) reported that alginates extracted from *Sargassum duplicatum* originated from the South coast of Yogyakarta (Indonesia) that was applied to cake as an emulsifier and stabilizers produced a smooth texture and prevented the cake from hardening. Some extracts rich in alginates obtained from *Sargassum* sp. introduced to animal feed had increased the phagocytosis activity of the shrimp *L. vannamei*. Thus, alginates contribute to the defense system of shrimp (Ridlo & Pramesti, 2005). Moreover, interested by phlorotannins, Widowati, Susanto, Puspita, Stiger-Pouvreau, and Bourgougnon (2012) demonstrated the interest of *S. polycystum, S. duplicatum* and *Sargassum echinocarpum* from Jepara, Central Java as a source of antibacterial and antioxidant compounds.

5.3 Fertilizer and bio-stimulants

As fertilizer, algae have been applied to soils as manure and as conditioning agents (Hong, Hien, & Son, 2007, pp. 817—826). Extract of *Sargassum* also stimulates the germination, improve the growth and biochemical constituents and enhance the immunity of certain plants, like lentil (*Vigna mungo*), mung bean (*Vigna radiata*), mustard (*Brassica juncea* L.), garden cress (*Lepidium sativium*), and tomato (*Solanum lycopersicum* L.) (Godlewska, Michalak, Tuhy, & Chojnacka, 2016; Hernández-Herrera, Santacruz-Ruvalcaba, Ruiz-López, Norrie, & Hernández-Carmona, 2014; Kalaivanan & Venkatesalu, 2012; Kavipriya, Dhanalakshmi, Jayashree, & Thangaraju, 2011; Siregar, Lubis, & Supriyadi, 2014).

5.4 Cosmetic applications

Regarding the existing products, *Sargassum* extracts have been commercially formulated in cosmetic products. As an example, the CAS Number 223751-78-8 states that *S. filipendula* extract is classified as a skin protective ingredient. *S. filipendula* extract has been introduced as one of the compositions in clay mask and hand lotion by the NU SKIN®.

Table 4 summarizes several researches regarding the potential of Indonesian *Sargassum* bioactive compounds from the past few years.

6. *Sargassum* exploitation

Despite the massive biomass of *Sargassum,* this genus has been under the radar for so many years. *Sargassum* production has not been as significant as the other commercial algae as previously mentioned. Therefore, detail on *Sargassum* production is highly scarce. The stranded massive biomass of *Sargassum* in Indonesia by certain coastal communities is used as fodder for the ruminant. Economic value of *Sargassum* is not as interesting as the red algae since local communities are not yet aware the remarkable potential hidden by this genus. Until present day, the price of *Sargassum* dry material remains in 0.25 US $, approximately (Puspita, Pers. Comm).

Their exploitation is still limited to their raw material either fresh or dried collected from wild stocks since *Sargassum* has not been cultivated yet. Coastal community, especially fishermen, considers this species as 'weeds'

Table 4 List of *Sargassum* bioactive compounds presenting potential bioactivities and collected from different locations in Indonesia.

No	Species	Location	Bioactive compounds	Potential	References
1	*S. fillipendula*	Talango Island, Sumenep Madura	Polyphenol	Antioxidant	Khotimah, Darius, & Sasmito, 2013
2	*S. duplicatum*		Polyphenol	Anti-inflammatory Bowl Disease	Aulanni'am, Rosdiana, & Rahmah, 2011
3	*S. echinocarpum*		Polyphenol	Antidiabetic	Firdaus and Astawan, 2010
4	*S. duplicatum*	Pantai Raca, Nusa Kambangan	Polyphenol	Antioxidant	Septiana & Asnani, 2013
5	*S. crassifolium*	Sumba Island	Polyphenol	Antioxidant	Supriyono, 2007
6	*S. hystrix*	South Coast of Yogyakarta	Crude phlorotannin	Antioxidant	Harnita et al., 2013
7	*Sargassum* sp.	Gresik	Polyphenol	Antibacterial	Bachtiar, Tjahjaningsih, & Sianita, 2012
8	*S. binderi*	Gunung Kidul, Yogyakarta	Alginate	Antidiabetic	Mushollaeni & Rusdiana, 2013
9	*Sargassum* sp.			Decreasing malondialdehide level	Fauziah & Mahdi, 2013
10	*S. crassifolium, S. polycystum, S. echinocarpum, S. duplicatum, S. binderi*	Craggy Coast of Gunung Kidul, Yogyakarta	Alginate	Emulsifier and stabilizer	Mushollaeni & Endang Rusdiana, 2012
11	*S. polycystum, S. duplicatum, S. echinocarpum*	Teluk Awur, Jepara	Polyphenol	Antioxidant and antibacterial	Widowati et al., 2014

(Continued)

Table 4 List of *Sargassum* bioactive compounds presenting potential bioactivities and collected from different locations in Indonesia.—cont'd

No	Species	Location	Bioactive compounds	Potential	References
12	*S. filipendula, S.duplicatum, S. crassifolium, S. binderi*	Sumenep, Madura	Polyphenol	Antioxidant	Bambang, Kumalaningsih, Susinggih, & Hardoko, 2013
13	*S. polyceratium, S. angustifolium, S. filipendula*	Jepara, Teluk Awur	Polyphenol	Antioxidant	Budhiyanti, Raharjo, Marseno, & Lelana, 2012
14	*S. hystrix*	Gunung Kidul, Yogyakarta	Polyphenol	Free radical scavenging, metal chelating and singlet oxygen quenching	(Budhiyanti, Raharjo, Marseno, & Lelana, 2011)
15	*Sargassum* sp.	Geger Beach and Sawangan Nusa Dua Beach	Polyphenol	Antibacterial and antifungal	(Hutasoit, Suada, & Susrama, 2013)
16	*S. cristaefolium*	Sumenep Beach, Madura	Polyphenol	Antioxidant	(Lailiyah, Adi, Hakim, & Yusnawan, 2014)
17	*S. echinocarpum*		Protein	Antibacterial and antifungal	(Arifudin et al., 2001)
18	*S. aquifolium, S. ilicifolium, S. polycystum*	Telukawur and Panjang Island, Jepara	Polyphenol	Antioxidant and antibacterial	(Puspita et al., 2012)

particularly when they grow in the same sites as the aquaculture activity. Thus, the presence of *Sargassum* will harm the cultivated species since they might block the penetration of light caused by their floating thallus. Furthermore, their thallus might get stuck in the small boat propeller and provoke a serious damage to the machine since the thallus can grow more than 2 m long. These are common problems experienced by fishermen, notably those who perform the aquaculture activity. To our knowledge, alginate industry in Indonesia has not yet well developed (Setyawidati, 2017).

Certain small industries start to take interest in *Sargassum*. They have begun to process *Sargassum* into valuable products mostly from processed dry material. Fertilizer and bio-stimulant have been mentioned, as the well-known featured product from *Sargassum* and it is just one step closer to industrialization. Another outstanding derivative product from *Sargassum* is alginate. Setyawidati et al. (2018) estimated the potential alginate stock within the Ekas Bay (Lombok, Indonesia) and Sargassaceae species represents a timely source of alginates. Nevertheless, based on the information obtained from local industry that has been working with *Sargassum*, the cost-production to process *Sargassum* to alginate is expensive. As a consequence, it is yet impossible to commercialize alginate production. Until present, the acknowledgments of *Sargassum*-based products basically come from the anglers or coastal communities that perform aquaculture activity. As a result, the need of alginate in Indonesia for food industries depends on foreign market such as China. One of local industry that requires alginate is the 'Batik' textile industry. Alginate is used to bind the coloring agent for Batik. Unfortunately, there has not been any recorded data officially published by related authorities regarding the exact amount of alginate that is imported and used in Indonesia for this industry.

7. Challenge for *Sargassum* sustainable development

The genus *Sargassum* is ubiquitous and pseudo-perennial, which means that thalli are present along the year. However, considering the industrialization of *Sargassum,* massive biomass from natural populations might not be adequate in certain extent. This genus might be perennial, however, the abundance of some species is seasonal (Setyawidati et al., 2018). Therefore, aquaculture is surely the answer if this genus is meant for commercial purpose. It is to anticipate the high demand in the future. Since the aquaculture of *Sargassum* has not been performed in Indonesia, particularly for the commercialization, it is important to comprehend not only about the aquaculture but also about the species as well, especially about the life cycle, the reproduction process and the ecology. Moreover, to achieve a sustainable exploration, integration and collaboration of multi-disciplinary groups of biologists, chemists, engineers and oceanographer is crucial (Pomponi, 1999). Furthermore, governmental supports as well as reliable collaboration of inter-institutional research are also required (Dewi, Tarman, & Uria, 2008).

Government might support through the infrastructural development, the provision of required technology and regulations or policies. Recently, government support for algae exploitation through the Ministry of Marine and Fisheries has announced their new policy that will be implemented in 2018. They encourage all related stakeholders to gradually decrease the export of algae raw material instead they have to transform them into products. The purpose is to increase the economic value of Indonesian macroalgae. The Ministry will facilitate this decision by constructing new factories and by providing the required technology to support the production of seaweed-based products. Akrim (2014) stated that in the interest of developing the coastal community, the Indonesian government has come with some strategic programs, which are (1) science and technology based seaweed industry, (2) conservation development, (3) integrated approach, (4) community development and (5) close cooperation between government and stakeholders. In addition, based on presidential directive order, Inpres No. 7 Tahun 2016 (Instruction of Indonesian President No. 7 / 2016), macroalgae are considered as one of national commodities included for further development. This presidential directive order emphasizes the acceleration of fisheries industries through the improvement of fisheries production, cultivation and fisheries processing.

In addition, an understanding and reliable collaboration between government, universities/institutes and stakeholders - privates sectors or industries, will bring remarkable output for the interest of marine algae bioprospecting. In the end, comprehensive and firm regulations will cover everything mentioned above as well as protect the future prospect of marine algal bioprospecting, regulations that include the Intellectual Property Right and Patent Law as well as conservation issues (Dewi et al., 2008; Siswandi, 2013).

To develop seaweed industries in Indonesia, it is important to assess and analyze the strength weakness, opportunity and threats influencing the global value chain. It might further serve as basic information for the involved parties to issue feasible decisions in the interest of developing seaweed industries. To summarize, the strength, the weakness, the opportunity, and the threat in seaweed value chain are presented in Table 5.

Table 5 Balance sheet in terms of strength, weakness, opportunity and threat of Indonesian seaweeds value chain, issued of an analysis of scientific articles (Duijn, Beukers, & Pijl, 2012; Mulyati & Geldermann, 2016; Neish, 2007; Ngabalin, 2013; Sunadji et al., 2014; Wibowo, 2014).

Strength	**Weakness**
1. Quality of seed	1. Skill
2. Labor	2. Availability of seed
3. Production of seaweed	3. Quality of seaweed
4. Quality of area	4. Post-harvest handling
5. Area width	5. Infrastructure
6. Capital	6. Education
7. Seaweed marketing	7. Competitiveness
8. Diverse end products	8. Technology and innovation
9. Respectful attitude toward the environment	9. Poor spatial planning
10. As national commodity with high economic value, Algae could maintain the unity of Indonesia as an archipelagic state	10. Lack of strategic market projection
11. Seaweed's farmers upholds the Traditional Rights law conserving surrounding ecosystem	11. Domestic demand regarding Algae products is still low
	12. Lack of well-qualified human resources
	13. Characteristic of seaweed cultivation is not based on the corporate system
Opportunity	**Threat**
1. Amounts of merchants	1. Price of medicaments
2. Institutional	2. Education costs
3. Roles of government	3. Health costs
4. Affordable technology for seaweed cultivation	4. Food prices
5. High market demand	5. Seaweed price
6. Hydrocolloid industry wants cultivated material	6. Other cultivators
7. IMTA well developed in Asia	7. Regulations
8. Growing demand for seaweed	8. Disease
9. Societal interest in cooking, new ingredients, functional foods, healthy products and nutraceuticals	9. Climate change
10. Cultivated macroalgae is considered as one of the largest un-exploited global biomass resources for a sustainable production of food and replacement of fossil industries	10. Pollution from mainland
	11. Competition with marine eco-tourism
	12. There is no guarantee of sustainability raw material supply

Good quality of seed, post-harvest handling, infrastructure or access, and technology are the most up-rizing issues affecting the production of Indonesian seaweeds. As an example, in the case of red macroalgae *Kappaphycus* and *Eucheuma,* the quality and quantity of dry material produced over decades have been declining leading to an unstable price. Over decades, seaweed farming in Indonesia has been mostly conducted in distant or even remote islands, thus, the transportation cost to bring the seaweeds to mainland are quite high. In addition, the characteristics of seaweed farming in Indonesia are not based on 'company-farming' but more 'traditional and family farming' where the farmers themselves are the owner of their farm. Therefore, the farmers use the cheapest and simplest method since technology is not an option as it costs a fortune for them.

In order to perform bioprospecting based program, all parts involved, *i.e.* government, universities/institutes and other stakeholders as well as media, need to cooperate in order to overcome the challenges that might be faced during the process of marine bioprospecting. As shown with Fig. 7, each involved part has its own role to play and need to coordinate and communicate in the process. In addition, the government will also have to protect every step of the marine algal bioprospecting through their policies and regulations.

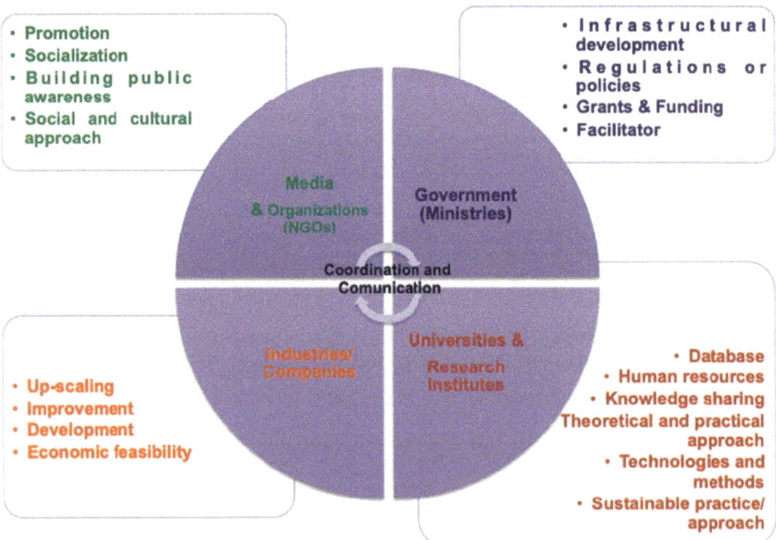

Fig. 7 Inter-linkages between government, institutes/universities and other stakeholders in marine bioprospecting.

8. Legal status of marine bioprospecting in Indonesia

The significance of the ocean to humankind is definitely indisputable. Experts have estimated that marine biological diversity is much higher than tropical rain forests, particularly in certain marine ecosystems, like coral reefs and deep coral floor (Reaka-Kudla, 1997). In addition, marine organisms, and in particular marine algae, are exclusive since evolution has equipped them with appropriate mechanisms to survive and develop complex biological and chemical mechanisms as their defense systéms (Lalegerie et al., 2019; Pawlik et al., 2013). As a consequence, it is not astonishing that marine species have been center of attention for many years especially for their chemical compounds as source of natural products.

From 1963, there have been 24.662 new compounds that have been isolated (Blunt et al., 2015). At least 300 patents have been issued on marine natural products (Demunshi & Chugh, 2009; Hunt and Vincent, 2006). Still from the same authors, it was reported that as for algae, there have been 522 patents carried out in WIPS® for pharmaceutical, nutraceutical, cosmeceutical, medicinal research, food additives and beverages (182 patents) and others (111 patents). Fig. 8 presents four phases that have to be followed before selecting a candidate, i.e. *Sargassum* species in our case, with market potential. In Indonesia, all researches concerning the potential of *Sargassum* have actually reached the third phase and remain in this phase. In other

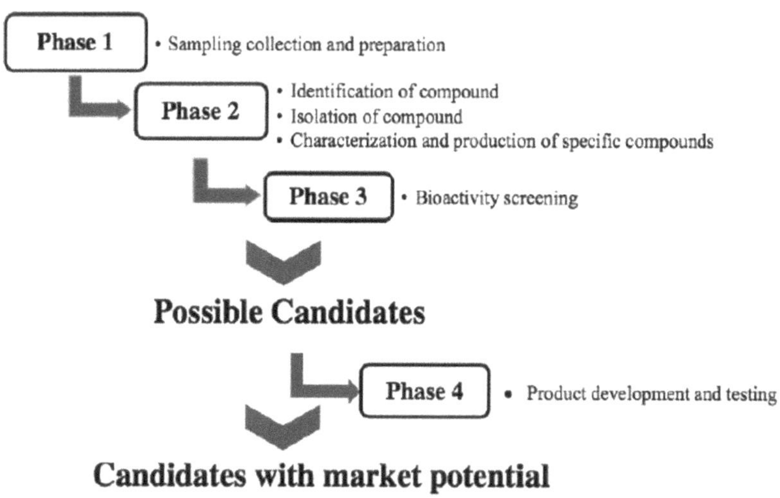

Fig. 8 Illustration of bioprospecting process based on Bhatia and Chugh (2015).

word, the continuation of phase 3 to the phase 4 where product development and testing are scarce. It is mostly due to the lack of financial support and necessary technology to develop the designated product.

In legal aspect, compared to bioprospecting activity in terrestrial organisms, the exploration and exploitation of marine biodiversity is particularly unique. This is due to the necessity to consider different legal regimes across various maritime zones established under the framework of the maw of the sea (Siswandi, 2013). The United Nations Convention on Biological Diversity (CBD) regulating the utilization of genetic resources does not contain particular provisions with regard to marine genetic resources. Then, the United Nations Convention on the Law of the Sea (LOSC) does not explicitly address marine bioprospecting as a part of it (Siswandi, 2013). Thus, legal agreement mostly applied to marine bioprospecting activity is the *access and benefit-sharing* (ABS) regime as contained in the CBD (Dewi et al., 2008; Martinez and Biber-Klemm, 2010). This issue illustrates the details about access to genetic resources, technology transfer, exchange of information, technical and scientific co-operation, handling the biotechnology and the benefit distribution, financial resources and mechanisms (Arico & Sapin, 2005). As in Indonesia, beside the implementation of ABS regime, current institutional framework will fall under the *Research and Development Permits Regulation* as issued by the Ministry of Research and Technology (Siswandi, 2013).

9. Conclusion

Over the past few years, Indonesian *Sargassum* species have gradually changed its status from merely "weeds" to "valuable assets" with many hidden potential waiting to be further revealed. Coastal communities benefit from part of the algal biomass by transforming it into products such as fodder and tea. However, their production is still limited, and based on home production. It is more realistic since the fact has revealed that *Sargassum* potential is more than being a source of alginate.

In term of marine algae bioprospecting, Indonesian *Sargassum* is still far behind the two red genera *Kappaphycus* and *Gracilaria* either from their production or economic value. Nevertheless, their promising potential either in the food, cosmetic, or pharmacology industries, will later put them in the same grade as these two algae. Many researches have been conducted related to the potential of *Sargassum*. Bioprospecting based on sustainable use of marine biodiversity of this genus is important to be acknowledged and considered, so the futures commercial use of *Sargassum* species will not only bring economic benefits but will also promote the protection and conservation of the Indonesian biodiversity.

Acknowledgment

This study was a collaboration research between Université Bretagne Sud (UBS), Vannes, France and Diponegoro University (UNDIP), Semarang, Indonesia under Double-Degree program supported by the Directorate of Higher Education (DIKTI), together with the Université de Bretagne Occidentale (UBO). The author would like to thank all members of the Laboratoire de Biotechnologie et Chimie Marines (LBCM) who had been very helpful and supporting during the research. Authors would like to express their gratitude to Indonesian Seaweed Association (ARLI) and Indonesian Seaweed Industries Association (ASTRULI) for several data that they provided.

References

Abdala-Diaz, R. T., Cabello-Pasini, A., Perez-Rodriguez, E., Conde Alvarez, R. M., & Figueroa, F. L. (2006). Daily and seasonal variations of optimum quantum yield and phenolic compounds in *Cystoseira tamariscifolia* (Phaeophyta). *Marine Biology, 148*, 459–465.

Adhikari, U., Mateu, C. G., Chattopadhyay, K., Pujol, C. A., Damonte, E. B., & Ray, B. (2006). Structure and antiviral activity of sulfated fucans from *Stoechospermum marginatum*. *Phytochemistry, 67*, 2474–2482.

Agatonovic-Kustrin, S., & Morton, D. W. (2013). Cosmeceuticals derived from bioactive substances found in marine algaes. *Journal of Oceanography and Marine Research, 1*, 1–11.

Ahn, M. J., Yoon, K. D., Min, S. Y., Lee, J. S., Kim, J. H., Kim, T. G., et al. (2004). Inhibition of HIV-1 reverse transcriptase and protease by phlorotannins from the brown alga *Ecklonia cava*. *Biological and Pharmaceutical Bulletin, 27*, 544–547.

Aizawa, M., Asaoka, K., Atsumi, M., & Sakou, T. (2007). Seaweed bioethanol production in Japan – the ocean sunrise project. In *Oceans conference record (IEEE)* (pp. 1–5). IEEE.

Akrim, D. (2014). The seaweed business development in Indonesia. *Journal of Shipping and Ocean Engineering, 4*, 198–202.

Alekseyenko, T. V., Zhanayeva, S. Y., Venediktova, A. A., Zvyagintseva, T. N., & Kuznetsova, T. A. (2007). Antitumor and antimetastatic activity of fucoidan, a sulfated polysaccharide isolated from the Okhotsk sea *Fucus evanescens* brown algae. *Bulletin of Experimental Biology and Medicine, 143*, 730–732.

Ale, M., & Meyer, A. (2013). Fucoidans from brown seaweeds: An update on structures, extraction techniques and use of enzymes as tools for structural elucidation. *RSC Advances, 3*, 8131–8141.

Ananthi, S., Raghavendran, H. R., Sunil, A. G., Gayathri, V., Ramakrishnan, G., & Vasanthi, H. R. (2010). In vitro antioxidant and in vivo anti-inflammatory potential of crude polysaccharide from *Turbinaria ornata* (marine brown alga). *Food and Chemical Toxicology, 48*, 187–192.

Ang, P. O. (2006). Phenology of *Sargassum* spp. in Tung Ping Chau marine park, Hong Kong SAR, China. *Journal of Applied Phycology, 18*, 629–636.

Ank, G., da Gama Perez, B. A., Pereira, R. C., da Gama, B. A. P., & Pereira, R. C. (2013). Polyphenols from *Stypopodium zonale* (Phaeophyceae): Intrapopulational variation, induction by simulated herbivory, and epibiosis effects. *Aquatic Botany, 111*, 125–129.

Apoya, M., Ogawa, H., & Nanba, N. (2002). Alginate content of farmed *Undaria pinnatifida* (Harvey) Suringar from the three bays of Iwate, Japan during harvest period. *Botanica Marina, 45*, 445–452.

Arico, S., & Sapin, C. (2005). *Bioprospecting of genetic resources in the deepsea bed: Scientific, legal and policy aspects*. UNU-IAS Report. United Nations University. Japan, 76 pp.

Arifudin, Patong, R., & Ahmad, A. (2001). The investigation of bioactive proteins in macroalgae as antibacterial and antifungus agents. *Mar Chim Acta, 2*, 11–18.

Arnold, T. M., Tanner, C. E., & Hatch, W. I. (1995). Phenotypic variation in polyphenolic content of the tropical brown alga *Lobophora variegata* as a function of nitrogen availability. *Marine Ecology Progress Series, 123*, 177–184.

Arnold, T. M., & Targett, N. M. (2000). Polyphenolics in tropical brown algae. *Journal of Chemical Ecology, 26*, 1393–1410.

Arnold, T. M., & Targett, N. M. (2002). Mini review marine tannins : The importance of a mechanistic framework for predicting ecological roles. *Journal of Chemical Ecology, 28*, 1919–1934.

Athukorala, Y., Lee, K. W., Kim, S. K., & Jeon, Y. J. (2007). Anticoagulant activity of marine green and brown algae collected from Jeju island in Korea. *Bioresource Technology, 98*, 1711–1716.

Atmadja, W. S., & Prudhomme van Reine, W. F. (2014). *Checklist of the seaweed specie biodiversity of Indonesia with their distribution and classification: Green algae (Chlorophyta) and brown algae (Phaeophyceae, Ochrophyta) Indonesia*. Naturalis Biodiversity Center, Indonesian Institute of Sciences (LIPI), 59 pp.

Aulanni'am, Rosdiana, A., & Rahmah, N. L. (2011). Potency of brown seaweed (*Sargassum duplicatum* Bory) ethanol and ethyl acetic fraction to malondialdehyde concentration decreasing and histological retriveal of IBD (inflammatory bowel disease) rat small intestinal jejunum. *Journal of Ilmiah Kedokteran, 4*, 57–64.

Azizi, W. A., Ekantari, N., & Husni, A. (2019). Inhibitory activity of Sargassum hystrix extract and its methanolic fractions on inhibiting α-glucosidase activity. *Indonesian Journal of Pharmacy, 30*, 36.

Bachtiar, S. Y., Tjahjaningsih, W., & Sianita, N. (2012). Pengaruh ekstrak alga coklat (*Sargassum* sp.) terhadap bakteri *Eschericia coli* (effect of *Sargassum* sp. extracts towards *Eschericia coli*). *Journal of Marine and Coastal Sciences, 1*, 53–60.

Balboa, E. M., Conde, E., Moure, A., Falque, E., & Dominguez, H. (2013). In vitro antioxidant properties of crude extracts and compounds from brown algae. *Food Chemistry, 138*, 1764–1785.

Bambang, B. S., Kumalaningsih, S., Susinggih, W., & Hardoko. (2013). Polyphenol content and antioxidant activities of crude extract from brown algae by various solvents. *Journal of Life Science and Biomedicine, 3*, 439–443.

Barsanti, L., & Gualtieri, P. (2006). *Algae anatomy, biochemistry, and biotechnology*. Boca Raton: Taylor and Francis, 361 pp.

Bedoux, G., Hardouin, K., Burlot, A. S., & Bourgougnon, N. (2014). Bioactive components from seaweed: Cosmetics applications and future development. *Advances in Botanical Research, 71*, 345–378.

Bhatia, P., & Chugh, A. (2015). Role of marine bioprospecting contracts in developing access and benefit sharing mechanism for marine traditional knowledge holders in the pharmaceutical industry. *Global Ecology and Conservation, 3*, 176–187.

Bourgougnon, N., & Stiger-Pouvreau, S. (2012). Chemodiversity and bioactivity within red and brown macroalgae along the French coasts, metropole and overseas departments and territories. In S. K. Kim (Ed.), *Handbook of marine macroalgae: Biotechnology and applied phycology* (1st ed.). United Kingdom: Wiley-Blackwell, 557 pp.

Budhiyanti, S. A., Raharjo, S., Marseno, D. W., & Lelana, I. Y. B. (2011). Free radical scavenging, metal chelating and singlet oxygen quenching activity of fractionated brown seaweed *Sargassum hystrix* extract. *Journal of Biological Sciences*, 288–298.
Budhiyanti, S. A., Raharjo, S., Marseno, D. W., & Lelana, I. Y. B. (2012). Antioxidant activity of brown algae *Sargassum* species extract from the coastline of Java island. *American Journal of Agricultural and Biological Sciences, 7*, 337–346.
Buschmann, A. H., Camus, C., Infante, J., Neori, A., Israel, Á., Hernández-González, M. C., et al. (2017). Seaweed production: Overview of the global state of exploitation, farming and emerging research activity. *European Journal of Phycology, 52*, 391–406.
Cardozo, K. H. M., Guaratini, T., Barros, M. P., Falcao, V. R., Tono, A. P., Lopes, N. P., et al. (2007). Metabolites from algae with economical impact. *Comparative Biochemistry & Physiology, 146*, 60–78.
Chakraborty, K., Praveen, N. K., Vijayan, K. K., & Rao, G. S. (2013). Evaluation of phenolic contents and antioxidant activities of brown algae belonging to *Turbinaria* spp. (Phaeophyta, Sargassaceae) collected from Gulf of Mannar. *Asian Pacific Journal of Tropical Biomedicine, 3*(1), 8–16.
Chandia, N. P., Matsuhiro, B., Mejias, E., & Moenne, A. (2004). Alginic acids in *Lessonia vadosa*: Partial hydrolysis and elicitor properties of the polymannuronic acid fraction. *Journal of Applied Phycology, 16*, 127–133.
Chan, Y. Y., Kim, K. H., & Cheah, S. H. (2011). Inhibitory effects of *Sargassum polycystum* on tyrosinase activity and melanin formation in B16F10 murine melanoma cells. *Journal of Ethnopharmacology, 137*, 1183–1188.
Chattopadhyay, N., Ghosh, T., Shinha, S., Chattodpadhyay, K., Karmakar, P., & Ray, B. (2010). Polysaccharides from *Turbinaria conoides*: Structural features and antioxidant capacity. *Food Chemistry, 118*, 823–829.
Chen, B., & Zou, D. (2014). Growth and photosynthetic activity of *Sargassum henslowianum* (Fucales, Phaeophyta) seedlings in responses to different light intensities, temperatures and CO_2 levels under laboratory conditions. *Marine Biology Research, 10*, 1019–1026.
Choi, J. S., Lee, K. Y., Lee, B. B., Kim, Y. C., & Kim, Y. D. (2014). Antibacterial activity of phlorotannins dieckol and phlorofucofuroeckol-A from botanical sciences *Ecklonia cava* against *Propionibacterium acnes*. *Botanical Sciences, 92*(3), 425–431.
Chojnacka, K., Saeid, A., Witkowska, Z., & Tuhy, Ł. (2012). Biologically active compounds in seaweed extracts – the prospects for the application. *The Open Conference Proceedings Journal, 3*, 20–28.
Connan, S., Goulard, F., Stiger, V., Deslandes, E., & Gall, E. A. (2004). Interspecific and temporal variation in phlorotannin levels in an assemblage of brown algae. *Botanica Marina, 47*, 410–416.
Costa, L. S., Fidelis, G. P., Cordeiro, S. L., Oliveira, R. M., Sabry, D. A., Câmara, R. B. G., et al. (2010). Biological activities of sulfated polysaccharides from tropical algae. *Biomedicine & Pharmacotherapy, 64*, 21–28.
Cox, S., Gupta, S., & Abu-Ghannam, N. (2012). Effect of different rehydration temperatures on the moisture, content of phenolic compounds, antioxidant capacity and textural properties of edible Irish brown seaweed. *Food Science and Technology, 47*, 300–307.
Critchley, A. T., Peddemors, V., & Pienaar, R. (1991). Reproduction and establishment of *Sargassum heterophyllum* (Turner) c. Ag.(phaeophyceae, Fucales). *European Journal of Phycology, 26*, 303–314.
Cumashi, A., Ushakova, N. A., Preobrazhenskaya, M. E., D'Incecco, A., Piccoli, A., Totani, I., et al. (2007). Comparative study of anti-inflammatory, anticoagulant, antiangiogenic, and antiadhesive activities of nine different fucoidans from brown algae. *Glycobiology, 17*, 541–552.

DeBoer, J. A., & Ryther, J. H. (1977). Potential yields from a waste-recycling algal mariculture system. In R. Krauss (Ed.), *The marine plant biomass of the Pacific Norwest coast* (pp. 231–249). Corvallis, Oregon: Oregon State University Press.

Demunshi, Y., & Chugh, A. (2009). Patenting trends in marine bioprospecting based pharmaceutical sector. *Journal of Intellectual Property Rights, 14*, 122–130.

Deniaud-Bouët, E., Hardouin, K., Potin, P., Kloareg, B., & Hervé, C. (2017). A review about brown algal cell walls and fucose-containing sulfated polysaccharides: Cell wall context, biomedical properties and key research challenges. *Carbohydrate Polymers, 175*, 395–408.

Dewi, A. S., Tarman, K., & Uria, A. R. (2008). Marine natural products : Prospects and impacts on the sustainable development in Indonesia. In *Proceeding of Indonesian students' scientific meeting* (pp. 54–63).

Dore, C. M., Faustino Alves, M. G. D. C., Will, L. S., Costa, T. G., Sabry, D. A., De Souza Rego, L. A., et al. (2013). A sulfated polysaccharide, fucans, isolated from brown algae *Sargassum vulgare* with anticoagulant, antithrombotic, antioxidant and anti-inflammatory effects. *Carbohydrate Polymers, 91*, 467–475.

Draget, K., Smidsrød, O., & Skjåk-Bræk, G. (2005). Alginates from algae. *Biopolymers*, 1–30.

Van Duijn, A. P., Beukers, R., & Van Der Pijl, W. (2012). *The Indonesian seafood sector a value chain analysis*. Report. Wagenigen, 72 pp.

El-Shoubaky, G. A., Moustafa, A. M. Y., & Salem, E. A. E. (2008). Comparative phytochemical investigation of beneficial essential Fatty Acids on a variety of marine seaweeds algae. *Research Journal of Phytochemistry, 2*, 18–26.

Eom, S. H., Kang, Y. M., Park, J. H., Yu, D. U., Jeong, E. T., Lee, M. S., et al. (2011). Enhancement of polyphenol content and antioxidant activity of brown alga *Eisenia bicyclis* extract by microbial fermentation. *Journal of Fisheries and Aquatic Sciences, 14*, 192–197.

Eom, S. H., Kim, Y. M., & Kim, S. K. (2012). Antimicrobial effect of phlorotannins from marine brown algae. *Food and Chemical Toxicology, 50*, 3251–3255.

Ertesvag, A. G. H., Valla, S., & Skjak-Brak, G. (1996). Genetics and biosynthesis of alginates. *European Carbohydrate, 14*, 6–13.

FAO. (2014). *The state of world fisheries and aquculture 2014*. Rome. 223 pp.

FAO. (2018). *The state of world fisheries and aquaculture 2018 – meeting the sustainable development goals*. Rome. Licence: CC BY-NC-SA 3.0 IGO. 227 pp.

Farvin, K. H. S., & Jacodsen, C. (2013). Phenolic compounds and antioxidant activities of selected species of algae from Danish coast. *Food Chemistry, 138*, 1670–1681.

Fauziah, F., & Mahdi, C. (2013). A Study on brown seaweed therapy (*Sargassum* sp.) toward MDA levels and histological improvement on rat foot suffering rheumatoid arthritis. *Journal of Pure and Applied Chemistry Research, 2*, 102–107.

Ferreres, F., Lopes, G., Gil-izquierdo, A., & Andrade, P. B. (2012). Phlorotannin extracts from Fucales characterized by inhibitory capacity and antioxidant properties. *Marine Drugs, 10*, 2766–2781.

Firdaus, M., Astawan, M., Muschtadi, D., Wresdiyati, T., Waspadji, S., & Karyono, S. K. (2010). Prevention of endothelial dysfunction in streprozotocin-induced diabetic rats by *Sargassum echinocarpum* extract. *Medical Journal of Indonesia, 19*, 32–35.

Fujii, Y., Tanaka, R., Miyake, H., Tamaru, Y., Ueda, M., & Shibata, T. (2013). Evaluation for antioxidative properties of phlorotannins isolated from the brown alga *Eisenia bicyclis*, by the H-ORAC method. *Food and Nutrition Sciences, 4*, 78–82.

Glombitza, K. W., & Gerstberger, G. (1985). Phlorotannins with dibenzodioxin structural elements from the brown alga *Eisenia arborea*. *Phytochemistry, 24*, 543–551.

Glombitza, K. W., & Grosse-Damhues, J. (1985). Antibiotics from algae XXXIII1: Phlorotannins of the brown alga *Himanthalia elongata*. *Planta Medica, 51*, 42–46.

Glombitza, K. W., Hauperich, S., & Keusgen, M. (1997). Phlorotannins from the brown algae *Cystophora torulosa* and *Sargassum spinuligerum*. *Natural Toxins, 5*, 58–63.

Glombitza, K., & Keusgen, M. (1995). Fuhalols and deshydroxyfuhalols from the brown alga *Sargassum spinuligerum*. *Phytochemistry, 38*, 987–995.

Glombitza, K., Keusgen, M., & Hauperich, S. (1997). Fucophlorethols from the brown algae *Sargassum spinuligerum* and *Cystophora torulosa*. *Phytochemistry, 46*, 1417–1422.

Glombitza, K. W., Rauwald, H. W., & Eckhardt, G. (1975). Fucols, polyhydroxyoligo- phenyls from *Fucus vesiculosus*. *Phytochemistry, 14*, 1403–1405.

Glombitza, K. W., & Schmidt, A. (1999). Trihydroxyphlorethols from the brown alga *Carpophyllum angustifolium*. *Phytochemistry, 51*, 1095–1100.

Glombitza, K. W., & Vogels, H. P. (1985). Antibiotics from algae. XXXV. Phlorotannins from *Ecklonia maxima*. *Planta Medica, 51*, 308–312.

Godlewska, K., Michalak, I., Tuhy, L., & Chojnacka, K. (2016). Plant growth biostimulants based on different methods of seaweed extraction with water. *BioMed Research International, 2016*, 1–11.

Goo, H. R., Choi, J. S., & Na, D. H. (2010). Quantitative determination of major phlorotannins in *Ecklonia stolonifera*. *Archives of Pharmacal Research, 33*, 539–544.

Grosse-Damhues, J., & Glombitza, K.-W. (1984). Isofuhalols, a type of phlorotannin from the brown alga *Chorda filum*. *Phytochemistry, 23*, 2639–2642.

Guiry, M. D., & Guiry, G. M. (2018). *AlgaeBase*. Galway: World-Wide Electronic Publication. National University of Ireland. http://www.algaebase.org.

Gupta, S., & Abu-ghannam, N. (2011). Recent developments in the application of algae or seaweed extracts as a means for enhancing the safety and quality attributes of foods. *Innovative Food Science and Emerging Technologies, 12*, 600–609.

Hafezieh, M., Ajdari, D., Ajdehakosh Por, A., & Hosseini, S. H. (2014). Using Oman Sea *Sargassum illicifolium* meal for feeding white leg shrimp *Litopenaeus vannamei*. *Iranian Journal of Fisheries Sciences, 13*, 73–80.

Ham, Y. M., Baik, J. S., Hyun, J. W., & Lee, N. H. (2007). Isolation of a new phlorotannin, fucodiphlorethol G, from a brown alga *Ecklonia cava*. *Bulletin of the Korean Chemical Society, 28*, 1595–1597.

Harden, E. A., Falshaw, R., Carnachan, S. M., Kern, E. R., & Mark, N. (2010). Virucidal activity of polysaccharide extracts from four algal species against herpes simplex virus. *Antiviral Research, 83*, 282–289.

Harnita, A. N. I., Santosa, I. E., Martono, S., Sudarsono, Widyarini, S., & M Harren, F. J. (2013). Inhibition of lipid peroxidation induced by ultraviolet radiation by crude phlorotannins isolated from brown algae *Sargassum hystrix* v. buxifolium C. Agardh. *Indonesian Journal of Chemistry, 13*, 14–20.

Harvey, A. L., & Gericke, N. (2011). Bioprospecting: Creating a value for biodiversity. In *Research in biodiversity – models and applications* (pp. 322–339).

Heesch, S., Serrano-Serrano, M., Luthringer, R., Peters, A. F., Destombe, C., Cock, J. L., et al. (2019). *Evolution of life cycles and reproductive traits: Insights from the brown algae*. bioRxiv, 530477.

Hemmi, V., & Jormalainen, A. (2004). Genetic and environmental variation in performance of a marine isopod: Effects of eutrophication. *Oecologia, 140*, 302–311.

Hemmi, A., Mäkinen, A., Jormalainen, V., & Honkanen, T. (2005). Responses of growth and phlorotannins in *Fucus vesiculosus* to nutrient enrichment and herbivory. *Aquatic Ecology, 39*, 201–211.

Hemmingson, J. A., Falshaw, R., Furneaux, R. H., & Thompson, K. (2006). Structure and antiviral activity of the galactofucan sulfates extracted from *Undaria pinnatifida*. *Journal of Applied Phycology, 18*, 185–193.

Henry, G. E., Momin, R. A., Nair, M. G. L., & Dewitt, D. L. (2002). Antioxidant and cyclooxygenase activities of fatty acids found in food. *Journal of Agricultural and Food Chemistry, 50*, 2231–2234.

Heo, S. J., & Jeon, Y. J. (2008). Radical scavenging capacity and cytoprotective effect of enzymatic digests of *Ishige okamurae*. *Journal of Applied Phycology, 20*, 1087–1095.
Heo, S. J., Ko, S. C., Cha, S. H., Kang, D. H., Park, H. S., Choi, Y. U., et al. (2009). Effect of phlorotannins isolated from *Ecklonia cava* on melanogenesis and their protective effect against photo-oxidative stress induced by UV-B radiation. *Toxicology in Vitro, 23*, 1123–1130.
Hernandez-Carmona, G., Carrillo-Domınguez, S., Arvizu-Higuera, D. L., Rodriguez-Montesinos, Y. E., Murillo-Alvarez, J. I., Munoz-Ochoa, M., et al. (2009). Monthly variation in the chemical composition of *Eisenia arborea* J.E. Areschoug. *Journal of Applied Phycology, 21*, 607–616.
Hernández-Herrera, R. M., Santacruz-Ruvalcaba, F., Ruiz-López, M. A., Norrie, J., & Hernández-Carmona, G. (2014). Effect of liquid seaweed extracts on growth of tomato seedlings (*Solanum lycopersicum* L.). *Journal of Applied Phycology, 26*, 619–628.
He, W. F., Yao, L. G., Liu, H. L., & Guo, Y. W. (2014). Thunberol, a new sterol from Chinese brown alga *Sargassum thunbergii*. *Journal of Asian Natural Products Research, 16*, 685–689.
Hidayati, J. R., Yudiati, E., Pringgenies, D., Arifin, Z., & OktaviyantI, D. T. (2019). Antioxidant activities, total phenolic compound and pigment contents of tropical *Sargassum* sp. extract, macerated in different solvents polarity. *Jurnal Kelautan Tropis, 22*, 73.
Holdt, S., & Kraan, S. (2011). Bioactive compounds in seaweed: Functional food applications and legislation. *Journal of Applied Phycology, 23*, 543–597.
Hong, D. D., Hien, H. M., & Son, P. N. (2007). Seaweeds from Vietnam used for functional food, medicine and biofertilizer. *Journal of Applied Phycology, 19*(6), 817–826. Springer Netherlands.
Hunt, B., & Vincent, A. C. J. (2006). Scale and sustainability of marine bioprospecting for pharmaceuticals. *Ambio, 35*, 57–64.
Hutasoit, S., Suada, I. K., & Susrama, I. G. K. (2013). Uji aktivitas antijamur ekstrak beberapa jenis biota laut terhadap *Aspergillus flavus* LINK. dan *Penicillium* sp LINK. Study of antifungi extract of some types of marine biota to the Aspergillus flavus LINK. and Penicillium LINK *E-Jurnal Agroekoteknologi Tropika, 2*, 27–38.
Huynh, T. G., Yeh, S. T., Lin, Y. C., Shyu, J. F., Chen, L. L., & Chen, J. C. (2011). White shrimp *Litopenaeus vannamei* immersed in seawater containing *Sargassum hemiphyllum* var. chinense powder and its extract showed increased immunity and resistance against *Vibrio alginolyticus* and white spot syndrome virus. *Fish & Shellfish Immunology, 31*, 286–293.
Hwang, R. L., Tsai, C. C., & Lee, T. M. (2004). Assessment of temperature and nutrient limitation on seasonal dynamics among species of *Sargassum* from a coral reef in Southern Taiwan. *Journal of Phycology, 40*, 463–473.
Imbs, T. I., Krasovskaya, N. P., Ermakova, S. P., Makarieva, T. N., Shevchenko, N. M., & Zvyagintseva, T. N. (2009). Comparative study of chemical composition and antitumor activity of aqueous–ethanol extracts of brown algae *Laminaria cichorioides, Costaria costata,* and *Fucus evanescens*. *Russian Journal of Marine Biology, 35*, 164–170.
Jégou, C., Kervarec, N., Cérantola, S., Bihannic, I., & Stiger-Pouvreau, V. (2015). NMR use to quantify phlorotannins: The case of Cystoseira tamariscifolia, a phloroglucinol-producing brown macroalga in Brittany (France). *Talanta, 135*, 1–6.
Jormalainen, V., & Honkanen, T. (2008). Macroalgal chemical defenses and their roles in structuring temperate marine communities. In C. D. Amsler (Ed.), *Algal chemical defence* (pp. 57–89). Berlin, Heidelberg: Springer Berlin Heidelberg.
Jung, H. A., Hyun, S. K., Kim, H. R., & Choi, J. S. (2006). Angiotensin-converting enzyme I inhibitory activity of phlorotannins from *Ecklonia stolonifera*. *Fisheries Science, 72*, 1292–1299.
Jung, H. A., Yoon, N. Y., Woo, M. H., & Choi, J. S. (2008). Inhibitory activities of extracts from several kinds of algae and phlorotannins from the brown alga *Ecklonia stolonifera* on

glucose-mediated protein damage and rat lens aldose reductase. *Fisheries Science, 74*, 1363–1365.

Kadi, A. (2005). Beberapa catatan kehadiran marga *Sargassum*. *Oseana, XXX*, 19–29.

Kalaivanan, C., & Venkatesalu, V. (2012). Utilization of seaweed *Sargassum myriocystum* extracts as a stimulant of seedlings of *Vigna mungo* (L.) Hepper. *Spanish Journal of Agricultural Research, 10*, 466–470.

Kang, M.-C., Kim, E.-A., Kang, S.-M., Wijesinghe, W. A. J. P., Yang, X., Kang, N.-L., et al. (2012). Thermostability of a marine polyphenolic antioxidant dieckol, derived from the brown seaweed *Ecklonia cava*. *Algae, 27*, 205–213.

Karaki, N., Sebaaly, C., Chahine, N., Faour, T., Zinchenko, A., Rachid, S., et al. (2013). The antioxidant and anticoagulant activities of polysaccharides isolated from the brown algae Dictyopteris polypodioides growing on the Lebanese coast. *Journal of Applied Pharmaceutical Science, 3*, 043–051.

Kavipriya, R., Dhanalakshmi, P. K., Jayashree, S., & Thangaraju, N. (2011). Seaweed extract as a biostimulant for legume crop, green gram. *Journal of Ecobiotechnology, 3*, 16–19.

Keusgen, M., & Glombitza, K. W. (1997). Pseudofuhalols from the brown alga *Sargassum spinuligerum*. *Phytochemistry, 46*, 1403–1415.

Keyrouz, R., Abasq, M. L., Le Bourvellec, C., Blanc, N., Audibert, L., Ar Gall, E., et al. (2011). Total phenolic contents, radical scavenging and cyclic voltammetry of algae from Brittany. *Food Chemistry, 126*, 831–836.

Khotimah, K., Darius, & Sasmito, B. B. (2013). Uji aktivitas senyawa aktif alga coklat *Sargassum fillipendulla* sebagai antioksidan pada minyak ikan lemuru *Sardinella langiceps*. *THPI Student Journal, I*, 1–9.

Kim, Y. C., An, R. B., Yoon, N. Y., Nam, T. J., & Choi, J. S. (2005). Hepatoprotective constituents of the edible brown alga *Ecklonia stolonifera* on tacrine-induced cytotoxicity in hep G2 cells. *Archives of Pharmacal Research, 28*, 1376–1380.

Kim, K. N., Heo, S. J., Song, C. B., Lee, J., Heo, S. M., Yeo, I. K., et al. (2006). Protective effect of *Ecklonia cava* enzymatic extracts on hydrogen peroxide-induced cell damage. *Process Biochemistry, 41*, 2393–2401.

Kim, S. M., Kang, K., Jeon, J.-S., Jho, E. H., Kim, C. Y., Nho, C. W., et al. (2011). Isolation of phlorotannins from *Eisenia bicyclis* and their hepatoprotective effect against oxidative stress induced by tert-butyl hyperoxide. *Applied Biochemistry and Biotechnology, 165*, 1296–1307.

Kim, S. M., Kang, S. W., Jeon, J. S., Jung, Y. J., Kim, W. R., Kim, C. Y., et al. (2013). Determination of major phlorotannins in *Eisenia bicyclis* using hydrophilic interaction chromatography: Seasonal variation and extraction characteristics. *Food Chemistry, 138*, 2399–2406.

Kim, M. M., & Kim, S. K. (2010). Effect of phloroglucinol on oxidative stress and inflammation. *Food and Chemical Toxicology, 48*, 2925–2933.

Kita, N., Fujimoto, K., Nakajima, I., Hayashi, R., & Shibuya, K. (1990). Screening test for deodorizing substances from marine algae and identification of phlorotannins as the effective ingredients in *Eisenia bicyclis*. *Journal of Applied Phycology, 2*, 155–162.

Kloareg, B., & Quatrano, R. S. (1988). Structure of the cell walls of marine algaes and ecophysiological functions of the matrix polysaccharides. *Oceanography and Marine Biology an Annual Review, 26*, 259–315.

Koch, M., Glombitza, K. W., & Eckhardt, G. (1980). Phlorotannins of phaeophyceae *Laminaria ochroleuca*. *Phytochemistry, 19*, 1821–1823.

Koivikko, R., Eranen, J. K., Loponen, J., & Jormalainen, V. (2008). Variation of phlorotannins among three populations of *Fucus vesiculosus* as revealed by HPLC and colorimetric quantification. *Journal of Chemical Ecology, 34*, 57–64.

Kraan, S. (2012). *Algal polysaccharides: Novel applications and outlook*. https://doi.org/10.5772/51572.

Kursar, T. A., Caballero-George, C. C., Capson, T. L., Cubilla-Rios, L., Gerwick, W. H., Gupta, M. P., Ibanez, A., Linington, R. G., McPhail, K. L., Ortega-Barria, E., Romero, L. I., Solis, P. N., & Coley, P. D. (2006). Securing economic benefits and promoting conservation through bioprospecting. *BioScience, 56*(12), 1005–1012.

Kursar, T. A., Capson, T. L., Emmen, D. A., Gerwick, W., Romero, L. I., Solis, P. N., et al. (2011). Linking insights from ecological research with bioprospecting to promote conservation, enhance research capacity, and provide economic uses of biodiversity. *Tropical Forest Community Ecology*, 429–441.

Laffoley, D. d'A., Roe, H. S. J., Angel, M. V., Ardon, J., Bates, N. R., Boyd, I. L., et al. (2011). *The protection and management of the Sargasso Sea: The golden floating rainforest of the Atlantic Ocean.* Summary Science and Supporting Evidence Case.

Lailiyah, A., Adi, T. K., Hakim, A., & Yusnawan, E. (2014). Kapasitas antioksidan dan kandungan total senyawa fenoli ekstrak kasar *Sargassum cristaefolium* Dari pantai Sumenep Madura. *Alchemy, 3*, 18–30.

Lalegerie, F., Gager, L., Stiger-Pouvreau, V., & Connan, S. (2019). The stressfull life of red and brown seaweeds on the temperate intertidal zone: Impact of abiotic and biotic parameters on the physiology of macroalgae and variability of the content of particular metabolites. *Advances in Botanical Research.*

Le Lann, K. (2009). *Etude de la biodiversité de Sargassaceae (Fucales, Phaeophyceae) en milleux tempéré et tropical: ecologie,chimiotaxonomie et source de composés bioacifs.* Université de Bretagne Occidentale, 367 pp.

Le Lann, K., Connan, S., & Stiger-Pouvreau, V. (2012). Phenology, TPC and size-fractioning phenolics variability in temperate *Sargassaceae* (*Phaeophyceae*, Fucales) from western Brittany: Native versus introduced species. *Marine Environmental Research, 80*, 1–11.

Le Lann, K., Ferret, C., VanMee, E., Spagnol, C., Lhuillery, M., & Stiger-Pouvreau, V. (2012). Total phenolic, size-fractionated phenolics and fucoxanthin content of tropical *Sargassaceae* (Fucales, *Phaeophyceae*) from the South Pacific ocean: Spatial and specific variability. *Phycological Research, 60*(1), 37–50.

Le Lann, K., Jégou, C., & Stiger-Pouvreau, V. (2008). Effect of different conditioning treatments on total phenolic content and antioxidant activities in two *Sargassacean* species: Comparison of the frondose *Sargassum muticum* (yendo) Fensholt and the cylindrical *Bifurcaria* bifurcata R. Ross. *Phycological Research, 56*, 238–245.

Le Lann, K., Surget, G., Couteau, C., Coiffard, L., Cérantola, S., Gaillard, F., et al. (2016). Sunscreen, antioxidant, and bactericide capacities of phlorotannins from the brown macroalga *Halidrys siliquosa. Journal of Applied Phycology, 28*, 3547–3559.

Lee, J. J., Eom, S. H., Lee, E. H., Jung, Y. J., Kim, H. J., Jo, M. R., et al. (2014). *In vitro* antibacterial and synergistic effect of phlorotannins isolated from edible brown seaweed *Eisenia bicyclis* against acne-related bacteria. *Algae, 29*, 47–55.

Lee, M. H., Lee, K. B., Oh, S. M., Lee, B. H., & Chee, H. Y. (2010). Antifungal activities of dieckol isolated from the marine brown alga *Ecklonia cava* against *Trichophyton rubrum. Food Science and Biotechnology, 53*, 504–507.

Lee, J. B., Takeshita, A., Hayashi, K., & Hayashi, T. (2011). Structures and antiviral activities of polysaccharides from *Sargassum trichophyllum. Carbohydrate Polymers, 86*, 995–999.

Lee, J. M., You, S. G., & Kim, S. M. (2005). Functional activities of low molecular weight peptides purified from enzymatic hydrolysates of algae. *Journal of the Korean Society of Food Science and Nutrition, 34*, 1124–1129.

Li, S.-M., & Glombitza, K.-W. (1991). Carmalols and phlorethofuhalols from the brown alga *Carpophyllum maschalocarpum. Phytochemistry, 30*, 3417–3421.

Li, H., Huang, J., Xin, Y., Zhang, B., Jin, Y., & Zhang, W. (2009). Optimization and scale-up of a new photobleaching agar extraction process from Gracilaria lemaneiformis. *Journal of Applied Phycology, 21*, 247–254.

Liu, L., Heinrich, M., Myers, S., & Dworjanyn, S. A. (2012). Towards a better understanding of medicinal uses of the brown seaweed *Sargassum* in traditional Chinese medicine: A phytochemical and pharmacological review. *Journal of Ethnopharmacology, 142*, 591–619.

Lopes, G., Sousa, C., Silva, L. R., pinto, E., Andrade, P. B., Bernardo, J., et al. (2012). Can phlorotannins purified extracts constitute a novel pharmacological alternative for microbial infections with associated inflammatory conditions? *PLoS One,* 7(2), e31145.

Manley, S. L., & North, W. J. (1984). Phosphorus and the growth of juvenile *Macrocystis pyrufera* (Phaeophyta) sporophytes. *Journal of Phycology, 20*, 389–393.

Martinez, S. I., & Biber-Klemm, S. (2010). Scientists-take action for access to biodiversity. *Current Opinion in Environmental Sustainability, 2*, 27–33.

Martinez, J. H. I., & Castaneda, H. G. T. (2013). Preparation and chromatographic analysis of phlorotannins. *Journal of Chromatographic Science, 51*, 825–838.

Maruyama, H., Tamauchi, H., Iizuka, M., & Nakano, T. (2006). The role of NK cells in antitumor activity of dietary fucoidan from *Undaria pinnatifida* sporophylls (mekabu). *Planta Medica, 72*, 1415–1417.

Matanjun, P., Mohamed, S., Mustapha, N. M., & Muhammad, K. (2009). Nutrient content of tropical edible seaweeds *Eucheuma cottonii, Caulerpa lentillifera* and *Sargassum polycystum. Journal of Applied Phycology, 21*, 75–80.

Mayer, A. M. S., Rodriguez, A. D., Taglialatela-Scafati, O., & Fusetani, N. (2013). Marine pharmacology in 2009–2011: Marine compounds with antibacterial, antidiabetic, antifungal, anti-inflammatory, antiprotozoal, antituberculosis, and antiviral activities; affecting the immune and nervous system, and other miscellaneous mechanism of action. *Marine Drugs, 11*, 2510–2573.

Meslet-Cladière, L., Delage, L., Leroux, C. J.-J., Goulitquer, S., Leblanc, C., Creis, E., et al. (2013). Structure/function analysis of a type iii polyketide synthase in the brown alga *Ectocarpus siliculosus* reveals a biochemical pathway in phlorotannin monomer biosynthesis. *The Plant Cell, 25*, 3089–3103.

Michalak, I., & Chojnacka, K. (2015). Algae as production systems of bioactive compounds. *Engineering in Life Science*, 1–17, 00.

Milledge, J. J., Nielsen, B. V., & Bailey, D. (2015). High-value products from macroalgae: The potential uses of the invasive brown seaweed, *Sargassum muticum. Reviews in Environmental Science and Biotechnology, 15*, 67–88.

Ministry of Marine and Fisheries. (2013). *Indonesia as the word largest cottonii producer (Indonesia sebagai produsen rumput laut cottonii terbesar dunia).* http://kkp.go.id. (Accessed November 2014).

Moghadamtousi, S. Z., Karimian, H., Khanabdali, R., Razavi, M., Firoozinia, M., Zandi, K., et al. (2014). Anticancer and antitumor potential of fucoidan and fucoxanthin, two main metabolites isolated from brown algae scientific. *WorldatWork Journal.* https://doi.org/10.1155/2014/768323.

Mulyati, H., & Geldermann, J. (2016). Managing risks in the Indonesian seaweed supply chain. *Clean Technologies and Environmental Policy*, 1–15. https://doi.org/10.1007/s10098-016-1219-7.

Mushollaeni, W., & Rusdiana, E. (2013). Test of hypoglikemic effect of brown algae extract (*Padina* and *Sargassum binderi*) on mice induced by alloxan. *Journal of Agriculture and Food Technology, 3*, 5–11.

Mushollaeni, W., & Endang Rusdiana, S. (2012). Optimizing the use of alginate from *Sargassum* and *Padina* as natural emulsifier and stabilizer in cake. *Journal of Agriculture and Food Technology, 2*, 108–112.

Mwangi, H. M., Westhuizen, J. V. D., Marnewick, J., Mabusela, W. T., Kabanda, M. M., & Ebenso, E. E. (2013). *Isolation, identification and radical scavenging activity of phlorotannin derivatives from brown algae,* Ecklonia maxima*: An experimental and theoretical study.* https://doi.org/10.1016/j.fra.2013.10.006.

Nagayama, K., Shibata, T., Fujimoto, K., Honjo, T., & Nakamura, T. (2003). Algicidal effect of phlorotannins from the brown alga *Ecklonia kurome* on red tide microalgae. *Aquaculture, 218*, 601–611.

Nakai, M., Kageyama, N., Nakahara, K., & Miki, W. (2006). Phlorotannins as radical scavengers from the extract of *Sargassum ringgoldianum*. *Marine Biotechnology, 8*, 409–414.

Namvar, F., Mohammad, R., Baharara, J., Zafar-Balanejad, S., Fargahi, F., & Rahman, H. S. (2013). Antioxidant, antiproliferative, and antiangiogenesis: Effects of polyphenol-rich seaweed (*Sargassum muticum*). *BioMed Research International*. https://doi.org/10.1155/2013/604787.

Narayan, B., Kazuo, M., & Masashi, H. (2004). Comparative evaluation of fatty acid composition of different *Sargassum* (Fucales, Phaeophyta) species harvested from temperate and tropical waters. *Journal of Aquatic Food Product Technology, 13*, 53–70.

Neish, I. C. (2007). *Assessment of the seaweed value chain in Indonesia*. Jakarta: USAID Report, 35 pp.

Neish, I. C. (2013). Social and economic dimensions of carrageenan seaweed farming in Indonesia. In D. Valderrama, J. Cai, N. Hishamunda, & N. Ridler (Eds.), *Social and economic dimensions of carrageenan seaweed farming. Fisheries and aquaculture technical paper no. 580* (pp. 61–89). Rome: FAO, 204 pp.

Ngabalin, A. M. (2013). *Analysis of value chain system and marketing strategy seaweed in southeast maluku regency, Maluku province*. Master Thesis. Agricultural Institute of Bogor, 139 pp.

Nwosu, F., Morris, J., Lund, V. A., Stewart, D., Ross, H. A., & McDougall, G. J. (2011). Anti-proliferative and potential anti-diabetic effects of phenolic-rich extracts from edible marine algaes. *Food Chemistry, 126*, 1006–1012.

Ogawa, H. (1976). Antheridium development of *Sargassum fulvellum* and *S. kjellmanianum*. *Marine Biology, 38*, 163–168.

O'Kelley, J. C. (1974). Inorganic nutrients. In W. D. P. Stewart (Ed.), *Algal physiology and biochemistry* (pp. 610–635). New York: Blackwell Scientific Publications.

Paduch, R., Kandefer-Szerszea, M., Trytek, M., & Fiedurek, J. (2007). Terpenes: substances useful in human healthcare. *Archivum Immunologiae et Therapiae Experimentalis* (Warsz), *55*, 315–327.

Pangestuti, R., & Kim, S. K. (2015). An overview of phycocolloids: The principal commercial seaweed extracts. *Marine Algae Extracts*, 319–330. https://doi.org/10.1002/9783527679577.ch19.

Park, Y. H., Jang, D. S., & Kim, S. B. (1997). *Utilization of marine products* (Chapter 4), Seaweed composition (2nd ed.). Hyoungsul press.

Parys, S., Kehraus, S., Krick, A., Glombitza, K.-W., Carmeli, S., Klimo, K., et al. (2010). In vitro chemopreventive potential of fucophlorethols from the brown alga *Fucus vesiculosus* L. by anti-oxidant activity and inhibition of selected cytochrome P450 enzymes. *Phytochemistry, 71*, 221–229.

Pavia, H., & Åberg, P. (1996). Spatial variation in polyphenolic content of *Ascophyllum nodosum* (Fucales, Phaeophyta). *Hydrobiology*, 199–203, 326–327.

Pavia, H., Toth, G., & Åberg, P. (1999). Trade-offs between phlorotannin production and annual growth in natural populations of the brown seaweed *Ascophyllum nodosum*. *Journal of Ecology, 87*, 761–771.

Pawlik, J. R., Loh, T. L., McMurray, S. E., & Finelli, C. M. (2013). Sponge communities on Caribbean coral reefs are structured by factors that are top-down, not bottom-up. *PLoS One, 8*, 1–7.

Percival, E., & McDowell, R. H. (1967). *Chemistry and enzymology of marine algal polysaccharides*. London and New York: Academic Press.

Pérez-Larrán, P., Balboa, E. M., Torres, M. D., & Domínguez, H. (2018). Antioxidant and antitumoral properties of aqueous fractions from frozen *Sargassum muticum*. *Waste Biomass Valorization*, 1–9.
Perez, C. M. T., Pajares, I. G., Alcantara, V. A., & Simbahan, J. F. (2018). Bacterial laminarinase for application in ethanol production from brown algae *Sargassum* sp. using halotolerant yeast. *Biofuel Research Journal, 5*, 792–797.
Pillai, S., Oresajo, C., & Hayward, J. (2005). Ultraviolet radiation and skin aging: roles of reactive oxygen species, inflammation and protease activation, and strategies for prevention of inflammation-induced matrix degradation—a review. *International Journal of Cosmetic Science, 27*, 17–34.
Plouguerné, E., Georgantea, P., Ioannou, E., Vagias, C., Roussis, V., Hellio, C., et al. (2010). Anti-microfouling activity of lipidic metabolites from the invasive brown alga *Sargassum muticum* (yendo) Fensholt. *Marine Biotechnology, 12*(1), 52–61.
Plouguerné, E., Hellio, C., Deslandes, E., Véron, B., & Stiger-Pouvreau, V. (2008). Anti-microfouling activities in extracts of two invasive algae: *Grateloupia turuturu* and *Sargassum muticum*. *Botanica Marina, 51*, 202–208.
Pomin, V. H., & Mourao, P. A. S. (2008). Structure, biology, evolution, and medical importance of sulfated fucans and galactans. *Glycobiology, 18*, 1016–1027.
Pomponi, S. A. (1999). The bioprocess-technological potential of the sea. *Journal of Biotechnology, 70*, 5–13.
Puspita, M. (2017). *Enzyme-assisted extraction of phlorotannins from sargassum and biological activities*. PhD. Thesis. Université Bretagne Sud and Diponegoro University, 334 pp.
Puspita, M., Deniel, M., Widowati, I., Radjasa, O. K., Douzenel, P., Bedoux, G., et al. (2012). Antioxidant and antibacterial activity of solid-liquid and enzyme-assisted extraction of phenolic compound from three species of tropical *Sargassum*. *Journal of Physics: Conference Series, 365*, 11001. https://doi.org/10.1088/1742-6596/365/1/011001.
Rahbari, H., Akram, A., Pazoki, M., & Aghbashlo, M. (2019). Bio-oil production from *Sargassum* macroalgae: A green and healthy source of energy. *JundiShapur Journal of Health Sciences, 11*(1), e84301. https://doi.org/10.5812/jjhs.84301.
Reaka-Kudla, M. L. (1997). The global biodiversity of coral reefs: A comparison with rain forest. In *Biodiversity II: Understanding and protecting our biological resources* (pp. 83–108). The National Academy of Sciences.
Rebours, C., Marinho-Soriano, E., Zertuche-Gonzalez, J. A., Hayashi, L., Vasquez, J. A., Kradolfer, P., et al. (2014). Algae: An opportunity for wealth and sustainable livelihood for coastal community. *Journal of Applied Phycology, 26*, 1939–1951.
de Reviers, B. (2002). *Biologie et Phylogénie des Algues, tome 1*. Paris: Belin, 352 pp.
Ridlo, A., & Pramesti, R. (2005). Application of seaweed extracts as immunostimulant agents in shrimp (*Litopennaeus vannamei*) non-specific defense system. *Ilmu Kelautan, 14*, 133–137.
Rinaudo, M. (2014). Biomaterials based on a natural polysaccharide: Alginate. *TIP Rev Espec en Ciencias Quim, 17*, 92–96.
Rioux, L. E., Turgeon, S. L., & Beaulieu, M. (2009). Effect of season on the composition of bioactive polysaccharides from the brown seaweed *Saccharina longicruris*. *Phytochemistry, 70*, 1069–1075.
Sailler, B., & Glombitza, K. W. (1999). Halogenated phlorethols and fucophlorethols from the brown alga *Cystophora retroflexa*. *Natural Toxins, 7*, 57–62.
Samarakoon, K., & Jeon, Y. J. (2012). Bio-functionalities of proteins derived from marine alage – a review. *Food Research International, 48*, 948–960.
Santoso, J., Podungge, F., & Sumaryanto, H. (2013). Chemical composition and antioxidant activity of tropical brown algae *Padina australis* from pramuka island, district of Seribu island, Indonesia. *Jurnal Ilmu dan Teknologi Kelautan Tropis, 5*, 287–297.

Saraswathi, S. J., Babu, B., & Rengasamy, R. (2003). Seasonal studies on the alginate and its biochemical composition I: *Sargassum polycystum* (Fucales), *phaeophyceae. Phycological Research, 51*, 240–243.

Scott, P. (2001). Bioprospecting as a conservation tool: History and background. In *Crossing boundaries in park management: Proceedings of the 11th conference on research and resource management in parks and on public lands*. Michigan, USA: The George Wright Society.

Septiana, A. T., & Asnani, A. (2013). Antioxidan activity of *Sargassum duplicatum* seaweed extract. *Jurnal Teknologi Pertanian, 14*, 79–86.

Setyawidati, N. A. R. (2017). *Marine seaweeds with ecological and economical potentials in Indonesia: Spatio-temporal analysis, valorization, and aquaculture*. PhD. Thesis. Université de Bretagne Occidentale, 301 pp.

Setyawidati, N., Kaimuddin, A., Wati, I.,P., Widowati, I., Rossi, N., Liabot, P.,O., et al. (2017). Percentage cover, biomass, distribution, and potential habitat mapping of natural macroalgae, based on high-resolution satellite data and in situ monitoring, at Libukang island, Malasoro Bay, Indonesia. *Journal of Applied Phycology, 30*(1), 159–171.

Setyawidati, N. A. R., Liabot, P. O., Perrot, T., Radiarta, N., Deslandes, E., Bourgougnon, N., et al. (2017). *In situ* evolution of carrageenan content and biomass estimation by satellite image processing for the red macroalgae *Kappaphycus alvarezii* cultivated in Malasoro Bay, Indonesia. *Journal of Applied Phycology, 29*(5), 2307–2321.

Setyawidati, N., Puspita, M., Kaimuddin, A., Widowati, I., Delandes, E., Bourgougnon, N., et al. (2018). Seasonal biomass and alginate stock assessment of three abundant genera of brown macroalgae using multispectral high resolution satellite remote sensing: A case study at Ekas bay (Lombok, bay). *Marine Pollution Bulletin, 131*(B), 40–48. https://doi.org/10.1016/j.marpolbul.2017.11.068.

Shibata, T., Ishimaru, K., Kawaguchi, S., Yoshikawa, H., & Hama, Y. (2008). Antioxidant activities of phlorotannins isolated from Japanese *Laminariaceae. Journal of Applied Phycology, 20*, 705–711.

Shibata, T., Kawaguchi, S., Hama, Y., Inagaki, M., Yamaguchi, K., & Nakamura, T. (2004). Local and chemical distribution of phlorotannins in brown algae. *Journal of Applied Phycology, 16*, 291–296.

Shin, T., Ahn, M., Hyun, J. W., Kim, S. H., & Moon, C. (2014). Antioxidant marine algaes phlorotannins and radioprotection: A review of experimental evidence. *Acta Histochemica*. https://doi.org/10.1016/j.acthis.2014.03.008.

Sinha, S., Astani, A., Ghosh, T., Schnitzler, P., & Ray, B. (2010). Polysaccharides from *Sargassum tenerrimum*: Structural features, chemical modification and anti- viral activity. *Phytochemistry, 71*, 235–242.

Siregar, T. R. P., Lubis, A., & Supriyadi. (2014). Utilization of two type of brown algae compost (*Sargassum polycystum*) in improving the fertility of ultisol soil and the production of mustard plant (*Brassica juncea* L.). *Journal of Online Agroekoteknol, 2*, 1259–1268.

Siriwardhana, N., Lee, K. W., & Jeon, Y. J. (2005). Radical scavenging potential of hydrophilic phlorotannins of *Hizikia fusiformis. Algae, 20*, 69–75.

Siswandi, A. G. C. (2013). *Marine bioprospecting: International law, Indonesia and sustainable development* (p. 581). Dissertation. ANU College of Law. Australia: Australian National University. Retrieved from https://openresearch-repository.anu.edu.au/bitstream/1885/9730/1/02Whole_Siswandi.pdf.

Skriptsova, A., Khomenko, V., & Isakov, V. (2004). Seasonal changes in growth rate, morphology and alginate content in *Undaria pinnatifida* at the northern limit in the Sea of Japan (Russia). *Journal of Applied Phycology, 16*, 17–21.

Skriptsova, A. V., Shevchenko, N. M., Zvyagintseva, T. N., & Imbs, T. I. (2010). 'Monthly changes in the content and monosaccharide composition of fucoidan from *Undaria pinnatifida* (*Laminariales*, Phaeophyta). *Journal of Applied Phycology, 22*, 79–86.

Soares, A. R., Robaina, M. C. S., Mendes, S., Silva, T. S. L., Gestinari, L. M. S., Pamplona, S., et al. (2012). Antiviral activity of extracts from Brazilian seaweeds against herpes simplex virus. *The Brazilian Journal of Pharmacognosy, 22*, 714–723.

Steen, H. (2004). Effects of reduced salinity on reproduction and germling development in *Sargassum muticum* (*Phaeophyceae*, Fucales). *European Journal of Phycology, 39*, 293–299.

Steinberg, P. D. (1989). Biogeographical variation in brown algal polyphenolics and other secondary metabolites: Comparison between temperate Australasia and North America. *Oecologia, 78*, 373–382.

Stengel, D., Connan, S., & Popper, Z. A. (2011). Algal chemodiversity and bioactivity: Sources of natural variability and implications for commercial application. *Biotechnology Advances, 29*, 483–501.

Stiger-Pouvreau, V., Bourgougnon, N., & Deslandes, E. (2016). Chapter 8. Carbohydrates from seaweeds. In J. Fleurence, & I. Levine (Eds.), *Seaweed in health and disease prevention* (pp. 223–274). Elsevier. https://doi.org/10.1016/B978-0-12-802772-1.00008-7.

Stiger-Pouvreau, V., Jégou, C., Cérantola, S., Guérard, F., & Le Lann, K. (2014). Chapitre 13. Phlorotannins in *Sargassaceae* species from Brittany (France): Interesting molécules for ecophysiological and valorisation purposes. In N. Bourgougnon (Ed.), *Advances in botanical research* (Vol. 71). Academic Press, Elsevier. https://doi.org/10.1016/B978-0-12-408062-1.00013-5.

Stiger, V., Deslandes, E., & Payri, C. E. (2004). Phenolic contents of two brown algae, *Turbinaria ornata* and *Sargassum mangarevense* on Tahiti (French Polynesia): Interspecific, ontogenic and spatio-temporal variations. *Botanica Marina, 47*, 402–409.

Sudaryono, A., Aditomo, A. H. C., & Isnansetyo, A. (2015). Evaluation of dietary supplementation of aqueous extract of brown algae *Sargassum cristaefolium* on growth performance and feed utilization of juvenile white shrimp *Litopenaeus vannamei*. *AACL Bioflux, 8*, 142–149.

Sugiura, Y., Kunio, I., & Amano, H. (2011). The anti-allergic and anti-inflammatory effects of seaweed polyphenols (phlorotannin). *Abstr.* Food and Food J Japan., *216*(1).

Sunadji, M. S., & Tjahjono, A. (2014). Development strategy of seaweed aquaculture business in kupang regency, East Nusa Tenggara province , Indonesia. *Journal of Agricultural Studies, 2*, 27–39.

Supriyono, A. (2007). Aktivitas antioksidan beberapa spesies rumput laut Dari pulau sumba. *Jurnal Sains Dan Teknologi Indonesia, 9*, 34–38.

Surget, G., Roberto, V., Le Lann, K., Mira, S., Laizé, V., Guérard, F., et al. (2017). Green algae: A source of natural compounds with proliferative, mineralogenic and antioxidant activities. *Journal of Applied Phycology, 29*(1), 575–584.

Tang, H. F., YI, Y. H., YAO, X. S., XU, Q. Z., Zhang, S. Y., & Lin, H. W. (2002). Bioactive steroids from the brown alga *Sargassum caryophyllum*. *Journal of Asian Natural Products Research, 4*, 95–105.

Tanniou, A., Serrano Léon, E., Vandanjon, L., Ibanez, E., Mendiola, J. A., Cérantola, S., et al. (2013). Green improved processes to extract bioactive phenolic compounds from brown macroalgae using *Sargassum muticum* as model. *Talanta, 104*, 44–52.

Tanniou, A., Vandajon, L., Goncalves, O., Kervarec, N., & Stiger-Pouvreau, V. (2014). Spectroscopy, rapid geographical differentiation of the European spread brown macroalga *Sargassum muticum* using HRMAS NMR and Fourier-transform infrared. *Talanta*. https://doi.org/10.1016/j.talanta.2014.09.002.

Tanniou, A., Vandanjon, L., Incera, M., Serrano Léon, E., Husa, V., Le Grand, J., et al. (2014). Assessment of the spatial variability of phenolic contents and associated bioactivities in the invasive alga *Sargassum muticum* sampled along its European range from Norway to Portugal. *Journal of Applied Phycology, 26*, 1215–1230.

Targett, N. M., Coen, L. D., Boettcher, A. A., & Tanner, C. E. (1992). Biogeographic comparisons of marine algal polyphenolics: Evidence against a latitudinal trend. *Oecologia, 89*, 464–470.

Thomas, N. V., & Kim, S. K. (2011). Potential pharmacological applications of polyphenolic derivatives from marine brown algae. *Environmental Toxicology and Pharmacology, 32*, 325–335.

Tierney, M. S., Soler-Vila, A., Croft, A. K., & Hayes, M. (2012). Antioxidant activity of the brown macro algae *Fucus spiralis* Linnaeus harvested from the west coast of Ireland. *Current Research Journal of Biological Sciences, 5*, 81–90.

Topinka, J. A., & Robbins, J. V. (1976). Effects of nitrate and ammonium enrichment on growth and nitrogen physiology in *Fucus spiralis*. *Limnology, 21*, 659–664.

Truus, K., Vaher, M., Koel, M., Mähar, A., & Taure, I. (2004). Analysis of bioactive ingredients in the brown alga *Fucus vesiculosus* by capillary electrophoresis and neutron activation analysis. *Analytical and Bioanalytical Chemistry, 379*, 849–852.

Valderrama, D., Cai, J., Hishamunda, N., & Ridler, N. (2013). *Social and economic dimensions of carrageenan seaweed farming*. Fisheries and Aquaculture Technical Paper No. 580. Rome: FAO, 204 pp.

Vo, T. S., Ngo, D. H., & Kim, S. K. (2012). Marine algae as a potential pharmaceutical source of anti-allergic therapy. *Process Biochemistry, 47*, 386–394.

Wang, W., Wang, S., & Guan, H. (2012). The antiviral activities and mechanisms of marine: An overview. *Marine Drugs, 10*, 2795–2816.

Wibowo, A. E. (2014). *Overview of agribusiness and factors that influence the Indonesian seaweed*. Agricultural Institute of Bogor, 73 pp.

Widowati, I., Susanto, A. B., Puspita, M., Stiger-Pouvreau, V., & Bourgougnon, N. (2012). Potentiality of using spreading *Sargassum* species from Jepara as an interesting source of antibacterial and antioxidant compounds: A preliminary study. *International Journal of Marine and Aquatic Resource Conservatian and Co-existence, 1*, 63–67.

Widowati, I., Susanto, A. B., Puspita, M., Stiger-Pouvreau, V., & Bourgougnon, N. (2014). Potentiality of using spreading *Sargassum* species from Indonesia as an interesting source of antibacterial and radical scavenging compounds: A preliminary study. *International Journal of Marine and Aquatic Resource Conservation and Co-existence, 1*, 63–67.

Wijengsihe, W. A. J. P., & Jeon, Y. J. (2011). Biological activities and potential cosmeceutical applications of bioactive components from brown seaweed: A review. *Phytochemistry Reviews, 10*, 431–443.

Wijesekara, I., Yoon, N. Y., & Kim, S. K. (2010). Phlorotannins from *Ecklonia cava* (*Phaeophyceae*): Biological activities and potential health benefits. *BioFactors*, 408–414.

Wijesinghe, W. A. J. P., & Jeon, dan Y. J. (2012). Enzyme-assistant extraction (EAE) of bioactive components: A useful approach for recovery of industrially important metabolites from algae: A review. *Fototerapia, 83*, 6–12.

Wynberg, R. (2016). Marine genetic resources and bioprospecting in the western Indian ocean. In *Regional state of the coast report* (pp. 397–408). New York: Western Indian Coast.

Yamasaki-Miyamoto, Y., Yamasaki, M., Tachibana, H., & Yamada, K. (2009). Fucoidan induces apoptosis through activation of caspase-8 on human breast cancer MCF-7 cells. *Journal of Agricultural and Food Chemistry, 57*(18), 8677–8682.

Yang, E.-J., Ham, Y. M., Lee, W. J., Lee, N. H., & Hyun, C. G. (2013). Anti-inflammatory effects of apo-9'-fucoxanthinone from the brown alga, *Sargassum muticum*. *Daru Journal of Pharmaceutical Sciences, 21*, 62.

Yang, H. C., Zeng, M. Y., Dong, S. Y., Liu, Z. Y., & Li, R. X. (2010). Anti-proliferative activity of phlorotannin extracts from brown algae *Laminaria japonica* Aresch. *Chinese Journal of Oceanology and Limnology, 28*, 122–130.

Yang, L., & Zhang, L. M. (2009). Review: Chemical structural and chain conformational characterization of some bioactive polysaccharides isolated from natural sources. *Carbohydrate Polymers, 76*, 349–361.

Yeong, B. M.-L., & Wong, C.-L. (2013). Seasonal growth rate of *Sargassum* species at teluk kemang, port Dickson, Malaysia. *Journal of Applied Phycology, 25*, 805–814.

Yoon, N. Y., Eom, T. K., Kim, M. M., & Kim, S. K. (2009). Inhibitory effect of phlorotannins isolated from *Ecklonia cava* on mushroom tyrosinase activity and melanin formation in mouse B16F10 melanoma cells. *Journal of Agricultural and Food Chemistry, 57*, 4124–4129.

Yoon, N. Y., Lee, S.-H., & Kim, S.-K. (2009). Phlorotannins from *Ishige okamurae* and their acetyl- and butyrylcholinesterase inhibitory effects. *The Journal of Functional Foods, 1*, 331–335.

Yuan, H., Zhang, W., Li, X., Lü, X., Li, N., & Gao, X. (2005). Preparation and in vitro antioxidant activity of carrageenan oligosaccharides and their oversulfated, acetylated, and phosphorylated derivates. *Carbohydrate Research, 340*, 685–692.

Zamannejad, N., Emadi, H., & Hafezieh, M. (2016). Effects of supplementation of algae (Sargassum ilicifolium) on growth, survival and body composition of rainbow trout *Oncorhynchus mykiss*. *Iranian Journal of Fisheries Sciences, 15*, 194–205.

Zhu, W., Chiu, L. C. M., Ooi, V. E. C., Chan, P. K. S., & Ang, P. O. (2004). Antiviral property and mode of action of a sulphated polysaccharide from *Sargassum patens* against herpes simplex virus type 2. *International Journal of Antimicrobial Agents, 24*, 81–85.

Zou, Y., Qian, Z.-J., Li, Y., Kim, M.-M., Lee, S.-H., & Kim, S.-K. (2008). Antioxidant effects of phlorotannins isolated from *Ishige okamurae* in free radical mediated oxidative systems. *Journal of Agricultural and Food Chemistry, 56*, 7001–7009.

Zou, X. X., Xing, S. S., Su, X., Zhu, J., Huang, H. Q., & Bao, S. X. (2017). The effects of temperature, salinity and irradiance upon the growth of Sargassum polycystum C. Agardh (Phaeophyceae). *Journal of Applied Phycology, 30*, 1207–1215.

Zoysa, M., Nikapitiya, C., Jeon, Y. J., Jee, Y., & Lee, J. (2008). Anticoagulant activity of sulfated polysaccharide isolated from fermented brown seaweed *Sargassum fulvellum*. *Journal of Applied Phycology, 20*, 67–74.

Zubia, M., Fabre, M. S., Kerjean, V., Lann, K. L., Stiger-Pouvreau, V., Fauchon, M., et al. (2009). Antioxidant and antitumoural activities of some Phaeophyta from Brittany coasts. *Food Chemistry, 116*, 693–701.

Zubia, M., Thomas, O. P., Soulet, S., Demoy-Schneider, M., Saulnier, D., Connan, S., et al. (2019). Potential of tropical macroalgae from French Polynesia for biotechnological applications. *Journal of Applied Phycology*. ISS19 (in press) org/10-1007/s10811-019-01920-8.

CHAPTER SIX

Dictyotaceae (Dictyotales, Phaeophyceae) species from French Polynesia: current knowledge and future research

Tohei Theophilus[a], Christophe Vieira[b], Gérald Culioli[c], Olivier P. Thomas[d], Antoine De Ramon N'Yeurt[e], Serge Andréfouët[f], Lydiane Mattio[g,h], Claude E. Payri[f] and Mayalen Zubia[a,*]

[a]University of French Polynesia, UMR-EIO, LabEx CORAIL, BP6570, 98702 Faa'a, Tahiti, French Polynesia
[b]Phycology Research Group and Centre for Molecular Phylogenetics and Evolution, Ghent University, Krijgslaan 281 (S8), Ghent, Belgium
[c]Université de Toulon, MAPIEM, Toulon, France
[d]Marine Biodiscovery, School of Chemistry and Ryan Institute, National University of Ireland Galway, University Road, Galway, Ireland
[e]Pacific Centre for Environment and Sustainable Development, The University of the South Pacific, Suva, Fiji
[f]French National Research Institute for Sustainable Development, UMR 9220/250 ENTROPIE (IRD-UR-CNRS), Noumea, New Caledonia
[g]blue[c]weed, Locmaria-Plouzané, France
[h]Department of Biological Sciences, University of Cape Town, Rondebosch, South Africa
*Corresponding author: E-mail: mayalen.zubia@upf.pf

Contents

Advances in Botanical Research, Volume 95
ISSN 0065-2296
https://doi.org/10.1016/bs.abr.2019.12.001

Abstract

The coral reefs of French Polynesia (FP) have experienced repeated macroalgal blooms over the last decades. These events have prompted intense efforts in fundamental and applied research on macroalgae in this ecoregion, especially regarding species of the order Fucales (*Turbinaria ornata* and *Sargassum pacificum*). Recently, however, these proliferations have occurred with a higher frequency, and they now involve additional species. Specifically, over the past decade, the abundance of species belonging to the Dictyotaceae family (e.g., *Dictyota bartayresiana* and *Spatoglossum asperum*) has increased on coral reefs around Tahiti Island, the largest and most inhabited island in FP. On the course of evolution, these species have developed physical and chemical defenses to deter grazers, including the production of a wide array of specialized metabolites. These molecules are of particular interest for their promising biological activities as well as for the new Blue Economy opportunities they can offer to FP. We review the current state of knowledge on the diversity, ecology, and potential uses of Dictyotaceae species present in FP. The first section focuses on the diversity and distribution of the family Dictyotaceae in FP. The second part examines the ecological dynamics of Dictyotaceae species in the coral reef ecosystem and their response to various environmental factors. The third and final part reviews the metabolites known from Dictyotaceae species that are present in FP, their associated biological activities, and potential for the development of biotechnological applications in FP.

1. Introduction

French Polynesia (FP) is a vast territory (5.5 million km^2) in the South Pacific Ocean (7°S-28°S; 134°W-155°W) consisting of 118 islands (34 high volcanic islands and 84 atolls) that belong to five archipelagos (Austral, Gambier, Marquesas, Society, and Tuamotu; Andréfouët & Adjeroud, 2019, Fig. 1). It harbors a wide variety of habitats and climates, ranging from high volcanic islands with fringing and barrier reefs in the Society Islands to low-lying coral atolls in the Tuamotus, and from equatorial conditions in the Marquesas to sub-tropical environments in the Austral Islands (Andréfouët & Adjeroud, 2019).

A remarkable floristic effort has been conducted to record the marine macroalgae in FP, and more than 40 relevant studies have been published since the 19th-century pioneering work made by Bory De Saint-Vincent

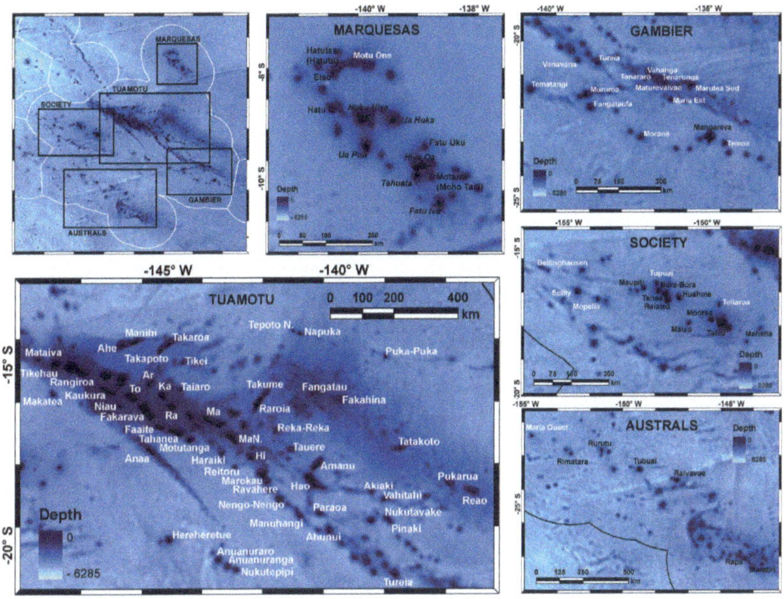

Fig. 1 Map showing the five archipelagos of French Polynesia.

(1828) (Table S1). The most comprehensive contributions were by Setchell (1926), Payri (1982), Payri and Meinesz (1985), Payri and Denizot (1993), Payri and N'Yeurt (1997), and N'Yeurt and Payri (2006; 2007; 2010). Based on previously published records, Payri and N'Yeurt (1997) provided a checklist of 425 taxa including 42 Phaeophyceae, 96 Chlorophyceae, 170 Rhodophyceae, and 117 Cyanophyceae. Using more in-depth analyses and extensive new field collections, N'Yeurt and Payri (2006; 2007; 2009; 2010) then revised the total number of marine macroalgal species known from FP to 312, including 32 Phaeophyceae, 82 Chlorophyceae, and 198 Rhodophyceae (Table S1).

The largest macroalgal diversity is recorded for the Society and the Austral Archipelagos, with the latter harboring 20% of endemic species (N'Yeurt & Payri, 2006). The remoteness of these islands is likely to explain this high level of endemism (Gillespie, Foody, Rocchini, Giorgi, & Saatchi, 2008; Kier et al., 2009). In their study, N'Yeurt and Payri (2006) also stressed the role of water temperature and the availability of specific habitats in shaping species distribution across the French Polynesian archipelago. For instance, in the Marquesas Islands, the absence of barrier reefs and lagoons, and the exposure to oceanic swell have been put forward to explain the dominance of crustose coralline algae and the low Phaeophyceae specific richness (Payri, N'Yeurt, Fiat, & Andréfouët, 2016). Because the Society, Austral, and Gambier archipelagos harbor more substantial habitat diversity (e.g., lagoons, fringing, and barrier reefs) compared to the Tuamotus, only consisting of atolls, it is expected that they harbor higher biodiversity (Sanciangco, Carpenter, Etnoyer, & Moretzsohn, 2013). They also happen to have received most of the sampling effort (Table S1). Since 2003, a series of scientific expeditions and research programs using both morphological and molecular approaches (the *Biocode* project in Moorea, the *Pakaihi i te Moana* expedition in the Marquesas, the *Global Reef Expedition of the Khaled bin Sultan Living Ocean Foundation* in the Gambier Islands, and the *AlgalReef* project in the Tuamotus) have been carried out to increase sampling efforts and better understand the distribution of macroalgal diversity in FP. The first published studies to use a molecular-based approach for the taxonomic revision of FP macroalgae focused on the Sargassaceae family (Fucales, Phaeophyceae; Rohfritsch, Payri, Stiger, & Bonhomme, 2007, 2010; Mattio, Payri, & Stiger-Pouvreau, 2008). Other molecular investigations of FP species appeared sporadically as a component of regional or global studies, for example for the red encrusting algal Order Corallinales (Rhodophyceae; Bittner et al., 2010; Caragnano et al. 2018), the red algal genus *Asparagopsis* (Bonnemaisoniales, Rhodophyceae; Dijoux, Viard, & Payri, 2014), the green algal genus *Halimeda* (Verbruggen, De Clerck, N'Yeurt, Spalding, et al., 2006) or the brown algal genera *Dictyota* (Dictyotales, Phaeophyceae; De Clerk et al., 2006) and *Lobophora* (Dictyotales, Phaeophyceae; Vieira, Camacho et al., 2016).

In general, Phaeophyceae is the group that has received the most attention in FP. Indeed, despite being the least diverse lineage compared to Chlorophyceae and Rhodophyceae, Phaeophyceae species often represent the most abundant biomass, are particularly significant ecologically (e.g., *Sargassum* beds provide food, habitat and nursery grounds for a wide array of marine organisms; Chaves, Pereira, & Feitosa, 2013), and have also been put forward for their economic potential (Zubia, Payri, Deslandes, & Guezennec, 2003; Zubia, Payri, & Deslandes, 2008; Zubia, Andréfouët, & Payri, 2014). Currently, taking into account the latest molecular work on *Lobophora* (Vieira, Camacho et al., 2016; 2017), the diversity of Phaeophyceae in FP accounts for 44 species (26 Dictyotales, 8 Ectocarpales, 4 Fucales, 1 Ralphsiales, 1 Scytothamnales, 2 Sphacelariales, 2 Tilopteridales) (Table 1). The highest Phaeophyceae diversity is reported for the Society Archipelago with 22 species. *Turbinaria ornata* and *Sargassum pacificum* (Fucales) are the most prevalent species, particularly in the case of abnormal proliferation, during which they represent the main species involved (Stiger & Payri, 1999a, 1999b; Payri & Stiger, 2001). Their molecular diversity was assessed in detail (Mattio et al., 2008; Rohfritsch et al., 2007, 2010). In contrast, the more species-rich order Dictyotales has received little or no attention

Table 1 Phaeophyceae species of French Polynesia and their distribution across the five archipelagos (Mattio et al., 2008; N'Yeurt & Payri, 2006; 2010; Vieira, Camacho et al., 2016; 2017).

	Austral	Gambier	Marquesas	Society	Tuamotu
Dictyotales Bory					
Dictyopteris repens (Okamura) Børgesen				+	
Dictyota acutiloba★ J.Agardh	+				
Dictyota bartayresiana J.V. Lamouroux	+			+	
Dictyota ceylanica Kützing		+		+	
Dictyota friabilis Setchell		+		+	+
Dictyota hamifera Setchell			+	+	
Lobophora obscura (Dickie) C.W.Vieira, De Clerck & Payri				+	
Lobophora pacifica (Setch.) C.W.Vieira, De Clerck & Payri				+	
Lobophora sp.6		+			
Lobophora sp.11	+	+			
Lobophora sp.13		+			
Lobophora sp.17					+
Lobophora sp.19		+		+	
Lobophora sp.31			+		
Lobophora sp.32			+		
Lobophora sp. 57			+		
Lobophora sp.70			+		
Lobophora sp.73			+		
Lobophora sp.74			+		
Lobophora variegata (J.V.Lamouroux) Womersley ex E.C.Oliveira	+	+		+	+
Padina boryana Thivy		+		+	
Padina melemele★ I.A.Abbott & Magruder	+				
Padina pavonica (Linnaeus) Thivy				+	
Padina stipitata Tanaka & Nozawa	+				

(Continued)

Table 1 Phaeophyceae species of French Polynesia and their distribution across the five archipelagos (Mattio et al., 2008; N'Yeurt & Payri, 2006; 2010; Vieira, Camacho et al., 2016; 2017).—cont'd

	Austral	Gambier	Marquesas	Society	Tuamotu
Spatoglossum asperum J.Agardh		+		+	
Stypopodium australasicum★ (Zanardini) Allender & Kraft	+				
Ectocarpales Bessey					
Chnoospora implexa J.Agardh	+	+		+	
Chnoospora minima (Hering) Papenfuss	+		+	+	
Cladosiphon novae-caledoniae★ Kylin	+				
Colpomenia sinuosa (Mertens ex Roth) Derbès & Solier	+	+		+	+
Hincksia mitchelliae (Harvey) P.C.Silva	+				+
Hydroclathrus clathratus (C.Agardh) M.Howe	+	+		+	
Hydroclathrus tumulis★ Kraft & Abbott	+				
Rosenvingea intricata (J.Agardh) Børgesen	+			+	+
Fucales Bory					
Sargassum echinocarpum J.Agardh	+	+			
Sargassum obtusifolium J.Agardh	+				
Sargassum pacificum Bory		+		+	
Turbinaria ornata (Turner) J.Agardh	+			+	
Ralfsiales Nakamura ex P.-E.Lim & H.Kawai					
Ralfsia expansa (J.Agardh) J.Agardh				+	

Table 1 Phaeophyceae species of French Polynesia and their distribution across the five archipelagos (Mattio et al., 2008; N'Yeurt & Payri, 2006; 2010; Vieira, Camacho et al., 2016; 2017).—cont'd

	Austral	Gambier	Marquesas	Society	Tuamotu
Scytothamnales A.F.Peters & M.N.Clayton					
Asteronema breviarticulatum (J.Agardh) Ouriques & Bouzon		+		+	
Sphacelariales Migula					
Sphacelaria rigidula Kützing	+			+	
Sphacelaria tribuloides Meneghini				+	
Tilopteridales Bessey					
Cutleria irregularis I.A.Abbott & Huisman	+				
*Cutleria mollis** Allender & Kraft	+				
Number of species per archipelago	**21**	**15**	**8**	**22**	**6**

(*) record is for Rapa Island only.

although a significant increase in their abundance has been observed on the reefs surrounding the island of Tahiti over the past decade (Zubia, pers. obs.). The present study reports the first detailed assessment of the family Dictyotaceae in FP and, as a background, provides a review of the current state of knowledge on the species diversity, ecology, and potential uses of Dictyotaceae species present in FP.

2. Diversity and distribution of the family Dictyotaceae in French Polynesia

The order Dictyotales is composed of only one family, the Dictyotaceae, which is subdivided into 19 genera (Bittner et al., 2008; Guiry & Guiry, 2019). Most Dictyotaceae belong to three genera, namely *Dictyota*, *Lobophora*, and *Padina*, mainly present in tropical regions (Silberfeld,

Fig. 2 Illustrations of the most abundant Dictyotaceae species in Tahiti Island: (A) *Spatoglossum asperum*; (B) *Dictyota bartayresiana*; (C) *Lobophora pacifica*; (D) *Padina boryana*. @: Zubia M., UPF.

Rousseau, & De Reviers, 2014; Vieira et al., 2017). Based on the available literature, we listed a total of 26 Dictyotaceae species recorded from FP that belong to six genera: *Dictyopteris* (1 species), *Dictyota* (5), *Lobophora* (14), *Padina* (4), *Spatoglossum* (1) and *Stypopodium* (1) (Table 1; Fig. 2). The molecular work carried out on the genus *Lobophora* by Vieira, Camacho et al. (2016; 2017), at the Indo-Pacific scale, extends the known diversity to more than 14 taxa in FP, including two new species, *L. pacifica* and *L. obscura,* and 12 undescribed species (Table 1). Molecular analyses also identified three *Dictyota* species in FP: *D. ceylanica*, *D. hamifera*, and *D. friabilis* (Tronholm, Steen et al., 2010; Tronholm, Sansón, Afonso-Carillo, Verbruggen, De Clerck, 2010). Except for *Lobophora* and *Dictyota*, none of the other Dictyotaceae genera have been re-assessed using molecular data, and the French Polynesian Dictyotaceae diversity is probably largely underestimated.

Compared to other archipelagos, the Society Islands have been the subject of more detailed studies, and 11 Dictyotaceae species have been recorded (Table 1). Between November 2017 and September 2018, we collected a total of 49 specimens of Dictyotaceae at three sites in Tahiti (Arue, Faa'a, and Punaauia). The methods are detailed in the Supplementary Material S2. The resulting species list is presented in Table 2 and includes, in particular, the new species *L. pacifica* described by Vieira, Camacho et al. (2016). This species was previously recorded as *L. variegata*, which, as a result of the revision, is now geographically restricted to the Caribbean region. Our observations confirmed the presence of most of the species listed by N'Yeurt and Payri (2006), except *Dictyopteris repens*. The molecular analyses carried out on our samples also revealed the presence of species previously unrecorded from Tahiti, as well as previous misidentifications. This is the case of two *Padina* species that were previously overlooked in FP: *P. boergesenii* (previously misidentified in FP as *P. pavonica*, a widely distributed Atlantic and Mediterranean species; Gómez Garreta, Rull, & Ribera Siguán, 2007) and *P. minor* (previously subsumed in FP under *P. boryana*) (Table 2).

3. Ecological dynamics of Dictyotaceae in Tahiti

To better understand the ecological dynamics of the Dictyotaceae species around Tahiti Island, we conducted several surveys during a period of one year (March 2018 to March 2019) at two sites, on the north coast (Arue 17.51°S-149.54°W) and on the west coast (Punaauia 17.66°S-149.60°W). Methods are detailed in the supplementary material S2. This section updates the previous knowledge on the ecology of these species using data obtained during this monitoring (percent cover, water temperature, swell, and nutrients). A bibliographical synthesis of the biotic factors that influence these populations is also presented. Finally, we discuss the risk of Dictyotaceae proliferation in FP.

3.1 Abundance, seasonality, and habitat of Dictyotaceae in Tahiti

All species, genera, and sites included, the percent cover of Dictyotales varied from 3.39 ± 9.68% in October to 14.55 ± 5.16% in March 2018 (Table S2). The maximum abundances were observed on reefs of Punaauia. We provide the results for each genus in Table 3 and discuss them below.

Lobophora. *Lobophora* species were found in a wide diversity of habitats. Because of limited morphological differences between *Lobophora* species,

Table 2 Dictyotaceae species collected at different sites around Tahiti Island during previous studies.

Species	Afaahiti	Arue	Faa'a	Paea	Papeari	Punaauia	Tata'a	Markers used
Dictyopteris repens (Okamura)Børgesen	+					+		–
Dictyota bartayresiana J.V.Lamouroux		★	★		+	+/★		cox3/psbA
Dictyota ceylanica Kützing		★	✧			+/★		cox3/psbA/cox1/26S
Dictyota friabilis Setchell		+/★/✧				+/★	+	psbA
Dictyota hamifera Setchell	+/✧	+/★	✧	+		+/★		psbA/cox1/cox3/nad1
Lobophora pacifica (Setch.) C.W.Vieira, De Clerck & Payri	+		✧			+/✧		cox3/rbcL
Padina boryana Thivy	+		★		+	+/★	+	cox3
Padina boergesenii Allender & Kraft						★		cox3
Padina minor Yamada			★			★		cox3/psbA
Spatoglossum asperum J.Agardh	+	+/★				+/★		psbA

(+): N'Yeurt & Payri, 2006; (✧): Vieira, Camacho et al., 2016, 2017; Tronholm, Sansón, Afonso-Carillo, Verbruggen, De Clerck, 2010; Tronholm, Steen et al., 2010; and (★) new collections between November 2017 and September 2018; the DNA markers used in each study are indicated.

Table 3 Percent cover (mean ± sd) of Dictyotaceae species measured during our survey in 2018 at Arue and Punaauia (Tahiti Island).

	Arue			Punaauia		
	March	**July**	**October**	**March**	**July**	**October**
Dictyotaceae	4.48 ± 4.18	5.22 ± 10.82	3.39 ± 9.68	14.55 ± 5.16	7.42 ± 8.80	6.43 ± 6.53
Dictyota bartayresiana	3.84 ± 2.69	2.53 ± 2.22	1.72 ± 1.45	6.63 ± 4.82	2.59 ± 2.43	1.68 ± 1.46
Dictyota hamifera	1.53 ± 1.43	1.40 ± 1.89	2.04 ± 1.48	–	–	1.26 ± 1.53
Dictyota friabilis	0.62 ± 0.52	0.27 ± 0.10	–	0.57 ± 0.50	0.50 ± 0.38	0.34*
Dictyota ceylanica	0.83 ± 0.54	–	–	–	–	0.60*
Lobophora spp.	2.13 ± 1.91	4.20 ± 6.68	3.18 ± 3.99	0.29 ± 0.22	1.71 ± 1.44	1.57 ± 1.45
Padina spp.	2.03 ± 3.36	0.63 ± 0.41	1.16 ± 1.54	11.80 ± 13.36	5.15 ± 4.57	5.37 ± 7.76
Spatoglossum asperum	–	–	–	3.24 ± 2.44	4.96 ± 5.01	2.63 ± 2.15

(–) Species absent; (*) Species found only once.

identification in the field is difficult. This was a limiting factor for our monitoring, and we only recorded them as *Lobophora* spp. At all sites surveyed, *Lobophora* spp. cover was low and ranged from 0.29 ± 0.22% (March) to 4.20 ± 6.68% (July) (Table 3). The highest abundances were observed during the austral winter (July) at both sites. While these results tend to suggest that *Lobophora* species are not abundant on Tahitian reefs, they do not account for the cryptic biomass growing under the canopy of other seaweeds (e.g., *Sargassum, Turbinaria*), or on the underside of rocks and crevices.

Dictyota. *Dictyota* species can grow on various substrates and can epiphytize numerous organisms (Beach et al., 2003; Herren, Walters, & Beach, 2006). During our survey, *Dictyota bartayresiana* was found growing in dense patches or balls of up to 15 cm in diameter, on coral rubbles, dead corals, and pavement. Sometimes this species was epiphytic on corals, sponges, and other benthic organisms. The most abundant *Dictyota* species was *D. bartayresiana*; although its percent cover was under 10% (Table 3). Monthly surveys found a similar seasonality at both Arue and Punaauia, with the highest abundance occurring in March (3.84 ± 2.69% and 6.63 ± 4.82%, respectively) and the lowest abundance in October (1.72 ± 1.45% and

1.68 ± 1.46%, respectively) (Table 3). *Dictyota hamifera* was growing mainly as an epiphyte on *Sargassum pacificum* or *Turbinaria ornata*. This species is easy to recognize and distinguish from other species in the genus thanks to its branchlets bearing hooks. At Punaauia, *D. hamifera* was rare and only found in October 2018 (Table 3). At Arue, this species was found throughout the year within *Sargassum* beds on the algal crest, with the highest abundant observed on October 2018 (2.04 ± 1.48%) (Table 3). The species *D. friabilis* and *D. ceylanica* were found in low abundances (<1% cover) in all habitats and at both sites (Table 3).

Padina. The most common *Padina* species found in Tahiti was *P. boryana*. *Padina* species were found throughout the year at Arue and Punaauia mostly on coral rubbles in sandy habitats or located at the base of dead corals. The abundance of *Padina* was much larger at Punaauia (5–12%) in comparison to Arue (1–2%) (Table 3), probably because more sandy sites were surveyed at Punaauia. The abundance of *Padina* was the lowest during the austral winter (July; 0.63 ± 0.41% at Arue, and 5.15 ± 4.57% at Punaauia) and reached its maximum during the austral summer (March; 2.03 ± 3.36% at Arue and 11.80 ± 13.36% at Punaauia) (Table 3).

Spatoglossum. *Spatoglossum asperum* was commonly found at the base of massive coral colonies (such as *Porites*) in the lagoon (0.5–2 m deep) or forming large beds at depths of 15–20 m on the outer slope of Taharaa and Punaauia. This species was also found in large patches at 100 m deep outside of Papeete harbor (Payri, com.pers.). In lagoons, older plants can reach large sizes (50 cm) and are heavily epiphytized by crustose coralline algae and turf. During our survey, *S. asperum* was only found on the reef of Punaauia, where its abundance increased from the beginning of the year to reach a maximum in July (4.96 ± 5.01%), and then declined to reach a minimum abundance in October (2.63 ± 2.15%) (Table 3).

3.2 Main factors controlling the abundance of Dictyotaceae species

The combination of bottom-up (e.g., nutrient supply, light availability) and top-down control (e.g., herbivory) usually maintains macroalgal abundance and prevents macroalgal proliferation (Lapointe, Littler, & Littler, 1997; Littler, Littler, & Brooks, 2006; Smith, Hunter, & Smith, 2010). The loss of this ecosystem balance can lead to a proliferation of macroalgae, and in the long-term to ecological phase-shifts (Fong, Frazier, Tompkins-Cook, Muthukrishnan, & Fong, 2016; Hughes, 1994; Hughes, Graham, Jackson,

Mumby, & Steneck, 2010; Rasher, Hoey, & Hay, 2013). Dictyotales species including members of the genera *Lobophora, Dictyota,* and *Stypopodium* have been reported to proliferate in several regions of the world (Barott et al., 2011; Bonaldo & Hay, 2014; Hughes et al., 2010; Longo & Hay, 2015; 2017; Rasher et al., 2013). Following reef degradation, damaged corals are initially overgrown by small ephemeral algae, which are quickly replaced by longer-lived taxa, notably *Sargassum, Lobophora, Dictyota* and *Padina,* some of which could represent a threat to coral reefs (Brown et al., 2018; Fong & Paul, 2011, pp. 241–272; Ho & Carpenter, 2017; Mumby, Steneck, Adjeroud, & Arnold, 2016; Paul et al., 2011). In the following paragraphs, we discuss some of the main ecological processes known to control the abundance and distribution of Dictyotaceae species globally.

3.2.1 Abiotic factors

3.2.1.1 Nutrients

Macroalgae need macronutrients (e.g., carbon, nitrogen, and phosphorus), micronutrients (e.g., iron, zinc, etc.), and vitamins (e.g., vitamin B12) to grow (Lobban & Harrison, 1994, p. 366). An excess of nutrients can favor the overgrowth of certain species of macroalgae (Zubia, Depetris, Flores, Turquet, & Cuet, 2018). As a bottom-up factor, nutrification can lead to coral reef degradation (Fabricius, 2005; Lapointe, 1997). Several sources of nutrients have been described including groundwater flow, terrestrial run-off (Camoin, Gautret, Montagioni, & Cabioch, 1999), upwelling (Camoin et al., 2006), and dissolved organic matter (DOM) derived from reef organisms (Heindel et al., 2012). In Tahiti, during our surveys, nitrogen concentration reached a maximum during the hot and rainy season (November to March; >1 μmol.L^{-1}) and a minimum during the cold and dry season (April to October; <1 μmol. L^{-1}, Fig. S2). In Tahiti, annual precipitations can reach up to 10,000 mm/year (Pheulpin, Recking, Sichoix, & Barriot, 2016). Therefore, the riverine input is particularly significant during the rainy season (November to March), during which large amounts of sediments are released into the lagoon, increasing nutrient concentrations (Pheulpin, Sichoix, Barriot, & Recking, 2014). In our study, the abundance of *D. bartayresiana* appeared to follow nitrogen concentrations, with the highest percent cover occurring during the rainy/hot season (Table 3; Fig. S2). In Moorea, Clausing and Fong (2016) also reported a higher growth rate in *D. bartayresiana* after a rainy period followed by high light intensity. These observations tend to support the hypothesis that nutrient inputs enhance the growth of *D. bartayresiana*. Besides, experiments in the Caribbean reefs have demonstrated that the growth and proliferation of *Dictyota* spp. were influenced mainly by nutrient enrichment and herbivore exclusion (Sotka & Hay, 2009).

3.2.1.2 Water temperature

Water temperature has a direct effect on metabolic and photosynthetic rates, which control growth, reproduction, and decay of seaweeds (Lüning & Tom Dieck, 1989). It is considered a key driver of seasonal fluctuations of algal biomass (Fulton et al., 2014). During our survey, the minimum sea surface temperature (SST) was recorded at Arue in September 2018 (26.4 °C) and the highest in May 2019 (31.9 °C) (Fig. S3). The cover of *Dictyota bartayresiana* and *Spatoglossum asperum* seemed to follow STT fluctuations in different ways over the survey period (Table 3). During the warm season (November to April), *S. asperum* was less abundant, and thalli started to deteriorate, while this was the period when *D. bartayresiana* was the most abundant. These observations illustrate the different thermal preferences of *S. asperum* and *D. bartayresiana*, and, therefore, their distinct seasonality.

3.2.1.3 Swell

Swell is a determinant factor which regulates algal abundance in coral reefs by ripping macroalgae off the reef and exporting them to other coral reefs (Herren et al., 2006). For example, strong swells can rip off macroalgae and contribute to the formation of drifting rafts (Goldberg & Kendrick, 2004; Zubia et al., 2014). The eastern and northern coasts of Tahiti are exposed to swell mostly during the hot season (October to April), while south swell is almost constant throughout the year (Laurent, Maamaatuaiahutapu, Maiau, & Varney, 2004). Swells generated by hurricanes and lows, between December and April, may also come from the west (Andréfouët & Adjeroud, 2019). Data collected during our surveys follow these general patterns (Fig. S3). The swell energy at Punaauia was high during the cold season (May to October), especially in June, September, and October (>30 $kW.m^{-1}$), which also coincided with lower abundances of *D. bartayresiana* (1.68 ± 1.46%) (Table 3; Fig. S3). *Spatoglossum asperum* was more abundant in July (4.96 ± 5.01%) when the swell energy was the highest at Punaauia (Table 3; Fig. S3). *Spatoglossum asperum* appeared less influenced by the swell, probably because it occurs far from the reef crest and is protected by massive coral colonies at the base of which it grows.

3.2.2 Biotic factors

3.2.2.1 Herbivory

Herbivory is one of the critical factors that control macroalgal abundance. The lack of grazing due to overfishing, for example, can favor the proliferation of macroalgal species, sometimes leading to phase-shifts where macroalgae-dominated reefs replace coral-dominated reefs (Done, 1992). Macroalgae are also known to naturally produce allelopathic compounds that deter herbivores (Hay, 1997; Rasher & Hay, 2014; Sotka, Bell, Lowry, & Poore, 2017). Some herbivores have adapted to tolerate these defenses, but intense grazing pressure can favor macroalgae with a wide range of chemical defenses (Sotka, Bell, Lowry, & Poore, 2017).

Herbivore assemblages in Moorea, an island of the Society Archipelago near Tahiti, are mainly composed of species that belong to the families Acanthuridae, Scaridae, and Pomacanthidae with six fish species responsible for over 80% of the total macroalgae biomass consumption: *Ctenochaetus striatus, Scarus psittacus, Chlorurus spilurus, Acanthurus nigrofuscus, Zebrasoma scopas,* and *Naso lituratus* (Holbrook, Schmitt, Adam, & Brooks, 2016). According to Holbrook et al. (2016), sea urchins and other invertebrates contribute little to macroalgal herbivory in Moorea, although Zizka (2012) showed that *Dictyota bartayresiana* is the most consumed species by the sea urchins *Echinometra.* Fong (2015) and Fong et al. (2016) reported that grazing is an effective biomass regulator for *Dictyota bartayresiana* in Moorea, especially by large grazers when there is little influence of nutrients. *Dictyota bartayresiana* was also rapidly consumed by herbivorous fish species in Fiji (up to 86–100% after 48 h; Rasher et al., 2013). The efficiency of *Dictyota* chemical defenses varies depending on herbivorous species (Sotka, Bell, Lowry, & Poore, 2017). For example, *Dictyota pfafii* showed significant activity against generalist herbivorous fishes and the sea urchin *Lytechinus variegatus,* while no effect was observed against the crab *Pachygrapsus transversus* (Barbosa et al., 2004). Pereira, Cavalcanti, & Texeira, 2000; Sotka, Bell, Lowry, & Poore (2017) report that herbivores highly appreciate *Dictyota* species despite their high level of secondary metabolites.

Grazing is a critical top-down process to maintain low *Lobophora* cover as illustrated by the shift in assemblages that occurred following herbivore die-offs in the Caribbean (Van Steveninck & Breeman, 1987) and the Canary Islands (Sangil, Sansón, Clemente, Afonso-Carrillo, & Hernández, 2014). In Palau, Mumby et al. (2016) showed that the primary grazers of *Lobophora* were Acanthuridae. Vieira, Stenger, Moleana, De Clerck, and Payri (2019) demonstrated limited variation in grazing sensitivity in *Lobophora* species by the sea-urchin *Diadema setosum* and the rabbitfish *Siganus lineatus*. These authors concluded that spatial and associational refuges are primary defense strategies against herbivores. For instance, in Tahiti, *Lobophora* spp. are commonly found growing under the canopy of *Sargassum* and *Turbinaria* or encrusting reef crevices.

Padina species are calcifying macroalgae that consistently rank low in herbivorous fish preference (Hay, 1991). They are chemically defended by secondary metabolites (Renaud, Hay, & Schmitt, 1990). For instance, unlike parrotfish species, Siganidae species actively avoid calcified macroalgae such as *Padina* due to the high acidity of their stomachs, (Mantyka & Bellwood, 2007; Schupp & Paul, 1994). Herbivory also plays an essential role in controlling the morphology of some *Padina* species. For instance, *Padina sanctae-crucis* and *P. boergesenii* can shift from a vegetative turf morphology (*Vaughaniella* stage) to a foliose and fan-shaped morphology when grazing pressure is low, allowing them to persist in habitats subject to various grazing pressures (Diaz-Pulido, Villamil, & Almanza, 2007; Lewis, Norris, & Searles, 1987). No data is currently available on the effect of grazing on *Padina* and *Spatoglossum* species in FP.

3.2.2.2 Interactions with corals

While little is known about the coral reef health in Tahiti due to the absence of surveys, in the neighboring island of Moorea, located 25 km from Tahiti, the situation is better informed. Since 1979, severe degradations of the coral reefs of Moorea have been observed, and several human and natural disturbances have been incriminated. These include organic and chemical pollution, terrestrial runoff, dredging, coral exploitation, microbial diseases, coral bleaching, and *Acanthaster planci* infestations (Faure, 1989; Gleason & Wellington, 1993; Kayal et al., 2012; Trapon, Pratchett, & Penin, 2011). These disturbances are not only affecting coral species abundance and assemblages (i.e., the decline in *Acropora* and increasing prevalence of *Pocillopora* and *Porites*; Trapon et al., 2011), but they also impact macroalgal assemblages and cover (e.g., *Turbinaria ornata*; Bittick, Clausing, Fong, Scoma, & Fong, 2019). As corals decline and macroalgae proliferate, their interactions intensify and potentially affect the survival, growth, reproduction, and recruitment of coral larvae (e.g., Diaz-Pulido, Harii, McCook, & Hoegh-Guldberg, 2010; Ritson-Williams, Arnold, Fogarty, Steneck, et al., 2009). Macroalgae then directly or indirectly impact coral health through shading, abrasion, transmitting coral diseases or microbes, and allelochemicals (Hughes et al., 2007; McCook, Jompa, & Diaz-Pulido, 2001; Nugues,

Smith, Van Hooindonk, Seabra, & Bak, 2004; Vieira, Engelen et al., 2016). However, macroalgae-coral interactions are known to be species-specific, which means that different coral species will not be equally sensitive and affected by the same macroalgal species. The genus *Lobophora* is known for its ability to outcompete corals through direct contact and associated microbiome carrying coral pathogens leading to coral bleaching and death (Vieira, Payri, & De Clerck, 2015; Vieira, Camacho et al., 2016; Vieira, Engelen et al., 2016). The situation is similar for the genus *Dictyota* which can induce coral bleaching and death through abrasion and the effect of their associated microbiomes (Barott et al., 2011; Rasher & Hay, 2010, 2011), and inhibit larval settlement (Diaz-Pulido et al., 2010; Paul et al., 2011). For instance, in Moorea, Longo and Hay (2017) showed that *D. bartayresiana* lipid-soluble extract could decrease the photochemical efficiency of *Pocillopora verrucosa* by 20–30%. Similarly, *Padina* species have been shown to impact the settlement and metamorphosis of *Pocillopora damicornis* larvae in the Philippines (Maypa & Raymundo, 2004), and Kaullysing et al. (2016) suggested that *Padina boryana* overgrowth is potentially stressful to the *Acropora muricata* colonies. To the best of our knowledge, there is no study reporting negative or positive interactions between *Spatoglossum* and corals.

3.3 Is there a risk for further macroalgae proliferation in Tahiti?

The coral reefs of FP have experienced repeated macroalgal blooms over the last decades. These events mainly involved two species of Sargassaceae (Fucales): *Sargassum pacificum* and *Turbinaria ornata*, which also happen to be the most prevalent species in terms of biomass in normal conditions (Payri & Stiger, 2001; Stiger & Payri, 1999a, 1999b). These proliferation events have prompted intense efforts in fundamental and applied research on macroalgae in the region. Recently, however, these proliferations have occurred with a higher frequency, and they now seem to involve additional species of the family Dictyotaceae. Indeed, over the past decade, two species of the family Dictyotaceae (*Dictyota bartayresiana* and *Spatoglossum asperum*) have become increasingly abundant in the lagoons of Tahiti (Zubia, pers. obs.). However, the percent cover measured during our survey is low at both sites (<10%) and does not currently support the hypothesis of an ongoing proliferation of these species. In the absence of previous baseline data, it is not possible to tell whether the abundance of these species has indeed significantly increased over the past decade. In the neighboring island of Moorea, Adam et al. (2011) and Holbrook et al. (2016) have demonstrated that intense grazing could prevent the establishment and proliferation of macroalgae. Although this biotic factor was not assessed during our surveys, we recommend that it is included in further studies and monitoring in Tahiti, where inhabitants and environmental pressures are high in comparison to other islands in FP (Andréfouët & Adjeroud, 2019). With increasing global ocean acidification, this is particularly worrying because naturally acidified reefs tend to host Dictyotaceae species such as *Dictyota* (Del Monaco, Hay, Gartrell, Mumby, et al., 2017), *Lobophora* (Bouraké, New Caledonia; Payri, com. Pers.), and *Spatoglossum* species (Enochs et al., 2015). For instance, Enochs et al. (2015) observed a clear shift from coral to fleshy macroalgae ecosystems at low pH with *Spatoglossum stipitatum* covering more than 50% of the substrate.

4. Potential for the development of biotechnological applications

Through evolution, macroalgae have adapted to external pressures (e.g., desiccation, UV radiations, infections, grazing) and to mitigate these pressures some species are producing a wide array of bioactive metabolites (Bischof et al., 2006; Duffy & Harvilicz, 2001; Steinberg & De Nys, 2002). With more than 1500 metabolites described, the class Phaeophyceae (brown algae) represents a rich source of natural products that belong to several chemical families such as terpenoids, phenols, fatty acids derivatives, and polysaccharides among others. Within the brown algae, the families Dictyotaceae and Sargassaceae, in particular, are recognized for the production of phenolics (mainly phlorotannins) and terpenoids (mainly sesquiterpenes, diterpenes and meroterpenes) (Carroll, Copp, Davis, Keysers, & Prinsep, 2019; and previous reports in the series) which are widely used in the industry. Here, we review the metabolites known for the species of Dictyotaceae recorded in FP.

4.1 Chemodiversity & biological activities

Dictyotaceae species are well-known for their outstanding richness in terpenoids (De Paula, Vallim, & Teixeira, 2011; Vieira et al., 2017), which represents more than 33% of all the specialized metabolites described from the class Phaeophyceae (Carroll et al., 2019; and previous reports in the series). More specifically, several Dictyotaceae species can biosynthesize a wide array of sesquiterpenes (e.g., *Dictyopteris* and *Taonia* spp.), cyclic diterpenes (e.g., *Dictyota* and *Spatoglossum* spp.) or meroterpenes, also called terpenoids of mixed biogenesis (e.g., *Dictyopteris* and *Stypopodium* spp.). Over 200 diterpenes have been isolated from 35 species of Dictyotaceae worldwide, mainly from the genus *Dictyota* (Chen et al., 2018; Vallim, De Paula, Pereira, & Teixeira, 2005). Besides their use in taxonomic determination and biogeographic markers, these molecules have been found to exhibit interesting biological activities. Table 4 provides an overview of the compounds and biological activities recorded in the literature for Dictyotaceae species recorded in Tahiti.

Dictyopteris. Aside from two studies on the content of sulfuric acid and inorganic ions of Dictyotaceae species (Sasaki, Kataoka, Kamiya, & Kawai, 1999; Sasaki, Kataoka, Murakami, & Kawai, 2004) and on the antineoplastic activity of algae where its extracts were found to be inactive (Kashiwagi, Mynderse, Moore, & Norton, 1980), no comprehensive work has been published on the chemical characterization of the species *Dictyopteris repens*. This lack of information is unexpected as *Dictyopteris* spp. are known to biosynthesize a broad array of specific compounds such as C_{11} hydrocarbons, sesquiterpenes, merosesquiterpenes or sulfur containing derivatives (Philippus et al., 2018; Zatelli, Philippus, & Falkenberg, 2018).

Dictyota. Species of the genus *Dictyota* are mostly known for their ability to produce a wide range of cyclic diterpenes (Chen et al., 2018; Hay, 1997; Kelecom & Teixeira, 1986; Teixeira & Kelecom, 1988; Vallim et al., 2005). Cyclic diterpenes found in Dictyotaceae can be divided into three structural groups (Fig. 3). Group I includes compounds obtained through a first C-1/C-10 cyclization from geranylgeranylpyrophosphate (GGPP) including the prenylated guaiane subgroup among others. Group II contains diterpenes formed from GGPP through a first C-1/C-11 cyclization, belonging mainly to the dolabellane subgroup among others. Finally, group III contains compounds derived from a first C-2/C-10 cyclization of the GGPP precursor belonging to the crenulane, xeniane, and dichotomane families subgroups among others (Chen et al., 2018). Among the *Dictyota* species found in FP, *Dictyota bartayresiana* appeared as the most studied with several diterpenes isolated (Table 4, Fig. 4). *Dictyota bartayresiana* was also investigated for its fatty acids (Gosch, Paul, De Nys, & Magnusson, 2015) and lipid hydroperoxydes contents (Kumari, Singh, Bijo, Reddy, & Jha, 2012), and it was found to be rich in cellulose which could be interesting for further economic applications (Siddhanta et al., 2009). Several extracts of these species have been assayed for their antibacterial activity (Antonysamy, Velayutham, Mani, Thangaiah, & Irullappan, 2015; Parmar et al., 2016) and their antiviral properties against *Vaccinia* virus (Kamat et al., 1992). In

Table 4 Overview of the biological activities recorded in the literature for Dictyotaceae species present in French Polynesia with indications of the type of molecules reported and the type of extraction used.

Taxa	Molecules reported	Extraction methods	Biological activities	Location	References
Dictyotaceae	Dictyols (diterpenes)	–	–	Worldwide	Hay, 1997
Dictyotaceae	Diterpenes	–	Herbivory deterrent & use in chemotaxonomy	Worldwide	Kelecom & Teixeira, 1986
Dictyopteris	C11-hydrocarbons, sulfur compounds, terpenes, meroditerpenes, halogenated compounds and sterols	–	Therapeutic and ecological activities	Worldwide	Zatelli et al., 2018
Dictyopteris	No compound could be identified in the extracts	DCM:MeOH (2:1)	–	São Paulo, Brazil	Philippus et al., 2018
Dictyopteris repens	–	–	No Antineoplastic activity	Palau, Western Caroline Islands	Kashiwagi et al., 1980
Dictyopteris repens	Sulfuric acid and inorganic ions content	Crude extract distilled water	–	Ishigaki Island, Okinawa	Sasaki et al., 1999
Dictyopteris repens	Sulfuric acid and inorganic ions content	Crude extract distilled water	–	Ishigaki Island, Okinawa	Sasaki et al., 2004
Dictyota bartayresiana	–	Ethanol extract	Antiviral (*Vaccinia* virus)	Gujarat, India	Kamat et al., 1992
Dictyota bartayresiana	A known dolastane, five known dolabellanes and five new diterpenes	DCM:MeOH (3:1)	–	Mannar, India	Bheemasankara Rao et al., 1994

Dictyota bartayresiana	–	Acetone, ethanol and water extracts	Ichtyotoxic activity	Mexico	De Lara-Isassi et al., 2000
Dictyota bartayresiana	A known dolastane, five known dolabellane, five new compounds (3 dolabellanes, 2 dolastanes)	–	–	Mannar, India	Sarma, Krishna, & Kawai, 2006
Dictyota bartayresiana	Cellulose rich	Methanol extract	–	India	Siddhanta et al., 2009
Dictyota bartayresiana	–	Methanol extract	Antioxidant activities	India	Palanisamy et al., 2010
Dictyota bartayresiana	Terpenes	Methanol extract	Coral bleaching	Viti Levu, Fiji	Rasher, Stout, Engel, Kubanek, & Hay, 2011
Dictyota bartayresiana	Fatty acids and lipid hydroperoxydes contents	Chloroform:MeOH (1:2)	–	Saurashtra coast, Gujarat, India	Kumari et al., 2012
Dictyota bartayresiana	Synthesize silver nanoparticules (Ag-NPs)	Methanol extract	Antifungal activity	Tamilnadu, India	Kumar & Sudha, 2013
Dictyota bartayresiana	–	Methanol extract	Highest antidiatom and antifouling	Mannar, India	Deepa et al., 2014
Dictyota bartayresiana	–	Methanol extract	No haemogglutinin activity (blood cell agglutination)	Manoli island	Mangaiyarkarasi, Kanna, Girija, & Gnanamurthy, 2014

(Continued)

Table 4 Overview of the biological activities recorded in the literature for Dictyotaceae species present in French Polynesia with indications of the type of molecules reported and the type of extraction used.—cont'd

Taxa	Molecules reported	Extraction methods	Biological activities	Location	References
Dictyota bartayresiana	–	Methanol extract	*In vivo* lethality test *Artemia salina* (shrimp)	Tamilnadu, India	Antonysamy, Thangiah et al., 2015
Dictyota bartayresiana	–	Methanol extract	Larvicidal effect on *Culex quinquefasciatus*	Tamilnadu, India	Antonysamy et al., 2015a
Dictyota bartayresiana	–	Methanol extract	Antibacterial screening	Tamilnadu, India	Antonysamy, Velayutham et al., 2015
Dictyota bartayresiana	–	Aqueous extract	Cytotoxicity and anticancer potential	Tamilnadu, India	Antonysamy, Thangiah et al., 2015
Dictyota bartayresiana	–	Crude extract	Antibacterial (*Staphylococcus aureus* and *Escherichia coli*) and antifungal (*Candida albicans*) activity	Mtwapa, Kenya	Kaaria et al., 2015
Dictyota bartayresiana	–	Ethanol extract	low antibacterial	Gujarat, India	Parmar et al., 2016
Dictyota bartayresiana	Lipid soluble extracts	Methanol extract	Coral bleaching of *Porites* spp.	Votua Reef, Viti Levu, Fij	Rasher & Hay, 2010
Dictyota bartayresii	Diterpenoids	Crude extract	Antifungal activity	Belize	Norris & Fenical, 1982

Dictyota bartayresii	Pachydictyol-A	DCM:MeOH (3:1)	Host and herbivory test	Belize and Bahamas	Hay, Duffy& Fenical, 1990
Dictyota bartayresii	β-dimethylsulfonio-propionate content	25% NaOH (w/v) and DMS: acrylic acid (1:1)	–	Meixia, China	Bischoff et al., 1994
Dictyota bartayresii	Fatty acids content	Trans-esterification method	–	North Queensland, Australia	Gosh, Paul, De Nys, & Magnusson, 2015
Dictyota ceylanica	Phenolic concentrations	Methanol extract	–	Thevenard island, Western Australia	Van Hees et al., 2017
Dictyota friabilis	Dolabelladienetriol	CH_2Cl_2 extract	Strong potential against HIV-1 in cervical tissue explants	Atol das Rocas reef, Rio Grande do Norte State, Brazil	Stephen et al., 2017
Dictyota friabilis	Phlorotannins	Methanol extract	Prevent melanosis and extend shelf-life of *Litopenaeus vannamei* shrimps	Iran	Sharifian, Shabanpour, Taheri, & Kordjazi, 2019
Dictyota pfaffii	10,18-diacetoxy-8-hydroxy-2, 6-dolabelladiene, 10-acetoxy-8,18_dihydroxy-2, 6-dolabelladiene, 8,10,48-trihydroxy-2,6-dolabelladiene	MeOH:DCM (3:7), MeOH extracts	Antiviral (HIV) and anti-herpes virus activities	Rio Grande, Brazil	Barbosa et al., 2004
Dictyota pfaffii	Dolabellane diterpene: 10,18-diacetoxy-8-hydroxy-2,6-dolabelladiene	CH_2Cl_2: MeOH 7:3 and MeOH (100%)	Herbivory deterrent against sea-urchins and fish	Atol das Rocas reef, Rio Grande do Norte State, Brazil	Barbosa, Teixaira, Villaça, Pereira, et al., 2003

(Continued)

Table 4 Overview of the biological activities recorded in the literature for Dictyotaceae species present in French Polynesia with indications of the type of molecules reported and the type of extraction used.—cont'd

Taxa	Molecules reported	Extraction methods	Biological activities	Location	References
Dictyota pfaffii	Dolabellane diterpene	Crude extract	Herbivore deterrent against the sea-urchin *Lytechinus variegatus*, generalist fishes and not against crab *Pachygrapsus transversus*	Atol das Rocas reef, Rio Grande do Norte State, Brazil	Barbosa et al., 2004
Dictyota pfaffii	Diterpene	Hexane extract	Strong antifouling activity against the mussel *Perna perna*	Niterói, Rio de Janeiro, Brazil	Barbosa, Fleury, Da Gama, Teixeira, & Pereira, 2007
Dictyota pfaffii	Three new dolabellane diterpenes: Dolabelladienols A–C	CH_2Cl_2 extract	Antiviral activity (HIV) and low cytotoxic against MT-2 lymphocyte tumor cells	Atol das Rocas reef, Rio Grande do Norte State, Brazil	Pardo-Vargas et al., 2014
Dictyota pfaffii	Dolabelladienetriol	CH_2Cl_2 extract	Potential against HIV-1 on infected female mice	Atol das Rocas reef, Rio Grande do Norte State, Brazil	Garrido et al., 2017
Dictyota pfaffii	Dolabelladienetriol	CH_2Cl_2 extract	Low activity potential against HIV-1 on infected female mice	Atol das Rocas reef, Rio Grande do Norte State, Brazil	Garrido et al., 2011
Dictyota pfaffii	Dolabelladienetriol	—	Inhibition of HIV-1 replication in human primary cells	Brazil	Cirne-Santos et al., 2006
Dictyota pfaffii	Dolabelladienetriol	Dolabelladienetriol	Inhibition of HIV-1 on peripheral blood mononuclear cells	Brazil	Cirne-Santos et al., 2008

Dictyota pfaffii	Dolabelladienetriol	Two diterpenes	Inhibition of HSV-1 infection in Vero cells	Brazil	Abrantes et al., 2010
Dictyota pfaffii	Dolabelladienetriol	Dolabelladienetriol	Inhibit Leishmania in macrophages	Brazil	Soares et al., 2012
Dictyota spp.	Dictyoepoxide, pachydictyol-A, pachydictyol-A epoxide, acetals 2a and 2b, dictyolactone, acetyl dictyolal, dictyoxide, dictyol B acetate	CH_2Cl_2 extract and Trimethylpentane	Inhibition of vasopressin binding to pig kidney	Bahia, Los Angeles	Patil et al., 1993
Dictyota spp.	pachydictyol-A acetals 2a and 2b, dictyolactone	CH_2Cl_2 extract and Trimethylpentane	Vasopressin antagonists	Bahia, Los Angeles	Patil et al., 1993
Dictyota spp.	Diterpene alcohols pachydictyol A and dictyol E	Lipid-soluble extract	Inhibit invertebrate larvae settlement	North Carolina, USA	Schmitt, Hay, & Lindquist, 1995
Dictyota spp.	—	—	Inhibit invertebrate larvae settlement	Hawai'i	Walters, Ruban, & Horton, 1996
Dictyota spp.	Rich in terpenoid	—	Defenses against generalist herbivores	Worldwide	Hay, 1996
Dictyota spp.	Lipid-soluble chemical	—	Defenses that deter feeding by reef herbivores	Worldwide	Hay, 1997
Dictyota spp.	Rich in terpenoid	—	Defenses against generalist herbivores	Worldwide	Fong & Paul, 2011
Dictyota spp.	Dictyols	—	Herbivory deterrent	Worldwide	Paul & Puglisi, 2004

(Continued)

Table 4 Overview of the biological activities recorded in the literature for Dictyotaceae species present in French Polynesia with indications of the type of molecules reported and the type of extraction used.—cont'd

Taxa	Molecules reported	Extraction methods	Biological activities	Location	References
Dictyota spp.	Rich in terpenoid	—	Defenses against generalist herbivores	Tropical Atlantic American region	Vallim et al., 2005
Dictyota spp.	—	—	Increased mortality rates of coral recruits	Long Key, Florida	Kuffner et al., 2006
Dictyota spp.	93 diterpenes and 15 skeletal classes	—	Herbivory deterrent & use in chemotaxonomy	Worldwide	Kelecom & Teixeira, 1986
Dictyota spp.	Over 90 diterpenes and 17 skeletal classes	—	Chemotaxonomy	Worldwide	Teixeira & Kelecom, 1988
Dictyota spp.	233 diterpenes	—	—	Worldwide	Chen et al., 2018
Padina boergesenii	Squalene	N-hexane defatted methanol (99.5%) crude extract	Antioxidant and anti-inflammatory	Gulf of Mannar, India	Rajamani et al., 2018
Padina boergesenii	Triterpene	Methanol extract	Enzyme inhibitory activities	Gulf of Oman	Ali et al., 2017
Padina boryana	Fucose, galactose, mannose, glucose and uronic acid residues	Sulfated and Acetylated fraction	Anticancer activity *in vitro*	Nhatrang Bay, Vietnam	Usoltseva et al. 2018
Padina boryana	Polysaccharides especially galactofucans	Aqueous EtOH (70%) with added $NaHCO_3$	Non-toxic and antitumor activity *in vitro* against DLD-1 human colon carcinoma cells	Vietnam	Shevchenko et al., 2017
Padina boryana	High productivity of lipids and fatty acid methyl esters	DCM:MeOH (2:1) with added NaCl	Candidate for investigation of biodiesel production	Red Sea at Jeddah Corniche, Saudi Arabia	Abomohra et al., 2018

Padina boryana	Polyphenolic and flavonoid contents	Ethanol, acetone, and hexane extract	Antimicrobial activity against Gram-negative and positive strains and a pathogen fungus	Red Sea in Al-Leith Provence, Saudi Arabia	Sameeh et al., 2016
Padina boryana	Biochemical composition	—	Antioxidant activity and potential food uses	Abu Qir Bay, Egypt	Ismail et al., 2017
Padina boryana	Lipid soluble extract	Methanol extract	Allelopathic effect on coral *Porites porites*	Votua Reef, Viti Levu, Fij	Rasher & Hay, 2010
Padina minor	β-dimethylsulfonio-propionate content	25% NaOH (w/v) and DMS: acrylic acid (1:1)	—	Ximaozhu, China	Bischoff et al., 1994
Padina pavonica	—	Methanol extract	Antibiotic activities against *Bacillus subtilis,* (B*) cereus* and *Staphylococcus aureus/* Reported as inactive for antimicrobial activities in previous Indian and Mediterranean studies	Yucatan, Mexico	Salvador, Gómez Garreta, Lavelli, & Ribera, 2007
Padina pavonica	Chemical composition (total phenolic, flavonoid, tannin, lipid and sterol profiles)	Acetone extract	Anti-proliferative and pro-apoptotic activity on human osteosarcoma cells and antioxidant activity	Moorea, French Polynesia	Bernardini et al., 2018

(Continued)

Table 4 Overview of the biological activities recorded in the literature for Dictyotaceae species present in French Polynesia with indications of the type of molecules reported and the type of extraction used.—cont'd

Taxa	Molecules reported	Extraction methods	Biological activities	Location	References
Padina sp.	Contained additional non-polyphenolic non-polar secondary metabolites	Aqueous methanol extract	Herbivory deterrent	Australia	Steinberg & Paul, 1990
Spatoglossum asperum	—	Ethanol extract	Cytotoxic activity on *Artemia salina* (Brine shrimp)	Karachi, Pakistan	Ara et al., 1999
Spatoglossum asperum	—	Ethanol extract	Hypolipidaemic activity, effect on cardiac and liver enzymes	Karachi, Pakistan	Ara et al., 2002
Spatoglossum asperum	Polyunsaturated fatty acids	Chloroform:MeOH (1:2)	Nutritional significance	Gujarat, India	Kumari, Kumar, Gupta, Reddy, & Jha, 2010
Spatoglossum asperum	Polyphenols	Methanol extract	Antioxidant activity (relatively high DPPH activity)	Maharashtra, India	Vinayak et al., 2011
Spatoglossum asperum	Polyphenols	Methanol extract	Cytotoxic activity (highly toxic and caused the death of the brine shrimp)	Maharashtra, India	Vinayak et al., 2011
Spatoglossum asperum	Fatty acids and lipid hydroperoxydes contents	Chloroform:MeOH (1:2)	—	Saurashtra coast, Gujarat, India	Kumari et al., 2012

Spatoglossum asperum	–	Ethanol extract	Hypolipidemic activities/decreases in the serum total cholesterol	–	Gade et al., 2013
Spatoglossum asperum	–	Methanol extract	No antidiatom/ antifouling activity	Mannar, India	Deepa et al., 2014
Spatoglossum asperum	–	Chloroform:MeOH (1:1)	Maximum activity against *V.cholera* and *K.pneumonia*	Tamilnadu, India	Sivakumar & Vignesh, 2014
Spatoglossum asperum	–	Methanol extract	Free radical scavenging activity	Tamilnadu, India	Pandithurai & Murugesan, 2014
Spatoglossum asperum	Polyphenols	Sequential extraction: hexane, dichloromethane, ethyl acetate, and methanol	antitumor activity against pancreatic cancer	India	Aravindan et al., 2015
Spatoglossum asperum	–	Crude extract	Antibacterial (*Staphylococcus aureus* and *Escherichia coli*) and antifungal (*Candida albicans*) activity	Mkomani, Kenya	Kaaria et al., 2015
Spatoglossum asperum	Synthesis of silver nanoparticules (Ag-NPs), terpenoids and phenols	Chloroform:MeOH (1:1)	Antimicrobial activity against plant diseases	Mannar, India	Jothirethinam, Prathilba, Shanti, & Arunkumar, 2015
Spatoglossum asperum	–	Methanol extract	Antimicrobial activity, antibacterial activity against *S. aureus*	Mandapam coast, India	Pandithurai et al., 2015

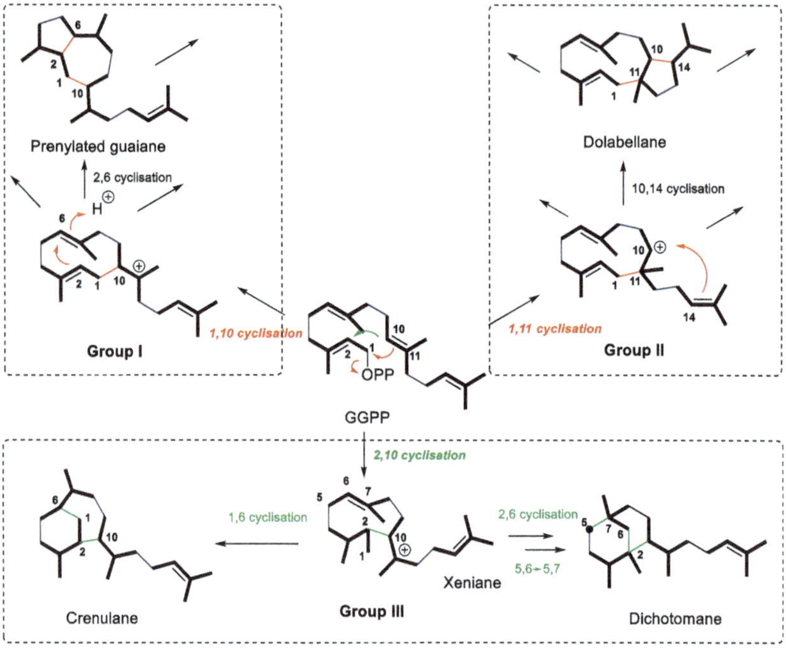

Fig. 3 Biosynthetic pathways of the three main groups of diterpenes isolated from various Dictyotaceae species.

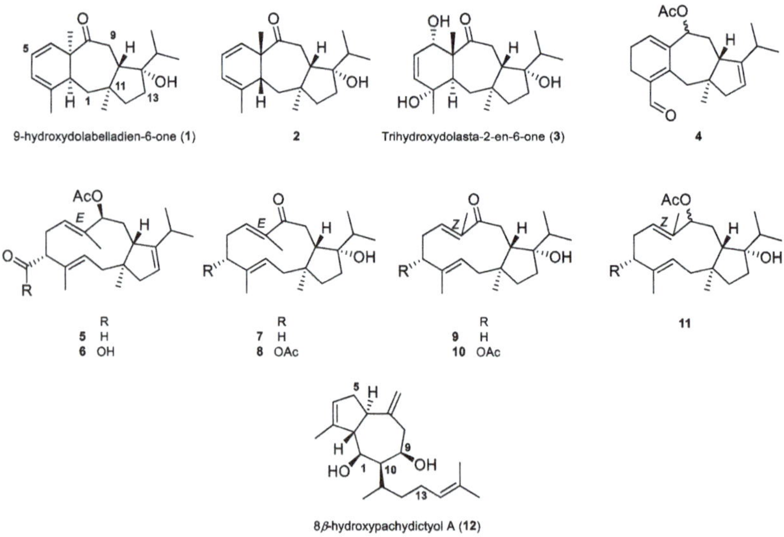

Fig. 4 Structures of diterpenoids isolated from *Dictyota bartayresiana*.

addition, *D. bartayresiana* extracts have shown a lethal activity against *Artemia salina* and *Culex quinquefasciatus* larvae (Antonysamy, Velayutham et al., 2015; Antonysamy, Thangiah, & Irulappan, 2015), an ichthyotoxic effect against the fish *Carassius auratus* (De Lara-Isassi, Álvarez-Hernández, & Collado-Vides, 2000), and antifouling properties against the diatom *Navicula subinflata* (Deepa, Srikumar, & Padmakumar, 2014). It was also suggested that *D. bartayresiana* could be used as a natural healing drug and for general good health as some of its extracts were active against cancer cell lines (Antonysamy, Thangiah et al., 2015) and showed antioxidant activity (Palanisamy, Sellappa, & Stella, 2010).

In the specific case of *Dictyota ceylanica*, only the phenolic contents (Van Hees, Olsen, Wernberg, Van Alstyne, & Kendrick, 2017) and overall biochemical composition (total carbohydrates, total proteins, pigments, vitamins) (Chakraborty & Santra, 2008) have been described. For *Dictyota friabilis* (previously known as *Dictyota pfaffii*), analyses of Brazilian specimens allowed the isolation of several dolabellane diterpenes displaying a broad range of biological activities (antiviral, cytotoxic, antifouling, etc.) (Table 4, Fig. 5). More specifically, the known dolabellane diterpene (group II) (1*R**, 2*E*, 4*R**, 6*E*, 8*S**, 10*S**, 11*S**, 12*R**)-10,18-diacetoxy-8-hydroxy-2,6-dolabelladiene was the first compound described (Fig. 5) as the main component of the organic extract (Barbosa, Teixaira, Villaça, Pereira, et al., 2003). Further studies on this species allowed to isolate four new dolabellanes (1*R**, 2*E*, 4*R**, 6*E*, 8*S**, 10*S**, 11*S**, 12*R**-10-acetoxy-

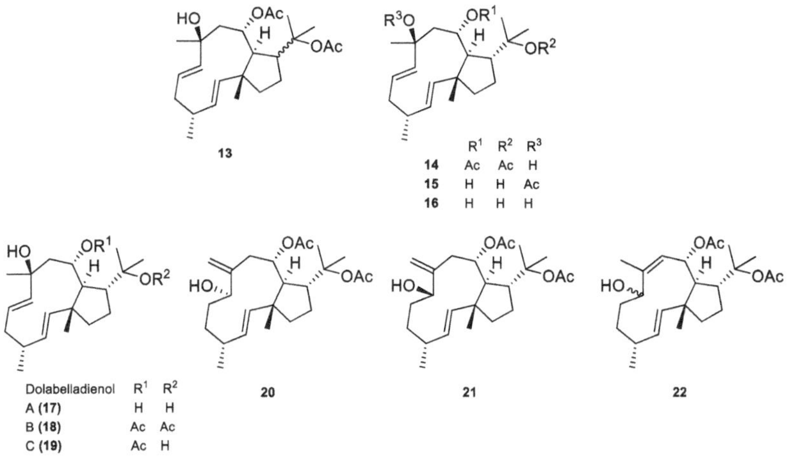

Fig. 5 Structures of diterpenoids isolated from *Dictyota friabilis* (syn. *Dictyota pfaffii*).

8,18-dihydroxy-2,6-dolabelladiene, Dolabelladienol A, B, and C) as well as the known Dolabelladienetriol, and previously described from the digestive gland of *Dolabella californica* (Ireland & Faulkner, 1977; Barbosa et al., 2004; Pardo-Vargas et al., 2014) (Fig. 5). Dolabelladienetriol, 10-acetoxy-8,18-dihydroxy-2,6-dolabelladiene, and 10,18-diacetoxy-8-hydroxy-2,6-dolabelladiene showed potent *in vitro* antiviral activity against HIV-1, but Dolabelladienetriol was the only compound which inhibited the reverse transcriptase enzyme of the virus. The antiviral properties of Dolabelladienetriol were then deeply investigated (Abrantes et al., 2010; Cirne-Santos, Teixeira, Castello-Branco, Frugulhetti, & Bou-Habib, 2006; Cirne-Santos et al., 2008; Garrido et al., 2017; Garrido et al., 2011; Stephens et al., 2017) and this diterpene was also shown to significantly inhibit the growth of *Leishmania* in macrophages (Soares et al., 2012). More recently, Dolabelladienols A and B were evaluated in the antiviral test against HIV-1 and were seen to be more active than Dolabelladienetriol (Pardo-Vargas et al., 2014). In another field of application, *D. friabilis* phlorotannins were assessed for their capacity to prevent the melanosis, quality changes and to prolong the shelf-life of iced Pacific white shrimps (*Litopenaeus vannamei*) (Sharifian, Shabanpour, Taheri, & Kordjazi, 2019). Regarding *Dictyota hamifera,* we found no studies dedicated to chemodiversity or biological properties.

Lobophora. There are no studies dedicated to the evaluation of the phytochemical content and the subsequent biological properties of *Lobophora pacifica.* Considering the vast array of bioactivities exhibited by *Lobophora* spp. metabolites in general (Gaubert, Greff, Thomas, & Payri, 2019; Vieira et al., 2015), it would be extremely relevant to investigate the chemodiversity associated to the species *Lobophora* in FP.

Padina. A minimal number of studies have isolated specialized metabolites from *Padina.* Fucoidans extracted from Vietnamese samples of *Padina boryana* were evaluated for their anticancer properties against colorectal carcinoma cells (Shevchenko et al., 2017; Usoltseva et al., 2018). Samples of *Padina boryana* from the Red Sea have been assessed for their potential in biodiesel production (Abomohra, El-Naggar, & Baeshen, 2018) polyphenolic contents, and antimicrobial activity (Sameeh, Mohamed, & Elazzazy, 2016), but also to determine their gross biochemical composition and their antioxidant properties (Ismail, El Zokm, & El-Sayed, 2017). Rasher and Hay (2010) assessed the allelopathic effects of the lipid-soluble fraction of the methanolic extract of *P. boryana* against the coral *Porites porites.* An antioxidant and anti-inflammatory compound identified as squalene was isolated from *Padina boergesenii* samples collected in the Gulf of Mannar (India) (Rajamani, Balasubramanian, & Thirugnanasambandan, 2018), while a new triterpene with several enzyme-inhibitory activities was described in samples of the same species from Oman (Ali et al., 2017). Many other studies were dedicated to the evaluation of a large number of biological activities (antioxidant, antidiabetic, hepatoprotective, etc.) of the extracts of *P. boergesenii*, mainly from samples from Mannar (Table 4). Dimethylsulfoniopropionate (DMSP) content was evaluated from Chinese samples of *Padina minor* (Bischoff et al., 1994), while several studies dealt with the biological properties of extracts coming from various collecting sites (Table 4). In Moorea, *Padina pavonica* (not recorded from Tahiti during our survey) was collected and tested for its pro-apoptotic activity on human osteosarcoma cells, a rare bone cancer condition (Bernardini, Minetti, Polizzotto, Biazzo, & Santucci, 2018). The chemical composition of this species revealed interesting pharmacological and health-protecting activities and a great richness in flavonoids (Bernardini et al., 2018).

Spatoglossum. *Spatoglossum asperum* has been studied in the Indian Ocean exclusively (Pakistan, India, and Kenya) (Table 4). Ara, Sultana, Ehteshamul-Haque, Qasim, and Ahmad (1999) showed a positive response to the cytotoxic activity of *S. asperum* from Pakistan. This cytotoxicity was confirmed by Vinayak, Sabu, and Chatterji (2011) on brine shrimps. Ara, Sultana, Qasim, and Ahmad (2002) showed a hypolipidaemic effect and effects on cardiac and liver enzymes. *Spatoglossum asperum* has numerous other interesting biological activities including antibacterial activity against *Staphylococcus aureus* (Pandithurai, Murugesan, & Sivamurugan, 2015), *Escherichia coli* (Kaaria, Wakibia, Matiru, Ndung'u, & Bii, 2015), *Bacillus cereus, Bacillus subtilis, Klebsiella pneumonia, Pseudomonas aeruginosa,* and *Salmonella typhi* (Pandithurai et al., 2015; Sivakumar & Vignesh, 2014), antifouling activities (Deepa et al., 2014), and antioxidant activities (Pandithurai & Murugesan, 2014; Vinayak et al., 2011). Antimicrobial activities have been tested against plant diseases for agricultural uses (Jothirethinam, Prathilba, Shanti, & Arunkumar, 2015). In 2015, *S. asperum* extracts revealed positive antitumoral effects (Aravindan, Ramraj, Somasundaram, Herman, & Aravindan, 2015). Among the publications listed in Table 4, none gives precise chemical compositions of *S. asperum*. Vinayak et al. (2011) and Aravindan et al. (2015) believe that polyphenols are responsible for the cytotoxic and anti-tumoral activities; however, they did not refer to isolated molecules.

4.2 Applications

4.2.1 Human consumption

The consumption of algae as food is the most significant market, representing about eighty percent of the world's algal production (FAO, 2016). According to a review by Zubia and Mattio (2019), four Dictyotaceae genera are used for food, mainly in the South Pacific Islands, including *Dictyota*, *Dictyopteris* (*Dictyopteris plagiogramma* "limu lipoa"; Hart, Ticktin, Kelman, Wright, & Tabandera, 2014), *Padina*, and *Spatoglossum*. Other Dictyotaceae such as *Padina* and *Dictyota* are used for their medicinal properties and as a source of food in many Pacific Islands (Abbott, 1984; Conte & Payri, 2002; Novaczek, 2001, pp. 1–48; South, 1993). *Dictyota* is eaten in various ways (seasoning, soup, etc.), and is also recognized as preventive medicine for heart diseases, stroke (algin compounds), and goiter (iodine content) (Novaczek, 2001, pp. 1–48). *Padina* is used to add flavor to dishes and is a good source of calcium (Novaczek, 2001, pp. 1–48). *Dictyopteris plagiogramma* is consumed traditionally in Hawaii (Abbott, 1984) and has a particularly strong and distinctive fragrance (due to dictyoterpenes; Moore, 1974), but this species is disappearing from the coast due to pollution (Hart et al., 2014).

In FP, seaweeds are still consumed today, particularly in the Marquesas and Austral islands with 11 species recorded (Conte & Payri, 2002, 2006; De Gaillande, Payri, Remoissenet, & Zubia, 2017), six of them belonging to the genus *Caulerpa*. In most of the Islands, this tradition has been lost, except in the Marquesas and the Austral archipelagos where *Caulerpa chemnitzia* (= *C. racemosa* var. *turbinata)* known as sea grapes ("remu" or "imu topua" in polynesian language) is consumed regularly (De Gaillande et al., 2017). In recent years, seaweed consumption has become more popular, and the French Polynesian Government wants to diversify the offer through the development of seaweed farming for *Caulerpa* spp. and a few red algae (*Gracilaria* spp., *Gelidiella acerosa*, *Acanthophora spicifera*). The development of *Dictyotaceae* species consumption like *Dictyota*, *Padina* or *Spatoglossum* could also be an exciting option to diversify the market.

4.2.2 Animal nutrition

A large proportion of methane emitted into the atmosphere derives from agriculture and specifically ruminant enteric fermentation which contributes approximately to 28% of global anthropogenic CH_4 emissions (Beauchemin, Kreuzer, O'Mara, & McAllister, 2008). Feed management and additives such as high-quality forages, grains, ionophores, fats, yeasts, enzymes, microbes, plant extracts, and algae have the potential for CH_4 abatement (Beauchemin, Mc Allister, & McGinn, 2009). Algae products can improve ruminant health and productivity, increase feed quality, and inhibit methanogenesis (Holdt & Kraan, 2011; Kinley, De Nys, Vucko, Machado, & Tomkins, 2016). Kinley et al. (2016) showed that *Asparagopsis* induces near elimination of methane *in vitro*, and Machado, Magnusson, Paul, De Nys, and Tomkins (2014) found that the brown algae *Dictyota* spp. can reduce methane production by over 92% compared to controls.

Many experiments are currently underway in the field of aquaculture, for the incorporation of algae in shrimp and fish feeds (Nelson, Glenn, Conn, Walsh, & Akutagawa, 2001; Reverter, Bontemps, Lecchini, Banaigs, & Sasal, 2014). These algae or seaweed extracts are used either to replace proteins of animal origin either for their probiotic or antimicrobial effects, usually in the context of Integrated Multi-Trophic Aquaculture (IMTA) projects. In FP, emerging *Platax orbicularis* (orbicular batfish) aquaculture is affected by bacterial infections causing severe mortality episodes. Effective solutions are investigated to find natural bioactive compounds to counteract the infections caused principally by *Tenacibaculum maritimum* and *Vibrio harveyi*, and Reverter et al. (2016) demonstrated the promising results obtained with the Rhodophyta *Asparagopsis taxiformis*. Due to their antiviral and antimicrobial activities, their abundance, availability, and accessibility, Dictyotaceae species represent an equally interesting group to consider. Our research unit in Tahiti (UMR-EIO) is currently investigating the efficiency of French Polynesian *Dictyota bartayresiana, Dictyota hamifera,* and *Spatoglossum asperum* extracts against the pathogens *V. harveyi* and *T. maritimum.*

4.2.3 Health care, medicinal, pharmaceutical, and cosmetic properties

The use of macroalgae in cosmetics is quickly expanding as they represent a natural and active environmental resource. Several species are already used by several cosmetic industries in hair care, skin whitening, anti-aging, moisturizers, or in sunscreens (Ariede et al., 2017), but there is no reference on the use of tropical species for cosmetics in the literature. As previously discussed in section 4.1 and presented in Table 4, Dictyotaceae species synthesize a wide range of molecules with specific biological activities that could be used as active ingredients in cosmetics or health industries. For instance, Fayad et al. (2017) highlighted the promising activity of *Padina pavonica* for the inhibition of hyalurosidase activity, which is expected to delay skin aging. *Spatoglossum asperum* also showed hypolipidic effects interesting against cholesterol disease (Gade, Tulasi, & Bhai, 2013). Other species of Dictyotaceae could be relevant for the health industry for their antiviral, antimicrobial, antioxidant, anti-inflammatory, and anti-cancer activities (Table 4), which are significant medical research challenges. According to our literature review, it is clear that the Dictyotaceae species of FP are a potential source of new exciting and potentially unique molecules that could be extremely useful for the cosmetic and health industry. However, although the potentials are high, only limited research is available that has explored specifically the diversity of the French Polynesian Dictyotaceae molecules and their associated biological potential.

4.2.4 Agriculture

Seaweeds can be used for many agricultural applications including biofertilizers, soil conditioners, and enhancers because of their high quantities of micro- and macronutrients, vitamins, amino acids, and growth regulators (Arioli, Mattner, & Winberg, 2015; Kumar & Sahoo, 2011; N'Yeurt & Iese, 2015a). Seaweed extracts represent a major opportunity to significantly enhance crop production and resistance to stress and disease (Arioli et al., 2015). Brown algae are particularly useful fertilizers (Arioli et al., 2015). Experiments in Fiji have demonstrated the potential of both *Sargassum polycystum* and *Gracilaria edulis* as excellent liquid fertilizers for a variety of common Pacific crops (unpublished results, N'Yeurt, pers. com.). Brown-algae based fertilizers contain alginates and sulfated polysaccharides, their chelating properties combined with metal ions in the soil form chelates that absorb moisture and improve the growth of soil bacteria (Cardozo et al., 2007; Khan et al., 2009). In FP, agronomical enrichment trials were conducted using drifting brown algae (*S. pacificum* and *T. ornata*) as organic additives (Zubia et al., 2014). This study demonstrated that low supplements of drift algae (1 and 3%) to plant compost significantly improved maize growth (stems length, aerial plants dry mass, and roots dry mass). In Tuvalu, farmers successfully use additives made of dried *Sargassum polycystum* on their crops (N'Yeurt & Iese, 2015b). Although we did not find studies on potential uses of Dictyotaceae species for agricultural purposes during our review, they could represent good candidates to develop eco-friendly fertilizer and additives of farming practices, in particular, to fight against phytopathogens due to their strong antimicrobial activities (Table 4).

5. Scope for future research

This paper represents the first detailed assessment of the Dictyotaceae family in FP. It reviews and supplements the existing knowledge on the diversity, ecology, chemodiversity, and potential economic applications of these species. Molecular analyses of the different genera are underway and will undoubtedly increase the known diversity of the Dictyotaceae family (44 species across all archipelagos to date). These results will be crucial for a better understanding of the diversity patterns in FP, and biogeographical studies in the Indo-Pacific, which are currently constrained by overlooked cryptic diversity (Vieira et al., 2017). Our research project was motivated by the observation of increasing amounts of Dictyotaceae species (especially *Dictyota bartayresiana* and *Spatoglossum asperum*) on the reefs of Tahiti Island over the past decade. However, our results do not indicate the occurrence of proliferation events, as the abundances we measured were relatively low (cover <10%) on all reefs surveyed throughout the year. Despite this result, and in light of our review, we argue here that monitoring Dictyotaceae species should be continued because of the potential negative impact they may have on coral health, particularly during periods of intense bleaching (see Section 3.2.2). Proliferations events were mainly triggered by increased temperatures, nutrients, and/or pH or biotic factors such as a decrease in herbivores which negatively impacted coral cover and favored the more resistant and opportunistic Dictyotaceae species (see Section 3). As Tahiti's coral reefs are subject to significant anthropogenic and environmental pressures (Andréfouët & Adjeroud, 2019), we believe that further studies are needed to better characterize the ecological dynamics of Dictyotacae, particularly biotic factors. These data will be critical to design appropriate management policies and avoid future additional proliferations on the reefs of Tahiti. Finally, Dictyotaceae species represent promising resources for many biotechnological applications (human and animal nutrition, agriculture, health, cosmetic), most of which could be developed in FP. However, further studies are needed to characterize the chemodiversity of local populations and assess which economic sectors would be most suited for FP.

Acknowledgments

This work is part of the Ph.D. thesis of Tohei Theophilus. We are very thankful to the University of French Polynesia and the Délégation à la Recherche de Polynésie française for their financial support without which the research program DIVECOBAL would not have been possible.

Appendix A. Supplementary data

Supplementary data to this article can be found online at https://doi.org/10.1016/bs.abr.2019.12.001.

References

Abbott, I. A. (1984). *Limu: An ethnobotanical study of some Hawaiian seaweeds*. National Tropical Botanical Garden.

Abomohra, A. E.-F., El-Naggar, A. H., & Baeshen, A. A. (2018). Potential of macroalgae for biodiesel production: Screening and evaluation studies. *Journal of Bioscience and Bioengineering, 125*(2), 231–237.

Abrantes, J. L., Barbosa, J., Cavalcanti, D., Pereira, R. C., Fontes, F. C. L., Teixeira, V. L., et al. (2010). The effects of the diterpenes isolated from the Brazilian brown algae *Dictyota pfaffii* and *Dictyota menstrualis* against the herpes simplex type-1 replicative cycle. *Planta Medica, 76*, 339–344.

Adam, T. C., Schmitt, R. J., Holbrook, S. J., Brooks, A. J., Edmunds, P. J., Carpenter, R. C., et al. (2011). Herbivory, connectivity, and ecosystem resilience: Response of a coral reef to a large-scale perturbation. *PLoS One, 6*(8), e23717.

Ali, L., Khan, A. L., Al-Broumi, M., Al-Harrasi, R., Al-Kharusi, L., Hussain, J., et al. (2017). New enzyme-inhibitory triterpenoid from marine macro brown alga *Padina boergesenii* Allender & Kraft. *Marine Drugs, 15*(19), 1–8.

Andréfouët, S., & Adjeroud, M. (2019). French Polynesia. In *World seas: An environmental evaluation* (pp. 827–854). Academic Press.

Antonysamy, J. M. A., Thangiah, S., & Irulappan, R. (2015). Green synthesis of silver nanoparticles using *Dictyota bartayresiana* J.V. Lamouroux and their cytotoxic potentials. *International Biological and Biomedical Journal, 1*(3), 112–118.

Antonysamy, J. M. A., Velayutham, K., Mani, N., Thangaiah, S., & Irullappan, R. (2015). Antibacterial, cytotoxic and larvicidal potential of *Dictyota bartayresiana* Lamour. *Journal of Coastal Life Medicine, 3*(5), 352–355.

Ara, J., Sultana, V., Ehteshamul-Haque, S., Qasim, R., & Ahmad, V. U. (1999). Cytotoxic activity of marine macroalgae on *Artemia salina* (brine shrimp). *Phytotherapy Research, 13*, 304–307.

Ara, J., Sultana, V., Qasim, R., & Ahmad, V. U. (2002). Hypolipidaemic activity of seaweed from Karachi coast. *Phytotherapy Research, 16*(5), 479–483.

Aravindan, S., Ramraj, S. K., Somasundaram, S. T., Herman, T. S., & Aravindan, N. (2015). Polyphenols from marine brown algae target radiotherapy-coordinated EMT and stemness-maintenance in residual pancreatic cancer. *Stem Cell Research and Therapy, 6*(182), 1–14.

Ariede, M. B., Candido, T. M., Jacome, A. L. M., Valesco, M. V. R., De Carvalho, J. C. M., & Baby, A. R. (2017). Cosmetic attributes of algae – A review. *Algae Research, 25*, 483–487.

Arioli, T., Mattner, S. W., & Winberg, P. C. (2015). Applications of seaweed extracts in Australian agriculture: Past, present and future. *Journal of Applied Phycology, 27*(5), 2007–2015.

Barbosa, J. P., Fleury, B. G., Da Gama, B. A. P., Teixeira, V. L., & Pereira, R. C. (2007). Natural products as antifoulants in the Brazilian brown alga *Dictyota pfaffii* (Phaeophyta, Dictyotales). *Biochemical Systematics and Ecology, 35*, 549–553.

Barbosa, J. P., Pereira, R. C., Abrantes, J. L., Cirne-Santos, C. C., Rebello, M. A., Frugulhetti, I. C. P., et al. (2004). *In vitro* antiviral diterpenes from the Brazilian brown alga *Dictyota pfaffii*. *Planta Medica, 70*, 856–860.

Barbosa, J. P., Teixeira, V. L., Villaça, R. C., Pereira, H. S., Abrantes, J. L., & Frugulhetti, I. C. N. P. (2003). A dolabellane diterpene from de Brazilian brown alga *Dictyota pfaffii*. *Biochemical Systematics and Ecology, 31*, 1451–1453.

Barott, K. L., Rodriguez-Mueller, B., Youle, M., Marhaver, K. L., Vermeij, M. J. A., Smith, J. E., et al. (2011). Microbial to reef scale interactions between the reef-building coral *Montastraea annularis* and benthic algae. *Proceedings of the Royal Society B: Biological Sciences, 279*, 1655–1664.

Beach, K., Walters, L., Borgeas, H., Smith, C., Coyer, J., & Vroom, P. (2003). The impact of *Dictyota* spp. on *Halimeda* populations of Conch Reef, Florida Keys. *Journal of Experimental Marine Biology and Ecology, 297*, 141–159.

Beauchemin, K. A., Kreuzer, M., O'Mara, F., & McAllister, T. A. (2008). Nutritional management for enteric methane abatement: A review. *Australian Journal of Experimental Agriculture, 48*, 21–27.

Beauchemin, K. A., McAllister, T. A., & McGinn, S. M. (2009). Dietary mitigation of enteric methane from Cattle. *CAB Reviews: Perspectives in Agriculture, Veterinary Science, Nutrition and Natural Resources, 4*, 1–18.

Bernardini, G., Minetti, M., Polizzotto, G., Biazzo, M., & Santucci, A. (2018). Pro-apoptotic activity of French Polynesian *Padina pavonica* extract on human osteosarcoma cells. *Marine Drugs, 16*(12), 504.

Bheemasankara Rao, C., Trimurtulu, G., Sreedhara, C., Ventaka Rao, D., Bobzin, S. C., & Faulkner, D. J. (1994). Diterpenes from the brown alga *Dictyota bartayresiana*. *Phytochemistry, 37*(2), 509–513.

Bischoff, B., Karsten, U., Daniel, C., Kuck, K., Xia, B., & Wiencke, C. (1994). Preliminary assessment of the β-dimethylsulfoniopropionate (DMSP) content of macroalgae from the tropical island of Hainan (People's Republic of China). *Australian Journal of Marine and Freshwater Research, 45*(7), 1329–1336.

Bischof, K., Gómez, I., Molis, M., Hanelt, D., Karsten, U., Lüder, U., et al. (2006). Ultraviolet radiation shapes seaweed communities. *Reviews in Environmental Science and Biotechnology, 5*, 141–166.

Bittick, S. J., Clausing, R. J., Fong, C. R., Scoma, S. R., & Fong, P. (2019). A rapidly expanding macroalga acts as a foundational species providing trophic support and habitat in the South Pacific. *Ecosystems, 22*(1), 165–173.

Bittner, L., Halary, S., Payri, C., Cruaud, C., De Reviers, B., Lopez, P., et al. (2010). Some considerations for analyzing biodiversity using integrative metagenomics and gene networks. *Biology Direct, 5*, 47.

Bittner, L., Payri, C. E., Couloux, A., Cruaud, C., De Reviers, B., & Rousseau, F. (2008). Molecular phylogeny of the Dictyotales and their position within the brown algae, based on nuclear, plastidial and mitochondrial sequence data. *Molecular Phylogenetics and Evolution, 49*, 211–226.

Bonaldo, R. M., & Hay, M. E. (2014). Seaweed-coral interactions: Variance in seaweed allelopathy, coral susceptibility, and potential effects on coral resilience. *PLoS One, 9*(1). e85786, 1-12.

Bory De Saint-Vincent, J. B. G. M. (1828). Cryptogamie. In L. I. Duperrey (Ed.), *Voyage Autour du Monde exécuté par ordre du Roi, sur la corvette de sa Majesté, La Coquille, pendant les années 1822, 1823, 1824 et 1825* (pp. 1–136). Paris: Arthus Bertrand.

Brown, K. T., Bender-Champ, D., Kubicek, A., van der Zande, R., Achlatis, M., Hoegh-Guldberg, O., et al. (2018). The dynamics of coral-algal interactions in space and time on the southern Great Barrier Reef. *Frontiers in Marine Science, 5*, 181.

Camoin, G., Cabioch, G., Eisenhauer, A., Braga, J.-C., Hamelin, B., & Lericolais, G. (2006). Environmental significance of microbialites in reef environments during the last deglaciation. *Sedimentary Geology, 185*, 277–295.

Camoin, G. F., Gautret, P., Montaggioni, L. F., & Cabioch, G. (1999). Nature and environmental significance of microbialites in Quaternary reefs: The Tahiti paradox. *Sedimentary Geology, 126*, 271–304.

Caragnano, A., Foetisch, A., Maneveldt, G. W., Millet, L., Liu, L. C., Lin, S. M., et al. (2018). Revision of Corallinaceae (Corallinales, Rhodophyta): Recognizing *Dawsoniolithon* gen. nov., *Parvicellularium* gen. nov. and Chamberlainoideae subfam. nov. containing *Chamberlainium* gen. nov. and Pneophyllum. *Journal of Phycology, 54*(3), 391–409.

Cardozo, K. H. M., Guaratini, T., Barros, M. P., Falcao, V. R., Tonon, A. P., Lopes, N. P., et al. (2007). Metabolites from algae with economical impact. *Comparative Biochemistry Physiology C-Toxicology Pharmacology, 146*, 60–78.

Carroll, A. R., Copp, B. R., Davis, R. A., Keysers, R. A., & Prinsep, M. R. (2019). Marine natural products. *Natural Product Reports, 36*, 122–173.

Chakraborty, S., & Santra, S. C. (2008). Biochemical composition of eight benthic algae collected from Sunderban. *Indian Journal of Marine Sciences, 37*(3), 329–332.

Chaves, L. T. C., Pereira, P. H. C., & Feitosa, J. L. L. (2013). Coral reef fish association with macroalgal beds on a tropical reef system in North-eastern Brazil. *Marine and Freshwater Research, 64*(12), 1101–1111.

Chen, J., Li, H., Zhao, Z., Xia, X., Li, B., Zhang, J., et al. (2018). Diterpenes from the marine algae of the genus *Dictyota*. *Marine Drugs, 16*(159), 1–25.

Cirne-Santos, C. C., Souza, T. M. L., Teixeira, V. L., Fontes, C. F. L., Rebello, M. A., Castello-Branco, L. R. R., et al. (2008). The dolabellane diterpene Dolabelladienetriol is a typical noncompetitive inhibitor of HIV-1 reverse transcriptase enzyme. *Antiviral Research,* 77, 64–71.

Cirne-Santos, C. C., Teixeira, V. L., Castello-Branco, L. R., Frugulhetti, I. C. P. P., & Bou-Habib, D. C. (2006). Inhibition of HIV-1 Replication in human primary cells by a dolabellane diterpene isolated from the marine algae *Dictyota pfaffii*. *Planta Medica,* 72, 295–299.

Clausing, R. J., & Fong, P. (2016). Environmental variability drives rapid and dramatic changes in nutrient limitation of tropical macroalgae with different ecological strategies. *Coral Reefs, 35*, 669–680.

Conte, E., & Payri, C. E. (2002). La consommation des algues en Polynésie française : Premiers résultats d'une enquête. *Le Journal de la Société des Océanistes, 114–115*, 165–172.

Conte, E., & Payri, C. E. (2006). Present day consumption of edible algae in French Polynesia: A study of the survival of pre-European practices. *Journal of the Polynesian Society, 115*(1), 77–93.

De Clerck, O., Leliaert, F., Verbruggen, H., Lane, C. E., De Paula, J. C., Payo, D. A., et al. (2006). A revised classification of the dictyotaceae (Dictyotales, Phaeophyceae) based on *rbc*L and 26S Ribosomal DNA sequence analyses. *Journal of Phycology, 42*(6), 1271–1288.

De Gaillande, C., Payri, C., Remoissenet, G., & Zubia, M. (2017). Caulerpa consumption, nutritional value and farming in the Indo-Pacific region. *Journal of Applied Phycology, 29*(5), 2249–2266.

De Lara-Isassi, G., Álvarez-Hernández, S., & Collado-Vides, L. (2000). Ichtyotoxic activity of extracts from Mexican marine macroalgae. *Journal of Applied Phycology, 12*, 45–52.

De Paula, J. C., Vallim, M. A., & Teixeira, V. L. (2011). What are and where are the bioactive terpenoids metabolites from Dictyotaceae (Phaeophyceae). *Revista Brasileira de Farmacognosia Brazilian Journal of Pharmacognosy, 21*(2), 216–228.

Deepa, S., Srikumar, M., & Padmakumar, K. P. (2014). Antifouling potential of selected macroalgae from the Gulf of Mannar, India. *International Journal of Bioassays, 3*(11), 3479–3487.

Del Monaco, C., Hay, M. E., Gartrell, P., Mumby, P. J., & Diaz-Pulido, G. (2017). Effects of ocean acidification on the potency of macroalgal allelopathy to a common coral. *Scientific Reports,* 7(41053), 1–10.

Diaz-Pulido, G., Harii, S., McCook, L. J., & Hoegh-Guldberg, O. (2010). The impact of benthic algae on the settlement of a reef-building coral. *Coral Reefs, 29*, 203–208.
Diaz-Pulido, G., Villamil, L., & Almanza, V. (2007). Herbivory effects on the morphology of the brown alga *Padina boergesenii* (Phaeophyta). *Phycologia, 46*(2), 131–136.
Dijoux, L., Viard, F., & Payri, C. (2014). The more we search, the more we find: Discovery of a new lineage and a new species complex in the genus *Asparagopsis*. *PLoS One, 9*(7). e103826, 1-13.
Done, T. J. (1992). Phase shifts in coral reef communities and their ecological significance. *Hydrobiologia, 247*(1–3), 121–132.
Duffy, J. E., & Harvilicz, A. M. (2001). Species-specific impacts of grazing amphipods in an eelgrass-bed community. *Marine Ecology Progress Series, 223*, 201–211.
Enochs, I. C., Manzello, D. P., Donham, E. M., Kolodziej, G., Okano, R., Johnston, L., et al. (2015). Shift from coral to macroalgae dominance on a volcanically acidified reef. *Nature Climate Change, 5*(12), 1083.
Fabricius, K. E. (2005). Effects of terrestrial runoff on the ecology of corals and coral reefs: Review and synthesis. *Marine Pollution Bulletin, 50*, 125–146.
FAO. (2016). *The State of World Fisheries and Aquaculture 2016. Contributing to food security and nutrition for all.* Rome, 200 p.
Faure, G. (1989). Degradation of coral reefs at Moorea island (French Polynesia) by *Acanthaster planci*. *Journal of Coastal Research, 5*, 295–306.
Fayad, S., Nehmé, R., Tannoury, M., Lesellier, E., Pichon, C., & Morin, P. (2017). Macroalga *Padina pavonica* water extracts obtained by pressurized liquid extraction and microwave-assisted extraction inhibit hyaluronidase activity as shown by capillary electrophoresis. *Journal of Chromatography A, 1497*, 19–27.
Fong, C. R. (2015). An experimental assessment of herbivory and nutrient effects on a small-scale in a coral reef macroalgal community. *Aquatic Botany, 123*, 1–5.
Fong, P., Frazier, N. M., Tompkins-Cook, C., Muthukrishnan, R., & Fong, C. R. (2016). Size matters: Experimental partitioning of the strength of fish herbivory on a fringing coral reef in Moorea, French Polynesia. *Marine Ecology, 37*, 933–942.
Fong, P., & Paul, V. J. (2011). *Coral reef algae. Coral reefs: An ecosystem in transition.* New York.
Fulton, C. J., Depczynski, M., Holmes, T. H., Noble, M. M., Radford, B., Wernberg, T., et al. (2014). Sea temperature shapes seasonal fluctuations in seaweed biomass within the Ningaloo coral reef ecosystem. *Limnology and Oceanography, 59*(1), 156–166.
Gade, R., Tulasi, M. S., & Bhai, V. A. (2013). Seaweeds: A novel biomaterial. *International Journal of Pharmacy and Pharmaceutical Sciences, 5*(2), 40–44.
Garrido, V., Barros, C., Melchiades, V. A., Fonseca, R. R., Pinheiro, S., Ocampo, P., et al. (2017). Subchronic toxicity and anti-HSV-1 activity in experimental animal of dolabelladienetriol from the seaweed, *Dictyota pfaffii*. *Regulatory Toxicology and Pharmacology, 86*, 193–198.
Garrido, V., Teixeira, G. A. P. B., Teixeira, V. L., Ocampo, P., Ferreira, W. J., Cavalcanti, D. N., et al. (2011). Evaluation of the acute toxicity of dolabelladienotriol, a potential antiviral from the brown alga *Dictyota pfaffii*, in BALB/c mice. *Revista Brasileira de Farmacognosia, 21*(2), 209–215.
Gaubert, J., Greff, S., Thomas, O. P., & Payri, C. E. (2019). Metabolomic variability of four macroalgal species of the genus *Lobophora* using diverse approaches. *Phytochemistry, 162*, 165–172.
Gillespie, T. W., Foody, G. M., Rocchini, D., Giorgi, A. P., & Saatchi, S. (2008). Measuring and modelling biodiversity from space. *Progress in Physical Geography, 32*, 203–221.
Gleason, D. F., & Wellington, G. M. (1993). Ultraviolet radiation and coral bleaching. *Nature, 365*, 836–838.

Goldberg, N. A., & Kendrick, G. A. (2004). Effects of island groups, depth, and exposure to ocean waves on subtidal macroalgal assemblages in the Recherche Archipelago, Western Australia. *Journal of Phycology, 40*, 631–641.
Gómez Garreta, M., Rull, J., & Ribera Siguán, M. A. (2007). On the presence of fertile gametophytes of *Padina pavonica* (Dictyotales, Phaeophyceae) from the Iberian coasts. *Anales del Jardín Botánico de Madrid, 64*(1), 27–33.
Gosch, B. J., Paul, N. A., De Nys, R., & Magnusson, M. (2015). Spatial, seasonal, and within-plant variation in total fatty acid content and composition in the brown seaweeds Dictyota bartayresii and Dictyopteris australis (Dictyotales, Phaeophyceae). *Journal of Applied Phycology, 27*(4), 1607–1622.
Guiry, M. D., & Guiry, G. M. (2019). *AlgaeBase*. Galway: World-wide electronic publication, National University of Ireland.
Hart, G. M., Ticktin, T., Kelman, D., Wright, A. D., & Tabandera, N. (2014). Contemporary gathering practice and antioxidant benefit of wild seaweeds in Hawai'i. *Economy Botany, 68*(1), 30–43.
Hay, M. E. (1991). Fish-seaweed interactions on coral reefs: Effects of herbivorous fishes and adaptations of their prey. In *The ecology of fishes on coral reefs* (pp. 96–119). San Diego: Academic Press.
Hay, M. E. (1996). Marine chemical ecology: what's known and what's next? *Journal of Experimental Marine Biology and Ecology, 200*, 103–134.
Hay, M. E. (1997). The ecology and evolution of seaweed-herbivore interactions on coral reefs. *Coral Reefs, 16*, 67–76.
Hay, M. E., Duffy, J. E., & Fenical, W. (1990). Host-plant specialization decreases Predation on a marine amphipod: An herbivore in plant's clothing. *Ecology, 71*(2), 733–743.
Heindel, K., Birgel, D., Brunner, B., Thiel, V., Westphal, H., Gischler, E., et al. (2012). Post-glacial microbialite formation in coral reefs of the Pacific, Atlantic, and Indian Oceans. *Chemical Geology, 304–305*, 117–130.
Herren, L. W., Walters, L. J., & Beach, K. S. (2006). Fragmentation generation, survival, and attachment of *Dictyota* spp. at Conch Reef in the Florida Keys, USA. Coral Reefs, *25*, 287–295.
Ho, M., & Carpenter, R. C. (2017). Differential growth responses to water flow and reduced pH in tropical marine macroalgae. *Journal of Experimental Marine Biology and Ecology, 491*, 58–65.
Holbrook, S. J., Schmitt, R. J., Adam, T. C., & Brooks, A. J. (2016). Coral reef resilience, tipping points and the strength of herbivory. *Nature, 6*(35817), 1–11.
Holdt, S. L., & Kraan, S. (2011). Bioactive compounds in seaweed: Functional food applications and legislation. *Journal of Applied Phycology, 23*, 543–597.
Hughes, T. P. (1994). Catastrophes, phase shifts, and large-scale degradation of a -Caribbean coral reef. *Science, 265*(5178), 1547–1551.
Hughes, T. P., Graham, N. A. J., Jackson, J. B. C., Mumby, P. J., & Steneck, R. S. (2010). Rising to the challenge of sustaining coral reef resilience. *Trends in Ecology and Evolution, 25*(11), 634–642.
Hughes, T. P., Rodrigues, M. J., Bellwood, D. R., Ceccarelli, D., Hoegh-Guldberg, O., McCook, L., et al. (2007). Phase shifts, herbivory, and the resilience of coral reefs to climate change. *Current Biology, 17*, 360–365.
Ireland, C., & Faulkner, D. J. (1977). Diterpenes from *Dolabella californica*. *Journal of Organic Chemistry, 42*(19), 3157–3162.
Ismail, M. M., El Zokm, G. M., & El-Sayed, A. M. (2017). Variation in biochemical constituents and master elements in common seaweeds from Alexandria coast, Egypt, with special reference to their antioxidant activity and potential food uses: Prospective equations. *Environmental Monitoring and Assessment, 189*, 175–648.

Jothirethinam, A., Prathiba, S., Shanti, N., & Arunkumar, K. (2015). Green synthesized silver nanoparticles prepared from the antimicrobial Crude extracts of two brown seaweeds against plant pathogens. *American Journal of Nanotechnology, 6*(2), 31–39.

Kaaria, P., Wakibia, J., Matiru, V., Ndung'u, M., & Bii, C. (2015). Antimicrobial screening of marine endophytes and epiphytes isolated from marine algae of Kenyan Indian Ocean. *Journal of Applied and Environmental Microbiology, 3*(3), 70–74.

Kamat, S. Y., Wahidulla, S., D'Souza, L., Naik, C. G., Ambiye, V., Bhakuni, D. S., et al. (1992). Bioactivity of marine organisms VI. Antiviral evaluation of marine algal extracts from the Indian coast. *Botanica Marina, 35*, 161–164.

Kashiwagi, M., Mynderse, J.-S., Moore, R. E., & Norton, T. R. (1980). Antineoplastic evaluation of pacific basin marine algae. *Journal of Pharmaceutical Sciences, 69*(6), 735–738.

Kaullysing, D., Gopeechund, A., Mattan-Moorgawa, S., Taleb-Hossenkhan, N., Kulkarni, B., & Bhagooli, R. (2016). Increased density of the corallivore *Drupella cornus* on *Acropora muricata* colonies overgrown by Padina boryana. In *Proceedings of the 13th international coral reefs symposium, Hawaii, USA* (pp. 288–304).

Kayal, M., Vercelloni, J., De Loma, T. L., Bosserelle, P., Chancerelle, Y., Geoffroy, S., et al. (2012). Predator crown-of-thorns starfish (*Acanthaster planci*) outbreak, mass mortality of corals, and cascading effects on reef fish and benthic communities. *PLoS One*, 7(10), e47363, 1-9.

Kelecom, A., & Teixeira, V. L. (1986). Diterpenes of marine brown algae of the family dictyotaceae: Their possible role as defense compounds and their use in chemotaxonomy. *The Science of the Total Environment, 58*(1–2), 109–115.

Khan, W., Rayirath, U. P., Subramanian, S., Jithesh, M. N., Rayorath, P., Hodges, D. M., et al. (2009). Seaweed extracts as biostimulants of plant growth and development. *Journal of Plant Growth Regulation, 28*(4), 386–399.

Kier, G., Kreft, H., Lee, T. M., Jetz, W., Ibisch, P. L., Nowicki, C., et al. (2009). A global assessment of endemism and species richness across island and mainland regions. *Proceedings of the National Academy of Sciences, 106*(23), 9323–9327.

Kinley, R. D., De Nys, R., Vucko, M. J., Machado, L., & Tomkins, N. W. (2016). The red macroalgae *Asparagopsis taxiformis* is a potent natural antimethanogenic that reduces methane production during in vitro fermentation with rumen fluid. *Animal Production Science, 56*(3), 282–289.

Kuffner, I. B., Walters, L. J., Becerro, M. A., Paul, V. J., Ritson-Williams, R., & Beach, K. S. (2006). Inhibition of coral recruitment by macroalgae and cyanobacteria. *Marine Ecology Progress Series, 33*, 107–117.

Kumari, P., Kumar, M., Gupta, V., Reddy, C. R. K., & Jha, B. (2010). Tropical marine macroalgae as potential sources of nutritionally important PUFAs. *Food Chemistry, 120*, 749–757.

Kumari, P., Singh, R. P., Bijo, A. J., Reddy, C. R. K., & Jha, B. (2012). Estimation of lipid hydroperoxide levels in tropical marine macroalgae. *Journal of Phycology, 48*(6), 1362–1373.

Kumar, G., & Sahoo, D. (2011). Effect of seaweed liquid extract on growth and yield of *Triticum aestivum* var. Pusa Gold. *Journal of Applied Phycology, 23*(2), 251–255.

Kumar, P. S., & Sudha, S. (2013). Biosynthesis of silver nanoparticles from *Dictyota bartayresiana* extract and their antifungal activity. *Nano Biomedicine & Engineering, 5*, 72–75.

Lapointe, B. E. (1997). Nutrient thresholds for bottom-up control of macroalgal blooms on coral reefs in Jamaica and southeast Florida. *Limnology and Oceanography, 42*, 1119–1131.

Lapointe, B. E., Littler, M. M., & Littler, D. S. (1997). Macroalgal overgrowth of fringing coral reefs at Discovery bay, Jamaica: Bottom-up versus top-down control. In *Proceeding 8th international coral reefs symposium 1, Smithsonian Tropical Research Institute, Panama* (pp. 927–932).

Laurent, V., Maamaatuaiahutapu, K., Maiau, J., & Varney, P. (2004). *Atlas climatologique de la Polynésie française*. Papeete: Meteo France.
Lewis, S. M., Norris, J. N., & Searles, R. B. (1987). The regulation of morphological plasticity in tropical reef algae by herbivory. *Ecology, 68*(3), 636–641.
Littler, M. M., Littler, D. S., & Brooks, B. L. (2006). Harmful algae on tropical coral reefs: Bottom-up eutrophication and top-down herbivory. *Harmful Algae, 5*(5), 565–585.
Lobban, C. S., & Harrison, P. J. (1994). *Seaweed ecology and physiology*. New York: Press Syndicate Cambridge University.
Longo, G. O., & Hay, M. E. (2015). Does seaweed-coral competition make seaweeds more palatable? *Coral Reefs, 34*, 87–96.
Longo, G. O., & Hay, M. E. (2017). Seaweed allelopathy to corals: Are active compounds on, or in, seaweeds? *Coral Reefs, 36*, 247–253.
Lüning, K., & Tom Dieck, I. (1989). Environmental triggers in algal seasonality. *Botanica Marina, 32*(5), 389–398.
Machado, L., Magnusson, M., Paul, N. A., De Nys, R., & Tomkins, N. (2014). Effects of marine and freshwater macroalgae on *in vitro* total gas and methane production. *PLoS One, 9*(1). e85289, 1-11.
Mangaiyarkarasi, R., Kanna, L., Girija, M., & Gnanamurthy, S. (2014). Screening of Indian marine macro algae (Seaweeds) for haemagglutinin activity. *International Journal of Advanced Research, 2*(4), 43–52.
Mantyka, C. S., & Bellwood, D. R. (2007). Macroalgal grazing selectivity among herbivorous coral reef fishes. *Marine Ecology Progress Series, 352*, 177–185.
Mattio, L., Payri, C. E., & Stiger-Pouvreau, V. (2008). Taxonomic revision of *Sargassum* (Fucales, Phaeophyceae) from French Polynesia based on morphological and molecular analyses. *Journal of Phycology, 44*, 1541–1555.
Maypa, A. P., & Raymundo, L. J. (2004). Algae-coral interactions: Mediation of coral settlement, early survival and growth by macroalgae. *Silliman Journal, 45*, 76–95.
McCook, L. J., Jompa, J., & Diaz-Pulido, G. (2001). Competition between corals and algae on coral reefs: A review of evidence and mechanisms. *Coral Reefs, 19*, 400–417.
Moore, J. W. (1974). Benthic algae of southern baffin island: II. The epipelic communities in temporary ponds. *The Journal of Ecology*, 809–819.
Mumby, P. J., Steneck, R. S., Adjeroud, M., & Arnold, S. N. (2016). High resilience masks underlying sensitivity to algal phase shifts of Pacific coral reefs. *Oikos, 125*, 644–655.
N'Yeurt, A. D. R., & Payri, C. E. (2009). Four new species of Rhodophyceae from Fiji, Polynesia and Vanuatu, South Pacific. *Phycological Research, 57*(1), 12–24.
Nelson, S. G., Glenn, E. P., Conn, J., Walsh, T., & Akutagawa, M. (2001). Cultivation of *Gracilaria parvispora* (Rhodophyta) in shrimp-farm effluent ditches and floating cages in Hawaii: A two-phase polyculture system. *Aquaculture, 193*(3–4), 239–248.
Norris, J. N., & Fenical, W. (1982). Chemical defenses in tropical marine algae. Atlantic barrier reef ecosystem, Carrie Bow Cay Belize. *Smithsonian Contributions to the Marine Sciences, 12*, 417–431.
Novaczek, I. (2001). *A guide to the common edible and medicinal sea plants of the pacific islands, 3A*. Community Fisheries Training Pacific Series, University of the South Pacific, Secretariat of the Pacific Community.
Nugues, M. M., Smith, G. W., Van Hooindonk, R. J., Seabra, M. I., & Bak, R. P. M. (2004). Algal contact as a trigger for coral disease. *Ecology Letters, 7*, 919–923.
N'Yeurt, A. D. R., & Iese, V. (2015a). The proliferating brown alga *Sargassum polycystum* in Tuvalu, South Pacific: Assessment of the bloom and applications to local agriculture and sustainable energy. *Journal of Applied Phycology, 27*(5), 2037–2045.
N'Yeurt, A. D. R., & Iese, V. (2015b). Erratum to: The proliferating brown alga *Sargassum polycystum* in Tuvalu, South Pacific: Assessment of the bloom and applications to local agriculture and sustainable energy. *Journal of Applied Phycology, 5*(27), 2047.

N'Yeurt, A. D. R., & Payri, C. E. (2006). Marine algal flora of French Polynesia I. Phaeophyceae (Ochrophyta, brown algae). *Cryptogamie Algologie, 27*(2), 111–152.

N'Yeurt, A. D. R., & Payri, C. E. (2007). Marine algal Flora of French Polynesia II. Chlorophyceae (green algae). *Cryptogamie Algologie, 28*(1), 3–88.

N'Yeurt, A. D. R., & Payri, C. E. (2010). Marine algal flora of French Polynesia III. Rhodophyta, with additions to the Phaeophyceae and Chlorophyta. *Cryptogamie Bryologie, 31*(1), 3–205.

Palanisamy, S., Sellappa, S., & Stella, C. (2010). Antioxidant properties of methanolic extract of *Dictyota batresiana* from south east coast of India. *Journal of Pharmacy Research, 3*(12), 2974–2976.

Pandithurai, M., & Murugesan, S. (2014). Free radical scavenging activity of methanolic extract of brown alga *Spatoglossum asperum*. *Journal of Chemical and Pharmaceutical Research, 6*, 128–132.

Pandithurai, M., Murugesan, S., & Sivamurugan, V. (2015). Antibacterial activity of various solvent extracts of marine brown alga *Spatoglossum asperum*. *International Journal of Pharmacology Research, 5*(6), 133–138.

Pardo-Vargas, A., de Oliveira, I. B., Stephens, P. R. S., Cirne-Santos, C. C., Palmer Paixão, I. C. N., Ramos, F. A., et al. (2014). Dolabelladienols A–C, new diterpenes isolated from Brazilian Brown alga *Dictyota pfaffii*. *Marine Drugs, 12*, 4247–4259.

Parmar, P. V., Solanki, J. B., Nanjiyani, R. P., Kamaliya, N. L., Chavda, V. M., Parmar, H. V., et al. (2016). Antibacterial activity of seaweeds from Veraval coast. *The Bioscan, 11*(4), 2133–2135.

Patil, A. D., Berry, D., Brooks, D. P., Hemling, M. E., Kumar, N. V., Mitchell, M. P., et al. (1993). A diterpene epoxide from the marine brown alga *Dictyota* sp.: Possible vasopressin V1 receptor antagonist. *Phytochemistry, 33*(5), 1061–1064.

Paul, V. J., Kuffner, I. B., Walters, L. J., Ritson-Williams, R., Beach, K. S., & Becerro, M. A. (2011). Chemically mediated interactions between macroalgae *Dictyota* spp. and multiple life-history stages of the coral *Porites astreoides*. *Marine Ecology Progress Series, 426*, 161–170.

Paul, V. J., & Puglisi, M. P. (2004). Chemical mediation of interactions among marine organisms. *Natural Product Report, 21*, 189–209.

Payri, C. E. (1982). *Les Macrophytes du Lagon de Tiahura (Île de Moorea, Polynésie française)*. Thèse de troisième cycle. Montpellier: Université des Sciences et des Techniques du Languedoc.

Payri, C. E., & Denizot, M. (1993). Les Peuplements d'Algues. In ORSTOM (Ed.), *Atlas de la Polynésie française*. pl. 49. Paris.

Payri, C. E., & Meinesz, A. (1985). Algae. In B. Delesalle, R. Galzin, & B. Salvat (Eds.), *French polynesian coral reefs: Vol. 1. Proceedings of the 5th international coral reef congress, Tahiti, 27 May–1 June 1985* (pp. 498–518).

Payri, C. E., & N'Yeurt, A. D. R. (1997). A revised checklist of Polynesian benthic marine. Algae. *Australian Systematic Botany, 10*, 867–910.

Payri, C. E., N'Yeurt, A. D. R., Fiat, S., & Andréfouët, S. (2016). Macroflore marine des Marquises. *Biodiversité Marine et Terrestre des Marquises*, 207–219.

Payri, C. E., & Stiger, V. (2001). Macroalgal community changes on French Polynesian reefs, 1980–2000. *Phycologia, 40*(4), 111.

Pereira, R. C., Cavalcanti, D. N., & Teixeira, V. L. (2000). Effects of secondary metabolites from the tropical Brazilian brown alga *Dictyota menstrualis* on the amphipod *Parhyale hawaiensis*. *Marine Ecology Progress Series, 205*, 95–100.

Pheulpin, L., Recking, A., Sichoix, L., & Barriot, J.-P. (2016). Extreme floods regionalisation in the tropical island of Tahiti, French Polynesia. In *European Conference on Flood Risk Management, E3S Web of Conferences, 7, 01014*.

Pheulpin, L., Sichoix, L., Barriot, J.-P., & Recking, A. (2014). An example of flash-flood events in Tahiti, French Polynesia. In *International conference on Earth observations and societal impacts.*
Philippus, A. C., Zatelli, G. A., Wanke, T., Barros, M. G. A., Kami, S. A., Lhullier, C., et al. (2018). Molecular networking prospection and characterization of terpenoids and C_{15}-acetogenins in Brazilian seaweed extracts. *Royal Society of Chemistry, 8*, 29654–29661.
Rajamani, K., Balasubramanian, T., & Thirugnanasambandan, S. S. (2018). Bioassay-guided isolation of triterpene from brown alga *Padina boergesenii* possess anti-inflammatory and anti-angiogenic potential with kinetic inhibition of β-carotene linoleate system. *LWT-Food Science and Technology, 93*, 549–555.
Rasher, D. B., & Hay, M. E. (2010). Chemically rich seaweeds poison corals when not controlled by herbivores. *Proceedings of the National Academy of Sciences, 107*(21), 9683–9688.
Rasher, D. B., & Hay, M. E. (2014). Competition induces allelopathy but suppresses growth and anti-herbivore defence in a chemically rich seaweed. *Proceedings of the Royal Society B, 281*, 20132615, 1-9.
Rasher, D. B., Hoey, A. S., & Hay, M. E. (2013). Consumer diversity interacts with prey defences to drive ecosystem function. *Ecology, 94*(6), 1347–1358.
Rasher, D. B., Stout, E. P., Engel, S., Kubanek, J., & Hay, M. E. (2011). Chemically rich seaweeds poison corals when not controlled by herbivores. *Proceedings of the National Academy of Sciences, 108*(43), 17726–17731.
Renaud, P. E., Hay, M. E., & Schmitt, T. M. (1990). Interactions of plant stress and herbivory: Intraspecific variation in the susceptibility of a palatable versus an unpalatable seaweed to sea urchin grazing. *Oecologia, 82*(2), 217–226.
Reverter, M., Bontemps, N., Lecchini, D., Banaigs, B., & Sasal, P. (2014). Use of plant extracts in fish aquaculture as an alternative to chemotherapy: Current status and future perspectives. *Aquaculture, 433*, 50–61.
Reverter, M., Saulnier, D., David, R., Bardon-Albaret, A., Belliard, C., Tapissier-Bontemps, N., et al. (2016). Effects of local Polynesian plants and algae on growth and expression of two immune-related genes in orbicular batfish (*Platax orbicularis*). *Fish and Shellfish Immunology, 58*, 82–88.
Rita-Williams, R., Arnold, S. N., Fogarty, N. D., Steneck, R. S., Vermeij, M. J., & Paul, V. J. (2009). New perspectives on ecological mechanisms affecting coral recruitment on reefs. *Smithsonian Contributions to the Marine Sciences, 38*, 437–457.
Rohfritsch, A., Payri, C., Stiger, V., & Bonhomme, F. (2007). Molecular and morphological relationships between two closely related species, *Turbinaria ornata* and *T. conoides* (Sargassaceae, Phaeophyceae). *Biochemical Systematics and Ecology, 35*, 91–98.
Rohfritsch, A., Payri, C., Stiger, V., & Bonhomme, F. (2010). Habitat-related allelic variation revealed by an anonymous DNA locus in reef-dwelling *Turbinaria ornata* (Fucales, Phaeophyceae). *Botanica Marina, 53*, 189–192.
Salvador, N., Gómez Garreta, A., Lavelli, L., & Ribera, M. A. (2007). Antimicrobial activity of Iberian macroalgae. *Scientia Marina, 71*(1), 101–113.
Sameeh, M. Y., Mohamed, A. A., & Elazzazy, A. M. (2016). Polyphenolic contents and antimicrobial activity of different extracts of *Padina boryana* Thivy and *Enteromorpha* sp marine algae. *Journal of Applied Pharmaceutical Science, 6*(9), 87–92.
Sanciangco, J. C., Carpenter, K. E., Etnoyer, P. J., & Moretzsohn, F. (2013). Habitat availability and Heterogeneity and the Indo-pacific warm pool as predictors of marine species richness in the tropical Indo-pacific. *PLoS One, 8*(2). e56245, 1-18.
Sangil, C., Sansón, M., Clemente, S., Afonso-Carrillo, J., & Hernández, J. C. (2014). Contrasting the species abundance, species density and diversity of seaweed assemblages in alternative states: Urchin density as a driver of biotic homogenization. *Journal of Sea Research, 85*, 92–103.

Sarma, N. S., Krishna, M. S. R., & Pasha, S. G. (2006). Marine algal natural products: Contributions from India in the global context. *Indian Journal of Chemistry, Section B, 45B*, 433–449.

Sasaki, H., Kataoka, H., Kamiya, M., & Kawai, H. (1999). Accumulation of sulfuric acid in Dictyotales (Phaeophyceae): Taxonomic distribution and ion chromatography of cell extracts. *Journal of Phycology, 35*, 732–739.

Sasaki, H., Kataoka, H., Murakami, A., & Kawai, H. (2004). Inorganic ion compositions in brown algae, with special reference to sulfuric acid ion accumulations. *Hydrobiologia, 512*, 255–262.

Schmitt, T. M., Hay, M. E., & Lindquist, N. (1995). Constraints on chemically mediated coevolution: Multiple functions for seaweed secondary metabolites. *Ecology, 76*, 107–123.

Schupp, P. J., & Paul, V. J. (1994). Calcium carbonate and secondary metabolites in tropical seaweeds: Variable effects on herbivorous fishes. *Ecology, 75*(4), 1172–1185.

Setchell, W. A. (1926). *Tahitian Algae collected by W. A. Setchell, C. B. Setchell and H. E. Parks* (Vol. 12, pp. 61–142). University of California Publications in Botany.

Sharifian, S., Shahbanpour, B., Taheri, A., & Kordjazi, M. (2019). Effect of phlorotannins on melanosis and quality changes of Pacific white shrimp (*Litopenaeus vannamei*) during iced storage. *Food Chemistry, 298*, 124980.

Shevchenko, N. M., Usol'tseva, R. V., Ishina, I. A., Thinh, P. D., Ly, B. M., & Ermakova, S. P. (2017). Structural characteristics and in vitro antitumor activity of water-soluble polysaccharides from brown algae of the Russian Far East and Vietnam. *Chemistry of Natural Compounds, 53*(1), 1–5.

Siddhanta, A. K., Prasad, K., Meena, R., Prasad, G., Mehta, G. K., Chhatbar, M. U., et al. (2009). Profiling of cellulose content in Indian seaweed species. *Bioresource Technology, 100*, 6669–6673.

Silberfeld, T., Rousseau, F., & De Reviers, B. (2014). An updated classification of brown algae (Ochrophyta, Phaeophyceae). *Cryptogamie Algologie, 35*(2), 117–157.

Sivakumar, S. R., & Vignesh, A. (2014). *In vitro* activity of seaweed extracts collected from Gulf of Mannar coast islands, Tamilnadu on clinical isolates. *International Journal of Pharmaceutical and Biological Archives, 5*(3), 99–103.

Smith, J. E., Hunter, C. L., & Smith, C. M. (2010). The effects of top–down versus bottom–up control on benthic coral reef community structure. *Oecologia, 163*(2), 497–507.

Soares, D. C., Calegari-Silva, T. C., Lopes, U. G., Teixeira, V. L., De Palmer Paixão, I. C. N., Cirne-Santos, C., et al. (2012). Dolabelladienetriol, a compound from *Dictyota pfaffii* algae, inhibits the infection by *Leishmania amazonensis*. *PLoS One, Neglected Tropical Diseases, 6*(9), e1787, 1-12.

Sotka, E. E., Bell, T., Hughes, L. E., Lowry, J. K., & Poore, A. G. (2017). A molecular phylogeny of marine amphipods in the herbivorous family Ampithoidae. *Zoologica Scripta, 46*(1), 85–95.

Sotka, E. E., & Hay, M. E. (2009). Effects of herbivores, nutrient enrichment, and their interactions on macroalgal proliferation and coral growth. *Coral Reefs, 28*, 555–568.

South, G. R. (1993). Edible seaweeds of Fiji: An ethnobotanical study. *Botanica Marina, 36*, 335–349.

Steinberg, P. D., & De Nys, R. (2002). Chemical mediation of colonization of seaweed surfaces. *Journal of Phycology, 38*, 621–629.

Steinberg, P. D., & Paul, V. J. (1990). Fish feeding and chemical defences of tropical brown algae in Western Australia. *Marine Ecology Progress Series, 58*, 253–259.

Stephens, P. R. S., Cirne-Santos, C. C., De Souza Barros, C., Teixeira, V. L., Carneiro, L. A. D., Dos Santos De Amorim, L., et al. (2017). Diterpene from marine brown alga *Dictyota friabilis* as a potential microbicide against HIV-1 in tissue explants. *Journal of Applied Phycology, 29*(2), 775–780.

Stiger, V., & Payri, C. E. (1999a). Spatial and temporal patterns of settlement of the brown macroalgae *Turbinaria ornata* and *Sargassum mangarevense* in a coral reef on Tahiti. *Marine Ecology Progress Series, 191*, 91–100.

Stiger, V., & Payri, C. E. (1999b). Spatial and seasonal variations in the biological characteristics of two invasive brown algae, *Turbinaria ornata* (Turner) J. Agardh and *Sargassum mangarevense* (grunow) setchelll (Sargassaceae, Fucales) spreading on the reefs of Tahiti (French Polynesia). *Botanica Marina, 42*, 295–306.

Teixeira, V. L., & Kelecom, A. (1988). A chemotaxonomic study of diterpenes from marine brown algae of the genus *Dictyota*. *The Science of the Total Environment, 75*, 271–283.

Trapon, M. L., Pratchett, M. S., & Penin, L. (2011). Comparative effects of different disturbances in coral reef habitats in Moorea, French Polynesia. *Journal of Marine Biology, 807625*, 1–11.

Tronholm, A., Sansón, M., Afonso-Carrillo, J., Verbruggen, H., & De Clerck, O. (2010). Niche partitioning and the coexistence of two cryptic *Dictyota* (Dictyotales, Phaeophyceae) species from the canary islands. *Journal of Phycology, 46*(6), 1075–1087.

Tronholm, A., Steen, F., Tyberghein, L., Leliaert, F., Verbruggen, H., Antonia Ribera Siguan, M., et al. (2010). Species delimitation, taxonomy, and biogeography of Dictyota in Europe (Dictyotales, Phaeophyceae). *Journal of Phycology, 46*(6), 1301–1321.

Usoltseva, R. V., Anastyuk, S. D., Ishina, I. A., Isakov, V. V., Zvyagintseva, T. N., Thinh, P. D., et al. (2018). Structural characteristics and anticancer activity *in vitro* of fucoidan from brown alga *Padina boryana*. *Carbohydrate Polymers, 184*, 260–268.

Vallim, M. A., De Paula, J. C., Pereira, R. C., & Teixeira, V. L. (2005). The diterpenes from Dictyotacean marine brown algae in the Tropical Atlantic American region. *Biochemical Systematics and Ecology, 33*, 1–16.

Van Hees, D. H., Olsen, Y. S., Wernberg, T., Van Alstyne, K. L., & Kendrick, G. A. (2017). Phenolic concentrations of brown seaweeds and relationships to nearshore environmental gradients in Western Australia. *Marine Biology, 164*, 74.

Van Steveninck, E. D. R., & Breeman, A. M. (1987). Population dynamics of a tropical intertidal and deep-water population of Sargassum polyceratium (Phaeophyceae). *Aquatic Botany, 29*(2), 139–156.

Verbruggen, H., De Clerck, O., N'Yeurt, A. D., Spalding, H., & Vroom, P. S. (2006). Phylogeny and taxonomy of *Halimeda incrassata*, including descriptions of *H. kanaloana* and *H. heteromorpha* spp. nov.(Bryopsidales, Chlorophyta). *European Journal of Phycology, 41*(3), 337–362.

Vieira, C., Camacho, O., Sun, Z., Fredericq, S., Leliaert, F., Payri, C., et al. (2017). Historical biogeography of the highly diverse brown seaweed *Lobophora* (Dictyotales, Phaeophyceae). *Molecular Phylogenetics and Evolution, 110*, 81–92.

Vieira, C., Camacho, O., Wynne, M. J., Mattio, L., Anderson, R. J., Bolton, J. J., et al. (2016). Shedding new light on old algae: Matching names and sequences in brown algal genus *Lobophora* (Dictyotales, Phaeophyceae). *Taxon, 65*(4), 689–707.

Vieira, C., Engelen, A. H., Guentas, L., Aires, T., Houlbreque, F., Gaubert, J., et al. (2016). Species specificity of bacteria associated to the brown seaweeds *Lobophora* (Dictyotales, Phaeophyceae) and their potential for induction of rapid coral bleaching in *Acropora muricata*. *Frontiers in Microbiology*, 7(316), 1–13.

Vieira, C., Payri, C., & De Clerck, O. (2015). Overgrowth and killing of corals by the brown alga *Lobophora hederacea* (Dictyotales, Phaeophyceae) on healthy reefs in New Caledonia: A new case of the epizoism syndrome. *Phycological Research, 63*, 152–153.

Vieira, C., Stenger, P. L., Moleana, T., De Clerck, O., & Payri, C. (2019). Limited interspecific variation in grazing susceptibility of the brown alga Lobophora to herbivory. *Journal of Experimental Marine Biology and Ecology, 518*, 151175.

Vinayak, R. C., Sabu, A. S., & Chatterji, A. (2011). Bio-prospecting of a few brown seaweeds for their cytotoxic and antioxidant activities. *Evidence-based Complementary and Alternative Medicine*, 673083, 1-9.

Walters, R. G., Ruban, A. V., & Horton, P. (1996). Identification of proton-active residues in a higher plant light-harvesting complex. *Proceedings of the National Academy of Sciences of the United States of America, 93*, 14204–14209.

Zatelli, G. A., Philippus, A. C., & Falkenberg, M. (2018). An overview of odoriferous marine seaweeds of the *Dictyopteris* genus: Insights into their chemical diversity, biological potential and ecological roles. *Revista Brasileira de Farmacognosia, 28*(2), 243–260.

Zizka, M. D. (2012). Sea Urchins on the move: Distribution change of Echinometra in Mo'orea French Polynesia. *Berkeley Scientific Journal, 15*(1).

Zubia, M., Andréfouët, S., & Payri, C. (2014). Distribution and biomass evaluation of drifting brown algae from Moorea lagoon (French Polynesia) for eco-friendly agricultural use. *Journal of Applied Phycology, 27*, 1277–1287.

Zubia, M., Depetris, M., Flores, O., Turquet, J., & Cuet, P. (2018). Macroalgae as a tool for assessing the ecological status of coral reefs under the water framework directive: A case study on the reef flats of La Réunion (Indian Ocean). *Marine Pollution Bulletin, 137*, 339–351.

Zubia, M., & Mattio, L. (2019). *Macroalgues tropicales; une ressource durable d'avenir*. Techniques de l'ingénieur BIO 9040.

Zubia, M., Payri, C., & Deslandes, E. (2008). Alginate, mannitol, phenolic compounds and biological activities of two range-extending brown algae, *Sargassum mangarevense* and *Turbinaria ornata* (Phaeophyta: Fucales), from Tahiti (French Polynesia). *Journal of Applied Phycology, 20*, 1033–1043.

Zubia, M., Payri, C. E., Deslandes, E., & Guezennec, J. (2003). Chemical composition of attached and drift specimens of *Sargassum mangarevense* and *Turbinaria ornata* (Phaeophyta: Fucales) from Tahiti, French Polynesia. *Botanica Marina, 46*(6), 562–571.

CHAPTER SEVEN

Production and properties of mycosporine-like amino acids isolated from seaweeds

Gilles Bedoux[a,]*, Hugo Pliego-Cortés[a], Chrystèle Dufau[a], Kevin Hardouin[a], Romain Boulho[b], Yolanda Freile-Pelegrín[c], Daniel Robledo[c] and Nathalie Bourgougnon[a]

[a]Laboratoire de Biotechnologie et Chimie Marines, EA3884, UBS, IUEM, F-56000 Vannes, France
[b]Gilson Purification, Saint-Avé, France
[c]CINVESTAV Marine Resources Department, Mérida, Yucatan, Mexico
*Corresponding author: E-mail: gilles.bedoux@univ-ubs.fr

Contents

Abstract

Mycosporine-like amino acids (MAAs) belong to a family of secondary metabolites produced by various organisms. However their occurrence is limited to some terrestrial organisms and they appear to be more frequent in marine organisms, such as cyanobacteria, algae and invertebrates. These photoprotective compounds have been isolated from various macroalgal species, mainly in Rhodophyta. Their chemical structure corresponds to polysubstituted aminocyclohexenone or aminocyclohexenimine derivatives substituted with a wide variety of functional groups, such as amino acids. The major role of MAAs in photoprotection against UV radiation has been demonstrated, and is linked to their chemical structure. It would appear however

Advances in Botanical Research, Volume 95
ISSN 0065-2296
https://doi.org/10.1016/bs.abr.2019.11.004

that some MAAs may also have various functions such as antioxidant or nitrogen source allowing the macroalgae adaptation to stressful environments. The biosynthesis of MAAs is suggested through two possible pathways, the shikimate or the pentose-phosphate, which involve the production of key precursors, deoxygadusol and gadusol. Taxonomically, they have been studied in the orders Ceramiales, Gigartinales, Gracilariales, and Bangiales. MAAs are isolated from natural biomass including wild or cultivated sources. Its concentration depends on quality and quantity irradiation, and on inorganic nitrogen availability. In some species, MAAs have been induced under culture conditions. Production of MAAs can be achieved by scaling up the macroalgae biomass, especially under IMTA systems followed by solid/liquid extraction. Because of the low extraction yield from natural sources, the need for large scale production and standard molecules for chemical analysis motivates the attempt at chemical synthesis. Chemical analysis of MAAs requires different techniques such as liquid chromatography (LC) or capillary electrophoresis for the separation and, mass spectrometry or nuclear magnetic resonance for the structural characterization. LC chromatography is mostly applied for the purification of MAAs. Various applications are herein outlined with special emphasis on anti-UV, antioxidant, anti-aging and anti-inflammatory.

1. Introduction

1.1 Definition and description of mycosporine-like amino acids

The Mycosporine-like Amino Acids (MAAs) form a group of water-soluble secondary metabolites. They are the strongest UV-absorbing compounds identified in the environment but they also play a major role, as antioxidant, osmoregulator and thermoregulator, in the response to biotic and abiotic stresses (Oren & Gunde-Cimerman, 2007). They are produced by a wide variety of organisms as fungi, cyanobacteria, invertebrates, algae and lichens (Chollet-Krugler, Nguyen, Sauvager, Thüs, & Boustie, 2019; Kageyama & Waditee-Sirisattha, 2019). Historically, the first MAAs were discovered in 1965, by Charles M. Leach (Leach, 1965). He observed the presence of UV-absorbing compounds in fungi when sporulation was induced by UV radiations. Nowadays, at least 50 MAAs and derivatives have been reported (Chollet-Krugler et al., 2019) from the environment and among them, eleven were identified in red seaweeds (Kageyama & Waditee-Sirisattha, 2019; Lalegerie et al., 2019). The structure of MAAs consists in a cyclohexenone or a cyclohexenimine ring substituted by amino acids or imino-alcohol groups. The MAAs were differentiated by the position and the nature of substituents on the cyclic carbons C3 and C1 (Fig. 1). Monosubstituted MAAs with one substitution on the carbon C3, like

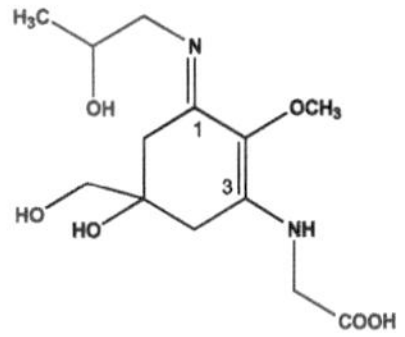

Aplysiapalythine A[1]
APA, M-Gly:Thr(-CO_2)
λ_{max}: 332 nm
Mw: 302 Da
ε: -

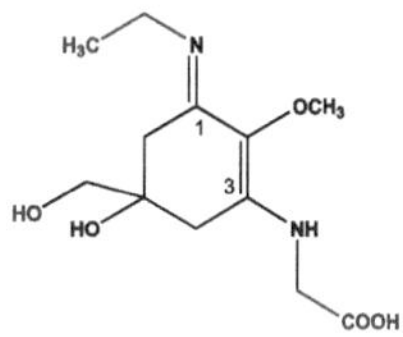

Aplysiapalythine B[1]
APB, M-Gly:Ala(-CO_2)
λ_{max}: 332 nm
Mw: 272 Da
ε: -

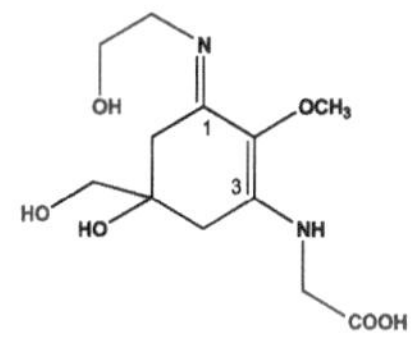

Asterina-330
A-330, M-Gly:Gly(-OH)
λ_{max}: 330 nm
Mw: 288 Da
ε: 43800

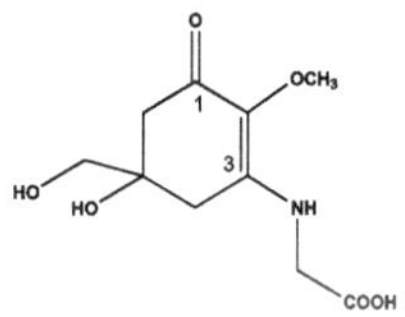

Mycosporine glycine
M-Gly
λ_{max}: 310 nm
Mw: 245 Da
ε: 28100

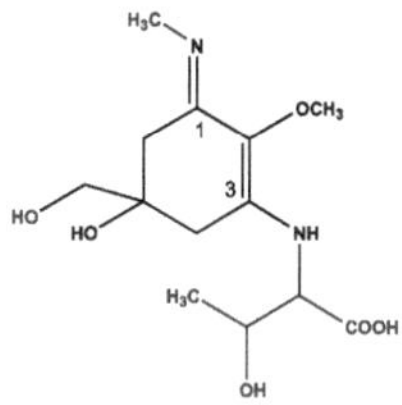

Mycosporine methylamine threonine
M-MA:Thr
λ_{max}: 330 nm
Mw: 303 Da
ε: 33300

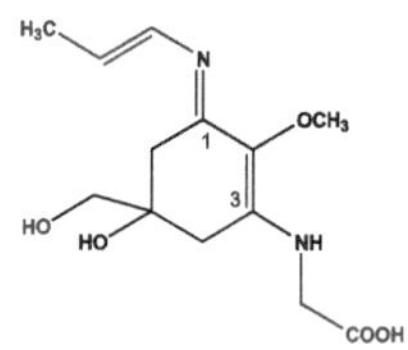

Palythene
Pe
λ_{max}: 360 nm
Mw: 284 Da
ε: 50000

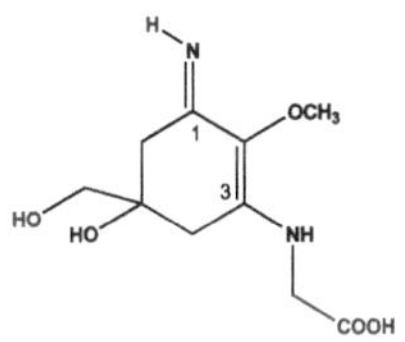

Palythine
Pi, M-NH_2-Gly
λ_{max}: 320 nm
Mw: 244 Da
ε: 36200

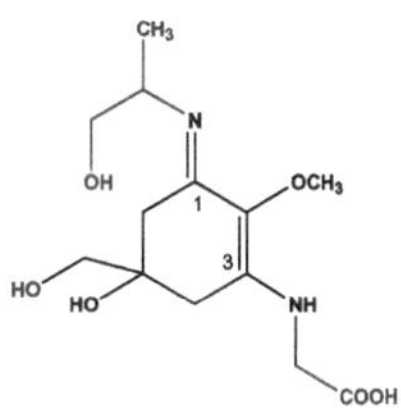

Palythinol
Pinol, M-Gly:Ala(-OH)
λ_{max}: 332 nm
Mw: 302 Da
ε: 43500

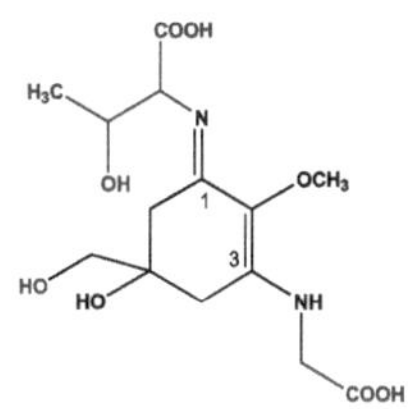

Porphyra-334
P-334, M-Gly:Thr
λ_{max}: 334 nm
Mw: 346 Da
ε: 42300

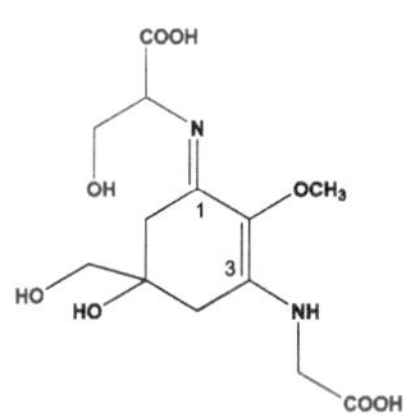

Shinorine
SH, M-Gly:Ser
λ_{max}: 334 nm
Mw: 332 Da
ε: 44668

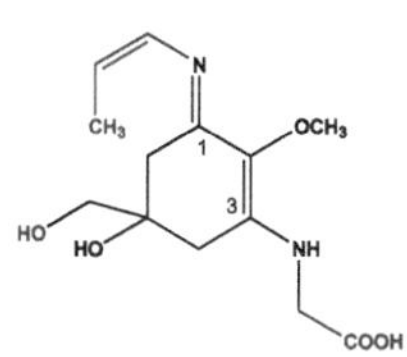

Usujirene
Ue
λ_{max}: 357 nm
Mw: 284 Da
ε: 50000

Fig. 1 Inventory and chemical properties of the eleven described mycosporine-like amino acids from red seaweeds Lalegerie et al., 2019; Kageyama et al., 2019; Orfanoudaki et al., 2019; La Barre et al., 2014.[1] mycosporine-glycine-threonine($-CO_2$), [2] mycosporine-glycine-alanine($-CO_2$) ε in $l\ mol^{-1}\ cm^{-1}$.

mycosporine-taurine and mycosporine-glycine, are oxocarbonyl-MAAs (Carreto & Carignan, 2011; Torres, Enk, Hochberg, & Srebnik, 2006). Mycosporine-glycine is a structural intermediate for the disubstituted MAAs with a second substitution on the C1 by an amino alcohol (ex: shinorine), or an amino acid (ex: mycosporine-2-glycine) or an enaminone (Muraoka et al., 2009). Additional modifications can occur on the amino acid group by oxidation, reduction, condensation, glycosylation, decarboxylation and dehydration (Arima et al., 2012). Fig. 1 inventories the MAAs isolated from red seaweeds. Chemically, they are characterized by the combination of three main characteristics, *1)* their maximum absorption wavelength (λ_{max}), *2)* their molecular weight (Mw) and *3)* their molar extinction coefficient (ε). MAAs have a molar extinction coefficient comprised between 28,100 and 50,000 L mol^{-1} cm^{-1}, an absorption maximum around 310 nm for oxo-mycosporine and 330 nm for imino-mycosporines, between 268 (deoxygadusol precursor) and 362 nm (palythene), and a molecular weight generally less than 400 Da.

1.2 Biosynthesis of mycosporine-like amino acids

So far it is thought that the original producer of the MAAs was a cyanobacterial ancestor. Studies of main steps in the MAAs biosynthesis and genes involved have been mainly elucidated in cyanobacteria, evidencing that MAAs biosynthesis proceeds from a branch-point intermediate of the Shikimate pathway or from the pentose phosphate pathway (Pathak et al., 2019; Pope et al., 2015). Fig. 2 shows the proposed pathways for MAAs biosynthesis, in the Shikimate pathway, the parent molecule 3-deoxy-D-arabino-heptulosinate 7-phosphate (DAHP) through 3-dehydroquinate synthase (DHQS) produce 3-dehydroquinate (3-DHQ) to be then transformed into gadusol and then into 4-deoxygadusol (4-DG). Whereas in the pentose phosphate pathway, the sedoheptulose 7-phosphate (SH 7P) yield 4-deoxyadiosol (4-DG). In *Anabaena variabilis*, the 2-epi-5-epi-valiolone synthase (EEVS) and the O-methyl-transferase (OMT) catalyzed this biosynthesis (Pope et al., 2015); OMT genes were also reported in the transcriptome of symbiotic dinoflagellates *Symbiodinium* (Rosic, 2012). In both pathways, 4-DG is the common precursor to produce MAAs (Balskus & Walsh, 2010; Favre-Bonvin, Bernillon, Salin & Arpin, 1987).

Conjugation with a glycine molecule through ATP-grasp amino acid ligase generates the MAA type mono-substituted cyclohexenone mycosporine-glycine (M-gly). M-gly is the common intermediate in the production of di-substituted aminocyclohexene imine-type MAAs such as shinorine

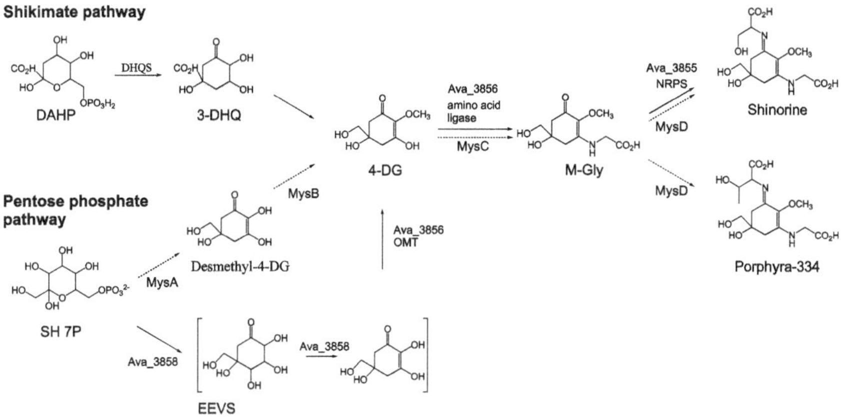

Fig. 2 Pathways involved in MAAs biosynthesis. Dotted arrows represent the biosynthetic pathway proposed for porphyra-334 in red algae according to Brawley et al. (2017). DAHP: 3-deoxy-D-arabinoheptulosinate 7-phosphate. DHQS: 3-dehydroquinate synthase; 3-DHQ: 3-dehydroquinate; SH 7P: sedoheptulose 7-phosphate; EEVS: 2-epi-5-epi-valiolone synthase; OMT: O-methyl-transferase; NRPS: nonribosomal peptide synthase; 4-DG: 4-deoxygadusol. *Modified from Pope et al. (2015) and Brawley et al. (2017).*

(Pope et al., 2015) which was yielded by the condensation with serine through a nonribosomal peptide synthase (NRPS) in *Chlorogloeopsis* sp. PCC 6912 (Portwich & Garcia-Pichel, 2003). In *A. variabilis* ATCC 29413, the shikimate and the pentose phosphate pathways converge for MAA biosynthesis in response to UV-A exposure (Pope et al., 2015). In *Chondrus crispus*, Krabs, Bischof, Hanelt, Karsten, and Wiencke (2002) reported that asterina-330, palythinol and palythene were strongly stimulated just under UV-B, the authors suggested that the different pathways to synthesize MAA might be induced by different wavelengths bands. Even though some MAAs can not be produced by the substitution of amino acids in 4-DG, it is thought that MAAs can be synthesized by modification of amino acid side chains by condensation (esterification and amidation), dehydration (double bond formation), decarboxylation (chain shortening), and oxidation and reduction (hydroxylation). Although, it is known that additional substitution reactions can occur (e.g., sulfonation and glycosylation) producing MAA sulfates in corals and MAA glycosides in cyanobacteria and fungi (Wada, Sakamoto, & Matsugo, 2015), these MAAs has not been identified in seaweeds.

Recently, Brawley et al. (2017) reported that the red macroalgae *Porphyra* and *Chondrus* genomes contain the genes which encode the MysA-MysB and the MysC—MysD protein fusion, present in other MAA synthesizer red algae and dinoflagellates as well (Waller, Slamovits, & Keeling, 2006), which were proposed to acquired the genes from a cyanobacterium through secondary or serial endosymbiosis. These genes were reported in *Nostoc punctiforme*, in which MysA, MysB, MysC, and MysD supported the biosynthesis of shinorine, while in *Anabaena* with homolog genes, MysD is replaced by a NRPS, MysD condense threonine instead serine onto mycosporine-glycine to yield porphyra-334 (Balskus & Walsh, 2010; Gao & Garcia-Pichel, 2011). Brawley et al. (2017) proposed that the production of porphyra-334 in red algae is done by the pentose pathway. Authors suggested that the arrangement of the gene-fusion and the conservation of the MAA gene cluster in *Porphyra* and *Chondrus* provide advantages and efficient MAA biosynthesis for red algae exposed to high quantities of UV irradiance. Studies in seaweed to elucidate the MAA biosynthesis are yet going on.

2. Ecophysiology

Attached to their substrate, marine seaweeds are confronted with various periodic stresses which play a decisive role in the physiology and determine their distribution on intertidal and subtidal zones. Each species has to face abiotic and biotic changes occurring at different scales and gradients to which it has adapted according to its own physiological parameters. Species settling in the supra-littoral zone are exposed to drought, high solar radiation, desiccation and temperature changes. The intertidal (eulittoral) zone is moreover characterized by regular and extreme changes in abiotic conditions (variations in salinity, temperature, high ultraviolet (UV) radiation, nutrient supply), and by herbivore pressure. Furthermore, mechanical stress due to tidal currents and wave exposure has a strong impact on intertidal communities. In the sub-littoral zone extending below the low tide level, seaweeds usually encounter a more stable habitat, as the water column above buffers against strong changes in abiotic parameters (Davison & Pearson, 1996). Climate change is imposing additional stresses on coastal ecosystems. The increase in sea surface temperature and irradiance, the ocean acidification, the storminess, the major grazing pressure are rapidly changing coastal environmental conditions (Quintano et al., 2019). Among Ultraviolet radiations (UVRs), UV-B radiations can induce formation of reactive oxygen species (ROS) altering DNA, lipids and protein molecules in macroalgae and it can also damage photosynthetic pigments such as chlorophyll and phycobiliproteins so that photosynthesis and growth become reduced (Barufi, Mata, Oliveira, & Figueroa, 2012; Bischof et al., 2006; Lalegerie et al., 2019; Sagert, Forster, Feuerpfeil, & Schubert, 1997). To counteract these negative effects, seaweeds have developed several photoprotective mechanisms and repair strategies for damaged molecules. These include a reduction of photosynthetic capacity, called photoinhibition, avoidance of DNA damage by synthesis of UV absorbing compounds and production of photosynthetic pigments (Bischof et al. 2006; Dring, 2006; Figueroa & Korbee, 2010; Korbee-Peinado, Abdala Dıaz, Figueroa, & Helbling, 2004). In parallel with pigments or phlorotannins from brown seaweeds, nitrogen-containing compounds, belonging to the family of MAAs, ensure photoprotective capabilities in red macroalgae (Figueroa & Korbee, 2010; García-Sánchez et al., 2012; Karsten et al., 1998; Korbee-Peinado et al., 2004).

3. Chemistry

3.1 Production

MAAs have been isolated from natural biomass including wild or cultivated sources. Currently, more than 100 species of seaweeds from temperate, tropical and polar regions have been studied and their content in MAAs varies from traces to more than 12 mg g^{-1} dw. So far, deep-sea species do not present MAAs or just in trace amounts and among the different species of seaweeds, red seaweed are the main producers. Total concentration of MAAs and its changes in the individual composition have been related to different parameters such as water depths, seasons of harvesting or environmental factors, irradiation (quality and quantity), nutrient availability (inorganic nitrogen), pH, temperature, or desiccation (Navarro et al., 2018; Pliego-Cortés et al., 2019).

3.1.1 Wild sources

In wild populations of *Digenia simplex* studied at different water depths (−1, −2.5 and −3.5 m), as the depth increased the MAAs concentration decreased, while a twofold times increased was observed from 11 to 13 h in thalli sampled at the shallow (~3 mg g^{-1} dw). The individual composition changed from the dominant MAA shinorine (33 %) in the shallow to mycosporine-glycine in the middle (37%) and deep (34%) level (Betancor, Dominguez, Tuya, Figueroa, & Haroun, 2015). A survey in 13 red intertidal seaweed from southern Chile during high solar irradiance showed that *Porphyra columbina* (10.6 mg g^{-1} dw), *Gelidium sp.* (6.8 mg g^{-1} dw), *Mazzaella laminarioides* (4.8 mg g^{-1} dw), and *Grateloupia doryphora* (3.5−4.4 mg g^{-1} dw) were species with the highest concentrations of MAAs, shinorine was the most common and present MAA in all species (Huovinen et al., 2004). In polar waters, MAAs were detected in 17 out of 28 red species and one green, *Porphyra endiviifolium* (9.0 mg g^{-1} dw), *Bangia atropurpurea*, (5.8 mg g^{-1} dw), *Curdiea racovitzae* (4.9 mg g^{-1} dw), and the green *Prasiola crispa subsp. Antarctica* (3.7 mg g^{-1} dw) showed the higher concentration (Hoyer, Karsten, Wiencke, 2002). In *C. racovitzae*, Karentz, McEuen, Land, and Dunlap (1991) reported 7.17 mg of palythine out of 9.3 mg g^{-1} dw of total.

Karsten, Sawall, and Wiencke (1998) screened 28 species of red seaweeds, in tropical and subtropical locations from China, Japan, America, Africa and Australia, the highest MAAs concentration was found in *Bostrychia radicans* (8.54 mg g^{-1} dw) collected in Mexico, the species presented up to 6.49 mg g^{-1} dw of palythinol. A positive correlation was observed between the total MAAs content and levels of PAR (Photosynthetically Active Radiation) in populations collected from different latitudinal gradients along the Chilean coast, *Chondracanthus chamissoi* (0.8−3.3 mg g^{-1} dw) and *Gelidium lingulatum* (2.6−6.8 mg.g^{-1}dw) concentrations changed among life cycle phases and sample places (Véliz, Chandía, Karsten, Lara, & Thiel, 2018). From 39 red seaweeds sampled in 23 collection points based on the abiotic variables along the Brazilian coast, in which 48.5% of samples presented MAAs concentrations lower than 0.3 mg g^{-1} dw, 31.6% between 0.3 and 1 mg g^{-1} dw, and 19% more than 1 mg g^{-1} dw, with a maximum of 5.89 mg in *Pyropia acanthophora*. A high UV index (>11), a basic pH (8.22) and concentrations of nitrate (0.95−1.21 μmol L^{-1}) and phosphate (0.12−0.24 μmol L^{-1}) influence MAA contents (Briani et al., 2018). While in thalli harvested on the intertidal coast of Brittany, France, during summer and winter, the authors reported 23 potential MAAs in 40 red seaweeds, in which the individual composition in some species were affected by the seasons, *Mastocarpus stellatus* contained a high concentration of shinorine in winter (3.12 mg g^{-1} dw) and lower in summer (1.62 mg g^{-1} dw). Furthermore, the high diversity of MAAs was not related to a phylogenetic pattern, anatomy and morphology, the position on the shore or the sampling site, even if a species was introduced or indigenous to Brittany (Lalegerie et al., 2019). This might be explained since no general MAAs induction patterns can be found; Hoyer et al. (2002) suggested that the induction and accumulation of MAAs is a very variable and a species-specific mechanism. For example, in *Pyropia plicata* collected in New Zealand over three seasons (April—November 2016), the highest MAA concentrations (14.5 mg g^{-1} dw) were found under low UV-B radiation during winter. On the contrary, at the end of spring (June) when the UV-B radiation increased, low concentrations of MAAs (<5 mg g^{-1} dw) were detected (Diehl, Michalik, Zuccarello, & Karsten, 2019).

These studies could be a reference to consider the MAAs concentration and individual composition in wild populations from different environments, in order to identify species with high amounts of MAAs or species from which MAAs concentration to be stimulated. In these sense, there are some studies that have managed to stimulate the production of MAAs concentration or individual composition in response to different abiotic factors. Most of these studies were conducted as experimental trials under controlled or semi-controlled conditions, including cultivation in bioreactors, mesocosms, tanks or ponds, in indoor and outdoor systems (Table 1).

3.1.2 Cultivated sources

The cultivation of *Pyropia columbina* under the interaction effects of full solar radiation exposure (PAR + UVA + UVB) and nitrate (NO_3^-) availability, reaching an amount of 10.4 mg g^{-1} dw compared to the initial value 2.2 mg g^{-1} dw. The individual composition of MAAs did not show a significant effect when the radiation treatments were used as an isolated factor (Navarro et al., 2014). This pattern was also observed in *Gracilaria longissima* under two different artificial lights and two nutrients concentrations, a three times increase in total MAAs content (0.69 mg g^{-1}dw) was observed, but no changes among the individual compositions were found (Álvarez-Gómez et al., 2017). The combined and/or interactive effects of radiation and nitrogen (N) concentrations on MAAs biostimulation have also been reported in *Gracilaria tenuistipitata* (Barufi, Korbee, Oliveira, & Figueroa, 2011; Barufi et al., 2012), *Porphyra columbina* (Korbee-Peinado et al. 2004), and *Grateloupia lanceola* (Huovinen, Matos, Pinto, & Figueroa, 2006). The MAAs concentration regarding the reproductive phase in cultivated *M. laminarioides* was higher in gametophytes than in tetrasporophytes (3.0 and 2.3 mg g^{-1} dw, respectively). A significant decrease in total MAAs content was observed in gametophytes under all solar radiations and irradiance treatments. After 14 days of cultivation, tetrasporophytes showed higher proportion of palythine and asterina-330, and gametophytes increased M-gly under PAR + UVA

Table 1 Induction of mycosporine-like amino acids in red seaweeds under different conditions including wild and cultivated sources.

		Total	Individual composition (%)									
Species	**Treatment**	**mg g dw**	**DG**	**M-Gly**	**SH**	**P-334**	**Pi**	**A-330**	**Pinol**	**Ue**	**Pe**	**Reference**
Rhodymenia pseudopalmata	Wild thalli	0.26 ± 0.1	0.01		35.04	57.17	3.55	4.13		0.09		Pliego-Cortés et al., 2019
IMTA cultivated	Full solar radiation + N				40.5	57.57	1.2	0.38	0.32	0.32		
	3 days	0.79 ± 0.2	0.01		44.5	50.37	2.06	1.84	0.50	0.50	0.48	
	7 days	1.24 ± 0.1										
Gracilaria cornea	5 m deep	~0.11			<5%		20–30	70–80				Quintano et al., 2019
Wild populations	Water deep transplantations	~0.15										
	Moved to 3 m deep											
G. vermiculophylla	Monthly harvested											Barceló-Villalobos, Figueroa, Korbee, Álvarez-Gómez, & Abreu, 2017
IMTA cultivated	Wild November	~3.8			★	★	★					
	Wild April	~4.8			★	★	★					
	IMTA April, 5 kg m^{-2}	~3.3			14.49	60.0	25.42					
	IMTA April, 7 kg m^{-2}	~3.8			12.58	65.32	22.09					
Gracilariopsis longissima	UVR and nutrients	0.33 ± 0.12			★	★	★	★	★			Álvarez-Gómez et al., 2017
Lab cultivated	Initial											
	PAB + High N	0.69 ± 0.05										

Digenia simplex	1 m deep, 11:00 h	~1.5	11.55	40.48	17.96	27.55	3.45		Betancor et al.,
Wild	1 m deep, 13:00 h	~3.0	31.17	33.10	7.53	25.72	2.51		2015
populations	2.5 m deep, 13:	~1.3	37.40	37.81	8.52	16.27			
	00 h	~0.7	34.18	22.95	13.59	29.28			
	3.5 m deep, 13:								
	00 h								
Ellisolandia	Increasing CO_2, N,								Figueroa et al.,
elongata	T°C	0.23 ±		57.3		24.76	2.7	15.25	2014
Mesocosm	Low CO_2, Low N	0.04							
cultivated	High CO_2, Noon	~0.25							
Pyropia columbina	Initial 8:00 h	2.2 ±	0	6.6	87.8	1.8	3.1		Navarro et al.,
Outdoor	Solar	0.6							2014
cultivated	radiation + N	~8.5							
	P, 12:30 h	10.4 ±							
	PAB, 12:30	1.1							
G. tenuistipitata	Initial	0.55 ±							Barufi et al.,
Lab cultivated	UVR + N	0.1							2012
	PAB, 3 d								
	PAB +	1.5							
	0.1 mM N, 3 d	2.5							
	PAB +	4.29							
	0.5 mM N, 7 d								

(Continued)

Table 1 Induction of mycosporine-like amino acids in red seaweeds under different conditions including wild and cultivated sources.—cont'd

		Total	Individual composition (%)									
Species	**Treatment**	**mg g dw**	**DG**	**M-Gly**	**SH**	**P-334**	**Pi**	**A-330**	**Pinol**	**Ue**	**Pe**	**Reference**
Hydropuntia cornea	Fishpond effluents	mg $m^{-2}d^{-1}$			★	★	★					Figueroa et al., 2012
IMTA cultivated	Outdoor, 7 d											
	Indoor, 7 d	10.2 ± 0.25										
	Natural seawater											
	Outdoor, 7 d	12.8 ± 0.09										
		12.9 ± 0.0										
G. conferta	Fishpond effluents											Figueroa et al., 2010
IMTA cultivated	High N	0.4			95	5						
	Low N + P	1.8			95	5						
	Low N + PA	~2.0			95	5						
Asparagopsis armata	Ammonium nitrogen flux (μM/h)							Traces				Figueroa et al., 2008
IMTA cultivated		~1.5			80–85		17–20					
		~2.7			Lower		Higher					
	17	~2			Lower		Higher					
	50											
	160											

Porphyra haitanensis	Control initial	2.3 ± 0.43	UV-absorbing compounds estimated at 336 nm peak					Jian et al., 2008
	PAB + desiccation							
Outdoor cultivated	Control, 4 d	~3.4						
	P + desiccated, 4 d	3.06						
	PAB + desiccated, 4 d	~7.0						
		~10.0						
Porphyra leucosticte	Initial	6.74		0.80^{+}	5.34^{+}	0.29^{+}	0.31^{+}	Korbee et al., 2005
	Seven-day cultures							
Lab cultivated	under:	10.4		2.67^{+}	7.27^{+}	0.22^{+}	0.24^{+}	
	White light	10.69		0.57^{+}	8.88^{+}	0.60^{+}	0.64^{+}	
	Blue	9.33		2.39^{+}	6.54^{+}	0.19^{+}	0.21^{+}	
	Green	8.0		1.96^{+}	5.75^{+}	0.14^{+}	0.15^{+}	
	Yellow	7.9		1.87^{+}	5.69^{+}	0.16^{+}	0.18^{+}	
	Red							
Porphyra columbina	Initial	~5.5	⋆	~20	~80	⋆	⋆	Korbee-Peinado et al., 2004
	PAB+300 mM NH4+, 6 d	~9						
Lab cultivated								
Chondrus crispus	Initial irradiance	Traces						Krabs et al., 2002
Outdor cultivated	under cut-off filters	~2.8		~0.8		~1.0	~0.3	
	320 nm, 5 d	~2.9		~0.3		~1.1	~0.7	
	335 nm, 7 d	~3		~0.5		~2.2	~0.5	
	420 nm, 9 d							

(Continued)

Table 1 Induction of mycosporine-like amino acids in red seaweeds under different conditions including wild and cultivated sources.—cont'd

		Total	Individual composition (%)									
Species	**Treatment**	**mg g dw**	**DG**	**M-Gly**	**SH**	**P-334**	**Pi**	**A-330**	**Pinol**	**Ue**	**Pe**	**Reference**
C. crispus	Initial	0.05										Franklin et al., 2011
Outdor cultivated	Irradiance under colored filters	~0.65			~0.1+		~0.4+	Traces			Traces	
	White light, 7 d	1.5			*		0.8+	Traces			Traces	
	Blue light 40 d											
C. crispus	Initial	0.14 ± 0.1			0		0.29+	0.08+			0.04+	Bischof et al., 2006
Mastocarpus stellatus	UV-B	0.67 ± 0.2			0.11+		0.28+	0.16+			0.09+	
Lab cultivated	Initial	2.6 ± 0.5			2.65+							
	UV-B	4.4 ± 1.0			4.40+							

DG: Deoxygadusol; M-Gly: Mycosporine-glycine; SH: Shinorine, P-334: Porphyra-334; Pi: Palythine; A-330: Asterina-330; Pinol: Palythinol; Ue: Usujirene; Pe: Palythene.
~ Approximate quantities based on the figures.
* Percentages of individual MAAs were presents in graphics or just identified without percentages.
+Individual composition in mg g^{-1} dw.
P: PAR (Photosynthetically Active Radiation); PA: PAR + UVA; PAB: PAR + UVA + UVB.
N: Nutrient such as NO_3^- or NH_4^+.

and PAR + UVA + UVB treatments (Navarro, Figueroa, & Korbee, 2017).

The effects of desiccation in *Porphyra haitanensis* maintained under indoor (artificial radiation) and outdoor (natural solar radiation) conditions, showed a significantly increase concentrations of UV-absorbing compounds (peak at 336 nm) about 119% and 200% respectively, compared to those without desiccation (Jiang, Gao, & Helbling, 2008). However, the authors did not found a UVR-dependent accumulation of UV-absorbing compounds during both long- and short-term experiments. Nevertheless, the quality of light has been reported to stimulate MAAs production. In *Porphyra leucosticta* an increase in white, blue and green light have been reported. The highest amount was obtain in the blue light, which also increase porphyra, palythine and asterina-330, and the lowest MAAs total content under yellow and red lights, decreasing the content of palythine and asterina-330 (Korbee, Figueroa, & Aguilera, 2005). In *C. crispus* exposed to filtered natural radiation during 40 days, palythine was increased under blue light. Under green or red light, the thalli showed just traces of MAAs (Franklin, Krabs, Kuhlenkamp et al., 2011). Whereas, in the same species exposed under different cut-off filters to natural solar radiation, UV-B exposition has shown a strong induction of asterina-330, palythinol and palythene, and UV-A stimulated shinorine and palythine (Krabs et al., 2002). In a trial to evaluate the vulnerability of seaweeds to climate change in short-terms (6 days) under cultivation, *Ellisolandia elongata* was exposed to increasing CO_2 (380 vs. 700 ppm), nitrate concentration (5 vs. 50 μM), temperature (ambient vs. ambient+4°C) and, harvesting at four times during the day. The interaction between two and three variables showed significant differences in both total MAAs content and individual composition, although no significant effect of nitrate as a single factor was found (Figueroa et al., 2014).

Nowadays, based on the information and data available about number of species and MAAs concentration in wild populations, and in spite of the fact that MAAs could be induced under certain conditions, it seems that a biotechnological production of MAAs is possible. However, new species should be studied in order to diversify MAA sources, especially under aquaculture systems, since these are excellent alternatives to wild harvesting. Aquaculture practice could be boosted by scaling up the macroalgae biomass for specific species, such as Integrated Multi-Trophic Aquaculture (IMTA) systems. Because the increase in MAAs concentration is generally nitrogen-demanding, the supply of nitrogen (N) could be further plenty covered from aquaculture wastewaters. Being also an effective way to reduce the eutrophication of the coastal areas, causes by traditional aquaculture, as seaweeds can eliminate efficiently the dissolved inorganic nutrients. Furthermore, biorefinery strategies should be considering maximizing biomass utilization.

Under IMTA system, Pliego-Cortés et al. (2019) recently showed that under high solar radiation and high ammonium concentrations from fishpond wastewaters, the total content of MAAs was stimulated in *Rhodymenia pseudopalmata*, compared with that from the wild populations from playa del Carmen, Mexico, and with that grown in lower irradiation conditions. The authors expected to obtain 1167 mg of MAAs in 941 g dw.100 m^{-2} d^{-1} of biomass. After seven days of solar exposition the individual composition increased from four to eight MAAs including the precursor deoxygadusol. While, in *Gracilaria vermiculophylla* collected in Ria de Aveiro, Portugal, and cultivated during eight months under IMTA conditions and using a stocking density of 3 kg fresh algae m^{-2}, its seems to be possible to obtain an average output of 212.05 mg MAAs m^{-2}, with a highest productivity in April and May (35.5% in April). The individual composition also changed according to stocking density and time. Comparing with wild thalli, higher contents of MAAs were detected in November and April (Barceló-Villalobos et al., 2017). According to Figueroa et al. (2012), some differences in MAAs production under cultivation could be explained by the species, the quantity of N supply, the biofiltration efficiency of algae, and the tank capacity. By growing *Hydropuntia cornea* in the Canary Islands, Spain, under outdoors conditions integrated to fishpond effluents, the authors obtained a MAAs concentration of 13 mg m^{-2} d^{-1} of dw, representing a production of 34.5 kg of MAAs from 19.7 T dw.y^{-1} of algal biomass using 100 m^2 surface for culture. Furthermore, in *Asparagopsis armata* grown with wastewaters from a fish farm located in southern Portugal, an average production of 1.75 mg MAAs per gram of seaweed dw, representing an expected annual production of 959 g of MAAs from 548 kg dw algal biomass was reported (Figueroa et al., 2008). While in *Gracilaria conferta*, cultivated in outdoor tanks in Eilat, Israel, under low nutrient and high nutrient (fishpond effluents), accumulation of MAAs was inversely related to quantities of N supply, the large MAA content was observed in the N-deficient algae (1.8 mg g^{-1} dw) compare to high N (0.4 mg g^{-1} dw), a production of 37 mg MAAs.m^{-2}.d^{-1} was reported (Figueroa et al., 2010). Again, this result showed no general MAAs induction, and accumulation of MAAs is a very variable and a species-specific mechanism.

3.1.3 Chemistry synthesis of mycosporine-like amino acids (MAAs) and analogs

The low extraction yield from natural sources, the need for large scale production and standard molecules for chemical analysis motivated the attempt at chemical synthesis. White, Cammack, and Sakuma (1989), White, Cammack, Sakuma, Rewcastle, Widener (1995) used D-(-)-quinic acid (1), an ubiquitous natural product, as an attractive starting material for asymmetric synthesis of mycosporins (Bandaranayake, 1998). Thus, (*S*)-mycosporine-glycine (2) and (*S*)-Mycosporin I (3) (Fig. 3) have been synthesized. However, this total organic synthesis including about 15 steps with an overall yield of 2% and using hazardous, explosive or flammable chemicals, can't be applied to large scale production. In addition, attempts to prepare iminomycosporines, palythine (4) and asterina-330 (5), were unsuccessful.

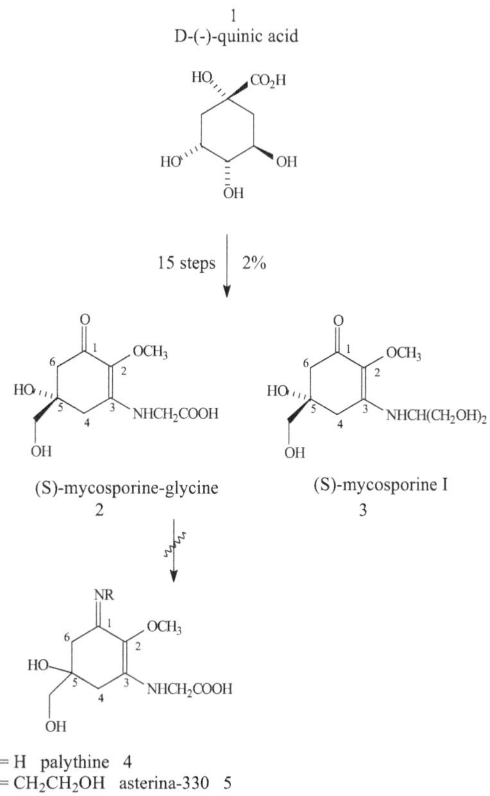

Fig. 3 Total synthesis of natural mycosporines.

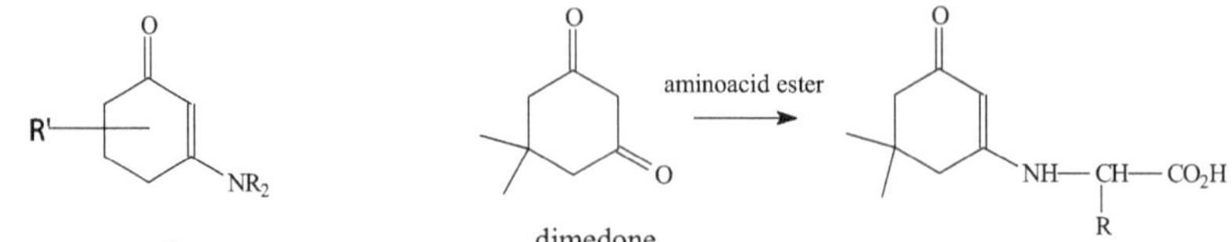

6
aminocyclohexenone
R=H, alkyl, aryl, aminoacid
R' = H or substituent(s)

dimedone
7

8

Fig. 4 Synthesis of aminocyclohexenone type derivatives.

This is the reason why it is interesting to look for more affordable chemical syntheses of simpler analogs retaining the same physico-chemical and biological properties (Bandaranayake, 1998; Cardozo et al. 2007; Chrapusta et al. 2017; Dunlap, Chalker, Bandaranayake, & Wu Won, 1998; Lawrence, Long, & Young, 2018; Nguyen 2013). At first it should be noted that the synthesis of aminocyclohexenone-type derivatives (6) has been described for several years (Fig. 4). For example, Halpern et al. (1964a,b), Halpern (1965) described the reaction of dimedone (7) with amino acid esters as a new protecting group in peptide synthesis to afford enaminone derivatives (8) after hydrolysis of the ester function.

Greenhill (1969, 1971, 1977) investigated the synthesis of enaminones (9) from derivatives of cylohexanedione or dimedone (Fig. 5) and reviewed their chemical and physical properties in order to demonstrate their potential uses in synthetic and medicinal chemistry.

Dunlap and Chalker (1986), Dunlap et al. (1998) prepared MAAs synthetic analogs (10) which preserve the UV-absorbing chromophore of

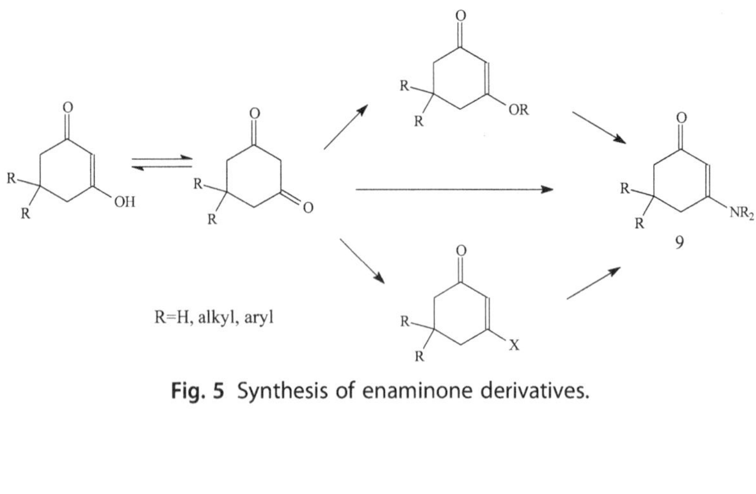

Fig. 5 Synthesis of enaminone derivatives.

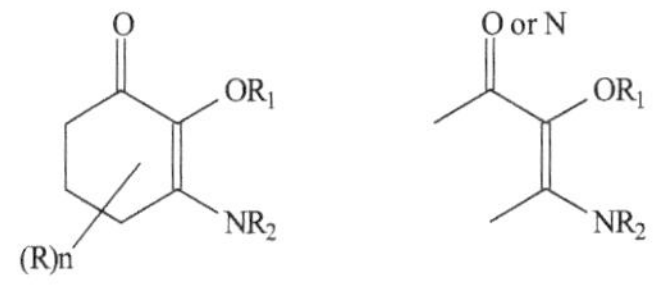

10

UV absorbing chromophore

Fig. 6 MAAs synthetic analogs which preserve the UV-absorbing chromophore.

MAAs due to electron conjugation of the free electron pair of the nitrogen at the 3-position and the carbonyl (or imino) unsaturation (Fig. 6).

Replacement of the amino acid substituent on the 3-position with alkylamino group reduces hydrophilic properties specific of highly water soluble MAAs. A substituent R on the ring can be chosen according to the hydrophilic/lipophilic balance. The authors presented this Structure Activity Relationship (SAR) and completed by the influence of substituents OR_1, NR_2 or five versus six-membered ring on UV-absorbance properties, hydrolytic reactivity and oxidative decomposition corresponding to the limited stability of MAAs once extracted from their natural origin. Indeed, useful UV agents for commercial purposes must have UV-absorbance excess of about 290 nm, acceptable stability and compatibility with media to which they will be applied or incorporated. Aminocyclohexenone were also prepared as intermediate products into the synthesis of aminocyclohexenimines analogs of MAAs (Chalker, Dunlap, & Bandaranayake, 1988; Sampedro Ruiz, Losantos Cabello, & Funes Ardoiz, 2015). Those analogous iminomycosporins (Fig. 7) have been synthesized (Chalker et al., 1988; Losantos et al., 2017; Sampedro Ruiz 2015). Replacement of carbonyle C=O by an imine C=N on 1-position increased thermal and photochemical stability as well as absorption. Iminoanalogs are also more photostable than natural MAAs. The

n = 0,1 or 2

R = different substituants

physiologically acceptable salt

$X^- = Cl^-, TsO^-, \ldots$

Fig. 7 Analogous iminomycosporins.

product may be in the form of physiologically acceptable salt. Changing substituants R allowed a SAR study about consequences on UV-absorption, solubility in different solvents and stability. Removing hydroxyl on 5—position substituents and the choice of the 3-position N-substituent allow a simpler synthesis with better yield. The methoxy on the 2-position can be replaced by alkyl, or S-alkyl. Finally, Losantos et al. (2017) concluded that they have prepared a new family of potential UV-filters, that the easy synthetic routes provide a suitable source for commercial applications for new generation of more efficient sunscreen. Their use in formulation including commercial sunscreens boosted the solar protection factor (SPF).

As aminocyclohexenone derivatives were too hydrolytically and oxidative unstable for practical applications, tetrahydropyridine derivatives (Fig. 8) were sufficiently stable for commercial application as suncare product and prolonged exposure to UV (Bandaranayake, 1998; Cardozo et al., 2007; Chrapusta et al., 2017; Dunlap et al., 1998).

In terms of synthesis, Wenkert et al. (1968) published a protocol describing the preparation of the compounds (11) (Fig. 9).

Chalmers et al. (1990), Bird, Fitzmaurice, Dunlap, Chalker, and Bandaranayake (1988, Bird, Chalmers, Fitzmaurice, & Thang, 1997) prepared this new generation of MAAs analogs for sunscreen compositions (Fig. 10); SAR studies were investigated to evaluate the influence of

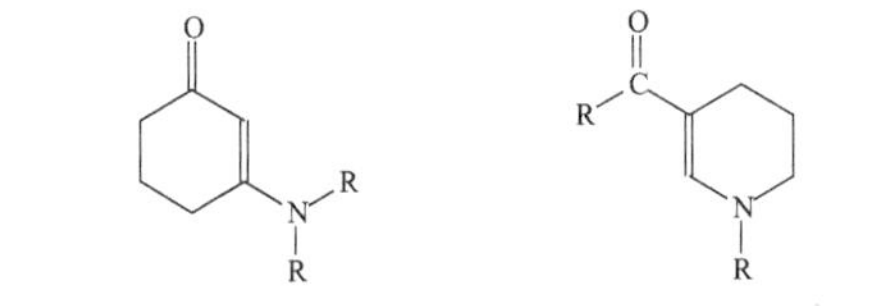

Fig. 8 Tetrahydropyridine derivatives.

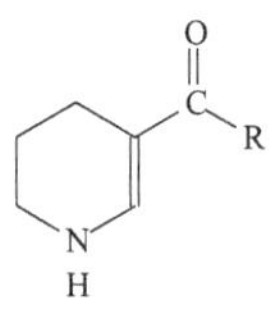

11

R = Me, H, OMe, OtBu, NH_2

Fig. 9 Preparation of the compounds (11).

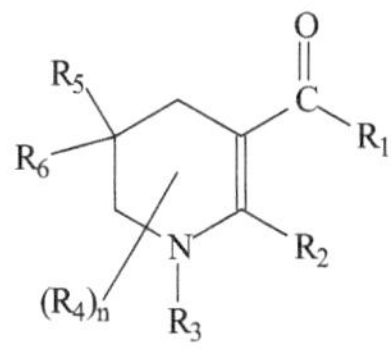

Fig. 10 New generation of MAAs analogs for sunscreen compositions.

different substituents R_1-R_6, to enhance stability and to find the best balance water resistance/UV absorption.

York and Ryan (2015) provide a new method for synthesizing such compounds with a good overall yield to enable their generation in commercial quantities. Then, they described the synthesis of various compounds (York & Ryan, 2018) and the SAR study in order to get a double absorption maximum in the UV A and UVB range to afford greater UV protection. Other proposals for syntheses of MAA analogs have been made. For instance Andreguetti, Stein, Pereira, Pinto, and Colepicolo (2013) also described the reaction of *m*-cyclohexanedione with amino acids by ultrasound and microwave promoted synthesis to produce the tautomeric structure (12) (Fig. 11); molar absorption coefficient and wavelengths of absorbance were lower than MAAs ones but these analogs showed remarkable antioxidant effect.

Abou-Khalil, Raeppel, and Raeppel (2013) also provided methods for the preparation of aminocylohexenimine compounds having the general formula (13) as protecting agents against UV radiations (Fig. 12).

Without mentioning the MAAs analogy, Lang and Malaval (1983) present a cosmetic composition containing synthesized β—aminoenone-like (14) as a filter agent for UV A radiation (Fig. 13).

Finally, Gouault, Nguyen, Tomasi, and Costuas (2016) and Nguyen et al. (2013) prepared analogs (15) (Fig. 14) with a high absorbance intensity in the UV A and/or UV B zone and without *in vitro* cytotoxicity.

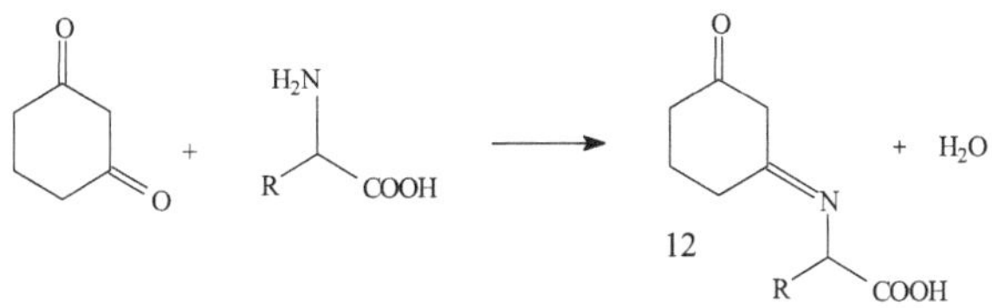

Fig. 11 Reaction of *m*-cyclohexanedione with amino acids by ultrasound and microwave promoted synthesis.

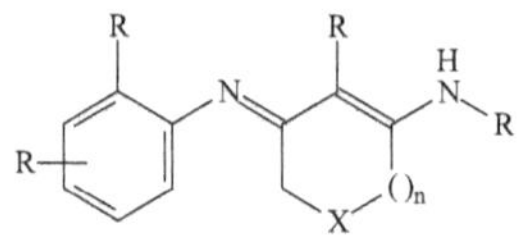

13

R = different substituents

Fig. 12 Aminocylohexenimine compounds having the general formula (13) as protecting agents against UV radiations.

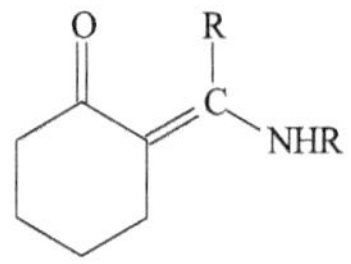

14

R = various substituents

Fig. 13 Synthesized β–aminoenone like as a filter agent for UV A radiation.

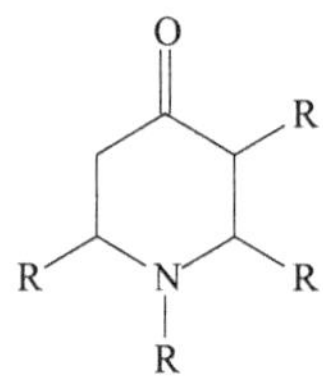

15

R = various sustituents

Fig. 14 Analogs that absorb with high intensity in the UV A and/or UV B zone and without cytotoxicity.

To conclude, varied synthesized chemical structures can be considered as "MAAs analogs"; some have interesting anti-UV and antioxidant properties. An alternative production of natural MAAs can also be considered through bioproduction by organisms or microorganisms under induced culture conditions.

3.2 Extraction, fractionation and purification

Due to their chemical properties, i.e. closely related in structures, polar, low molecular weight, the extraction of Mycosporine-like Amino Acids is generally performed by a solid/liquid extraction from the raw material in polar solvents, as aqueous ethanol or methanol. The conditions of temperature, pressure, pH and raw material pre-treatment have also been studies and sometimes induced variations in the extraction, i.e. qualitatively and quantitatively. After extraction, with or without purification step, the isolation by liquid chromatography-based (LC) applications coupled with a UV and/or Mass Spectrometry (MS), under positive ionization mode ($[M+H]^+$), detection is then the main way to analyze and identify MAAs in extracts.

In 2005, Volkmann & Gorbushina developed a broadly applicable method for the extraction and characterization of MAAs from terrestrial, marine and freshwater fungi. MAAs were isolated from the freeze-dried samples with a 0.2% acetic acid in 0.5% methanol solution for 12 h at 4°C. The extracts, recovered by centrifugation, were analyzed by LC-UV and LC-ESI-MS using a reverse phase C18 column, eluting with the extraction solvent and a double detection at 310 or 320 nm and by ESI-MS.

In their study, Alvarez-Gomez, Korbee, Casas-Arrojo, Abdala-Diaz, and Figueroa (2019) studied the photoprotection, cytotoxicity and immunology capacity of UV-absorbing red algae extracts. They extracted the MAAs from the dried raw material with 20% methanol prior an identification by HPLC-PDA and ESI-MS.

Bedoux et al. (2014) extracted the MAAs from freeze-dried samples of the red algae *Solieria chordalis* by using ethyl acetate or 2-octyldodecanol or octodecyl ester of L-pyrrolidone carboxylic acid (cosmetic emollient and moisturizer). They evaluated the efficiency of extraction by spectrophotometry and LC with a UV and MS detection.

In 2015 and 2017, Hartmann et al. described two different methods for the isolation and identification of mycosporine-like amino acids from a 75% methanolic extract of red and green algae. The first method was based on the hydrophilic interactions on HILIC resin with an elution by a mixture of acetonitrile and water with 5 mM ammonium acetate. MAAs were detected using UV and MS detectors (Hartmann, Becker, Karsten, Remias, & Ganzera, 2015). In 2017, they developed an innovative method by capillary electrophoresis (CE) for the qualitative and quantitative analysis of MAAs. The detection was also performed by DAD (diode array detector) (Hartmann, Murauer, Ganzezra, 2017). They conclude that the use of CE for the separation of MAAs lead to a lower analysis duration and better separation compared to HILIC and RP-C18 HPLC analyses, respectively.

Rosic, Braun, and Kvaskoff (2015) have published extraction and analytical methods of MAAs from marine algae. The standardized method was based on the sequential extraction of MAAs with ice-cold 80% methanol. They did the analyses on a RP-C18 column using a gradient of 1 mM ammonium acetate in 0.1% acetic acid and acetonitrile/isopropanol (4:1, v:v). In this protocol, MAAs were also detected by LC-MS.

Another parameter affecting the quality of isolation of MAAs by LC is the re-dissolution solvent. This parameter was studied by Karsten, Escoubeyrou, and Charles (2009), in order to set up a routine protocol for the analysis of MAAs. They compared three re-dissolution solvents, 100% methanol, water and the HPLC eluent, for dissolution of extracts prior to analysis. In this study, they also compared three reverse-phase HPLC column, C18 and two references of C8 columns, in order to evaluate the benefits of each condition. Samples were eluted in the same conditions by 2.5% methanol in 0.1% acetic acid. They highlighted that both the re-dissolution solvent and the column used for the analysis impact the separation of MAAs, the retention time and the intensity of peaks (Karsten et al., 2009). However, they also showed a great long-term reproducibility of analyses on C18 column in opposite to a gradual loss of resolution of analyses on C8 columns. Concerning the re-dissolution solvent, they concluded that the widely use of pure methanol may lead to analytical problems.

Finally, in 2019, Lalegerie et al. have screened the diversity of mycosporine-like amino acids for forty red seaweeds species from Brittany (France). The MAAs were extracted with 70% ethanol for 2 h at 45°C and under stirring. Extracts were analyzed by LC with a UV detection at 320, 330, 334 and 360 nm for the screening. They also analyzed the samples by LC-ESI-Q-TOF-MS, in order to identify the MAAs (Lalegerie et al., 2019).

To summarize, numerous extractions, isolation and separation methods and conditions have been presented in the literature. However, it appears that the more efficient extraction techniques are based on the use of water diluted ethanol or methanol solvents, at mild temperature and at short-term (few hours). The LC (RP-C18 or HILIC columns) with UV detection (between 300 and 340 nm) and LC-MS seems to be the universal method for the separation and identification of mycosporine-like amino acids in samples. This conclusion is relevant with the particularities of MAAs. The molecular weight, determined by LC-MS, and the maximum absorption wavelength (λ_{max}, determined by UV) being the two characteristics of MAAs.

To go further in the characterization of MAAs in samples, some studies reports the use of nuclear magnetic resonance (NMR) to analyze the samples and identify the MAAs (Hartmann et al., 2015; Orfanoudaki, Hartmann, Karsten, & Ganzera, 2019). This method has the advantage to be highly informative but also requires a pure sample in relatively large amounts. That conditions lead to develop a high yield purification technique for MAAs.

No many different purification methods were reported for MAAs. Some studies used Solid Extraction (SPE) on C-18 silica gel cartridge with an interesting yield and reproducibility (Shick, Romaine-Lioud, Ferrier-Pages, & Gattuso, 1999; Teai et al. 1997). This application was used for pre-purification on the crude extract to eliminate some compounds like pigments and lipids before HPLC analysis. Another alternative to purify mycosporine-like amino acid is ultrafiltration. The larger molecule content in the water-soluble extract could be removed by a membrane filtration with a cut-off rating according to the size of the MAAs. This method has shown high yield of recovering for shinorine, palythinol and usujirene purification (Carreto, Carignan, & Montoya, 2005).

3.3 Characterization

The use of analytical and spectroscopic techniques allows to solve structural elucidation problems. LC—MS had been the only hyphenated technique allowing a qualitative analysis of the molecular structure or a mixture until recent developments in NMR. Only LC-MS could fail for the analysis of highly polar MAAs. Whitehead, Karentz, and Hedges (2001) and Whitehead and Hedges (2002) developed an approach method to identify the individual MAAs. This method consisted to characterize the MAAs using a liquid chromatography equipped with an electrospray ionization mass spectrometry (LC/MS). This approach allowed to identify the MAAs differentiation with a low resolution on a HPLC column. An exception occurs with the palythenic acid, the usujirene and the palythene with an impossibility to differentiate the isomeric form without a tandem mass spectrometry (MS/MS) analysis. The studies of the MAAs identification showed similarity in the ion detection. Among these, the elimination of

Table 2 Mass spectrometry fragmentation patterns of mycosporine-like amino acids by positive mode ionization $[M+H]^+$.

MAAs	$[M+H]^+$	Fragment pattern	Mass
Shinorine	333	137, 168, 185, **186**, 197, 211, 230, 241, 255	312.12
Palythine	245	137, 150, 155, 162, 184, **186**, 199, 209, 230	244.11
Asterina 330	289	137, 150, 168, **186**, 197, 199, 213, 230, 243, 273, 289	288.13
Porphyra 334	365	137, 151, 168, 185, **186**, 197, 200, 243, 303	346.14
Mycosporine-serinol	262	**244**, 216, 194	261.12
Mycosporine-glutaminol-glucoside	465	447, 429, **303**, 285, 267, 235	464.20
Mycosporine-alanine	260	242, **171**, 147	259.11
Mycosporine-serine	276	**258**, 228, 236	275.10
Mycosporine-glutamicol	304	286, 268, **236**	303.13
Mycosporine-glutaminol-glucoside	466	448, 430, **304**, 286, 268, 236	465.18

the 15 Da (methyl radical) and the 44 Da (CO_2 loss) and the lateral chain (C-1) seems to be ubiquitous for all the MAAs. This elimination produced a m/z 186 ion which could be an additional information to prove the presence of MAAs in a complex sample. This m/z 186 fragment was not observed in the palythene and the palythenic acid forms because of the presence of a double bond within the lateral chain (Whitehead & Hedges, 2003). For palythine only, the hydrogen on the nitrogen at C-1 was substituted, leading to the formation of a m/z 192 fragment in the analysis with D_2O with the 1Da higher fragment compare to the other MAAs (m/z 191) (Cardozo, Carvalho, Pinto, & Colepicolo, 2006). Some fragmentation patterns, obtained by LC-MS analysis on mycosporine like amino acid were presented in Table 2 below.

To identify the compound structure, the nuclear magnetic resonance (NMR) was developed the last decade only with reasonably pure compounds. NMR spectra shown a ^{13}C-chemical shift typical for the assignment of the cyclohexenimine chromophore (C1, 2, 3, 4, 5 et 6) and the amino acid substituents (C9, 10, 11, 12, 13 et 14) (Nazifi et al., 2013) Table 3.

Table 3 NMR spectra recorded with a spectrometer in D_2O as a solvent. Chemical shifts (ppm) for porphyra-334, mycosporine-glycine (Nazifi et al., 2013) and shinorine (Volkmann & Gorbushina, 2005). The coupling patterns and constants (Hz) are shown in brackets. x, x′, Chemical shifts may be exchangeable for the 508-Da MAA.

	Porphyra-334		Shinorine		Mycosporine-glycine	
C	^{13}C	^{1}H	^{13}C	^{1}H	^{13}C	^{1}H
1	161.6	–	62.0	3.94	159.7	–
2	126.0	–	64.0	4.40	130.4	–
3	163.2	–	174.5	–	187.2	–
4	32.5	2.75	160	–	45.4	2.50, 2.73
5	71.3	–	125	–	7.9	–
6	33.0	2.77	158	–	33.8	2.72, 2.83
7	67.1	3.61 (s)	32.5	2.80	68.4	3.57 (s)
8	59.0	3.73 (s)	69.0	–	60.2	3.64 (s)
9	47.0	4.07 (s)	32.5	2.80	43.7	4.24 (s)
10	177.6	–	58.2	3.58	174.5	–
11	64.0	4.12 (d, 5.0)	67.0	3.46	–	–
12	178.0	–	43.5	4.25	–	–
13	68.0	4.33 (m, 5.0, 6.4)	170.6	–	–	–
14	19.0	1.26 (d, 6.4)	52.5	3.69	–	–

4. Concluding remarks

Red seaweeds synthesize specific MAAs characterized by an absorption maximum between 320 and 360 nm. Their production is directly or indirectly related to the absorption of solar energy and they are considered as photoprotective metabolites because they have maximum UV absorption, high molar extinction factors, and are photostable. In addition to the photoprotective and antioxidant activities, MAAs accumulation in the cells contributes to the maintenance of osmotic pressure. Under solar radiation and high ammonium concentrations from an IMTA system, the total content and individual composition of MAAs and antioxidant capacity are stimulated, compared with those from the wild populations and those grown in lower irradiation conditions. The ability of the species to conjugate the dynamic photoinhibition, the efficient use of its pigments, MAAs and the biochemical status have allowed them to establish in the intertidal zone under changing environmental conditions. Nowadays, it is necessary to diversify MAA sources produced under IMTA systems, since these are excellent alternatives to wild population harvesting, and to biorefinery strategies to maximize biomass utilization.

Acknowledgments

All co-authors acknowledge ECOS Nord funding (program M14A03, 2015–2018). Hugo Pliego-Cortés, gratefully acknowledges the Région Bretagne, France (SAD-Glycoalg-2017), and People Program (Marie Curie Actions) through PRESTIGE Campus France (PCOFUND-GA-2013-609102) for funding support. The co-authors (GB, NB, KH, RB) thank the Région Bretagne (France) for CP'Cosm'Actifs funding (CP'Cosm'Actifs, 2018).

References

Abou-Khalil, E., Raeppel, S., & Raeppel, F. (2013). *Imino compounds as protecting agents against ultraviolet radiations WO2013181741 (A1).*

Álvarez-Gómez, F., Bouzon, Z., Korbee, N., Celis-Plá, P. S. M., Schmidt, E. C., & Figueroa, F. L. (2017). Combined effects of UVR and nutrients on cell ultrastructure, photosynthesis and biochemistry in *Gracilariopsis longissima* (Gracilariales, Rhodophyta). *Algal Research, 26*, 190–202.

Alvarez-Gomez, F., Korbee, N., Casas-Arrojo, V., Abdala-Diaz, R. T., & Figueroa, F. L. (2019). UV photoprotection, cytotoxicity and immunology capacity of red algae extracts. *Molecules, 24*, 341.

Andreguetti, D., Stein, E. M., Pereira, C. M. P., Pinto, E., & Colepicolo, P. (2013). Antioxidant properties and UV absorbance pattern of mycosporine-like amino acids analogs synthesized in an environmentally friendly manner. *Journal of Biochemical and Molecular Toxicology, 27*, 305–312.

Arima, H., Horiguchi, N., Takaichi, S., Kofuji, R., Ishida, K., Wada, K., et al. (2012). Molecular genetics and chemotaxonomic characterization of the terrestrial cyanobacterium *Nostoc commune* and its neighboring species. *FEMS Microbiology Ecology, 79*, 34–45.

Balskus, E. P., & Walsh, C. T. (2010). The genetic and molecular basis for sunscreen biosynthesis in cyanobacteria. *Science, 329*, 1653–1656.

Bandaranayake, W. M. (1998). Mycosporines : Are they nature's sunscreen? *Natural Product Reports*, 159–172.

Barceló-Villalobos, M., Figueroa, F. L., Korbee, N., Álvarez-Gómez, F., & Abreu, M. (2017). Production of mycosporine-like amino acids from *Gracilaria vermiculophylla* (Rhodophyta) cultured through one year in an integrated multi-trophic aquaculture (IMTA) system. *Marine Biotechnology, 19*, 246–254.

Barufi, J., Korbee, N., Oliveira, M., & Figueroa, F. L. (2011). Effects of N supply on the accumulation of photosynthetic pigments and photoprotectors in *Gracilaria tenuistipitata* (Rhodophyta) cultured under UV radiation. *Journal of Applied Phycology, 23*, 457–466.

Barufi, B. J., Mata, M. T., Oliveira, M. C., & Figueroa, F. L. (2012). Nitrate reduces the negative effect of UV radiation on photosynthesis and pigmentation in *Gracilaria tenuistipitata* (Rhodophyta): The photoprotection role of mycosporine-like amino acids. *Phycologia, 51*, 636–648.

Bedoux, G., Hardouin, K., Marty, C., Taupin, L., Vandanjon, L., & Bourgougnon, N. (2014). Chemical characterization and photoprotective activity measurement of extracts from the red macroalga *Solieria chordalis*. *Botanica Marina, 57*, 291–301.

Betancor, S., Dominguez, B., Tuya, F., Figueroa, F. L., & Haroun, R. (2015). Photosynthetic performance and photoprotection of *Cystoseira humilis* (Phaeophyceae) and *Digenea simplex* (Rhodophyceae) in an intertidal rock pool. *Aquatic Botany, 121*, 16–25.

Bird, G., Chalmers, P. J., Fitzmaurice, N., Rigg, D. J., & Thang, S. H. (1997). *UV absorbing compounds US5637718 (A)*.

Bird, G., Fitzmaurice, N., Dunlap, W. C., Chalker, B. E., & Bandaranayake, W. M. (1988). *Sunscreen composition and compounds for use therein WO8802251*.

Bischof, K., Gómez, I., Molis, M., Hanelt, D., Karsten, U., Lüder, U., et al. (2006). Ultraviolet radiation shapes seaweed communities. *Reviews in Environmental Science and Biotechnology, 5*, 141–166.

Brawley, S., Blouina, N., Ficko-Bleanc, E., Wheelerd, G., Lohre, M., et al. (2017). Insights into the red algae and eukaryotic evolution from the genome of *Porphyra umbilicalis* (Bangiophyceae, Rhodophyta). *Proceedings of the National Academy of Sciences of the United States of America, 114*, E6361–E6370.

Briani, B., Sissini, M. N., Lucena, L. A., Batista, M. B., Costa, I. O., Nunes, J. M. C., et al. (2018). The influence of environmental features in the content of mycosporine-like amino acids in red marine algae along the Brazilian coast. *Journal of Phycology, 54*, 380–390.

Cardozo, K. H., Carvalho, V. M., Pinto, E., & Colepicolo, P. (2006). Fragmentation of mycosporine-like amino acids by hydrogen/deuterium exchange and electrospray ionisation tandem mass spectrometry. *Rapid Communications in Mass Spectrometry, 20*, 253–258.

Cardozo, K. H. M., Guaratini, T., Barros, M. P., Falcão, V. R., Tonon, A. P., Lopes, N. P., et al. (2007). Metabolites from algae with economical impact. *Comparative Biochemistry and Physiology – Part C, 146*, 60–78.

Carignan, M. O., & Carreto, J. I. (2013). Characterization of mycosporine-serine-glycine methyl ester, a major mycosporine-like amino acid from dinoflagellates: A mass spectrometry study. *Journal of Phycology, 49*, 680–688.

Carreto, J. I., & Carignan, M. O. (2011). Mycosporine-like amino acids (MAAs): Relevant secondary metabolites. Chemical and ecological aspects. *Marine Drugs, 9*, 387–446.

Carreto, J. I., Carignan, M. O., & Montoya, N. G. (2005). A high-resolution reverse-phase liquid chromatography method for the analysis of mycosporine-like amino acids (MAAs) in marine organisms. *Marine Biology, 146*, 237–252.

Chalker, B. E., Dunlap, W. C., & Bandaranayake, W. M. (1988). *U.V. absorbing amino ketimine compounds. AU-B-15291/88; (1994) sunscreen compounds 5352793(A)*.

Chalmers, P. J., Fitzmaurice, N., Rigg, D. J., Thang, S. H., & Bird, G. (1990). *UV absorbing compounds and compositions WO9009995*.

Chollet-Krugler, M., Nguyen, T. T. T., Sauvager, A., Thüs, H., & Boustie, J. (2019). Mycosporine-like amino acids (MAAs) in time-series of lichens specimens from natural history collections. *Molecules, 24*, 1070.

Chrapusta, E., Kaminski, A., Duchnik, K., Bober, B., Adamski, M., & Bialczyk, J. (2017). Mycosporine-like amino acids: Potential health and beauty ingredients. *Marine Drugs, 15*, 326.

Davison, I. R., & Pearson, G. A. (1996). Stress tolerance in intertidal seaweeds. *Journal of Phycology, 32*, 197–211.

Diehl, N., Michalik, D., Zuccarello, G., & Karsten, U. (2019). Stress metabolite pattern in the eulittoral red alga *Pyropia plicata* (Bangiales) in New Zealand – mycosporine-like amino acids and heterosides. *Journal of Experimental Marine Biology and Ecology, 510*, 23–30.

Dring, M. J. (2006). Stress resistance and disease resistance in seaweeds: The role of reactive oxygen metabolism. *Advanced Botanical Research, 43*, 175–207.

Dunlap, W. C., & Chalker, B. E. (1986). *Ultra violet agents. WO 8602350 (A1)*.

Dunlap, W. C., Chalker, B. E., Bandaranayake, W. M., & Wu Won, J. J. (1998). Nature's sunscreen from the Great Barrier Reef, Australia. *International Journal of Cosmetic Science, 20*, 41–51.

Favre-Bonvin, J., Bernillon, J., Salin, N., & Arpin, N. (1987). Biosynthesis of mycosporines: Mycosporine glutaminol in *Trichothecium roseum*. *Phytochemistry, 26*, 2509–2514.
Figueroa, F. L., Bonomi, J., Barufi, E., et al. (2014). Short-term effects of increasing CO_2, nitrate and temperature on three mediterranean macroalgae: Biochemical composition. *Aquatic Biology, 22*, 177–193.
Figueroa, F. L., Bueno, A., Korbee, N., Santos, R., Mata, L., & Schuenhoff, A. (2008). Accumulation of mycosporine-like amino acids in *Asparagopsis armata* grown in tanks with fishpond effluents of gilthead sea bream, *Sparus aurata*. *Journal of the World Aquaculture Society, 39*, 692–699.
Figueroa, F. L., Israel, A., Neori, A., Martínez, B., Malta, E. J., Put, A., et al. (2010). Effect of nutrient supply on photosynthesis and pigmentation to short-term stress (UV radiation) in *Gracilaria conferta* (Rhodophyta). *Marine Pollution Bulletin, 60*, 1768–1778.
Figueroa, F. L., & Korbee, N. (2010). Interactive effects of UV radiation and nutrients on ecophysiology: Vulnerability and adaptation to climate change. In A. Israel, R. Einvav, & J. Seckbach (Eds.), *Seaweeds and their role in globally changing environments* (pp. 157–182). Dordrecht: Springer.
Figueroa, F., Korbee, N., Abdala, R., Jerez, C. G., López-de la Torre, M., Güenaga, L., et al. (2012). Biofiltration of fishpond effluents and accumulation of N-compounds (phycobiliproteins and mycosporine-like amino acids) versus C-compounds (polysaccharides) in *Hydropuntia cornea* (Rhodophyta). *Marine Pollution Bulletin, 64*, 310–318.
Franklin, L., Krabs, G., & Kuhlenkamp, R. (2011). Blue light and UVA radiation control the synthesis of mycosporine-like amino acids in *Chondrus crispus* (Florideophyceae). *Journal of Phycology, 37*, 257–270.
Gao, Q., & Garcia-Pichel, F. (2011). An ATP-grasp ligase involved in the last biosynthetic step of the iminomycosporine shinorine in *Nostoc punctiforme* ATCC 29133. *Journal of Bacteriology, 193*, 5923–5928.
García-Sánchez, M., Korbee, N., Pérez-Ruzafa, I. M., Marcos, C., Domínguez, B., Figueroa, F. L., et al. (2012). Physiological response and photoacclimation capacity of *Caulerpa prolifera* (Forsskål) J.V. Lamouroux and *Cymodocea nodosa* (Ucria) Ascherson meadows in the Mar Menor lagoon (SE Spain). *Marine Environmental Research, 79*, 37–47.
Gouault, N., Nguyen, K. H., Tomasi, S., & Costuas, K. (2016). *Photoprotective compounds, compositions including same and uses thereof US 2016067162*.
Greenhill, J. V. (1969). Determination of the tautomeric ration in certain enaminones. *Journal of the Chemical Society B*, 299–300.
Greenhill, J. V. (1971). The reactions with aldehydes of enaminones derived from dimedone. *Journal of the Chemical Society C*, 2699–2703.
Greenhill, J. V. (1977). Enaminones. *Chemical Society Reviews, 6*, 277–294.
Halpern, B. (1965). The preparation of dimedone derivatives of amino acids and their use in peptide synthesis. *Australian Journal of Chemistry, 18*, 417–421.
Halpern, B., & James, L. B. (1964a). A new amino protecting group for amino-acids in peptide synthesis. *Nature, 202*, 592–593.
Halpern, B., & James, L. B. (1964b). Dimedone (5,5-dimethylcyclohexane-1,3-dione) as a protecting agent for amino groups in peptide synthesis. *Australian Journal of Chemistry, 17*, 1282–1287.
Hartmann, A., Becker, K., Karsten, U., Remias, D., & Ganzera, M. (2015). Analysis of mycosporine-like amino acids in selected algae and cyanobacteria by hydrophilic interaction liquid chromatography and a novel MAA from the red alga *Catenella repens*. *Marine Drugs, 13*, 6291–6305.
Hartmann, A., Murauer, A., & Ganzezra, M. (2017). Quantitative analysis of mycosporine-like amino acids in marine algae by capillary electrophoresis with diode-array detection. *Journal of Pharmaceutical and Biomedical Analysis, 138*, 153–157.

Hoyer, K., Karsten, U., & Wiencke, C. (2002). Induction of sunscreen compounds in Antarctic macroalgae by different radiation conditions. *Marine Biology, 141*, 619–627.
Huovinen, P., Gómez, I., Figueroa, F. L., Ulloa, N., Morales, V., & Lovengreen, C. (2004). Ultraviolet-absorbing mycosporine-like amino acids in red macroalgae from Chile. *Botanica Marina, 47*, 21–29.
Huovinen, P., Matos, J., Pinto, I. S., & Figueroa, F. L. (2006). The role of ammonium in photoprotection against high irradiance in the red alga *Grateloupia lanceola*. *Aquatic Botany, 84*, 308–316.
Jiang, H., Gao, K., & Helbling, E. W. (2008). UV-absorbing compounds in *Porphyra haitanensis* (Rhodophyta) with special reference to effects of desiccation. *Journal of Applied Phycology, 20*, 387.
Kageyama, H., & Waditee-Sirisattha, R. (2019). Mycosporine-like amino acids as multifunctional secondary metabolites in cyanobacteria: From biochemical to application aspects. *Studies in Natural Products Chemistry, 59*, 153–194.
Karentz, D., McEuen, F., Land, M., & Dunlap, W. C. (1991). Survey of mycosporine-like amino acid compounds in antarctic marine organisms: Potential protection from ultraviolet exposure. *Marine Biology, 108*, 157–166.
Karsten, U., Escoubeyrou, K., & Charles, F. (2009). The effect of re-dissolution solvents and HPLC columns on the analysis of mycosporine-like amino acids in the eulittoral macroalgae *Prasiola crispa* and *Porphyra umbilicalis*. *Helgoland Marine Research, 63*, 231–238.
Karsten, U., Sawall, T., Hanelt, D., Bischof, K., Figueroa, F. L., Flores-Moya, A., et al. (1998). An inventory of UV-absorbing mycosporine-like amino acids in macroalgae from polar to warm-temperate regions. *Botanica Marina, 41*, 443–453.
Karsten, U., Sawall, T., & Wiencke, C. (1998). A survey of the distribution of UV-absorbing substances in tropical macroalgae. *Phycological Research, 46*, 271–279.
Korbee-Peinado, N., Abdala Dıaz, R. T., Figueroa, F. L., & Helbling, E. W. (2004). Ammonium and UV radiation stimulate the accumulation of mycosporine-like amino acids in *Porphyra columbina* (Rhodophyta) from Patagonia (Argentina). *Journal of Phycology, 40*, 248–259.
Korbee, N., Figueroa, F. L., & Aguilera, J. (2005). Effect of light quality on the accumulation of photosynthetic pigments, proteins and mycosporine-like amino acids in the red alga *Porphyra leucosticta* (Bangiales, Rhodophyta). *Journal of Photochemistry and Photobiology A, 80*, 71–78.
Krabs, G., Bischof, K., Hanelt, D., Karsten, U., & Wiencke, C. (2002). Wavelength-dependent induction of UV-absorbing mycosporine-like amino acids in the red alga *Chondrus crispus* under natural solar radiation. *Journal of Experimental Marine Biology and Ecology, 268*, 69–82.
La Barre, S., Roullier, C., & Boustie, J. (2014). Mycosporine-Like Amino Acids (MAAs) in Biological Photosystems. *Outstanding Marine Molecules: Chemistry, Biology, Analysis*, 333–360.
Lalegerie, F., Lajili, S., Bedoux, G., Taupin, L., Stiger-Pouvreau, V., & Connan, S. (2019). Photoprotective compounds in red macroalgae from Brittany: A large diversity in mycosporine-like amino acids (MAAs). *Marine Environmental Research, 147*, 37–48.
Lang, G., & Malaval, A. (1983). *Composition cosmétique à base de béta amino énones comme agents filtrant les radiations UV-A, FR2520615 A1.*
Lawrence, K. P., Long, P. F., & Young, A. R. (2018). Mycosporine-like amino acids for skin photoprotection. *Current Medicinal Chemistry, 25*, 5512–5527.
Leach, C. M. (1965). Ultraviolet-absorbing substances associated with light-induced sporulation in fungi. *Revue Canadienne de Botanique, 43*, 185–200.
Losantos, R., Funes-Ardoiz, I., Aguilera, J., Herrera-Ceballos, E., Garcia-Iriepa, C., Campos, P. J., et al. (2017). Rational design and synthesis of efficient sunscreens to boost the solar protection factor. *Angewandte Chemie International Edition, 56*, 2632–2635.

Matsui, K., Nazifi, E., Kunita, S., Wada, N., Matsugo, S., & Sakamoto, T. (2011). Novel glycosylated mycosporine-like amino acids with radical scavenging activity from the cyanobacterium *Nostoc commune*. *Journal of Photochemistry and Photobiology B, 105*, 81–89.

Muraoka, O., Morikawa, T., Zhang, Y., Ninomiya, K., Nakamura, S., Matsuda, H., et al. (2009). Novel megastigmanes with lipid accumulation inhibitory and lipid metabolism-promoting activities in HepG2 cells from *Sedum sarmentosum*. *Tetrahedron, 65*, 4142–4148.

Navarro, N., Figueroa, F. L., & Korbee, N. (2017). Mycosporine-like amino acids vs carrageenan yield in *Mazzaella laminarioides* (Gigartinales; Rhodophyta) under high and low UV solar irradiance. *Phycologia, 56*, 570–578.

Navarro, N., Figueroa, F. L., Korbee, N., Bononi, J., Alvarez-Gomez, F., & de la Coba, F. (2018). Mycosporine-like amino acids from red algae to develop natural UV sunscreens. In R. Rastogi (Ed.), *Sunscreens sources, formul. effic. recomemendations* (pp. 99–129). New York: Nova Science Publishers, Inc.

Navarro, N., Mansilla, A., Figueroa, F. L., Korbee, N., Acevedo, J. J., & Plastino, E. M. (2014). Short-term effects of solar UV radiation and NO_3^- supply on the accumulation of mycosporine-like amino acids in *Pyropia columbina* (Bangiales, Rhodophyta) under spring ozone depletion in the Sub-Antarctic region, Chile. *Botanica Marina, 57*, 9–20.

Nazifi, E., Wada, N., Yamaba, M., Asano, T., Nishiuchi, T., Matsugo, S., et al. (2013). Glycosylated porphyra-334 and palythine-threonine from the terrestrial cyanobacterium *Nostoc commune*. *Marine Drugs, 11*, 3124–3154.

Nguyen, K. H., Tomasi, S., Le Roch, M., Toupet, L., Renault, J., Uriac, P., et al. (2013). Gold-mediated synthesis and functionalization of chiral halopyridones. *Journal of Organic Chemistry, 78*, 7809–7815.

Oren, A., & Gunde-Cimerman, N. (2007). Mycosporines and mycosporine-like amino acids: UV protectants or multipurpose secondary metabolites? *FEMS Microbiology Letters, 269*, 1–10.

Orfanoudaki, M., Hartmann, A., Karsten, U., & Ganzera, M. (2019). Chemical profiling of mycosporine-like amino acids in twenty-three red algal species. *Journal of Phycology, 55*, 393–403.

Pathak, J., Ahmeda, H., Shailendra, R., Singhb, P., Häderc, D., & Sinha, R. P. (2019). Genetic regulation of scytonemin and mycosporine-like amino acids (MAAs) biosynthesis in cyanobacteria. *Plant Gene, 17*, 100172.

Pliego-Cortés, H., Bedoux, G., Boulho, R., Taupin, L., Freile-Pelegrín, Y., Bourgougnon, N., et al. (2019). Stress tolerance and photoadaptation to solar radiation in *Rhodymenia pseudopalmata* (Rhodophyta) through mycosporine-like amino acids, phenolic compounds, and pigments in an Integrated Multi-Trophic Aquaculture system. *Algal Research, 41*, 101542.

Pope, M., Spence, E., Seralvo, V., Gacesa, R., et al. (2015). O-methyltransferase is shared between the pentose phosphate and shikimate pathways and is essential for mycosporine-like amino acid biosynthesis in *Anabaena variabilis* ATCC 29413. *ChemBioChem, 16*, 320–327.

Portwich, A., & Garcia-Pichel, F. (2003). Biosynthetic pathway of mycosporines (mycosporine-like amino acids) in the cyanobacterium. *Phycologia, 42*, 384–392.

Quintanoa, E., Celis-Plá, P. S. M., Martínez, B., Díez, I., Muguerza, N., Figueroa, F. L., et al. (2019). Ecophysiological responses of a threatened red alga to increased irradiance in an in situ transplant experiment. *Marine Environmental Research, 144*, 166–177.

Rosic, N. N. (2012). Phylogenetic analysis of genes involved in mycosporine-like amino acid biosynthesis in symbiotic dinoflagellates. *Applied Microbiology and Biotechnology, 94*, 29–37.

Rosic, N. N., Braun, C., & Kvaskoff, D. (2015). Extraction and analysis of mycosporine-like amino acids in marine algae. In D. B. Stengel, & S. Connan (Eds.), *Natural products from marine algae: Methods and protocols, methods in molecular biology* (Vol. 1308).

Sagert, S., Forster, R. M., Feuerpfeil, P., & Schubert, H. (1997). Daily course of photosynthesis and photoinhibition in *Chondrus crispus* (Rhodophyta) from different shore levels. *European Journal of Phycology, 32*, 363–371.

Sampedro Ruiz, D., Losantos Cabello, R., & Funes Ardoiz, I. (2015). *Compuestos fotoprotectores analogos de MAA, prodecimiento des sintesis y composicion que comprende los mismos. ES 2 550 374 A1.*

Shick, J. M., Romaine-Lioud, S., Ferrier-Pages, C., & Gattuso, J. P. (1999). Ultraviolet-B radiation stimulates shikimate pathway-dependent accumulation of mycosporine-like amino acids in the coral *Stylophora pistillata* despite decreases in its population of symbiotic dinoflagellates. *Limnology & Oceanography, 44*, 1667–1682.

Teai, T., Raharivelomanana, P., Bianchini, J. P., Faura, R., Martın, P. M. V., & Cambon, A. (1997). Structure de deux nouvelles imino-mycosporines isolées de *Pocillopora eydouxi*. *Tetrahedron Letters, 38*, 5799–5800.

Torres, A., Enk, C. D., Hochberg, M., & Srebnik, M. (2006). Porphyra-334, a potential natural source for UVA protective sunscreens. *Photochemical and Photobiological Sciences, 5*, 432–435.

Véliz, K., Chandía, N., Karsten, U., Lara, C., & Thiel, M. (2018). Geographic variation in biochemical and physiological traits of the red seaweeds *Chondracanthus chamissoi* and *Gelidium lingulatum* from the south east Pacific coast. *Journal of Applied Phycology, 31*, 1–18.

Volkmann, M., & Gorbushina, A. A. (2005). A broadly applicable method for extraction and characterization of mycosporines and mycosporine-like amino acids of terrestrial, marine and freshwater origin. *FEMS Microbiology Letters, 255*, 286–295.

Wada, N., Sakamoto, T., & Matsugo, S. (2015). Mycosporine-like amino acids and their derivatives as natural antioxidants. *Antioxidants, 4*, 603–646.

Waller, R. F., Slamovits, C. H., & Keeling, P. J. (2006). Lateral gene transfer of a multigene region from cyanobacteria to dinoflagellates resulting in a novel plastid-targeted fusion protein. *Molecular Biology and Evolution, 23*, 1437–1443.

Wenkert, E., Dave, K. G., Haglid, F., Lewis, R. G., Oishi, T., Stevens, R. V., et al. (1968). Tetrahydropyridines. *Journal of Organic Chemistry, 33*, 747–753.

White, J. D., Cammack, J. H., & Sakuma, K. (1989). The synthesis and absolute configuration of mycosporins. A novel application of the Staudinger reaction. *Journal of the American Chemical Society, 111*, 8970–8972.

White, J. D., Cammack, J. H., Sakuma, K., Rewcastle, G. W., & Widener, R. K. (1995). Transformations of quinic acid. Asymmetric synthesis and absolute configuration of mycosporine I and mycosporine-gly. *Journal of Organic Chemistry, 60*, 3600–3611.

Whitehead, K., & Hedges, J. I. (2002). Analysis of mycosporine-like amino acids in plankton by liquid chromatography electrospray ionization mass spectrometry. *Marine Chemistry, 80*, 27–39.

Whitehead, K., & Hedges, J. I. (2003). Tandem mass spectrometric and GCMS analysis of mycosporine-like amino acids (MAAs). *Rapid Communications in Mass Spectrometry, 17*, 2133–2138.

Whitehead, K., Karentz, D., & Hedges, J. I. (2001). Mycosporine-like amino acids (MAAs) in phytoplankton, a herbivorous pteropod (*Limacina helicina*), and its pteropod predator (*Clione antarctica*) in McMurdo Bay, Antarctica. *Marine Biology, 139*, 1013–1019.

York, M., & Ryan, J. (2015). *Synthesis of UV absorbing compounds WO 2014/082124 A1 - US 2015/0299124 A1.*

York, M., & Ryan, J. (2018). *UV absorbing compounds, compositions comprising same and uses thereof. US 9,920,009 B2.*

CHAPTER EIGHT

The stressful life of red and brown seaweeds on the temperate intertidal zone: effect of abiotic and biotic parameters on the physiology of macroalgae and content variability of particular metabolites

Fanny Lalegerie*,[1], Leslie Gager[1], Valérie Stiger-Pouvreau and Solène Connan
Univ Brest, CNRS, IRD, Ifremer, LEMAR, Plouzane, France
*Corresponding author: E-mail: fanny.lalegerie@univ-brest.fr

Contents

[1] Fanny Lalegerie and Leslie Gager contributed equally to this work.

Advances in Botanical Research, Volume 95
ISSN 0065-2296
https://doi.org/10.1016/bs.abr.2019.11.007

Abstract

Most marine macroalgae live attached to a substrate in coastal areas characterized by a very variable environment mainly due to the tide cycle which is more or less important depending on the geographical situation. Macroalgae spread over coastal zones according to their resistance to several abiotic and biotic factors. Some species are thus present on intertidal zones and are outside the water for a more or less long time each day; on the other hand, others are always underwater if they are living in the subtidal zone or in pools in which abiotic parameters can be strongly modified during a tide cycle. During these alternating immersion/emersion phases, macroalgae deal then with important modifications of many abiotic parameters, such as light, temperature, salinity, and biotic parameters, such as grazing, fouling, pathogens. To counteract these important variations, intertidal macroalgae develop resistance mechanisms such as the synthesis of special compounds, like photoprotectors, osmolytes, antifouling molecules as examples. In this chapter, after a presentation of the main environmental parameters facing intertidal seaweeds in temperate environment, some adaptation/protection mechanisms of these algae are presented, with an emphasis on brown (structural species whose biomass is very high on intertidal zones) and red seaweeds (whose diversity is very high).

1. Distribution of seaweeds on the rockyshore

Algae are a highly diverse group estimated at 50,370 species including ~7,300 red (Rhodophyta) and ~2,070 brown macroalgae (Ochrophyta, Phaeophyceae), playing a key role in marine ecosystems (Guiry & Guiry, 2019). Indeed, algae are at the base of the ocean food web, and contribute to primary photosynthetic production at a level almost equivalent to that of terrestrial plants (Falkowski & Raven, 2013). Among the different groups, brown macroalgae, *i.e.* Phaeophyceae, are the most easily observed seaweeds on rocky foreshores, including Fucales and Laminariales (Lüning, 1990). Red macroalgae, *i.e.* Rhodophyta, constitute a second group with the greatest specific diversity (Krause-Jensen & Duarte, 2016; Stengel, Connan, & Popper, 2011).

Although some species such as the red macroalga *Agarophyton vermiculophyllum* (previously *Gracilaria vermiculophylla*) live on sandy or muddy substrate (Surget et al., 2017), benthic macroalgae grow mainly on the rockyshore, and are thus directly subjected to changing environmental conditions (Lüning, 1990). By consequences, algal diversity and abundance is not the same everywhere in the ocean, and more particularly in polar, temperate, and tropical regions (Bolton, 1994; Hurd, Harrison, Bischof, & Lobban, 2014). On temperate rockyshores, macroalgae are affected daily

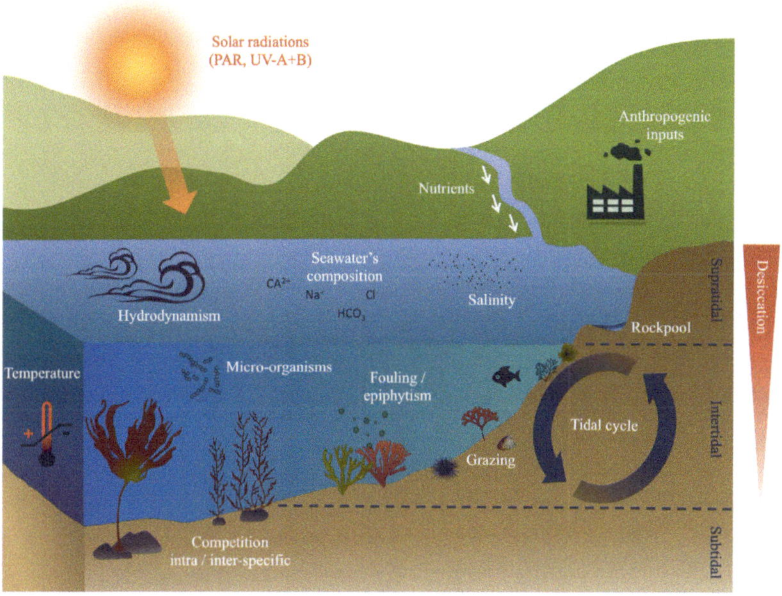

Fig. 1 Conceptual representation of the main abiotic and biotic factors presented in this review, influencing the ecology and physiology of seaweeds living on the rocky-shore and along the vertical zonation.

by the tidal cycle and have to deal with a whole range of stresses whose contribution varies at the different tidal levels (Fig. 1). One consequence of the alternation of emersion/immersion periods is the appearance of a vertical distribution (Lüning, 1990), which is supposed to be linked upwards to the resistance of the species to desiccation; but other various potentially stresses including high hydrodynamism, high light intensity, elevated temperature, nutrient limitation and osmotic stress resulting from emersion, also have a significant effect on intertidal macroalgae (Bird, Franklin, Smith, & Toonen, 2013; Contreras-Porcia, López-Cristoffanini, Meynard, & Kumar, 2017). Thus, rockyshores in temperate zones are divided into several levels described by Lüning (1990), in which algae will be affected by emersion for a more or less long period (Fig. 1). First of all, the supralittoral zone is the highest area of the foreshore with a dominance of lichens and halophytes but no seaweed. The mediolittoral zone is the area where the physico-chemical conditions are the most variable because it is subjected to daily emersion/immersion cycle with a tidal range up to 15 m. This level is divided into several distinct zones parallel to the shoreline at different shore heights, defined by the presence of particular species of brown seaweeds (belts) according to their resistance to different stresses (Ar Gall & Le Duff, 2014). Thus, on the upper mediolittoral zone, most macroalgae, such as the brown seaweed *Pelvetia canaliculata*, present a lower density and size than lower foreshore species to withstand long desiccation. At the mid-mediolittoral level, there is the brown seaweeds *Ascophyllum nodosum* and *Fucus vesiculosus* belt on sheltered sites, with a disappearance of the first species on exposed sites (Lüning, 1990). On the lower mediolittoral zone, there are the *Fucus serratus* belt and the brown seaweed *Himanthalia elongata* community, where the effects of the tide are less noticeable. Finally, at the bottom of the rockyshore, there is the subtidal level which is always submerged except in the event of very high tides. At this level, the light for photosynthesis is more limited and some species require near-permanent immersion, including *Laminaria* species.

Within this intertidal zone, seaweeds are subjected to a range of abiotic factors which can fluctuate during the day (tide/photoperiod), or with weather and seasons. Rocky foreshores are notably characterized by their wave exposure mode (*i.e.* hydrodynamism): sheltered, semi-sheltered, as opposed to exposed mode where wave intensity is much more noticeable. Communities are therefore different between shores, but there are also differences within each shore where species are positioned more or less high, depending on their resistance to different environmental parameters. As already mentioned, desiccation is the factor that influences mostly the vertical distribution of algae. Nevertheless, the impact of this factor is difficult to assess because it occurs at the same time as other stresses occurring during emersion. Desiccation is notably often associated with variation in salinity, especially in rockpools. In the Atlantic marine environment, the salinity of the water is around 35. However, this value can fluctuate, with for example a decrease in salinity in rockpools during rainy periods, or an increase during evaporation at low tide, such as in hot weather for example. As algae are photosynthetic organisms, the presence of light (quality and quantity), combined with sufficient nutrient concentration, is also an essential condition for their survival. Nevertheless, too much light and UV-radiations, combined with an increase in temperature, can be harmful. Finally, in addition to these abiotic stresses, macroalgae are subjected to biotic interactions such as competition (for light, space and available substrate), grazing and the presence of epiphytic bacteria, fungi, animals and/or algae. Among the great diversity of algae, some species are able to withstand extreme conditions, such as in rockpools on the intertidal zone, estuaries or salt marshes, where they experience severe variable salinity, temperature and nutrient supply conditions (Lüning, 1990).

To cope with their complex and highly variable environment, algae must constantly adapt, and this adaptation is now faster as the various environmental factors tend to change more rapidly due to the global change that currently weakens coastal ecosystems. Thus, to protect and adapt themselves against these abiotic and biotic stresses, macroalgae produce a wide range of metabolites in order to ensure effective photosynthesis, respiration, growth or reproduction (Stengel et al., 2011). These include photosynthetic, photoprotective, osmoregulatory, antioxidant, antifouling or particular compounds, which can be synthesized in response to different stresses, that can act separately or in synergy. In this context, the aim of this present chapter is to review the effect of the different abiotic and biotic factors on the physiology of brown and red macroalgae. For this purpose, each parameter is detailed one by one, presenting which kind of compounds is synthetized and under which circumstances.

2. Effect of abiotic parameters

2.1 Hydrodynamism

Hydrodynamism refers to water motions, due to the tidal cycle, and to waves or ocean currents resulting from winds and differences in the densities of water masses (Hurd et al., 2014). This factor influences the distribution and the vertical zonation of algae on the shore (Fig. 1), because: (1) currents and waves affect other environmental factors as light penetration, temperature, and the availability of nutrients and all elements necessary for algal development; (2) hydrodynamism influences the distribution of fauna, indirectly affecting algae through spatial competition and potential impact of herbivores; (3) moreover, to survive on exposed sites, macroalgae must be resistant enough to remain attached to their substrate and withstand wave action, giving rise in interesting adaptations which will be developed below (Hurd, 2000; Lüning, 1990). Resulting from hydrodynamism, macroalgal species are often acclimatized to a type of wave exposure, either sheltered (low hydrodynamism), exposed (high hydrodynamism) or the intermediate, semi-exposed environment.

Hydrodynamism can affect the growth, the morphology or the phenology of seaweeds. In this sense, Kraemer and Chapman (1991) showed that juveniles of the brown alga *Egregia menziesii* grown under high hydrodynamic conditions were about 2-times stronger and stiffer than those growing under low hydrodynamic conditions to resist hydrodynamism and not being swept away by waves and tide. In addition, seaweeds can adapt their size, shape and direction to thrive/disperse with the flow and increase their resistance to currents (Boller & Carrington, 2006). Thus, it has been shown that the red alga *Chondrus crispus* can realign its stipe and compact its thallus in response to the flow (Boller & Carrington, 2006). Moreover, nutrient uptake by macroalgae is influenced by water motion, impacting growth and photosynthesis (Gerard, 1982; Hurd, 2000). Indeed, with a high current, the nutrient absorption rate increases due to the decrease in the thickness of the diffusion boundary layer, and the acceleration of transport through it (Lüning, 1990). Thus, photosynthesis increases with hydrodynamism until enzyme saturation, as demonstrated in the brown alga *Macrocystis pyrifera* (Wheeler, 1980). In return, in order to compensate for low nutrient and gas absorption rates, algae in low hydrodynamic environments tend to have a higher surface-to-volume ratio (Stewart & Carpenter, 2003).

The reproduction can also be influenced by hydrodynamism. According to Gordon and Brawley (2004), water motion mainly affects gametes and spores release, and their dispersal distance. Thus, they have demonstrated an increase in the zoospores release of the brown alga *Alaria esculenta* under shaken conditions, and that, on the contrary, there were more gametes released by the antheridia under calm conditions.

The presence of some molecules in seaweeds could be influenced by hydrodynamism, especially polysaccharides in Phaeophyceae. In this sense, Craigie, Morris, Rees, and Thom (1984) found that the alginic acid composition in the brown alga *Saccharina longicruris* was related to the degree of wave exposure, although this hypothesis is controversial (Kraemer & Chapman, 1991). Moreover, Le Lann, Rumin, Cérantola, Culioli, and Stiger-Pouvreau (2014) showed that the brown alga *Bifurcaria bifurcata*, living in a semi-exposed/sheltered environment, synthetized more terpenes, including bifurcanone which is specific to sheltered environment. According to Rönnberg and Ruokolahti (1986), phenolic content of *F. vesiculosus* has been shown to depend on the degree of wave exposure of the habitat, with smaller fluctuations of phenolic content at exposed site compared to sheltered site, as also demonstrated in the brown alga *Sargassum muticum* by Plouguerné et al. (2006b). It is sometimes difficult to interpret the significance of the difference in composition or morphology with the exposure mode, as many other factors can interact, including desiccation and herbivory. In this sense, Kraemer and Chapman (1991) did not find any variation in alginic acid content in the Laminariales *E. menziesii* exposed to different wave velocities; and Fuller and Mathieson (1972) showed no significant difference in carrageenan content of *C. crispus* according to the wave exposure.

2.2 Desiccation

In the intertidal zone, seaweeds, as being sessile organisms living attached to the substrate, are faced with desiccation for a brief or longer period depending on the position on the shore, due to the tide and the alternation of immersion and emersion. Several preliminary studies agree that the ability of algae to withstand significant water loss and mostly to rehydrate quickly defines their tolerance to drought, and not their ability to retain water (Contreras-Porcia et al., 2017). Thus, the different algal species display variable desiccation tolerance, with the most resistant species at the top of the intertidal zone (Abe, Kurashima, Yokohama, & Tanaka, 2001; Lüning, 1990). Thanks to a slower water loss, a faster rehydration and a greater resistance to high water loss, macroalgae of the upper intertidal zone can loose up to 90% of their cellular water content (Contreras-Porcia et al., 2017). This rate can reach up to 96% for *Pyropia orbicularis* or *Porphyra columbina*, making these species models to study algal responses to desiccation (Contreras-Porcia, Thomas, Flores, & Correa, 2011; Fierro et al., 2017). Nevertheless, a whole range of factors can influence the rate of desiccation of each species, as for example weather and seasons, as high temperatures and wind favor a high rate of water evaporation (Lamote, Johnson, & Lemoine, 2012). Desiccation also seems to be related to the exposure mode, with the most exposed species being more likely to have a lower rate of desiccation due to thicker cell walls (Kristensen, 1968).

The loss of water induced by desiccation causes morphological changes such as retraction of protoplasts and chloroplasts disorganization due to an osmotic imbalance (Flores-Molina et al., 2014). To avoid these effects, some algae regularly subjected to emersion present a small specific surface to reduce desiccation, as species with a lower surface/volume ratio loose their water content less quickly (Lüning, 1990). In this way, macroalgae in the upper intertidal zone are smaller in size to better resist to water deprivation as demonstrated for *Fucus spiralis* (Schagerl & Möstl, 2011). The formation of mucilage on the surface of some particular algae is another way that has been suggested as resistance to desiccation (Bérard-Therriault & Cardinal, 1973). Indeed, the cell wall of macroalgae is composed of some unique polysaccharides (or phycocolloids), with a strong polyanionic character and which possess water retention properties, especially due to sulfatation. The main phycocolloids are fucoidans and alginates in brown seaweeds (Bérard-Therriault & Cardinal, 1973), whereas red seaweeds contain agars, xylans and carrageenans (Kloareg & Quatrano, 1988; reviewed by; Stiger-Pouvreau, Bourgougnon, & Deslandes, 2016). Contrary to freshwater algae and terrestrial plants, polysaccharides of marine macroalgae are highly sulfated, up to 20%–38% for carrageenans, which could be a specific adaptation to marine environment and desiccation (Kloareg & Quatrano, 1988).

From a physiological point of view, it has been shown that there is a decrease in photosynthesis and chlorophyll-*a* content during emersion (Contreras-Porcia et al., 2011; Flores-Molina et al., 2014; Gao & Wang, 2012). In this sense, López-Cristoffanini et al. (2015) demonstrated in *P. orbicularis* that a drying treatment could affect the expression of proteins involved in photosynthesis. The cyclic electron flow around photosystem I has been demonstrated to play a significant role in response to drought, by protecting the photosynthetic apparatus and modulating a pH gradient for the synthesis of ATP (Gao & Wang, 2012). Photosynthesis is then completely restored after re-immersion, except if a critical water content has been reached (Dring & Brown, 1982; Lamote et al., 2012).

In general, prolonged desiccation periods can affect a whole set of physiological activities due to the fact that being out of the water starves algae for nutrients and inorganic carbon (Lamote et al., 2012). For example, desiccation causes a decrease in phosphate absorption in *F. spiralis*, *A. nodosum* and *F. serratus* (Hurd & Dring, 1991). Macroalgae experiencing periodical emersion may benefit from increasing atmospheric CO_2 concentrations or light (Zou & Gao, 2002), which may explain why the rate of photosynthesis during emersion can sometimes exceed the one during immersion. In addition, a long period of desiccation induces oxidative stress with an over-supply of reactive oxygen species (ROS) (Contreras-Porcia et al., 2011; Sampath-Wiley, Neefus, & Jahnke, 2008). For example, concentrations of H_2O_2 increased up to 25-fold during desiccation in cortical cells of the brown macroalga *Lessonia spicata* (Flores-Molina et al., 2014). Molecular analysis demonstrated that some genes, such as those encoding the enzyme trehalose phosphate synthase (TPS) or catalase (CAT) which protect algae from oxidative stress, are overexpressed in *P. orbicularis* during desiccation (Fierro et al., 2017). As for higher plants, the enzyme glutathione reductase (Contreras-Porcia et al., 2017; Sampath-Wiley et al., 2008), and the hormone abscisic acid (ABA) (Guajardo, Correa, & Contreras-Porcia, 2016) regulate oxidative stress under water deficiency in algae.

One challenge in measuring the effect of desiccation is that it occurs naturally during periods of emersion, and is thus associated with other stresses like changes in light or temperature (Matta & Chapman, 1995; Sampath-Wiley et al., 2008), which can accelerate desiccation (Binet, 1956). In particular, it has been demonstrated in the Gigartinales *Endocladia muricata* that desiccation leads to better thermotolerance, protecting the alga from high heat during low tide (Hunt & Denny, 2008). Similarly, according to Ji et al. (2015) and López-Cristoffanini et al. (2015), heat shock proteins expression is altered during desiccation, due to high temperature.

2.3 Light

As algae are photosynthetic organisms, light is a key factor for their growth and distribution. More particularly, algae need to absorb enough Photosynthetically Active Radiation (PAR) in their thylakoids to transform light energy into chemical energy through photosynthesis, necessary for their development. In this sense, the growth of red macroalgae such as *Gracilaria domingensis* (Ramlov et al., 2011), *C. crispus* and *Palmaria palmata* (Manríquez-Hernández, Duston, & Garbary, 2016) have been enhanced under high photon flux intensity, indicating that light can be a limiting factor. There is a photosynthetic light compensation point, which corresponds to the photon fluence rate, for which the amount of oxygen produced by photosynthesis accurately compensates for the oxygen absorption related to respiration (Hurd et al., 2014; Lüning, 1990). On the other hand, if the light increases well beyond the compensation point, the light saturation can be reached and photosynthesis is thus maximum (P_{max}). At this point, the enzymatic reactions limit the photosynthesis, and if the light increased again, photosynthesis is then reduced (photoinhibition).

To achieve photosynthesis, all algal species have chlorophyll-*a* in their chloroplasts to capture light photon energy. Depending on the taxonomic group of algae, there are also other chlorophylls that absorb light at other wavelengths (Table 1). In addition, the different classes of seaweeds are characterized by their carotenoid composition including fucoxanthin, violaxanthin, antheraxanthin, lutein, zeaxanthin, α-carotene, and β-carotene (Lüning, 1990) (Table 1). The composition and quantities of carotenoids also varied widely between algal species, although no link has ever been demonstrated with phylogeny (Schubert, Garcia-Mendoza, & Pacheco-Ruiz, 2006). In red seaweeds, lutein is the most common

Table 1 Pigment composition and presence of phenolic compounds or mycosporine-like amino acids (MAAs) within the red and brown macroalgae according to Stengel, Connan, & Popper (2011).

Group	Chlorophylls	Other pigments	MAAs	Phenolic compounds
Rhodophyta	Chlorophyll-*a*	phycocyanin, phycoerythrin, allophycocyanin, α-carotene, β-carotene, violaxanthin, antheraxanthin, zeaxanthin, lutein	Yes	Few
Phaeophyceae	Chlorophyll-*a* Chlorophyll-c_1,c_2	β-carotene, fucoxanthin, violaxanthin, antheraxanthin, zeaxanthin, neoxanthin	No	Many

carotenoid and can represent up to 99% of the total carotenoid content (Lalegerie et al., 2019; Schubert, García-Mendoza, & Pacheco-Ruiz, 2006). In order to optimize photosynthesis in the presence of low light intensity, red algae also possess extra pigments called phycobiliproteins (*i.e.* phycocyanins, phycoerythrins and allophycocyanins in the phycobilisome) (Dumay, Morancais, Munier, Le Guillard, & Fleurence, 2014).

Irradiance in the ocean is highly variable, since it depends on weather, algal position on the shore, latitude, water turbitidy, seasons and time of the day (day/night and tidal cycles) (Hanelt & Figueroa, 2012; Hurd et al., 2014). However, any modification of the light, whether by the duration of irradiance, the intensity or the nature of the radiations, will have an effect on the physiology and distribution of seaweeds (Hurd et al., 2014). In particular, light is an environmental signal for reproduction and growth, since it has been demonstrated for example that the release and accumulation of gametes are light-induced (especially with the blue radiations) in several species like the brown macroalga *Dictyota dichotoma* (Hanelt & Figueroa, 2012). However, the amount of light first affects photosynthesis and the amount of chlorophyll-*a*. Thus, Torres, Chow, and Santos (2015) observed a 55% increase in chlorophyll-*a* content from 1000 to 60 μmol photon m^{-2} s^{-1} in *Gracilaria tenuifrons*, to enhance light energy capture at low irradiance. On the contrary, many studies have shown that chlorophyll-*a* level is lower under high light, particularly in summer due to the increasing irradiance. It has been observed, for example, a 50% decrease in chlorophyll-*a* in *A. nodosum* during summer (Stengel & Dring, 1997), and a threefold decrease from February to July in *P. palmata* (Guihéneuf, Gietl, & Stengel, 2018). Similarly, chlorophyll-*a* content increases with water depths with light attenuation and fluctuation in the composition of the light spectrum, as shown in *Porphyra umbilicalis* and *C. crispus* (Ramus, Beale, Mauzerale, & Howard, 1976).

Although light is vital for algae, too much radiation, especially in the UV-A and UV-B range, can be harmful, mainly for the early development stages (Karentz, 2001; Wiencke et al., 2000). UV-radiation can cause a decrease in growth, as observed in the red seaweeds *Crassiphycus birdiae* (Ayres-Ostrock & Plastino, 2014) and *Kappaphycus alvarezii* (Schmidt, Maraschin, & Bouzon, 2010). Above a certain threshold, extreme irradiances can cause photodamages such as DNA alterations, synthesis of reactive oxygen species (ROS), photo-oxidation, or degradation of photosynthetic pigments (Karentz, 2001; Karsten, 2008; van de Poll, Eggert, Buma, & Breeman, 2001). UV radiation is notably reported to be responsible for damage to chloroplasts, thylakoids and mitochondria in *K. alvarezii* (Schmidt et al., 2010) and *P. palmata* (Holzinger, Lütz, Karsten, & Wiencke, 2004). A very strong visible light can also lead to photoinhibition consisting in the reduction of the photosynthesis, which could be temporary or deleterious if very strong or long lasting (Franklin, Osmond, & Larkum, 2003; Hanelt & Nultsch, 2003). For example, photoinhibition can reach 95% in *C. crispus* (Yakovleva & Titlyanov, 2001). Under UV radiation, this phenomenon of inhibition of photosynthetis may be related to the activity of Rubisco and D_1 protein of photosystem II (Bischoff, Hanelt, & Wiencke, 2000; Schmidt et al., 2010).

For their own protection, algae are able to produce photoprotective compounds that will dissipate excessive energy (Bhatia, Sharma, Sharma, & Purohit, 2011; Rastogi, Sinha, Singh, & Häder, 2010). In this sense, carotenoids can play both photosynthetic and photoprotective roles (Hashimoto, Uragami, & Cogdell, 2016; Young & Frank, 1996). In particular, in some seaweeds, they protect the cells against excessive light energy in photosystem II and oxidative stress via the xanthophyll cycle (Fig. 2): when the amount of light is too high, violaxanthin is converted by de-epoxidation to antheraxanthin and then zeaxanthin, which reduces excess energy (Karentz, 2001; Ursi, Pedersen, Platino, & Snoeijs, 2003).

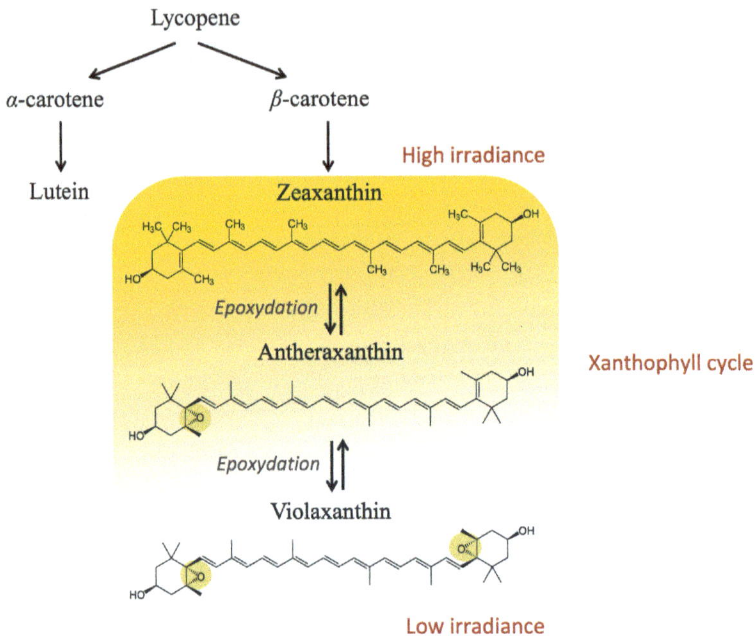

Fig. 2 The xanthophyll cycle in macroalgae according to Latowski, Kuczyńska, and Strzałka (2011) and Takaichi (2011).

Mycosporine-like amino acids (MAAs) are a second group of photoprotective compounds found in red macroalgae (Sinha, Singh, & Häder, 2007), which are reported to be effective against UV radiation (see Chapter 8). Yakovleva and Titlyanov (2001) for example, showed that repeated exposure to UV radiation resulted in the accumulation of MAAs, coupled with less photoinhibition in *C. crispus*. Some species such as *Furcellaria lumbricalis* or *Heterosiphonia plumosa* are not able to produce MAAs, while others like *P. palmata*, or *C. crispus* have been reported to synthesize up to 7 or 8 MAAs (Lalegerie et al., 2019). Shinorine, palythine and porphyra-334 are the MAAs most widely distributed among red algae. Several studies have suggested that MAAs could also be synthesized in response to other environmental stresses and more particularly with respect to nutrient limitation (Oren & Gunde-Cimerman, 2007).

Beyond the well-known photoprotective compounds mentioned above, it has also been demonstrated that phenolic compounds of brown seaweeds (phlorotannins) could play a role in the UV-protection (Schoenwaelder, 2002). For example, Pavia, Cervin, Lindgren, and Åberg (1997) have found that a 50% increase in UV radiation during 2 weeks resulted to a 30% increase in phlorotannin concentration of *A. nodosum*.

2.4 Temperature

Temperature is the major factor controlling survival, growth, reproduction and thus geographical distribution of seaweeds (Yarish, Breeman, & van den Hoeck, 1986). However, in temperate zones, temperature can change according to the latitude, depth, season, but also according to the tide, with emersion and immersion phases. Thus, according to Eggert (2012), seaweeds respond to the temperature in three different ways: genetic adaptation to the constant temperature of the living environment; phenotypic acclimatization; or short-term physiological regulation in response to the temporal temperature shocks. In summer and winter at low tide, the seawater temperature highly varies, forcing intertidal organisms to be euryceous and highly tolerant to any variation in temperature. Intertidal seaweeds living on the upper shore have thus been found to have temperature tolerance ranges broader than subtidal species, and can tolerate exposure to high temperatures for prolonged time periods (Einav, Breckle, & Beer, 1995). For example, the sublittoral brown alga *Laminaria* spp. can tolerate a temperature up to 20 °C, whereas *F. spiralis* from upper littoral is resistant up to 28 °C (Lüning, 1990). Moreover, the temperature tolerance of seaweed depends on latitude, with southern perennial kelp more resistant to short heat shock than the northern one (Pereira, Engelen, Pearson, Valero, & Serrão, 2015).

The temperature can affect the growth and reproduction, for which optimal temperature and timing must be particularly important (Molenaar & Breeman, 1997). Indeed, growth and reproduction can be under direct photoperiodic control or endogenous circannual rhythm (Lüning & Tom Dieck, 1989). For example, Molenaar and Breeman (1994) showed that the life history of the red alga *Phyllophora pseudoceranoides* is regulated by temperature and daylength, with the production of spores in winter. In addition, Yarish, Breeman and van den Hoeck (1986) showed that gametophytes of *Polyneura bonnemaisonii* (previously *P. hilliae*) have a smaller survival rate at low temperatures than tetrasporophytes, which limits its distribution.

Moreover, the temperature can impact the photosynthesis and the other metabolic pathways of seaweeds, as enzymatic activities are temperature-dependent. In this sense, the response of thermal acclimatation of photosynthesis appears to be highly comparable to that described for photoacclimatation (Huner, Öquist, & Sarhan, 1998). However, temperature and light have been shown to have interactive effects on photosynthesis and its regulation

(Savitch, Maxwell, & Huner, 1996). Moreover, temperature sensitivity affects proteins and membranes: quantitatively (concentration changing), qualitatively or through modulation (modify the protein environment) (Eggert, 2012). More particularly, it has been demonstrated that membrane fluidity decreases with a decrease in temperature (Eggert, 2012; Inaba et al., 2003; Szalontai, Nishiyama, Gombos & Murata, 2000). Heat stress affects membrane-associated processes because high temperatures cause fluidization of membranes and finally disintegration of the lipid bilayer (Los & Murata, 2004). In addition, Pakker et al. (2000) demonstrated that the reparation of DNA damages induced by UV-B radiation is dependent on the temperature in *P. palmata*.

The production of metabolites in macroalgae is also temperature dependant. Notably, high temperature induces the production of heat shock proteins. For example, Ireland et al. (2004) showed an induction of these molecules in *F. serratus* exposed to 42 °C, maximum after 2 h. Tanniou (2014) studied the impact of water temperature on total phenolic content during 30 days at 11° and 22 °C on *S. muticum*: a decrease in phenolic level was observed at 22 °C compared to those grown at 11 °C. According to Connan, Deslandes, and Ar Gall (2007), air-temperature correlated significantly with the phenolic content in uppershore *P. canaliculata* and midshore *A. nodosum,* but not for the lowshore *B. bifurcata.* Temperature can also influence pigment content of the algae. Indeed, Rothäusler, Gómez, Karsten, Tala, and Thiel (2011) showed the highest carotenoid level at the coolest temperature (12 °C) compared to ambient (17 °C) and warm (20 °C) temperatures and independently of irradiance, in *M. pyrifera* being afloat for 15 days. Vanitha and Chandra (2012) found no correlation between pigments (R-phycoerythrin, R-phycocyanin, allophycocyanin and chlorophyll-*a*) and temperature but some variations appeared maybe due to other factors (light or humidity) in red algae (*Gracilaria corticata, Grateloupia lithophila, Gelidium* sp. and *Bryocladia thwaistesii*). On the contrary, Kim, Kraemer, Neefus, Chung, and Yarish (2007) showed an influence of temperature on the phycoerythrin content in several *Porphyra* species (*P. leucosticta, P. linearis, P. umbilicalis*); with higher contents at low temperature (10 °C compared to 15 °C and 20 °C), but not for *P. amplissima.* In a similar manner, Guihéneuf, Gietl, and Stengel (2018) found an increase in chlorophyll-*a* and phycobiliprotein concentrations in *C. crispus* and *P. palmata* when temperature and irradiance decreased in autumn. Conversely, Ding, Ma, Huang, and Chen (2013) showed an increase in chlorophyll *a* content when water temperature increases from 15 until 25 °C in *Hypnea cervicornis.*

2.5 Salinity

The ranges of salinity encountered in marine habitats differ greatly with respect to space and time. Salinity is different according to the oceans and seas. More particularly, saltmarshes, estuaries and near-shore waters are environments with extreme salinity variations. In marine waters, intertidal algae are also subjected to variations in salinity according to the season, day, or even hours, as in the rockpools of the intertidal zone. Notably, at low tide seaweeds experienced extreme dilutions by rainwater or, on the contrary, extreme salinities by evaporation. In the intertidal zone, Fucales can withstand variations in salinity ranging from 0.1 to 3.5 times that of seawater, while subtidal algae, like the red macroalgal genera *Antithamnion* or *Plocamium*, are less tolerant, between 0.5 and 1.5 times that of seawater (Zaneveld, 1969). Macroalgae living in environments with variable salinity must maintain a higher osmotic pressure in their cell medium than in their environment; an osmotic pressure of 2.6–3 MPa is typical of seaweeds (Karsten, 2007).

Some algal species are euryhaline as they are able to tolerate a large salinity range, while stenohaline species cannot survive at a salinity too far from their salinity optimum (Bird & McLachlan, 1986). Karsten (2007) demonstrated that the euryhaline *Fucus distichus* was not affected by hypo- or hypersaline stresses (salinities ranging from 5 to 60) while *A. esculenta*, *Saccharina latissima* and *Laminaria solidungula* demonstrated a strong loss of pigments (bleaching) or even high mortality under hyposaline stress; intertidal species are thus mostly euryhaline while subtidal are stenohaline. Schubert, Feuerpfeil, Marquardt, Telesh, and Skarlato (2011) showed a decline in abundance of macroalgae with salinity until 7.2. Moreover, Stratil, Neulinger, Knecht, Friedrichts, and Wahl (2014) demonstrated that salinity was important in structuring algal-associated epibacterial communities: they showed a disappearance of epiphytic and epilithic betaproteobacteria with an increase in salinity, from 5 to 25.

Faced to variations in salinity, algae respond to osmotic shock by regulating either their concentrations of ions, such as K^+, Na^+, Cl^- or by synthesizing organic osmolytes, *i.e.* molecules with low molecular weight in order to be easily and rapidly exchange with the surrounding water to equilibrate osmotic pressure with their own tissues (Karsten, 2007). In general, osmolytes are concentrated in algal vacuoles and/or are located in the cytoplasm (Karsten, 2007). The last author drawed a schematic functioning of the osmotic adjustment in macroalgae, listing two phases in this phenomenon: a phase I which includes rapid water flows through the cell-wall, entering (hypoosmotic schock) or leaving (hyperosmotic schock) depending on the osmotic shock, and a phase II which represents metabolic control, including at the beginning an ionic regulation which is very rapid, followed by a slow ionic regulation involving osmolytes.

Variations in salinity affect different aspects of macroalgal metabolism, in particular photosynthesis, respiration, ionic fluxes and algal growth, as demonstrated in the brown macroalga *Dictyopteris membranacea* by Gessner (1969), with a photosynthetic optimum at a salinity between 28 and 55. During readjustment to hyperosmotic shock, rapid changes in mineral concentration often precede a slower synthesis of organic osmolytes, such as mannitol in brown algal species (Munda, 1978). Mannitol is the main product resulting from photosynthesis, which accumulates in Phaeophyceae where it represents a significant part of the dry matter of the algae: 23% and 19.5% in the Fucales *P. canaliculata* and *Halidrys siliquosa*, respectively (Reed, Davison, Chudek, & Foster, 1985). The accumulation of mannitol and volemitol has been shown in the uppershore *P. canaliculata* (Reed et al., 1985) while at the bottom of the mediolittoral zone, *H. elongata* synthesizes mannitol and altriol (Wright & Reed, 1985). Red algae produce mainly floridoside for their osmoregulation (Reed, 1990). In *Porphyra/Pyropia* spp., different forms of isofloridoside are present simultaneously with floridoside for the osmoregulation of each species (Reed, Collins, & Russell, 1980). The average floridoside content varies between 1.5 and 8% of the dry matter of the algae, but can reach 22%–30% in some species (Kerjean et al., 2007; Kirst, 1990). Floridoside is a small carbohydrate synthesized very early in photosynthesis, as demonstrated in the red alga *Solieria chordalis* (Goulard, Diouris, Quéré, Deslandes, & Floc'h, 2001). Simon-Colin, Bessieres, and Deslandes (2002) highlighted the major role of floridoside in the osmoregulation of the red macroalga *Grateloupia turuturu* (formerly *Grateloupia doryphora*), demonstrating an increase at high salinity and a decrease under hypohyaline stress, and returning at its initial state after 3 days of cultivation of the macroalga. Under controlled conditions, early after the hypo- or hyper-hyaline stress, Kerjean et al. (2007) demonstrated that the floridoside content can reach high values around 39 mg g^{-1} DW in *Mastocarpus stellatus*.

As a general rule, red algae do not synthesize alditol, *i.e.* non-cyclic polyol, except few Ceramiales, which accumulate mannitol, as *Caloglossa leprieuri*, whose mannitol contributes 23% of the internal osmotic pressure (Karsten et al., 1992a). *Bostrychia* (previously *Stictosiphonia*) *arbuscula* use a combination of D-dulcitol and D-sorbitol as organic osmolytes (Karsten, West, & Zuccarello, 1992b). Some amino acids are also involved in the osmoregulation of some species, generally characterized by low concentrations of basic amino acids and a dominance of alanine, aspartic and glutamic acids that can represent more than 50% of free amino acids (FAA; Fattorusso & Piattelli, 1980). Other amino acids are involved in the osmoregulation of red algae: proline (54.8% FAA in *Polysiphonia nigrescens* and 65.3% FAA in *Halopitys incurvus*; Fattorusso & Piatelli, 1980); taurine, a sulfur amino acid derivative, such as *A. vermiculophyllum* (Surget et al., 2017), *Gracilariopsis lemaneiformis* (Broberg, Kenne, & Pedersen, 1998), *Grateloupia* sp. (Impellizzeri et al., 1975) and *C. crispus* (Harnedy & Fitzgerald, 2011). In marine algae, there are also other compatible molecules that are accumulated in response to osmotic stress, such as quaternary ammoniums (betaines) found in some red algae, such as *M. stellatus, Calliblepharis jubata* and the brown alga *Dilophus fasciola* (Blunden et al., 1992), and dimethylsulfoniopropionate (DMSP), a tertiary sulfonium found in some Rhodophyta, such as *Porphyra, Polysiphonia* species and in minority in brown algae (Reed, 1983).

2.6 Seawater chemical composition

To achieve their photosynthesis and growth, algae need nutrients that are essential to their development. As a result, several experiments have shown that nutrient enrichment leads to higher growth and nutrient content in tissues, such as in *A. vermiculophyllum* (Abreu, Pereira, Buschmann, Sousa-Pinto, & Yarish, 2011). More particularly, many culture experiments have demonstrated that carbon (C), hydrogen (H), oxygen (O), nitrogen (N), magnesium (Mg), copper (Cu), manganese (Mn), and zinc (Zn) are essential for algae (Hurd et al., 2014). Nitrogen as nitrate ($NO_3{}^-$), nitrite ($NO_2{}^-$), ammonium ($NH_4{}^+$), or urea ($CO(NH_2)_2$), and phosphorus as phosphate ($H_3PO_4^{3-}$) are often limiting factors for primary production and algal growth in natural environment since they occur at low concentrations in seawater (Lüning, 1990). Some seaweeds could thus produce hyaline hairs under low nutrient conditions to increase nutrient uptake rate (Harrison & Hurd, 2001; Wallentinus, 1984). Different algal species do not have the same affinity for each form of nutrients available in the marine environment: NH_4^+ is taken up more preferentially and quickly by most seaweeds, such as *F. serratus* (Brenchley, Raven, & Johnston, 1997) and *Pelvetiopsis limitata* (Fujita, Wheeler, & Edwards, 1989) or *C. crispus* (Corey, Kim, Duston, Garbary, & Prithiviraj, 2013). On the other hand, *P. palmata* showed similar affinity for NH_4^+ and NO_3^- (Martínez & Rico, 2004).

Seaweed physiology depends on nutrients, whose concentrations are very variable in the marine environment: concentration of dissolved nutrients in surface waters are generally important in high latitudes, in coastal upwelling zones (Levitus, Conkright, Reid, Najjar, & Mantyla, 1993), and generally close to the coastlines and in estuaries or eutrophic sites. In addition, the nutrient availability can vary greatly depending on the time of the day or year. Indeed nutrient inputs are dependent on tidal cycle, since algae have no longer nitrogen source during emersion, affecting thus the nutrient seaweed content. For example, in *Porphyra umbilicalis*, the N content in tissues decreased during emersion (Kim, Kraemer, & Yarish, 2013). More specifically, species living at the upper intertidal zone need to be adapted to a longer nutrient deficiency, by modifying their uptake rate, uptake efficiency or enzymatic activities (Phillips & Hurd, 2003). In this sense, Benes and Bracken (2016) have demonstrated that the absorption rate of *F. vesiculosus* is 18% higher when living in the upper intertidal zone. In addition, algae are subjected to seasonal fluctuations in the nutrient content in the seawater. Macroalgae are particularly affected by very low ambient N and P concentrations during summer (Fujita et al., 1989). To cope with these variations, some species adapt their nutrient uptake, such as *Fucus gardneri* whose nitrate reductase activity increased sevenfold between February and March (Hurd, Berges, Osborne, & Harrison, 1995). Similarly, relative preferences for each N source vary according to the time of year in *Bostrychia arbuscula* and the two brown algae *Scytothamnus australis* and *Xiphophora gladiata*: in winter NH_4^+ is preferred over NO_3^- and urea, while in summer NH_4^+ and NO_3^- are equally absorbed (Phillips & Hurd, 2003). In addition, nutrient concentrations depend on several abiotic factors: it has been shown that nutrient uptake depends on water motion (*i.e.* hydrodynamism), since it decreases the size of the water boundary layer of the seaweed (Hurd, 2000).

To cope with the temporal variation of nutrients, some species have the ability to incorporate and store more or less important quantities of nutrients in their cells, for a longer or shorter time in N-rich molecules such as proteins, amino acids and pigments (Gagné, Mann, & Chapman, 1982). According to Naldi and Wheeler (1999), proteins can contribute up to 90% to total N storage in algae. In *P. palmata*, N from phycoerythrin represents up to 8.3% of the thallus N content (Martínez & Rico, 2002). Levels of these compounds increase with N concentration and are quickly degraded under N limitation, as demonstrated in *Porphyra/Pyropia* spp. (Korbee, Huovinen, Figueroa, Aguilera, & Karsten, 2005; Pereira, Kraemer, Yarish, & Sousa-Pinto, 2008). Recently, mycosporine-like amino acids have been also suggested as a N source for *P. palmata* in summer (Guihéneuf, Gietl, & Stengel, 2018). As a result, algal growth is not directly related to nutrient concentrations in the surrounding environment. Thus, to ensure their reproduction during summer, some species store nutrients in winter, when nutrients are high but not light and temperature. Thanks to these internal reserves, *Laminaria longicruris* can extend its growth several months after the decrease in N in the environment (Gagné et al., 1982).

The fast absorption or resistance to low nutrient contents can lead to competition between species: some species can take advantage during nutrient peaks, modifying the composition of algal communities since slow-growing benthic macrophytes are replaced by opportunistic ephemeral macroalgae with fast growth and high nutrient demands (Pedersen & Borum, 1996). If concentrations are too high, nutrients can be responsible for seasonal algal blooms (Heisler et al., 2008). This phenomenon of excessive algal biomass production is increasing from year to year due to eutrophication of the marine environment and nutrient inputs near agricultural lands or factories. Many algal blooms have been reported worldwide, mainly due to cyanobacteria, diatoms or some green algae, but there are also massive seasonal strandings of red and brown seaweeds (Teichberg, Martinetto, & Fox, 2012). Recently, nitrogen sources from the Amazon and Congo rivers are suspected to stimulate the growth and proliferation of pelagic *Sargassum* spp. in the North Atlantic Ocean (Ody et al., 2019; Oviatt, Huizenga, Rogers, & Miller, 2019). These developments modify community structures, impacting biodiversity and local economy (Teichberg et al., 2012).

Similar to nutrients, metals occur naturally in seawater at low concentrations; but elevated levels may occur in coastal waters due to contamination near mines, harbors, and industrial activity (*e.g.* Nriagu, 1990) which then can pose a threat to marine life if they are bioaccumulated. However, the metal bioavailability depends on many factors such as the salinity or the acidity of seawater. Among trace metals, copper is one of the most toxic: for example, for brown macroalgae, copper in excess inhibits photosynthesis (Connan & Stengel, 2011a), restricts growth (Nielsen, Brownlee, Coelho, & Brown, 2003), and disrupts gametophyte development (Garman, Pillai, & Cherr, 1994). Copper also affects the distribution of other algal compounds such as proteins (Contreras, Moenne, Gaillard, Potin, & Correa, 2010), free fatty acids (Ritter et al., 2008), phenolic compounds (Connan & Stengel, 2011b), and induces oxidative stress (Contreras, Mella, Moenne, & Correa, 2009).

3. Effect of biotic parameters

3.1 Microfouling

Any substrate immersed in aqueous solution is immediately colonized by organic molecules (forming the primary film) that are consumed by bacteria followed by microalgae and fungi (Lewis, 1998). These microorganisms constitute the microfouling that will be used by other benthic macroorganisms such as invertebrates or macroalgae for their settlement. In the marine environment, over 4000 organisms, including algae, are able to form fouling (Chambers, Stokes, Wahsh, & Wood, 2006; Yebra, Kiil, & Dam-Johansen, 2004). Macroalgae represent a living substrate on which other organisms, plant or animal, can attach themselves. The community of the epiphytic bacteria is mainly host specific (according to the algal species) and is changing according to the part of the thallus, site and/or season (for example: Bengtsson, Sjøtun, and Øvreås (2010) and Mancuso, D'Hondt, Willems, Airoldi, and De Clerck (2016) on brown seaweed; and de Oliveira et al. (2012) for red macroalgae). The physiological state of the basiphyte will also influence the settlement of epiphytic bacteria: Marzinelli et al. (2015) observed that microbial communities were strongly associated with the physiological condition of *Ecklonia radiata* (healthy vs. stressed). The epiphytic bacteria interact with seaweeds, thereby modulating the health of their host. They may have positive effect on the basiphyte as being involved in their morphology and physiology. For example, when cultured in an axenic environment, some seaweed species will develop abnormal morphologies which become normal again when bacteria are added, suggesting the synthesis by the bacteria of compounds necessary for the macroalgal morphology (Egan et al., 2013). Weinberger et al. (2007) have also observed that secondary compounds produce by the bacteria are important in the algal spore release and settlement. Moreover, cyanobacteria are often abundant on benthic macroalgae probably due to their nitrogen fixing function that will help the macroalga in relatively poor nitrogen environment (de Oliveira et al., 2012). This strong relationship was emphasized by Egan et al. (2013) who suggested the use of the term holobiont for the association seaweed-bacteria, similarly to the coral.

On the other hand, biofilms can reduce the access of their host to light, gases and nutrients and alter the interaction with other fouling epibionts, consumers and pathogens (Goecke, Labes, Wiese, & Imhoff, 2010; Wahl, Goecke, Labes, Dobretsov, & Weinberger, 2012). To limit or reduce the microfouling, macroalgae will produce metabolites. For example, some red and brown macroalgae are able to produce oxidative burst against epiphytic bacteria. Küpper, Kloareg, Guern, and Potin (2001), and Küpper, Muller, Peters, Kloareg, and Potin (2002) reported that brown macroalgae from the orders Laminariales and Desmarestiales and the species *Pylaiella littoralis* produce rapidly H_2O_2 in response to the degradation of cell wall alginates and thus the liberation of alginate oligomers by bacteria. Similarly, red macroalgae and especially *Gracilaria conferta*, produce a similar oxidative burst in response to agar oligomers, which resulted in a 60% reduction of agarolytic bacterial pathogens (Weinberger & Friedlander, 2000). Plouguerné et al. (2008) worked on two invasive species, *G. turuturu* and *S. muticum*, which presented clean surface, without epiphytes. They demonstrated the importance of lipidic molecules as potential defense against microfouling: cholesteryl formate and glycolipids in respectively *G. turuturu* (Plouguerné, Kikuchi, Oshima, Deslandes, & Stiger-Pouvreau, 2006a) and *S. muticum* (Plouguerné et al. 2010). Hellio, Bergé, Beaupoil, Le Gal, and Bourgougnon (2002) carried out screenings on French intertidal seaweeds and demonstrated that extracts of red and brown species, especially *Vertebrata* (formerly *Polysiphonia*) *lanosa*, *C. crispus*, *Undaria pinnatifida* and *S. muticum*, are a source of antifouling molecules.

3.2 Epiphytes and endophytes

Following the development of microorganisms at the surface of the macroalgae (first step of the biofouling), a secondary colonization occurs with macroorganisms such as other algae or animals, that settle on the algal surface. Other organisms will colonize the subsurface and enter into the seaweed thallus. The first group of species is called epiphytes as being epibiont of an alga (called basiphyte), and the other endophytes. The carrying of epiphytes is common on marine algae, ranging from obligate relationships (such as *Vertebrata lanosa* on *A. nodosum*; Garbary, Miller, & Scrosati, 2014), to species that colonize any available space. Five anatomical relationships between epiphyte and their host have been detected (reviewed in Potin, 2012) from the weakly attachment to the host surface with no tissue damage (Type I), up to the deep penetration of the epiphyte into the host cortex, reaching the medullary tissue, and causing the destruction of the host's cells in the area around the infection (Type V).

As for microfouling, the abundance and diversity of epiphytes follow a seasonal pattern (*e.g.* higher diversity and biomass of epiphytes on *F. vesiculosus* in spring-summer; Rindi & Guiry, 2004). Spatial variation in the epiphyte community on the stipes of *Laminaria hyperborea* was recorded in

Scotland, with differences with the depth along a stipe and according to the host position on the shore (Whittick, 1983). The age of the algal part also affected the epiphyte abundance, with epiphyte densities greatest on the oldest tissue; surface rugosity also affected fouling (Jennings & Steinberg, 1997). The epiphytes, as they live at the surface of the seaweeds, will reduce the amount of light reaching the basiphyte thus impacting the algal photosynthesis (Sand-Jensen, 1977), compete for nutrients and gases, and reduce water motion surrounding the host, thus impacting the host growth (Buschmann & Gómez, 1993). Epiphytes will add weight to the host alga promoting the detachment of the entire, or part of it (*e.g.* Buschmann & Gómez, 1993; Kraberg & Norton, 2007). Epiphytes also decrease the reproductive effort of algae (*e.g. V. lanosa* on *A. nodosum*; Kraberg & Norton, 2007). Finally, epiphytes will release metabolites in the surrounding water that can promote or be detrimental to the host algae (see Harlin, 1987 for a review).

On the other hand, epiphytes may also have a positive effect on the basiphyte by, for example, reducing the desiccation experienced at low tide by the host or protecting the basiphyte from grazers (Karez, Engelbrecht, & Sommer, 2000; see following section). On the other side, as for the primary foulers, the seaweed host can exhibit a variety of defense against the epiphyte attachment (Ducker & Knox, 1984), including the production of a mucilaginous covering, a rapid growth, a sloughing of outer cell walls (Filion-Myklebust & Norton, 1981; Moss, 1982), allelopathy, or the release of toxic chemicals such as phenolic compounds (Rönnberg & Ruokolahti, 1986). However, this activity of phenolic compounds is sometimes controversial (Jennings & Steinberg, 1997). The algal host will thus modify its metabolism and produce compounds to eliminate or control the epi- and endophytes growing on or in its thallus. For example, red and brown macroalgae are known to produce terpenes with antifouling activity against microorganisms (see previous section) but also against animals. Schmitt, Hay, and Lindquist (1995) identified the diterpenes pachydictyol A and dictyol E in *Dictyota menstrualis* which were active against the settlement of algal epiphyte and bryozoan larvae (*Bugula neritina*). The red macroalga *Delisea pulchra* produces halogenated furanones which were the most effective natural product against the settlement of barnacles (de Nys et al., 1995).

Many filamentous epiphytes can also develop into endophytes through notably the expansion of rhizoids or the filaments embedded into the host thallus (Chirapart, Praiboon, Boonprab, & Puangsombat, 2018; Leonardi et al., 2006). A benthic diatom, *Pseudogomphonema* sp. has also been found living and reproducing into the red macroalga *Neoabbottiella* sp. (Klochkova et al., 2014). Generally, endophytes are bacteria, fungi or other algae that live into their host, but these relationships are species-specific from being innocuous up to pathogenic (Schoenrock, Amsler, McClintock, & Baker, 2013). There is also many metabolite exchanges between the endophyte and the seaweed host. Endophytes induce morphological modifications (such as kelp twisted stipes and blades; Lein, Sjøtun, & Wakili, 1991) of the host up to the development of galls (Apt, 1988). The brown algal endophyte *Laminariocolax aecidioides* (previously *Streblonema aecidioides*) causes tissue thickening in *S. latissima* (Peters & Schaffelke, 1996), but has no impact on the sorus maturation of *U. pinnatifida* (Gauna, Parodi, & Cáceres, 2009). The red macroalga *C. crispus* and *Mazzaella laminarioides* both developed severe lesions and cellular damages when infected by the green alga *Ulvella* (previously *Acrochaete*) *operculata* (Correa & McLachlan, 1992) and *Undaria ramosa* (previously *Endophyton ramosum*) (Correa, Flores & Garrido, 1994), respectively. As against bacterial fouling (see previous section), *C. crispus* resisted to the infection by *Undaria operculata* through the setting up of an oxidative burst, resulting from the contact between carrageenan and the green algal filaments (Weinberger et al., 2005); burst which is stronger in the gametophyte than the sporophyte of *C. crispus* (Bouarab, Potin, Correa, & Kloareg, 1999). One positive reported effect of the presence of epiphyte or endophyte for their host is the decrease in algal palatability for grazers (Amsler, Amsler, McClintock, & Baker, 2009; Karez et al., 2000), although both can sometimes attract grazers to the basiphyte (Campbell, Vergés, & Steinberg, 2014; Wahl, Hay, & Enderlein, 1997).

3.3 Grazing

In the coastal communities, grazing is a structuring element because herbivores are able of significantly affecting the amount of seaweed biomass available (Lubchenco & Gaines, 1981). For example, uncontrolled urchin populations may impact a seaweed community up to completely remove the macroalgae and leave "urchin barrens" where only encrusting algae may survive such as in Norway or Alaska (Steneck et al., 2002). Grazers reduced the growth and reproduction of seaweeds (*e.g. F. gardneri*; Dethier, Williams, & Freeman, 2005) and induced the breaking down of seaweeds by removing parts that weakened their thalli and making them more vulnerable to hydrodynamism (reviewed in Iken, 2012).

Some macroalgae modify their morphology in response to a herbivore presence: for example, transplantation of the brown macroalga *Padina sanctae-crucis* (previously *P. jamaicensis*) into a grazing-protected area resulted in a rapid morphological shift from prostrate, highly branched turf to an erect foliose morphology (Lewis, Norris, & Searles, 1987). Another algal morphological change to decrease grazing is the inclusion of calcium carbonate into the thallus often concomitant with the synthesis of chemical defense (Paul & Hay, 1986). Brown macroalgae produce phenolic compounds whose role as anti-grazing molecule has been either observed (*e.g.* Geiselman & McConnell, 1981) or not (*e.g.* Kubanek, Lester, Fenical, & Hay, 2004) depending on the algal or herbivore species. Brown seaweeds belonging to Dictyotales are known to produce many defensive secondary metabolites with many having anti-herbivorous activity (reviewed in Amsler & Fairhead, 2005): for example, *Dictyota* spp. produce terpenes and notably pachydictyol-A (diterpene alcohol) which showed anti-grazing activity against many fishes, sea urchins and amphipods (Hay & Fenical, 1992). Flöthe, Molis, Kruse, Weinberger, and John (2014) studied the molecular response of *F. vesiculosus* to grazing by the isopod *Idotea baltica*: algal defense was induced by the grazing and they found up to ~500 genes that were up (*e.g.* lipid, carbohydrate metabolism, defense and stress response) or down-regulated (such as photosynthesis) and indicative thus of algal metabolic changes. These changes were also herbivore specific as the grazing by a periwinkle (*Littorea obtusata*) showed different transcriptome changes (Flöthe, Molis, & John, 2014). The same isopod produced also inducive defense in three red macroalgae, *F. lumbricalis*, *Delesseria sanguinea* and *P. pseudoceranoides,* in feeding experiments (Rohde & Wahl, 2008). *D. pulchra* produced halogenated polyketide metabolites or furanones (de Nys, Wright, Konig, & Sticher, 1993), which protect the alga not only against fouling (see previous section) but also grazers (Wright, de Nys, Poore, & Steinberg, 2004). A rapid response of two kelp species (*L. spicata* and *Laminaria digitata*) to grazing by their main herbivores was observed by Ritter et al. (2017): seaweeds produced first sulfur containing amino acids followed by free fatty acids liberation and eicosanoid oxylipins synthesis likely representing metabolites related to stress. Sometimes the production of defense metabolites by the seaweed does not need a physical contact between the seaweed and the grazer: for example, *A. nodosum* when grazed will produce and release compounds in the surrounding seawater that will enable neighboring *A. nodosum* thalli to produce defense compounds even if they were not grazed (Toth & Pavia, 2000).

4. Conclusion

Macroalgae in rockyshores are subjected to various environmental factors, such as hydrodynamism, desiccation, light, salinity and temperature (Fig. 1). However, these different disturbances vary during the day (tidal cycle or nycthemeral cycle), the season, or the geographical area, impacting the distribution of algae, whether between shores, or within the same shore (vertical zonation). To cope with this highly variable environment, seaweeds must therefore have a high adaptative ability to respond to significant changes that can occur in just a few hours. They are thus able to adapt their photosynthesis, respiration or reproduction; and to synthesize a wide range of compounds, including pigments (photosynthetic or photoprotective), mycosporine-like amino acids, lipids, proteins, polysaccharides, or phenolic compounds, specific to each species (Table 2). Nevertheless, different stresses

Table 2 Examples of molecules synthetized by marine macroalgae, with their chemical structures.

Compounds	Chemical structure
Pigments	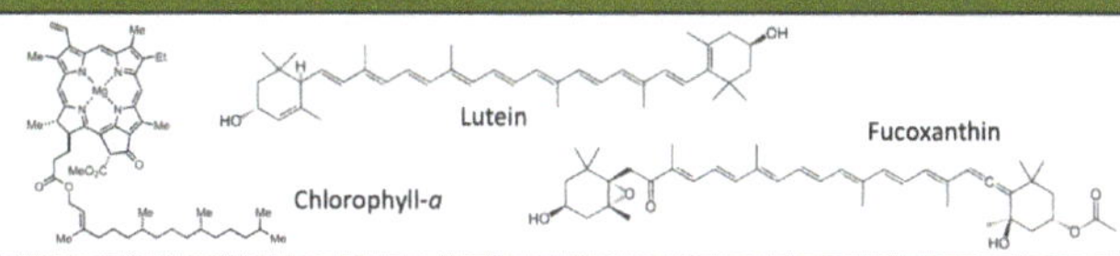
Carbohydrates	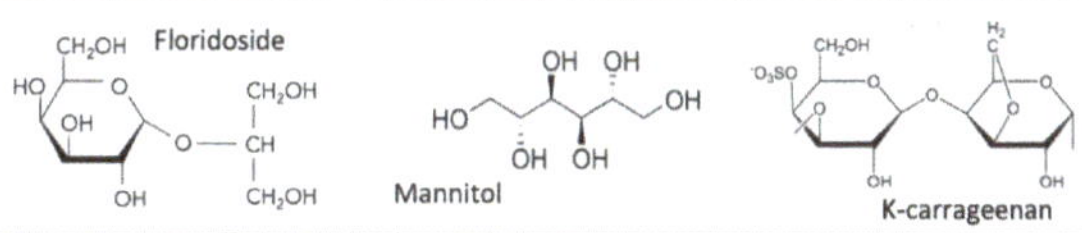
Lipids and fatty acids	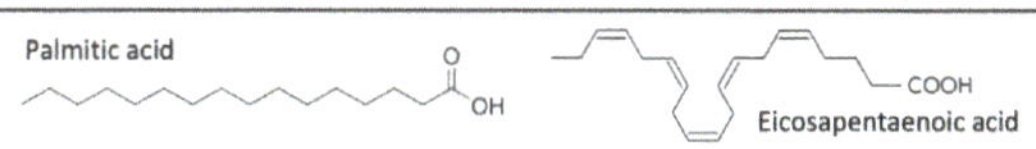
Terpenes	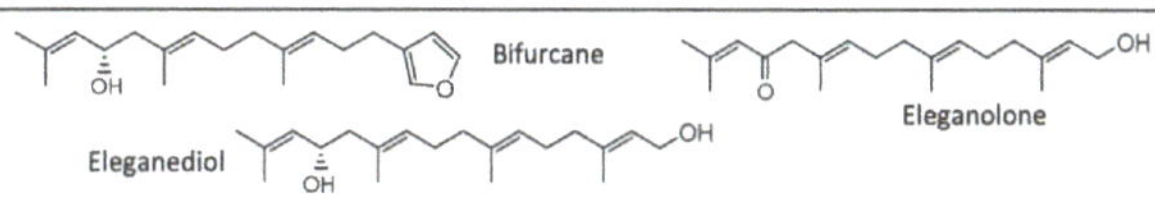
Phenolic compounds	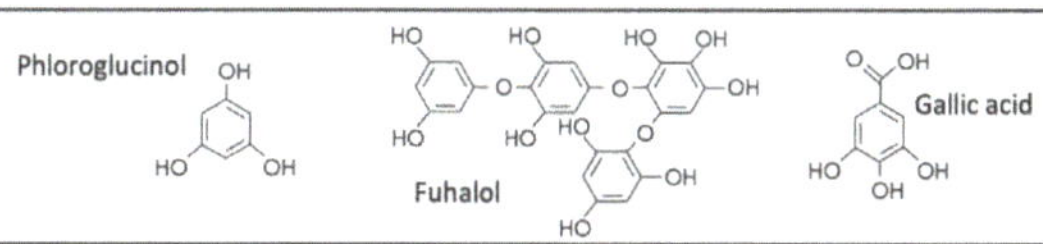
MAAs	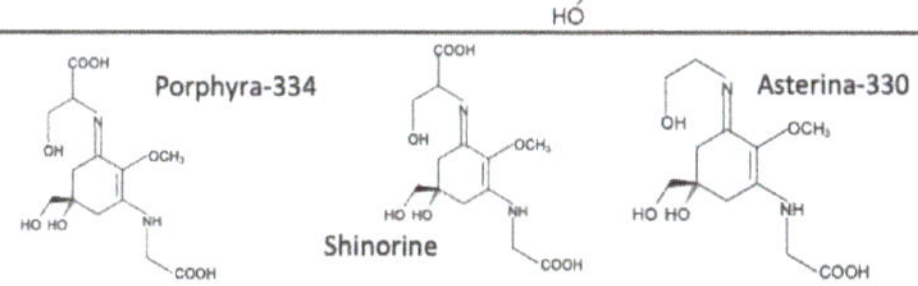

can act at the same time on the shore, and have a synergistic effect, making it complex to attribute the synthesis of a compound to a particular factor. In order to better understand the ecological and physiological responses of algae, it is therefore important to carry out field monitoring to study algae in their own environment with all factors; and to carry out culture experiments in the laboratory to test the impact of different stresses one by one or in combination.

In addition, by biometism, the study and understanding of the synthesis of these protective compounds can provide inspiration for valorization in various fields such as medicine, cosmetics (reviewed by Stiger-Pouvreau & Guérard, 2018), fertilizer, antifouling paints (Hellio, De La Broise, Dufosse, Le Gal, & Bourgougnon, 2001) or biofuel. Thus, algae have a high potential since they are a highly diversified resource that is much less studied than terrestrial plants. In this sense, there are, for example, cosmetic products with algal photoprotective compounds to reduce skin aging and the development of skin cancer: for example, the MAAs shinorine and porphyra-334 extracted from *Porphyra umbilicalis* in Helioguard 365 (Mibelle group, Switzerland), a commercial formulation used to protect the skin from UV radiation (La Barre, Roullier, & Boustie, 2014).

Finally, human activities are increasing, resulting in a stronger anthropogenic effect on the environment: they generate significant greenhouse gas emissions, which are mainly responsible of global warming and ocean acidification. These changes disrupt the physiology and ecology of many algal species (Walther, 2010). In particular, the geographical distribution of algae is guided by climatic limits; climate change induces recurrent exposition of species to extreme temperatures, variable in time and leading to not only a switch of species to northern latitudes, but also a succession of scattered local extinctions (*e.g.* in Portugal; Lima, Ribeiro, Queiroz, Hwakins, & Santos, 2007). Local communities are restructuring as species are added or deleted and some species may develop in areas where they could not previously survive or reproduce. In this sense, review of the different seaweed assemblages of each region have already been made (Garbary, 2001; Lüning, 1990; van den Hoeck, 1984), but it is important to carry out periodically local checklists as some species are expected to migrate or disappear with human activity and climate change. In particular, in recent years, the massive strandings of algae and the proliferation of invasive algae have increased, especially in Brittany (Stiger-Pouvreau & Thouzeau, 2015), weakening native algal fields. Climate change also affects interactions among marine organisms: for example, elevated temperature reduced herbivore defenses in *F. vesiculosus*

(Weinberger et al., 2011) and facilitate bacterial infections (Harley et al., 2012). Seaweed thus have to cope with these rapid changes in their environment.

References

Abe, S., Kurashima, A., Yokohama, Y., & Tanaka, J. (2001). The cellular ability of desiccation tolerance in Japanese intertidal seaweeds. *Botanica Marina, 44*, 125–131. https://doi.org/10.1515/BOT.2001.017.

Abreu, M. H., Pereira, R., Buschmann, A. H., Sousa-Pinto, I., & Yarish, C. (2011). Nitrogen uptake responses of *Gracilaria vermiculophylla* (Ohmi) Papenfuss under combined and single addition of nitrate and ammonium. *Journal of Experimental Marine Biology and Ecology, 407*, 190–199. https://doi.org/10.1016/j.jembe.2011.06.034.

Amsler, C. D., Amsler, M. O., McClintock, J. B., & Baker, B. J. (2009). Filamentous algal endophytes in macrophytic Antarctic algae: Prevalence in hosts and palatability to mesoherbivores. *Phycologia, 48*, 324–334. https://doi.org/10.2216/08-79.1.

Amsler, C. D., & Fairhead, V. A. (2005). Defensive and sensory chemical ecology of brown algae. *Advances in Botanical Research, 43*, 1–91.

Apt, K. E. (1988). Galls and tumor-like growth on marine macroalgae. *Diseases of Aquatic Organisms, 3*, 211–217.

Ar Gall, E., & Le Duff, M. (2014). Development of a quality index to evaluate the structure of macroalgal communities. *Estuarine, Coastal and Shelf Science, 139*, 99–109. https://doi.org/10.1016/j.ecss.2013.12.028.

Ayres-Ostrock, L. M., & Plastino, E. M. (2014). Effects of UV-B radiation on growth rates, pigment content and ultrastructure of red (wild type), greenish-brown and green strains of *Gracilaria birdiae* (Gracilariales, Rhodophyta). *European Journal of Phycology, 49*, 197–212. https://doi.org/10.1080/09670262.2014.904931.

Benes, K. M., & Bracken, M. E. S. (2016). Nitrate uptake varies with tide height and nutrient availability in the intertidal seaweed *Fucus vesiculosus*. *Journal of Phycology, 52*, 863–876. https://doi.org/10.1111/jpy.12454.

Bengtsson, M. M., Sjøtun, K., & Øvreås, L. (2010). Seasonal dynamics of bacterial biofilms on the kelp *Laminaria hyperborea*. *Aquatic Microbial Ecology, 60*, 71–83. https://doi.org/10.3354/ame01409.

Bérard-Therriault, L., & Cardinal, A. (1973). Importance de certains facteurs écologiques sur la résistance à la dessication des Fucacées (Phaeophyceae). *Phycologia, 12*, 41–52. https://doi.org/10.2216/i0031-8884-12-1-41.1.

Bhatia, S., Sharma, K., Sharma, A., & Purohit, A. P. (2011). Mycosporine and mycosporine-like amino acids: A paramount tool against ultra violet irradiation. *Pharmacognosy Reviews, 5*, 138–146. https://doi.org/10.4103/0973-7847.91107.

Binet, P. (1956). II. L'émersion des Algues marines : Étude physiologique [Emersion of marine algae: Physiological study]. *Bulletin de la Société Botanique de France, 103*, 386–400. https://doi.org/10.1080/00378941.1956.10837787.

Bird, C. E., Franklin, E. C., Smith, C. M., & Toonen, R. J. (2013). Between tide and wave marks: A unifying model of physical zonation on littoral shores. *PeerJ, 1*, e154. https://doi.org/10.7717/peerj.154.

Bird, C. J., & McLachlan, J. (1986). The effect of salinity on distribution of species of *Gracilaria* Grev. (Rhodophyta, Gigartinales): An experimental assessment. *Botanica Marina, 29*, 231–238. https://doi.org/10.1515/botm.1986.29.3.231.

Bischof, K., Hanelt, D., & Wiencke, C. (2000). Effects of ultraviolet radiation on photosynthesis and related enzyme reactions of marine macroalgae. *Planta, 211*, 555–562. https://doi.org/10.1007/s004250000313.

Blunden, G., Smith, B. E., Irons, M. W., Yang, M.-H., Roch, O. G., & Patel, A. V. (1992). Betaines and tertiary sulphonium compounds from 62 species of marine algae. *Biochemical Systematics and Ecology, 20*, 373–388. https://doi.org/10.1016/0305-1978(92)90050-N.

Boller, M. L., & Carrington, E. (2006). The hydrodynamic effects of shape and size change during reconfiguration of a flexible macroalga. *Journal of Experimental Biology, 209*, 1894–1903. https://doi.org/10.1242/jeb.02225.

Bolton, J. J. (1994). Global seaweed diversity: Patterns and anomalies. *Botanica Marina, 37*, 241–245. https://doi.org/10.1515/botm.1994.37.3.241.

Bouarab, K., Potin, P., Correa, J. A., & Kloareg, B. (1999). Sulphated oligosaccharides mediate the interaction between a marine red alga and its green algal pathogenic endopyhte. *Plant Cell, 11*, 1635–1650. https://doi.org/10.1105/tpc.11.9.1635.

Brenchley, J. L., Raven, J. A., & Johnston, A. M. (1997). Resource acquisition in two intertidal fucoid seaweeds, *Fucus serratus* and *Himanthalia elongata*: Seasonal variation and effects of reproductive development. *Marine Biology, 129*, 367–375. https://doi.org/10.1007/s002270050177.

Broberg, A., Kenne, L., & Pedersen, M. (1998). In-situ identification of major metabolites in the red alga *Gracilariopsis lemaneiformis* using high-resolution magic angle spinning nuclear magnetic resonance spectroscopy. *Planta, 206*, 300–307. https://doi.org/10.1007/s004250050404.

Buschmann, A. H., & Gómez, P. (1993). Interaction mechanisms between *Gracilaria chilensis* (Rhodophyta) and epiphytes. *Hydrobiologia, 260*, 345–351. https://doi.org/10.1007/BF00049039.

Campbell, A. H., Vergés, A., & Steinberg, P. D. (2014). Demographic consequences of disease in a habitat-forming seaweed and impacts on interactions between natural enemies. *Ecology, 95*, 142–152. https://doi.org/10.1890/13-0213.1.

Chambers, L. D., Stokes, K. R., Walsh, F. C., & Wood, R. J. K. (2006). Modern approaches to marine antifouling coatings. *Surface and Coatings Technology, 201*, 3642–3652. https://doi.org/10.1016/j.surfcoat.2006.08.129.

Chirapart, A., Praiboon, J., Boonprab, K., & Puangsombat, P. (2018). Epiphytism differences in the commercial species of *Gracilaria, G. fisheri, G. tenuistipitata*, and *G. salicornia*, from Thailand. *Journal of Applied Phycology, 30*, 3413–3423. https://doi.org/10.1007/s10811-018-1570-7.

Connan, S., Deslandes, E., & Ar Gall, E. A. (2007). Influence of day-night and tidal cycles on phenol content and antioxidant capacity in three temperate intertidal brown seaweeds. *Journal of Experimental Marine Biology and Ecology, 349*, 359–369. https://doi.org/10.1016/j.jembe.2007.05.028.

Connan, S., & Stengel, D. B. (2011a). Impacts of ambient salinity and copper on brown algae: 1. Interactive effects on photosynthesis, growth, and copper accumulation. *Aquatic Toxicology, 104*, 94–107. https://doi.org/10.1016/j.aquatox.2011.03.015.

Connan, S., & Stengel, D. B. (2011b). Impacts of ambient salinity and copper on brown algae: 2. Interactive effects on phenolic pool and assessment of metal binding capacity of phlorotannin. *Aquatic Toxicology, 104*, 1–13. https://doi.org/10.1016/j.aquatox.2011.03.016.

Contreras-Porcia, L., López-Cristoffanini, C., Meynard, A., & Kumar, M. (2017). Tolerance pathways to desiccation stress in seaweeds. In M. Kumar, & P. Ralph (Eds.), *Systems biology of marine ecosystems* (pp. 13–33). https://doi.org/10.1007/978-3-319-62094-7_2.

Contreras-Porcia, L., Thomas, D., Flores, V., & Correa, J. A. (2011). Tolerance to oxidative stress induced by desiccation in *Porphyra columbina* (Bangiales, Rhodophyta). *Journal of Experimental Botany, 62*, 1815–1829. https://doi.org/10.1093/jxb/erq364.

Contreras, L., Mella, D., Moenne, A., & Correa, J. A. (2009). Differential responses to copper-induced oxidative stress in the marine macroalgae *Lessonia nigrescens* and *Scytosiphon*

lomentaria (Phaeophyceae). *Aquatic Toxicology, 94*, 94–102. https://doi.org/10.1016/j.aquatox.2009.06.004.
Contreras, L., Moenne, A., Gaillard, F., Potin, P., & Correa, J. A. (2010). Proteomic analysis and identification of copper stress-regulated proteins in the marine alga *Scytosiphon gracilis* (Phaeophyceae). *Aquatic Toxicology, 96*, 85–89. https://doi.org/10.1016/j.aquatox.2009.10.007.
Corey, P., Kim, J. K., Duston, J., Garbary, D. J., & Prithiviraj, B. (2013). Bioremediation potential of *Palmaria palmata* and *Chondrus crispus* (basin head): Effect of nitrate and ammonium ratio as nitrogen source on nutrient removal. *Journal of Applied Phycology, 25*, 1349–1358. https://doi.org/10.1007/s10811-013-9977-7.
Correa, J. A., Flores, V., & Garrido, J. (1994). Green patch disease in *Iridaea laminarioides* (Rhodophyta) caused by *Endophyton* sp. (Chlorophyta). *Diseases of Aquatic Organisms, 19*, 203–213. https://doi.org/10.3354/dao019203.
Correa, J. A., & McLachlan, J. L. (1992). Endophytic algae of *Chondrus crispus* (Rhodophyta). IV. Effects on the host following infections by *Acrochaete operculata* and *A. heteroclada* (Chlorophyta). *Marine Ecology Progress Series, 81*, 73–87. https://doi.org/10.3354/meps081073.
Craigie, J. S., Morris, E. R., Rees, D. A., & Thom, D. (1984). Alginate block structure in Phaeophyceae from Nova Scotia: Variation with species, environment and tissue-type. *Carbohydrate Polymers, 4*, 237–252. https://doi.org/10.1016/0144-8617(84)90001-8.
Dethier, M. N., Williams, S. L., & Freeman, A. (2005). Seaweeds under stress: Manipulated stress and herbivory affect critical life-history functions. *Ecological Monographs, 75*, 403–418. https://doi.org/10.1890/03-4108.
Ding, L., Ma, Y., Huang, B., & Chen, S. (2013). Effects of seawater salinity and temperature on growth and pigment contents in *Hypnea cervicornis* J. Agardh (Gigartinales, Rhodophyta). *BioMed Research International*, 1–10. https://doi.org/10.1155/2013/594308.
Dring, M. J., & Brown, F. A. (1982). Photosynthesis of intertidal brown algae during and after periods of emersion: A renewed search for physiological causes of zonation. *Marine Ecology Progress Series, 8*, 301–308. https://doi.org/10.3354/meps008301.
Ducker, S. C., & Knox, R. B. (1984). Epiphytism at the cellular level with special references to algal epiphytes. In H.-F. Linskens, & J. Heslop-Harrison (Eds.), *Encyclopedia of plant physiology* (pp. 113–133). Berlin: Springer.
Dumay, J., Morancais, M., Munier, M., Le Guillard, C., & Fleurence, J. (2014). Phycoerythrins: Valuable proteinic pigments in red seaweeds. *Advances in Botanical Reseach, 71*, 321–343. https://doi.org/10.1016/B978-0-12-408062-1.00011-1.
Egan, S., Harder, T., Burke, C., Steinberg, P., Kjelleberg, S., & Thomas, T. (2013). The seaweed holobiont: Understanding seaweed–bacteria interactions. *FEMS Microbiology Reviews, 37*, 462–476. https://doi.org/10.1111/1574-6976.12011.
Eggert, A. (2012). Seaweed responses to temperature. In C. Wiencke, & K. Bischof (Eds.), *Seaweed biology. Ecological studies (analysis and synthesis)* (pp. 47–66). Berlin, Heidelberg: Springer.
Einav, R., Breckle, S., & Beer, S. (1995). Ecophysiological adaptation strategies of some intertidal marine macroalgae of the Israeli Mediterranean coast. *Marine Ecology Progress Series, 125*, 219–228.
Falkowski, P. G., & Raven, J. A. (2013). An introduction to photosynthesis in aquatic systems. In P. G. Falkowski, & J. A. Raven (Eds.), *Aquatic photosynthesis* (2nd ed., pp. 1–43). Oxford: Princeton University Press.
Fattorusso, E., & Piatelli, M. (1980). Amino acids from marine algae. In P. J. Scheuer (Ed.), *Marine natural products: Chemical and biological perspectives* (pp. 95–140). Elsevier.
Fierro, C., López-Cristoffanini, C., Meynard, A., Lovazzano, C., Castañeda, F., Guajardo, E., et al. (2017). Expression profile of desiccation tolerance factors in intertidal seaweed

species during the tidal cycle. *Planta, 245*, 1149–1164. https://doi.org/10.1007/s00425-017-2673-0.

Filion-Myklebust, C., & Norton, T. (1981). Epidermis shedding in the brown seaweed *Ascophyllum nodosum* (L.) Le Jolis, and its ecological significance. *Marine Biology Letters, 2*, 45–51.

Flores-Molina, M. R., Thomas, D., Lovazzano, C., Núñez, A., Zapata, J., Kumar, M., et al. (2014). Desiccation stress in intertidal seaweeds: Effects on morphology, antioxidant responses and photosynthetic performance. *Aquatic Botany, 113*, 90–99. https://doi.org/10.1016/j.aquabot.2013.11.004.

Flöthe, C. R., Molis, M., & John, U. (2014). Induced resistance to periwinkle grazing in the brown seaweed *Fucus vesiculosus* (Phaeophyceae): Molecular insights and seaweed-mediated effects on herbivore interactions. *Journal of Phycology, 50*, 564–576. https://doi.org/10.1111/jpy.12186.

Flöthe, C. R., Molis, M., Kruse, I., Weinberger, F., & John, U. (2014). Herbivore-induced defence response in the brown seaweed *Fucus vesiculosus* (Phaeophyceae): Temporal pattern and gene expression. *European Journal of Phycology, 49*, 356–369. https://doi.org/10.1080/09670262.2014.946452.

Franklin, L. A., Osmond, C. B., & Larkum, A. W. (2003). Photoinhibition, UV-B and algal photosynthesis. In A. W. D. Larkum, S. E. Douglas, & J. A. Raven (Eds.), *Photosynthesis in algae* (pp. 351–384). Dordrecht: Springer.

Fujita, R. M., Wheeler, P. A., & Edwards, R. L. (1989). Assessment of macroalgal nitrogen limitation in a seasonal upwelling region. *Marine Ecology Progress Series, 53*, 293–303. https://doi.org/10.3354/meps053293.

Fuller, S. W., & Mathieson, A. C. (1972). Ecological studies of eco- nomic red algae. IV. Variations of carrageenan concentration and properties in *Chondrus crispus* Stackhouse. *Journal of Experimental Marine Biology and Ecology, 10*, 49–58. https://doi.org/10.1016/0022-0981(72)90092-5.

Gagné, J. A., Mann, K. H., & Chapman, A. R. O. (1982). Seasonal patterns of growth and storage in *Laminaria longicruris* in relation to differing patterns of availability of nitrogen in the water. *Marine Biology, 69*, 91–101. https://doi.org/10.1007/2FBF00396965.

Gao, S., & Wang, G. (2012). The enhancement of cyclic electron flow around photosystem I improves the recovery of severely desiccated *Porphyra yezoensis* (Bangiales, Rhodophyta). *Journal of Experimental Botany, 63*, 4349–4358. https://doi.org/10.1093/jxb/ers082.

Garbary, D. J. (2001). Biogeography of marine algae. In *Encyclopedia of Life Sciences* (pp. 1–9).

Garbary, D. J., Miller, A. G., & Scrosati, R. A. (2014). *Ascophyllum nodosum* and its symbionts: XI. The epiphyte *Vertebrata lanosa* performs better photosynthetically when attached to *Ascophyllum* than when alone. *Algae, 29*, 321–331. https://doi.org/10.4490/algae.2014.29.4.321.

Garman, G. D., Pillai, M. M., & Cherr, G. N. (1994). Inhibition of cellular events during early algal gametophyte development: Effects of select metals and an aqueous petroleum waste. *Aquatic Toxicology, 28*, 127–144. https://doi.org/10.1016/0166-445X(94)90025-6.

Gauna, M. C., Parodi, E. R., & Cáceres, E. J. (2009). Epi-endophytic symbiosis between *Laminariocolax aecidioides* (Ectocarpales, Phaeophyceae) and *Undaria pinnatifida* (Laminariales, Phaeophyceae) growing on Argentinian coasts. *Journal of Applied Phycology, 21*, 11–18. https://doi.org/10.1007/s10811-007-9298-9.

Geiselman, J. A., & McConnell, O. J. (1981). Polyphenols in brown algae *Fucus vesiculosus* and *Ascophyllum nodosum*: Chemical defenses against the marine herbivorous snail, *Littorina littorea*. *Journal of Chemical Ecology, 7*, 1115–1133.

Gerard, V. A. (1982). *In situ* water motion and nutrient uptake by the giant kelp *Macrocystis pyrifera*. *Marine Biology, 69*, 51–54. https://doi.org/10.1007/BF00396960.

Gessner, F. (1969). Photosynthesis and ion loss in the brown algae *Dictyopteris membranacea* and *Fucus virsoides*. *Marine Biology, 4*, 349–351. https://doi.org/10.1007/BF00350365.
Goecke, F., Labes, A., Wiese, J., & Imhoff, J. F. (2010). Chemical interactions between marine macroalgae and bacteria. *Marine Ecology Progress Series, 409*, 267–300. https://doi.org/10.3354/meps08607.
Gordon, R., & Brawley, S. H. (2004). Effects of water motion on propagule release from algae with complex life histories. *Marine Biology, 145*, 21–29. https://doi.org/10.1007/s00227-004-1305-y.
Goulard, F., Diouris, M., Quéré, G., Deslandes, E., & Floc'h, J.-Y. (2001). Salinity effects on NDP-sugars, floridoside, starch, and carrageenan yield, and UDP-glucose-pyrophosphorylase and epimerase activities of cultivated *Solieria chordalis*. *Journal of Plant Physiology, 158*, 1387–1394. https://doi.org/10.1078/0176-1617-00569.
Guajardo, E., Correa, J. A., & Contreras-Porcia, L. (2016). Role of abscisic acid (ABA) in activating antioxidant tolerance responses to desiccation stress in intertidal seaweed species. *Planta, 243*, 767–781. https://doi.org/10.1007/s00425-015-2438-6.
Guihéneuf, F., Gietl, A., & Stengel, D. B. (2018). Temporal and spatial variability of mycosporine-like amino acids and pigments in three edible red seaweeds from western Ireland. *Journal of Applied Phycology, 30*, 2573–2586. https://doi.org/10.1007/s10811-018-1436-z.
Guiry, M. D., & Guiry, G. M. (2019). *AlgaeBase. World-wide electronic publication*. Galway: National University of Ireland. http://www.algaebase.org.
Hanelt, D., & Figueroa, F. L. (2012). Physiological and photomorphogenic effects of light on marine macrophytes. In C. Wiencke, & K. Bischof (Eds.), *Seaweed biology* (pp. 3–23). Berlin, Heidelberg: Springer.
Hanelt, D., & Nultsch, W. (2003). Photoinhibition in seaweeds. In G. Heldmaier, & D. Werner (Eds.), *Environmental signal processing and adaptation* (pp. 141–167). Berlin, Heidelberg: Springer.
Harley, C. D., Anderson, K. M., Demes, K. W., Jorve, J. P., Kordas, R. L., & Coyle, T. A. (2012). Effects of climate change on global seaweed communities. *Journal of Phycology, 48*, 1064–1078. https://doi.org/10.1111/j.1529-8817.2012.01224.x.
Harlin, M. M. (1987). Allelochemistry in marine macroalgae. *Critical Reviews in Plant Sciences, 5*, 237–249. https://doi.org/10.1080/07352688709382241.
Harnedy, P. A., & Fitzgerald, R. J. (2011). Bioactive proteins, peptides, and amino acids from macroalgae. *Journal of Phycology, 47*, 218–232. https://doi.org/10.1111/j.1529-8817.2011.00969.x.
Harrison, P., & Hurd, C. (2001). Nutrient physiology of seaweeds: Application of concepts to aquaculture. *Cahiers De Biologie Marine, 42*, 71–82.
Hashimoto, H., Uragami, C., & Cogdell, R. J. (2016). Carotenoids and photosynthesis. In C. Stange (Ed.), *Carotenoids in nature: Biosynthesis, regulation and function* (pp. 111–139). Switzerland: Springer.
Hay, M. E., & Fenical, W. (1992). Chemical mediation of seaweed-herbivore interactions. In V. M. John, S. J. Hawkins, & J. H. Price (Eds.), *Plant-animal interactions in the marine benthos* (pp. 317–337). Oxford, England: Clarendon Press.
Heisler, J., Glibert, P. M., Burkholder, J. M., Anderson, D. M., Cochlan, W., Dennison, W. C., et al. (2008). Eutrophication and harmful algal blooms: A scientific consensus. *Harmful Algae, 8*, 3–13. https://doi.org/10.1016/j.hal.2008.08.006.
Hellio, C., Bergé, J. P., Beaupoil, C., Le Gal, Y., & Bourgougnon, N. (2002). Screening of marine algal extracts for antisettlement activities against microalgae and macroalgae. *Biofouling, 18*, 205–215. https://doi.org/10.1080/08927010290010137.
Hellio, C., De La Broise, D., Dufosse, L., Le Gal, Y., & Bourgougnon, N. (2001). Inhibition of marine bacteria by extracts of macroalgae: Potential use for environmentally friendly

antifouling paints. *Marine Environmental Research, 52*, 231–247. https://doi.org/10.1016/S0141-1136(01)00092-7.
Holzinger, A., Lütz, C., Karsten, U., & Wiencke, C. (2004). The effect of ultraviolet radiation on ultrastructure and photosynthesis in the red macroalgae *Palmaria palmata* and *Odonthalia dentata* from Arctic waters. *Plant Biology, 6*, 568–577. https://doi.org/10.1055/s-2004-821003.
Huner, N. P. A., Öquist, G., & Sarhan, F. (1998). Energy balance and acclimation to light and cold. *Trends in Plant Science, 3*, 224–230. https://doi.org/10.1016/S1360-1385(98)01248-5.
Hunt, L. J. H., & Denny, M. W. (2008). Desiccation protection and disruption: A trade-off for an intertidal marine alga. *Journal of Phycology, 44*, 1164–1170. https://doi.org/10.1111/j.1529-8817.2008.00578.x.
Hurd, C. L. (2000). Water motion, marine macroalgal physiology, and production. *Journal of Phycology, 36*, 453–472. https://doi.org/10.1046/j.1529-8817.2000.99139.x.
Hurd, C. L., Berges, J. A., Osborne, J., & Harrison, P. J. (1995). An *in vitro* nitrate reductase assay for marine macroalgae: Optimization and characterization of the enzyme for *Fucus gardneri* (Phaeophyta). *Journal of Phycology, 31*, 835–843. https://doi.org/10.1111/j.0022-3646.1995.00835.x.
Hurd, C. L., & Dring, M. J. (1991). Desiccation and phosphate uptake by intertidal fucoid algae in relation to zonation. *British Phycological Journal, 26*, 327–333. https://doi.org/10.1080/00071619100650291.
Hurd, C. L., Harrison, P. J., Bischof, K., & Lobban, C. S. (2014). *Seaweed ecology and physiology*. Cambridge University Press, 551 pp.
Iken, K. (2012). Grazers on benthic seaweeds. In C. Wiencke, & K. Bishof (Eds.), *Seaweed biology* (pp. 157–175). Berlin, Heidelberg: Springer.
Impellizzeri, G., Mangiafico, S., Oriente, G., Piattelli, M., Sciuto, S., Fattorusso, E., et al. (1975). Amino acids and low-molecular-weight carbohydrates of some marine red algae. *Phytochemistry, 14*, 1549–1557. https://doi.org/10.1016/0031-9422(75)85349-0.
Inaba, M., Suzuki, I., Szalontai, B., Kanesaki, Y., Los, D. A., Hayashi, H., et al. (2003). Gene-engineered rigidification of membrane lipids enhances the cold inducibility of gene expression in *Synechocystis*. *Journal of Biological Chemistry, 278*, 12191–12198. https://doi.org/10.1074/jbc.M212204200.
Ireland, H. E., Harding, S. J., Bonwick, G. A., Jones, M., Smith, C. J., & Williams, J. H. H. (2004). Evaluation of heat shock protein 70 as a biomarker of environmental stress in *Fucus serratus* and *Lemna minor*. *Biomarkers, 9*, 139–155. https://doi.org/10.1080/13547500410001732610.
Jennings, J. G., & Steinberg, P. D. (1997). Phlorotannins versus other factors affecting epiphyte abundance on the kelp *Ecklonia radiata*. *Oecologia, 109*, 461–473. https://doi.org/10.1007/s004420050106.
Ji, D., Li, B., Xu, Y., Chen, C., & Xie, C. (2015). Cloning and quantitative analysis of five heat shock protein 70 genes from *Pyropia haitanensis*. *Journal of Applied Phycology, 27*, 499–509. https://doi.org/10.1007/s10811-014-0314-6.
Karentz, D. (2001). Chemical defenses of marine organisms against solar radiation exposure: UV-absorbing mycosporine-like amino acids and scytonemin. In J. McClintock, & B. J. Baker (Eds.), *Marine chemical ecology* (pp. 481–520). Boca Raton: CRC Press.
Karez, R., Engelbrecht, S., & Sommer, U. (2000). Co-consumption and protective coating: Two new proposed effects of epiphytes on their macroalgal hosts in mesograzer-epiphyte-host interactions. *Marine Ecology Progress Series, 205*, 85–93. https://doi.org/10.3354/meps205085.
Karsten, U. (2007). Salinity tolerance of Arctic kelps from Spitsbergen. *Phycological Research, 55*, 257–262. https://doi.org/10.1111/j.1440-1835.2007.00468.x.

Karsten, U. (2008). Defense strategies of algae and cyanobacteria against solar ultraviolet radiation. In C. D. Amsler (Ed.), *Algal chemical ecology* (pp. 273–296). Berlin: Springer.
Karsten, U., West, J. A., Mostaert, A. S., King, R. J., Barrow, K. D., & Kirst, G. O. (1992a). Mannitol in the red algal genus *Caloglossa* (Harvey) J. Agardh. *Journal of Plant Physiology, 140*, 292–297.
Karsten, U., West, J. A., & Zuccarello, G. (1992b). Polyol content of *Bostrychia* and *Stictosiphonia* (Rhodomelaceae, Rhodophyta) from field and culture. *Botanica Marina, 35*, 11–19. https://doi.org/10.1515/botm.1992.35.1.11.
Kerjean, V., Morel, B., Stiger, V., Bessieres, M.-A., Simon-Colin, C., Magné, C., et al. (2007). Optimization of floridoside production in the red alga *Mastocarpus stellatus*: Pretreatment, extraction and seasonal variations. *Botanica Marina, 50*, 59–64. https://doi.org/10.1515/BOT.2007.005.
Kim, J. K., Kraemer, G. P., Neefus, C. D., Chung, I. K., & Yarish, C. (2007). Effects of temperature and ammonium on growth, pigment production and nitrogen uptake by four species of *Porphyra* (Bangiales, Rhodophyta) native to the New England coast. *Journal of Applied Phycology, 19*, 431–440. https://doi.org/10.1007/s10811-006-9150-7.
Kim, J. K., Kraemer, G. P., & Yarish, C. (2013). Emersion induces nitrogen release and alteration of nitrogen metabolism in the intertidal genus *Porphyra*. *PLoS One, 8*, e69961. https://doi.org/10.1371/journal.pone.0069961.
Kirst, G. O. (1990). Salinity tolerance of eukaryotic marine algae. *Annual Review of Plant Physiology and Plant Molecular Biology, 41*, 21–53. https://doi.org/10.1146/annurev.pp.41.060190.000321.
Kloareg, B., & Quatrano, R. S. (1988). Structure of the cell walls of marine algae and ecophysiological functions of the matrix polysaccharides. *Oceanography and Marine Biology an Annual Review, 26*, 259–315.
Klochkova, T. A., Pisareva, N. A., Park, J. S., Lee, J. H., Han, J. W., Klochkova, N. G., et al. (2014). An endophytic diatom, *Pseudogomphonema* sp. (Naviculaceae, Bacillariophyceae), lives inside the red alga *Neoabbottiella* (Halymeniaceae, Rhodophyta). *Phycologia, 53*, 205–214. https://doi.org/10.2216/13-229.1.
Korbee, N., Huovinen, P., Figueroa, F. L., Aguilera, J., & Karsten, U. (2005). Availability of ammonium influences photosynthesis and the accumulation of mycosporine-like amino acids in two *Porphyra* species (Bangiales, Rhodophyta). *Marine Biology, 146*, 645–654. https://doi.org/10.1007/s00227-004-1484-6.
Kraberg, A. C., & Norton, T. A. (2007). Effect of epiphytism on reproductive and vegetative lateral formation in the brown, intertidal seaweed *Ascophyllum nodosum* (Phaeophyceae). *Phycological Research, 55*, 17–24. https://doi.org/10.1111/j.1440-1835.2006.00441.x.
Kraemer, G. P., & Chapman, J. C. (1991). Biomechanics and alginic acid composition during hydronynamic adaptation by *Egregia menziesii* (Phaeophyta) juveniles. *Journal of Phycology, 27*, 47–53.
Krause-Jensen, D., & Duarte, C. M. (2016). Substantial role of macroalgae in marine carbon sequestration. *Nature Geoscience, 9*, 737–742. https://doi.org/10.1038/ngeo2790.
Kristensen, I. (1968). Surf influence on the thallus of fucoids and the rate of desiccation. *Sarsia, 34*, 69–82. https://doi.org/10.1080/00364827.1968.10413372.
Kubanek, J., Lester, S. E., Fenical, W., & Hay, M. E. (2004). Ambigous role of phlorotannins as chemical defenses in the brown alga *Fucus vesiculosus*. *Marine Ecology Progress Series, 277*, 79–93. https://doi.org/10.3354/meps277079.
Küpper, F. C., Kloareg, B., Guern, J., & Potin, P. (2001). Oligoguluronates elicit an oxidative burst in the brown algal kelp *Laminaria digitata*. *Plant Physiology, 125*, 278–291. https://doi.org/10.1104/pp.125.1.278.
Küpper, F. C., Muller, D. G., Peters, A. F., Kloareg, B., & Potin, P. (2002). Oligoalginate recognition and oxidative burst play a key role in natural and induced resistance of

sporophytes of Laminariales. *Journal of Chemical Ecology, 28*, 2057–2081. https://doi.org/10.1023/A:1020706129624.
La Barre, S., Roullier, C., & Boustie, J. (2014). Mycosporine-like amino acids (MAAs) in biological photosystems. In S. La Barre, & J. M. Kornprobst (Eds.), *Outstanding marine molecules: Chemistry, biology, analysis* (pp. 333–360). Weinheim: Wiley-Blackwell.
Lalegerie, F., Lajili, S., Bedoux, G., Taupin, L., Stiger-Pouvreau, V., & Connan, S. (2019). Photo-protective compounds in red macroalgae from Brittany: Considerable diversity in mycosporine-like amino acids (MAAs). *Marine Environmental Research, 147*, 37–48. https://doi.org/10.1016/j.marenvres.2019.04.001.
Lamote, M., Johnson, L. E., & Lemoine, Y. (2012). Photosynthetic responses of an intertidal alga to emersion: The interplay of intertidal height and meteorological conditions. *Journal of Experimental Marine Biology and Ecology, 428*, 16–23. https://doi.org/10.1016/j.jembe.2012.05.021.
Latowski, D., Kuczyńska, P., & Strzałka, K. (2011). Xanthophyll cycle – a mechanism protecting plants against oxidative stress. *Redox Report, 16*, 78–90. https://doi.org/10.1179/174329211X13020951739938.
Le Lann, K., Rumin, J., Cérantola, S., Culioli, G., & Stiger-Pouvreau, V. (2014). Spatiotemporal variations of diterpene production in the brown macroalga *Bifurcaria bifurcata* from the western coasts of Brittany (France). *Journal of Applied Phycology, 26*, 1207–1214. https://doi.org/10.1007/s10811-013-0148-7.
Lein, T. E., Sjøtun, K., & Wakili, S. (1991). Mass-occurrence of a brown filamentous endophyte in the lamina of the kelp *Laminaria hyperborea* (Gunnerus) Foslie along the southwestern coast of Norway. *Sarsia, 76*, 187–193. https://doi.org/10.1080/00364827.1991.10413474.
Leonardi, P. I., Miravalles, A. B., Faugeron, S., Verónica, F., Beltrán, J., & Correa, J. A. (2006). Diversity, phenomenology and epidemiology of epiphytism in farmed *Gracilaria chilensis* (Rhodophyta) in northern Chile. *European Journal of Phycology, 41*, 247–257. https://doi.org/10.1080/09670260600645659.
Levitus, S., Conkright, M. E., Reid, J. L., Najjar, R. G., & Mantyl, A. (1993). Distribution of nitrate, phosphate and silicate in the world oceans. *Progress in Oceanography, 31*, 245–273. https://doi.org/10.1016/0079-6611(93)90003-V.
Lewis, J. A. (1998). Marine biofouling and its prevention on underwater surfaces. *Materials Forum, 22*, 41–61.
Lewis, S. M., Norris, J. N., & Searles, R. B. (1987). The regulation of morphological plasticity in tropical reef algae by herbivory. *Ecology, 68*, 636–641.
Lima, F. P., Ribeiro, P. A., Queiroz, N., Hawkins, S. J., & Santos, A. M. (2007). Do distributional shifts of northern and southern species of algae match the warming pattern? *Global Change Biology, 13*, 2592–2604. https://doi.org/10.1111/j.1365-2486.2007.01451.x.
López-Cristoffanini, C., Zapata, J., Gaillard, F., Potin, P., Correa, J. A., & Contreras-Porcia, L. (2015). Identification of proteins involved in desiccation tolerance in the red seaweed *Pyropia orbicularis* (Rhodophyta, Bangiales). *Proteomics, 15*, 3954–3968. https://doi.org/10.1002/pmic.201400625.
Los, D. A., & Murata, N. (2004). Membrane fluidity and its roles in the perception of environmental signals. *Biochimica et Biophysica Acta (BBA)-Biomembranes, 1666*, 142–157. https://doi.org/10.1016/j.bbamem.2004.08.002.
Lubchenco, J., & Gaines, S. D. (1981). A unified approach to marine plant-herbivore interactions. I. Populations and communities. *Annual Review of Ecology and Systematics, 12*, 405–437. https://doi.org/10.1146/annurev.es.12.110181.002201.
Lüning, K. (1990). *Seaweeds: Their environment, biogeography and ecophysiology* (p. 527pp). New York: John Wiley and Sons.

Lüning, K., & Dieck, I. T. (1989). Environmental triggers in algal seasonality. *Botanica Marina, 32*, 389–398. https://doi.org/10.1515/botm.1989.32.5.389.

Mancuso, F. P., D'Hondt, S., Willems, A., Airoldi, L., & De Clerck, O. (2016). Diversity and temporal dynamics of the epiphytic bacterial communities associated with the canopy-forming seaweed *Cystoseira compressa* (Esper) Gerloff and Nizamuddin. *Frontiers in Microbiology, 7*, 476. https://doi.org/10.3389/fmicb.2016.00476.

Manríquez-Hernández, J. A., Duston, J., & Garbary, D. J. (2016). Effect of irradiance on bioremediation capacity of the red algae *Chondrus crispus* and *Palmaria palmata. Aquaculture International, 24*, 39–55. https://doi.org/10.1007/s10499-015-9907-6.

Martínez, B., & Rico, J. M. (2002). Seasonal variation of P content and major N pools in *Palmaria palmata* (Rhodophyta). *Journal of Phycology, 38*, 1082–1089. https://doi.org/10.1046/j.1529-8817.2002.01217.x.

Martínez, B., & Rico, J. M. (2004). Inorganic nitrogen and phosphorus uptake kinetics in *Palmaria palmata* (Rhodophyta). *Journal of Phycology, 40*(4), 642–650. https://doi.org/10.1111/j.1529-8817.2004.03116.x.

Marzinelli, E. M., Campbell, A. H., Zozaya Valdes, E., Vergés, A., Nielsen, S., Wernberg, T., et al. (2015). Continental-scale variation in seaweed host- associated bacterial communities is a function of host condition, not geography. *Environmental Microbiology, 17*, 4078–4088. https://doi.org/10.1111/1462-2920.12972.

Matta, J. L., & Chapman, D. J. (1995). Effects of light, temperature and desiccation on the net emersed productivity of the intertidal macroalga *Colpomenia peregrina* Sauv. (Hamel). *Journal of Experimental Marine Biology and Ecology, 189*, 13–27. https://doi.org/10.1016/0022-0981(95)00005-C.

Molenaar, F. J., & Breeman, A. M. (1994). Ecotypic variation in *Phyllophora pseudoceranoides* (Rhodophyta) ensures winter reproduction througout its geographic range. *Journal of Phycology, 30*, 392–402. https://doi.org/10.1111/j.0022-3646.1994.00392.x.

Molenaar, F. J., & Breeman, A. M. (1997). Latitudinal trends in the growth and reproductive seasonality of *Delesseria sanguinea, Membranoptera alata,* and *Phycodrys rubens* (Rhodophyta). *Journal of Phycology, 33*, 330–343. https://doi.org/10.1111/j.0022-3646.1997.00330.x.

Moss, B. L. (1982). The control of epiphytes by *Halidrys siliquosa* (L.) Lyngb. (Phaeophyta, Cystoseiraceae). *Phycologia, 21*, 185–188. https://doi.org/10.2216/i0031-8884-21-2-185.1.

Munda, I. M. (1978). Salinity dependent distribution of benthic algae in estuarine areas of Icelandic fjords. *Botanica Marina, 21*, 451–468. https://doi.org/10.1515/botm.1978.21.7.451.

Naldi, M., & Wheeler, P. A. (1999). Changes in nitrogen pools in *Ulva fenestrata* (Chlorophyta) and *Gracilaria pacifica* (Rhodophyta) under nitrate and ammonium enrichment. *Journal of Phycology, 35*, 70–77. https://doi.org/10.1046/j.1529-8817.1999.3510070.x.

Nielsen, H. D., Brownlee, C., Coelho, S. M., & Brown, M. M. (2003). Inter-population differences in inherited copper tolerance involve photosynthetic adaptation and exclusion mechanisms in *Fucus serratus. New Phytologist, 160*, 157–165. https://doi.org/10.1046/j.1469-8137.2003.00864.x.

Nriagu, J. O. (1990). Global metal pollution - poisoning the biosphere? *Environment: Science and Policy for Sustainable Development, 32*, 7–33. https://doi.org/10.1080/00139157.1990.9929037.

de Nys, R., Steinberg, P. D., Willemsen, P., Dworjanyn, S. A., Gabelish, C. L., & King, R. J. (1995). Broad spectrum effects of secondary metabolites from the red alga *Delisea pulchra* in antifouling assays. *Biofouling, 8*, 259–271. https://doi.org/10.1080/08927019509378279.

de Nys, R., Wright, A. D., Konig, G. M., & Sticher, O. (1993). New halogenated furanones from the marine alga *Delisea pulchra* (cf. *fimbriata*). *Tetrahedron, 49*, 11213–11220. https://doi.org/10.1016/S0040-4020(01)81808-1.
Ody, A., Thibaut, T., Berline, L., Changeux, T., André, J.-M., Chevalier, C., et al. (2019). From *in situ* to satellite observations of pelagic *Sargassum* distribution and aggregation in the Tropical North Atlantic Ocean. *PLoS One, 14*, e0222584.
de Oliveira, L. S., Gregoracci, G. B., Silva, G. G. Z., Salgado, L. T., Filho, G. A., Alves-Ferreira, M., et al. (2012). Transcriptomic analysis of the red seaweed *Laurencia dendroidea* (Florideophyceae, Rhodophyta) and its microbiome. *BMC Genomics, 13*, 487. https://doi.org/10.1186/1471-2164-13-487.
Oren, A., & Gunde-Cimerman, N. (2007). Mycosporines and mycosporine-like amino acids: UV protectants or multipurpose secondary metabolites? *FEMS Microbiology Letters, 269*, 1–10. https://doi.org/10.1111/j.1574-6968.2007.00650.x.
Oviatt, C. A., Huizenga, K., Rogers, C. S., & Miller, W. J. (2019). What nutrient sources support anomalous growth and the recent *Sargassum* mass stranding on Caribbean beaches? A review. *Marine Pollution Bulletin, 145*, 517–525. https://doi.org/10.1016/j.marpolbul.2019.06.049.
Pakker, H., Martins, R. S. T., Boelen, P., Buma, A. G. J., Nikaido, O., & Breeman, A. M. (2000). Effects of temperature on the photoreactivation of ultraviolet-B-induced DNA damage in *Palmaria palmata* (Rhodophyta). *Journal of Phycology, 36*, 334–341. https://doi.org/10.1046/j.1529-8817.2000.99087.x.
Paul, V. J., & Hay, M. E. (1986). Seaweed susceptibility to herbivory: Chemical and morphological correlates. *Marine Ecology Progress Series, 33*, 255–264.
Pavia, H., Cervin, G., Lindgren, A., & Åberg, P. (1997). Effects of UV-B radiation and simulated herbivory on phlorotannins in the brown alga *Ascophyllum nodosum*. *Marine Ecology Progress Series, 157*, 139–146. https://doi.org/10.3354/meps157139.
Pedersen, M. F., & Borum, J. (1996). Nutrient control of algal growth in estuarine waters. Nutrient limitation and the importance of nitrogen requirements and nitrogen storage among phytoplankton and species of macroalgae. *Marine Ecology Progress Series, 142*, 261–272. https://doi.org/10.3354/meps142261.
Pereira, T. R., Engelen, A. H., Pearson, G. A., Valero, M., & Serrão, E. A. (2015). Response of kelps from different latitudes to consecutive heat shock. *Journal of Experimental Marine Biology and Ecology, 463*, 57–62. https://doi.org/10.1016/j.jembe.2014.10.022.
Pereira, R., Kraemer, G., Yarish, C., & Sousa-Pinto, I. (2008). Nitrogen uptake by gametophytes of *Porphyra dioica* (Bangiales, Rhodophyta) under controlled-culture conditions. *European Journal of Phycology, 43*, 107–118. https://doi.org/10.1080/09670260701763393.
Peters, A. F., & Schaffelke, B. (1996). *Streblonema* (Ectocarpales, Phaeophyceae) infection in the kelp *Laminaria saccharina* (Laminariales, Phaeophyceae) in the western Baltic. *Hydrobiologia, 326*, 111–116. https://doi.org/10.1007/BF00047795.
Phillips, J. C., & Hurd, C. L. (2003). Nitrogen ecophysiology of intertidal seaweeds from New Zealand: N uptake, storage and utilisation in relation to shore position and season. *Marine Ecology Progress Series, 265*, 31–48. https://doi.org/10.3354/meps264031.
Plouguerné, E., Georgantea, P., Ioannou, E., Vagias, C., Roussis, V., Hellio, C., et al. (2010). Anti-microfouling activity of lipidic metabolites from the invasive brown alga *Sargassum muticum* (Yendo) Fensholt. *Marine Biotechnology, 12*, 52–61. https://doi.org/10.1007/s10126-009-9199-9.
Plouguerné, E., Hellio, C., Deslandes, E., Véron, B., Bremer, G., & Stiger-Pouvreau, V. (2008). Anti-microfouling activities of extracts of two of invasive algae: *Grateloupia turuturu* and *Sargassum muticum*. *Botanica Marina, 51*, 202–208. https://doi.org/10.1515/BOT.2008.026.

Plouguerné, E., Kikuchi, H., Oshima, Y., Deslandes, E., & Stiger-Pouvreau, V. (2006a). Isolation of Cholest-5-en-3-ol formate from the red alga *Grateloupia turuturu* Yamada and its chemotaxonomic significance. *Biochemical Systematics and Ecology, 34*, 714–717. https://doi.org/10.1016/j.bse.2006.04.003.

Plouguerné, E., Le Lann, K., Connan, S., Jechoux, G., Deslandes, E., & Stiger-Pouvreau, V. (2006b). Spatial and seasonal variation in density, reproductive status, length and phenolic content of the invasive brown macroalga *Sargassum muticum* (Yendo) Fensholt along the coast of Western Brittany (France). *Aquatic Botany, 85*, 337–344. https://doi.org/10.1016/j.aquabot.2006.06.011.

van de Poll, W. H., Eggert, A., Buma, A. G. J., & Breeman, A. M. (2001). Effects of UV-B-induced DNA damage and photoinhibition on growth of temperate of marine red macrophytes: Habitat related differences in UV-B tolerance. *Journal of Phycology, 37*, 30–37. https://doi.org/10.1046/j.1529-8817.2001.037001030.x.

Potin, P. (2012). Intimate associations between epiphytes, endophytes, and parasites of seaweeds. In C. Wiencke, & K. Bischof (Eds.), *Seaweed biology* (pp. 203–234). Springer.

Ramlov, F., Souza, J. M. C., Faria, A. V. F., Maraschin, M., Horta, P. A., & Yokoya, N. S. (2011). Growth and accumulation of carotenoids and nitrogen compounds in *Gracilaria domingensis* (Kütz.) Sonder ex Dickie (Gracilariales, Rhodophyta) cultured under different irradiance and nutrient levels. *Revista Brasileira de Farmacognosia, 21*, 255–261. https://doi.org/10.1590/S0102-695X2011005000081.

Ramus, J., Beale, S. I., Mauzerall, D., & Howard, K. L. (1976). Changes in photosynthetic pigment concentration in seaweeds as a function of water depth. *Marine Biology, 37*, 223–229. https://doi.org/10.1007/BF00387607.

Rastogi, R. P., Sinha, R. P., Singh, S. P., & Häder, D. P. (2010). Photoprotective compounds from marine organisms. *Journal of Industrial Microbiology and Biotechnology, 37*, 537–558. https://doi.org/10.1007/s10295-010-0718-5.

Reed, R. H. (1983). The osmotic responses of *Polysiphonia lanosa* (L.) tandy from marine and estuarine sites: Evidence for incomplete recovery of turgor. *Journal of Experimental Marine Biology and Ecology, 68*, 169–170.

Reed, R. H. (1990). Solute accumulation and osmotic adjustment. In K. M. Cole, & R. G. Sheath (Eds.), *Biology of the red algae* (pp. 147–170). Cambridge: Cambridge University Press.

Reed, R. H., Collins, J. C., & Russell, G. (1980). The effects of salinity upon galactosyl-glycerol content and concentration of the marine red alga *Porphyra perf*orata (Roth) C. Ag. *Journal of Experimental Botany, 31*, 1539–1554. https://doi.org/10.1093/jxb/31.6.1539.

Reed, R. H., Davison, I. R., Chudek, J. A., & Foster, R. (1985). The osmotic role of mannitol in the Phaeophyta: An appraisal. *Phycologia, 24*, 35–47. https://doi.org/10.2216/i0031-8884-24-1-35.1.

Rindi, F., & Guiry, M. D. (2004). Composition and spatio temporal variability of the epiphytic macroalgal assemblage of *Fucus vesiculosus* Linnaeus at Clare Island, Mayo, western Ireland. *Journal of Experimental Marine Biology and Ecology, 311*, 233–252. https://doi.org/10.1016/j.jembe.2004.05.009.

Ritter, A., Cabioch, L., Brillet-Guéguen, L., Corre, E., Cosse, A., Dartevelle, L., et al. (2017). Herbivore-induced chemical and molecular responses of the kelps *Laminaria digitata* and *Lessonia spicata*. *PLoS One, 12*, e0173315. https://doi.org/10.1371/journal.pone.0173315.

Ritter, A., Goulitquer, S., Salaün, J.-P., Tonon, T., Correa, J. A., & Potin, P. (2008). Copper stress induces biosynthesis of octadecanoid and eicosanoid oxygenated derivatives in the brown algal kelp *Laminaria digitata*. *New Phytologist, 180*, 809–821. https://doi.org/10.1111/j.1469-8137.2008.02626.x.

Rohde, S., & Wahl, M. (2008). Antifeeding defense in baltic macroalgae: Induction by direct grazing versus waterborne cues. *Journal of Phycology, 44*, 85–90.

Rönnberg, O., & Ruokolahti, C. (1986). Seasonal variation of algal epiphytes and phenolic content of *Fucus vesiculosus* in a northern Baltic archipelago. *Finnish Zoological and Botanical Publishing Board, 23*, 317–323.

Rothäusler, E., Gómez, I., Karsten, U., Tala, F., & Thiel, M. (2011). Physiological acclimation of floating *Macrocystis pyrifera* to temperature and irradiance ensures long-term persistence at the sea surface at mid-latitudes. *Journal of Experimental Marine Biology and Ecology, 405*, 33–41. https://doi.org/10.1016/j.jembe.2011.05.018.

Sampath-Wiley, P., Neefus, C. D., & Jahnke, L. S. (2008). Seasonal effects of sun exposure and emersion on intertidal seaweed physiology: Fluctuations in antioxidant contents, photosynthetic pigments and photosynthetic efficiency in the red alga *Porphyra umbilicalis* Kützing (Rhodophyta, Bangiales). *Journal of Experimental Marine Biology and Ecology, 361*, 83–91. https://doi.org/10.1016/j.jembe.2008.05.001.

Sand-Jensen, K. (1977). Effect of epiphytes on eelgrass photosynthesis. *Aquatic Botany, 3*, 55–63. https://doi.org/10.1016/0304-3770(77)90004-3.

Savitch, L. V., Maxwell, D. P., & Huner, N. (1996). Photosystem II excitation pressure and photosynthetic carbon metabolism in *Chlorella vulgaris*. *Plant Physiology, 111*, 127–136. https://doi.org/10.1104/pp.111.1.127.

Schagerl, M., & Möstl, M. (2011). Drought stress, rain and recovery of the intertidal seaweed *Fucus spiralis*. *Marine Biology, 158*, 2471–2479. https://doi.org/10.1007/s00227-011-1748-x.

Schmidt, É. C., Maraschin, M., & Bouzon, Z. L. (2010). Effects of UVB radiation on the carragenophyte *Kappaphycus alvarezii* (Rhodophyta, Gigartinales): Changes in ultrastructure, growth, and photosynthetic pigments. *Hydrobiologia, 649*, 171–182. https://doi.org/10.1007/s10750-010-0243-6.

Schmitt, T. M., Hay, M. E., & Lindquist, N. (1995). Constraints on chemically mediated coevolution: Multiple functions for seaweed secondary metabolites. *Ecology, 76*, 107–123. https://doi.org/10.2307/1940635.

Schoenrock, K. M., Amsler, C. D., McClintock, J. B., & Baker, B. J. (2013). Endophyte presence as a potential stressor on growth and survival in Antarctic macroalgal hosts. *Phycologia, 52*, 595–599. https://doi.org/10.2216/13-188.1.

Schoenwaelder, M. E. (2002). The occurrence and cellular significance of physodes in brown algae. *Phycologia, 41*, 125–139. https://doi.org/10.2216/i0031-8884-41-2-125.1.

Schubert, H., Feuerpfeil, P., Marquardt, R., Telesh, I., & Skarlato, S. (2011). Macroalgal diversity along the Baltic Sea salinity gradient challenges Remane's species-minimum concept. *Marine Pollution Bulletin, 62*, 1948–1956. https://doi.org/10.1016/j.marpolbul.2011.06.033.

Schubert, N., García-Mendoza, E., & Pacheco-Ruiz, I. (2006). Carotenoids composition of marine red algae. *Journal of Phycology, 42*, 1208–1216. https://doi.org/10.1111/j.1529-8817.2006.00274.x.

Simon-Colin, C., Bessieres, M.-A., & Deslandes, E. (2002). An alternative HPLC method for the quantification of floridoside in salt-stressed cultures of the red alga *Grateloupia doryphora*. *Journal of Applied Phycology, 14*, 123–127. https://doi.org/10.1023/A:1019539225950.

Sinha, R. P., Singh, S. P., & Häder, D. P. (2007). Database on mycosporines and mycosporine- like amino acids (MAAs) in fungi, cyanobacteria, macroalgae, phytoplankton and animals. *Journal of Photochemistry and Photobiology B: Biology, 89*, 29–35. https://doi.org/10.1016/j.jphotobiol.2007.07.006.

Steneck, R. S., Graham, M. H., Bourque, B. J., Corbett, D., Erlandson, J. M., Estes, J. A., et al. (2002). Kelp forest ecosystems: Biodiversity, stability, resilience and future. *Environmental Conservation, 29*, 436–459. https://doi.org/10.1017/S0376892902000322.

Stengel, D. B., Connan, S., & Popper, Z. A. (2011). Algal chemodiversity and bioactivity: Sources of natural variability and implications for commercial application. *Biotechnology Advances, 29*, 483–501. https://doi.org/10.1016/j.biotechadv.2011.05.016.

Stengel, D., & Dring, M. (1997). Morphology and in situ growth rates of plants of *Ascophyllum nodosum* (Phaeophyta) from different shore levels and responses of plants to vertical transplantation. *European Journal of Phycology, 32*, 193–202. https://doi.org/10.1080/09670269710001737129.

Stewart, H. L., & Carpenter, R. C. (2003). The effects of morphology and water flow on photosynthesis of marine macroalgae. *Ecology, 84*, 2999–3012. https://doi.org/10.1890/02-0092.

Stiger-Pouvreau, V., Bourgougnon, N., & Deslandes, E. (2016). Carbohydrates from seaweeds. In J. Fleurence, & I. Levine (Eds.), *Seaweed in health and disease prevention* (pp. 223–274). Elsevier. https://doi.org/10.1016/B978-0-12-802772-1.00008-7.

Stiger-Pouvreau, V., & Guérard, F. (2018). Bio-inspired molecules extracted from marine macroalgae: A new generation of active ingredients for cosmetics and human health. In S. La Barre, & S. S. Bates (Eds.), *Blue biotechnology: Production and use of marine molecules, Part 2: Marine molecules for disease treatment/prevention and for biological research* (pp. 709–746), ISBN 978-3-527-34138-2.

Stiger-Pouvreau, V., & Thouzeau, G. (2015). Marine species introduced on the French channel-Atlantic coasts: A review of main biological invasions and impacts. *Open Journal of Ecology, 5*, 227–257. https://doi.org/10.4236/oje.2015.55019.

Stratil, S. B., Neulinger, S. C., Knecht, H., Friedrichts, A. K., & Wahl, M. (2014). Salinity affects compositional traits of epibacterial communities on the brown macroalga *Fucus vesiculosus*. *FEMS Microbiology Ecology, 88*, 272–279. https://doi.org/10.1111/1574-6941.12292.

Surget, G., Le Lann, K., Delebecq, G., Kervarec, N., Donval, A., Poullaouec, M.-A., et al. (2017). Seasonal phenology and metabolomics of the introduced red macroalga *Gracilaria vermiculophylla*, monitored in the Bay of Brest (France). *Journal of Applied Phycology, 29*, 2651–2666. https://doi.org/10.1007/s10811-017-1060-3.

Szalontai, B., Nishiyama, Y., Gombos, Z., & Murata, N. (2000). Membrane dynamics as seen by Fourier transform infrared spectroscopy in a cyanobacterium, *Synechocystis* PCC 6803. The effects of lipid unsaturation and the protein-to-lipid ratio. *Biochimica et Biophysica Acta (BBA) - Biomembranes, 1509*, 409–419. https://doi.org/10.1016/s0005-2736(00)00323-0.

Takaichi, S. (2011). Carotenoids in algae: Distribution, biosyntheses and functions. *Marine Drugs, 9*, 1101–1118. https://doi.org/10.3390/md9061101.

Tanniou, A. (2014). *Etude de la production de biomolecules d'intérêt (phlorotannins, pigments, lipides) d'algues brunes modèles par des approches combinées de profilage métabolique et d'écophysiologie* [*Study of the production of biomolecules of interest (phlorotannins, pigments, lipids) of brown algal models by combined metabolic profiling and ecophysiology approaches*]. PhD. France: Université de Bretagne Occidentale.

Teichberg, M., Martinetto, P., & Fox, S. E. (2012). Bottom-up versus top-down control of macroalgal blooms. In C. Wiencke, & K. Bischof (Eds.), *Seaweed biology* (pp. 449–467). Berlin Heidelberg: Springer. https://doi.org/10.1007/978-3-642-28451-9_21.

Torres, P. B., Chow, F., & Santos, D. Y. A. C. (2015). Growth and photosynthetic pigments of *Gracilariopsis tenuifrons* (Rhodophyta, Gracilariaceae) under high light *in vitro* culture. *Journal of Applied Phycology, 27*, 1243–1251. https://doi.org/10.1007/s10811-014-0418-z.

Toth, G. B., & Pavia, H. (2000). Water-borne cues induce chemical defense in a marine alga (*Ascophyllum nodosum*). *Proceedings of the National Academy of Sciences of the United States of America, 97*, 14418–14420. https://doi.org/10.1073/pnas.250226997.

Ursi, S., Pedersén, M., Plastino, E., & Snoeijs, P. (2003). Intraspecific variation of photosynthesis, respiration and photoprotective carotenoids in *Gracilaria birdiae* (Gracilariales: Rhodophyta). *Marine Biology, 142*, 997–1007. https://doi.org/10.1007/s00227-003-1031-x.

Vanitha, A., & Chandra, S. (2012). Studies on photosynthetic pigments of some red algae of Covelong, Chennai (India). *International Journal of Current Science*, 149–154.

van den Hoek, C. (1984). World-wide latitudinal and longitudinal seaweed distribution patterns and their possible causes, as illustrated by the distribution of Rhodophyta genera. *Helgoländer Meeresuntersuchungen, 38*, 227. https://doi.org/10.1007/BF01997483.

Wahl, M., Goecke, F., Labes, A., Dobretsov, S., & Weinberger, F. (2012). The second skin: Ecological role of epibiotic biofilms on marine organisms. *Frontiers in Microbiology, 3*, 292. https://doi.org/10.3389/fmicb.2012.00292.

Wahl, M., Hay, M. E., & Enderlein, P. (1997). Effects of epibiosis on consumer-prey interactions. *Hydrobiologia, 355*, 49–59.

Wallentinus, I. (1984). Comparisons of nutrient uptake rates for baltic macroalagae with different thallus morphologies. *Marine Biology, 80*, 215–225. https://doi.org/10.1007/BF02180189.

Walther, G.-R. (2010). Community and ecosystem responses to recent climate change. *Philosophical Transactions of the Royal Society, Biological Sciences, 365*, 2019–2024. https://doi.org/10.1098/rstb.2010.0021.

Weinberger, F., Beltran, J., Correa, J. A., Lion, U., Pohnert, G., Kumar, N., et al. (2007). Spore release in *Acrochaetium* sp. (Rhodophyta) is bacterially controlled. *Journal of Phycology, 43*, 235–241. https://doi.org/10.1111/j.1529-8817.2007.00329.x.

Weinberger, F., & Friedlander, M. (2000). Response of *Gracilaria conferta* (Rhodophyta) to oligoagars results in defense against agar-degrading epiphytes. *Journal of Phycology, 36*, 1079–1086. https://doi.org/10.1046/j.1529-8817.2000.00003.x.

Weinberger, F., Pohnert, G., Berndt, M. L., Bouarab, K., Kloareg, B., & Potin, P. (2005). Apoplastic oxidation of L-asparagine is involved in the control of the green algal endophyte *Acrochaete operculata* Correa & Nielsen by the red seaweed *Chondrus crispus* Stackhouse. *Journal of Experimental Botany, 56*, 1317–1326. https://doi.org/10.1093/jxb/eri132.

Weinberger, F., Rohde, S., Oschmann, Y., Shahnaz, L., Dobretsov, S., & Wahl, M. (2011). Effects of limitation stress and of disruptive stress on induced antigrazing defense in the bladder wrack *Fucus vesiculosus*. *Marine Ecology Progress Series, 427*, 83–94.

Wheeler, W. N. (1980). Effect of boundary layer transport on the fixation of carbon by the giant kelp *Macrocystis pyrifera*. *Marine Biology, 56*, 103–110. https://doi.org/10.1007/BF00397128.

Whittick, A. (1983). Spatial and temporal distributions of dominant epiphytes on the stipes of *Laminaria hyperborea* (Gunn.) Fosl. (Phaeophyta: Laminariales) in S.E. Scotland. *Journal of Experimental Marine Biology and Ecology, 73*, 1–10. https://doi.org/10.1016/0022-0981(83)90002-3.

Wiencke, C., Gómez, I., Pakker, H., Flores-Moya, A., Altamirano, M., Hanelt, D., et al. (2000). Impact of UV-radiation on viability, photosynthetic characteristics and DNA of brown algal zoospores: Implications for depth zonation. *Marine Ecology Progress Series, 197*, 217–229. https://doi.org/10.3354/meps197217.

Wright, J. T., de Nys, R., Poore, A. G. B., & Steinberg, P. D. (2004). Chemical defense in a marine alga: Heritability and the potential for selection by herbivores. *Ecology, 85*, 2946–2959. https://doi.org/10.1890/03-4041.

Wright, P. J., & Reed, R. H. (1985). The effects of osmotic stress on intracellular hexitols in the marine brown alga *Himanthalia elongata* (L.) S.F. Gray. *Journal of Experimental Marine Biology and Ecology, 93*, 183–190. https://doi.org/10.1016/0022-0981(85)90158-3.

Yakovleva, I. M., & Titlyanov, E. A. (2001). Effect of high visible and UV irradiance on subtidal *Chondrus crispus*: Stress, photoinhibition and protective mechanisms. *Aquatic Botany, 71*, 47–61. https://doi.org/10.1016/S0304-3770(01)00167-X.

Yarish, C., Breeman, A. M., & van den Hoek, C. (1986). Survival strategies and temperature responses of seaweeds belonging to different biogeographic distribution groups. *Botanica Marina, 29*, 215–230. https://doi.org/10.1515/botm.1986.29.3.215.

Yebra, D. M., Kiil, S., & Dam-Johansen, K. (2004). Antifouling technology—past, present and future steps towards efficient and environmentally friendly antifouling coatings. *Progress in Organic Coatings, 50*, 75–104. https://doi.org/10.1016/j.porgcoat.2003.06.001.

Young, A. J., & Frank, H. A. (1996). Energy transfer reactions involving carotenoids: Quenching of chlorophyll fluorescence. *Journal of Photochemistry and Photobiology B: Biology, 36*, 3–15. https://doi.org/10.1016/S1011-1344(96)07397-6.

Zaneveld, J. S. (1969). Factors controlling the delimitation of littoral benthic marine algal zonation. *American Zoologist, 9*, 367–391. https://doi.org/10.1093/icb/9.2.367.

Zou, D., & Gao, K. (2002). Effects of desiccation and CO_2 concentrations on emersed photosynthesis in *Porphyra haitanensis* (Bangiales, Rhodophyta), a species farmed in China. *European Journal of Phycology, 37*, 587–592. https://doi.org/10.1017/S0967026202003876.

CHAPTER NINE

Current knowledge and challenges in extraction, characterization and bioactivity of seaweed protein and seaweed-derived proteins

Hugo Pliego-Cortés[a,*], Isuru Wijesekara[b], Marie Lang[a], Nathalie Bourgougnon[a] and Gilles Bedoux[a]

[a]Laboratoire de Biotechnologie et Chimie Marines, EA3884, UBS, IUEM, F-56000 Vannes, France
[b]Department of Food Science & Technology, Faculty of Applied Sciences, University of Sri Jayewardenepura, Gangodawila, Nugegoda, Sri Lanka
*Corresponding author: E-mail: hugo-skyol.pliego-cortes@univ-ubs.fr

Contents

Advances in Botanical Research, Volume 95
ISSN 0065-2296
https://doi.org/10.1016/bs.abr.2019.11.008

Abstract

Given the current interest to reach sustainable protein supplies, seaweed proteins represent a potential source for pharmaceutical, nutraceutical, cosmeceutical, or food and feed applications. Some species are reported to contain high quantities of protein with original amino acid composition, which has been comparable to the values reported for other plant-based proteins. Further, seaweeds are rich in several sorts of valuable proteins, such as peptides, enzymes, phycobiliproteins, glycoproteins, cell wall-attached proteins and mycosporine-like amino acids. However, some challenges remain to be addressed concerning protein yield. The extractability is affected by the presence of the tough polysaccharide-rich cell wall and the occurrence of phenolic compounds. In addition, because the protein content depends on species, harvesting season, location, and growing conditions, aquaculture systems are proposed as an alternative to scaling up seaweed biomass and increasing protein production. An update to the current knowledge of the seaweed protein extraction is addressed over the conventional procedures, which have been improved using assisted and alternative methods, including enzymatic (EAE), microwave (MAE), ultrasound (UAE), pulsed electric fields (PEF), accelerated solvent extraction (ASE) and membrane filtration. The identification and characterization of protein could be done through electrophoresis, chromatography or spectrometry. The potential bioactivities such as antioxidant, antiviral, anti-inflammatory, or anticancer from seaweed-driven proteins, including novel glycoproteins and lectins, are discussed.

1. Introduction

1.1 Current knowledge in seaweed proteins

Macroscopic algae or seaweeds are an extraordinarily diverse group of marine organisms living in a wide range of habitats under highly variable environmental conditions. To ensure their survival they have developed complex metabolic pathways and producing molecules that represent a vast potential for use as natural products (Harnedy & FitzGerald, 2011). Seaweeds are mainly categorized in to three major groups such as green seaweeds (Chlorophyta), brown seaweeds (Phaeophyceae) and red seaweeds (Rhodophyta), depending on their chemical composition (*i.e.* pigments).

Seaweeds are considered a rich source of proteins, as well as nutraceutical and functional food ingredients for human and animal nutrition (Wijesekara & Kim, 2015), since they are also good sources of vitamins, minerals, dietary fibers, lipids and polyphenols. Seaweed contains several types of valuable proteins such as peptides, enzymes, glycoproteins, lectins, mycosporine-like amino acids, and phycobiliproteins in red seaweeds.

The protein content in seaweeds can reach up to 50% of dry weight (dw). For example, in the human edible red seaweeds *Porphyra* sp. ~44% dw (Marsham, Scott, & Tobin, 2007), *Palmaria palmata* 35%, and *Chondrus crispus* 19.5% (Fleurence, Morançais, & Dumay, 2018), while the green *Ulva* sp. 23%, and the brown *Undaria pinnatifida* 19% dw (Vieira et al., 2018). Generally, the protein concentration is higher in red species (20%–47% dw) than green (9%–26% dw) and brown (3%–15% dw) (Fleurence et al., 2018). However, seaweeds may contain non-proteinic nitrogen, obtained from nitrates, pigments or nucleic acids, and resulting in an over-estimation of their protein content, which usually is estimated by the general Nitrogen-to-Protein conversion factor of 6.25. Therefore, a Seaweed-Nitrogen-to-Protein (SNP) conversion factors of 5.0 is proposed based on the analysis of 103 species of seaweeds, across 44 studies including three phyla, multiple geographic regions and a range of nitrogen contents (Angell, Mata, de Nys, & Paul, 2016b). Specific SNP conversion factor for brown, red and green seaweeds of 5.38, 4.59 and 5.13 are also reported (Barbarino & Lourenço, 2005).

The composition of seaweed protein presented all amino acids, especially glycine, alanine, arginine, proline, glutamic, and aspartic acids. Total amino acids content have been reported up to 40 mg/100 mg dw (Table 1). Particularly, the content of Essential Amino Acids (EAA) may represent about 50% of total amino acids, as in *Fucus spiralis* 63.5%, *Porphyra* sp. 56.7%, *Osmundea pinnatifida* 41.6% (Paiva, Lima, Patarra, Neto, & Baptista, 2014), or *Rhodymenia pseudopalmata* 43% (Pliego-Cortés, Caamal-Fuentes, Montero-Muñoz, Freile-Pelegrín, & Robledo, 2017). The seaweed amino acid analysis have demonstrated profiles similar to ovalbumin (52.4% EEA) and leguminous plant (41.62% EAA), being comparable to those of the FAO/WHO requirement pattern (Fleurence et al., 2018; Paiva et al., 2014). In this sense, seaweeds have emerged as important field of research in food science and technology, because of their interesting nutritional value. The quality of proteins for human or animal nutrition, therefore, depends drastically on their digestibility and the availability of EAA. It is true that animal protein is generally considered complete because of their high

Table 1 Amino acid content and ratio AAA/NEAA of selected seaweeds species.

Species	Ratio EAA/ NEAA	Total AA (mg/100 mg of dry weight)	References
Rhodophyta			
Chondrus crispus	1.58 ± 0.05	31.2 ± 1.7	Vieira et al., 2018
Gracilaria sp.	1.74 ± 0.05	35.5 ± 1.2	Vieira et al., 2018
Porphyra spp.	1.32 ± 0.04	38.8 ± 0.9	Vieira et al., 2018
Porphyra spp.	0.54	17.2	Astorga-España et al., 2016
Gracilaria domingensis	0.90	7.6 ± 0.0	Gressler et al., 2010
Gracilaria birdiae	0.90	9.1 ± 0.0	Gressler et al., 2010
Chlorophyta			
Ulva spp.	1.32 ± 0.05	41.3 ± 1.5	Vieira et al., 2018
Ulva spp.	0.56	19.2	Astorga-España et al., 2016
Enteromorpha sp.	0.55	15.8	Astorga-España et al., 2016
Pheophyceae			
Ascophylum nodosum	1.06 ± 0.01	36.6 ± 1.2	Vieira et al., 2018
Fucus spiralis	1.08 ± 0.01	40.2 ± 1.2	Vieira et al., 2018
Undaria pinnatifida	0.87 ± 0.01	44.2 ± 1.2	Vieira et al., 2018
Laminaria filiformis	0.90	11.3 ± 0.0	Gressler et al., 2010
Laminaria intricata	0.80	6.7 ± 0.0	Gressler et al., 2010

AA, Amino acids; *EAA*, Essential amino acids; *NEAA*, Non-essential amino acids.

content in EAA, however it is recommended to limit its consumption due to the development of cardiovascular disease and diabetes (Bleakley & Hayes, 2017). Algal protein could be complemented with other food proteins to achieved a high quality intake of proteins (Fleurence et al., 2018). Amino acids also play a key role in taste aspects, aspartic acid and glutamic acids, are the two most represented amino acids in all species of seaweeds, and are important for the development of flavors. Glutamic acid is particularly responsible for the taste sensations of "umami" (Pangestuti & Kim, 2015). Seaweed proteins are valuable ingredients for animal feed as well (Angell, Angell, de Nys, & Paul, 2016a). The enzymatic hydrolysis of protein by-products from *Gracilaria fisheri* yielded a roasted seafood-like flavoring. This could be useful in the food industry to replace flavoring agents from animal protein sources (Laohakunjit, Selamassakul, & Kerdchoechuen, 2014). The protein concentrate of *Kappaphycus alvarezii* contained 62.3%

of proteins represents an inexpensive source of protein, and could be incorporated into several value-added food products (Suresh Kumar, Ganesan, Selvaraj, & Subba Rao, 2014).

1.2 Types of proteins

1.2.1 Peptides

Peptides are generally known to be protein fragments containing 3 to 40 amino acids. These fragments are generated from the parent protein through digestion process in the gastro-intestinal tract, but can also be artificially produced during fermentation or other processes like enzymatic hydrolysis (Bleakley & Hayes, 2017; Pangestuti & Kim, 2015). Thus, it is expected that the seaweed treatment by enzymatic hydrolysis could generate enriched peptide extracts. The production of algal peptides is a particularly developed sector because of the diverse biological activities. Indeed, if the specific amino acid sequences are not necessarily active in the protein, they become active after release in the form of peptides (Pangestuti & Kim, 2015). Successful production of bioactive peptides - as describe later in bioactivity section - originating from hydrolyzed proteins of *P. palmata, Solieria chordalis, Ulva lactuca* and *Saccharina longicruris* are reported (Bondu et al., 2015). While in *U. pinnatifida*, 10 dipeptides were characterized and, four of them, Tyr-His, Lys-Tyr, Phe-Tyr, and Ile-Tyr, presented strong bioactivity (Suetsuna, Maekawa, & Chen, 2004). An active peptide identified as Pro-Ala-Phe-Gly, purified from *Enteromorpha clathrata* proteins, showed resistance against gastrointestinal proteases (Pan, Wang, Jing, & Yao, 2016).

1.2.2 Enzymes

Seaweed are also rich sources of enzymes. Alkaline phosphatase is a Zn-containing metalloproteinase that catalyzes the non-specific hydrolysis of phosphate monoesters. This enzyme is widely distributed in seaweeds and have been purified and characterized from *Ulva pertusa* (Yang, Wang, Bao, & An, 2003). Alternative oxidases (AOX) proteins has been described for electron flow through electron transport chain and regulation of mitochondrial retrograde signaling pathway, in the green *Caulerpa cylindracea* two AOX encoding genes were described, which support the capacity of this species for invasion of new environments (Ünlü, Ünüvar, & Aydın, 2019). The fibrinolytic enzymes has been isolated from the green alga, *Codium latum* (Matsubara, Hori, Matsuura, & Miyazawa, 1999), and *Codium fragile* (Choi, Sapkota, Park, Kim, & Kim, 2013), which is a trypsin-like serine protease with a high substrate specificity. The bifunctional enzyme Rubisco, which is known to catalyze carbon dioxide fixation and oxygenation, was reported in *S. chordalis* (Bondu et al., 2015) and *K. alvarezii* (Tee, Yong, Rodrigues, & Yong, 2015).

1.2.3 Glycoproteins and lectins

Glycoproteins (GP) are carbohydrate-binding proteins, in which glycan is conjugated to peptide chains by two types of primary covalent linkages, N-glycosyl linkages and O-glycosyl linkages; some glycoproteins have both N-linked and O-linked forms. The carbohydrate is attached to the protein through glycosylation, which is a co-translational or post-translational modification (Senthilkumar & Jayanthi, 2016). According to Yoshiie et al. (2012), the high-mannose type N-glycans were linked to GP expressed in different species of seaweeds. GP are located on the cell wall, on the cell surfaces or they are secreted, and their functions in plants include recognition, intercellular interactions, and adhesion (Stiger-Pouvreau, Bourgougnon, & Deslandes, 2016). Currently, a few GP have been isolated from seaweeds (Table 2), and their structures and functional roles are yet to be investigated. In *Ulva* sp. three glycoprotein-rich fractions were obtained, UvGP-1 (hot water extraction), and UvGP-2-DA and UvGP-2-DS (cold-water extraction). GP-1 and GP-2-DA showed higher protein content (24.9% and 33.4%, respectively) than neutral sugars, while GP-2-DS contains a higher amount of neutral sugars (28.5%) than proteins (Wijesekara et al., 2017). While, in *U. lactuca* the high recovery of carbohydrates (42%) and protein (12.3%) in the extracts suggested the presence of GP (Abdel-fattah & Sary, 1987). Senthilkumar and Jayanthi (2016), isolated and partially characterized a GP from *Codium decorticatum* with a molecular mass of ~48 kDa, and Surendraraj, Habeebullah, and Jacobsen (2015), produces GP from the pH precipitated fractions after enzymatic extracts from the brown algae *Fucus serratus* and *Fucus vesiculosus*.

Lectins are proteins or glycoproteins of non-immune origin, containing at least one non-catalytic domain that binds reversibly to specific mono- or oligo-saccharides without modify the carbohydrates that they bind. Lectins precipitate glycoproteins and agglutinate cells (Peumans & Van Damme, 1995). In seaweeds, lectins are involved in gamete recognition and reproductive cell fusion (Ingram, 1985). Seaweeds are good sources of novel lectins (Table 2), such as griffithsin, a mannose-specific lectin isolated from the red alga *Griffithsia* sp. (Mori et al., 2005), or SfL-1 and SfL2 from *Solieria filiformis* (Chaves et al., 2018), or HRL40 from *Halimeda renschii* (Mu, Hirayama, Sato, Morimoto, & Hori, 2017). Singh and Walia (2018), and

Table 2 Glycoproteins and lectins extraction, identification and bioactivity in selected seaweeds species.

Species	Glycoprotein (GP) or lectin name	Extraction and purification	Identification	Mass (kDa)	Bioactivity	References
Chlorophyta						
Capsosiphon fulvescens	Cf-hGP	Sodium acetate Methanol Lectin wheat germ agglutinin resin	SDS-PAGE Immunoblotting	–	Reduce endoplasmic reticulum stress-induced cognitive dysfunction	Oh & Nam, 2019
Ulva sp.	UvGP-1 UvGP-2-DA UvGP-2-DS	Distilled Water TCA $(NH_4)_2SO_4$	SDS-PAGE MALDI-TOF	–	– Non cytotoxicity	Wijesekara et al., 2017
Codium decorticatum	GLPCD	Distilled Water $(NH_4)_2SO_4$ Ion exchange	SDS-PAGE FT-IR, CD, ^{1}H NMR, MALDI-TOF	~48	Anticancer activities on cell lines, MCF-7, Siha, and A549	Senthilkumar & Jayanthi, 2016
Codium decorticatum	GLPCD	$(NH_4)_2SO_4$ Ion exchange	SDS-PAGE	~48	Activity properties against human MDA-MB-231 breast cancer cells	Thangam et al., 2014
Halimeda renschii	Lectin HRL40	Phosphate buffer $(NH_4)_2SO_4$ gel-filtration Ion exchange	SDS-PAGE ESI-MS	46.5	Anti-Influenza Virus Activity	Mu et al., 2017
Caulerpa cupressoides var. lycopodium	Lectin CcL	Tris–HCl Ion exchange	–	–	Anti-nociceptive and anti-inflammatory	da Conceição Rivanor et al., 2014
Ulva lactuca	GP fraction G	Distilled Water NaOH DEAE-cellulose	Chromatographic paper	–	–	Abdel-fattah & Sary, 1987

(Continued)

Table 2 Glycoproteins and lectins extraction, identification and bioactivity in selected seaweeds species.—cont'd

Species	Glycoprotein (GP) or lectin name	Extraction and purification	Identification	Mass (kDa)	Bioactivity	References
Rhodophyta						
Solieria filiformis	Lectins SfL-1 SfL-2	Phosphate buffer $(NH_4)_2SO_4$ Ion exchange	SDS-PAGE LC-MS/MS ESI Protein and peptide sequencer	27.5 27.9	Anticancer effects on Human breast adenocarcinoma and dermal Fibroblasts	Chaves et al., 2018
S. filiformis	Lectin SfL	Tris-HCl buffer $(NH_4)_2SO_4$ Ion exchange	–		Antidepressant-like action	Abreu et al., 2018
Hypnea musciformis	Lectin HML	Phosphate buffer $(NH_4)_2SO_4$ Affinity and ion exchange	SDS-PAGE	10	Antimicrobial of biofilm formation from bacteria and yeast	Vasconcelos et al., 2014
Gracilaria ornata	Lectin GOL	Tris-HCl buffer $(NH_4)_2SO_4$	SDS-PAGE	17	effect on the development of the insect cowpea weevil	Leite et al., 2005
Griffithsia sp.	Lectin GRFT	$(NH_4)_2SO_4$ HPLC HPSEC	SDS-PAGE MALDI-TOF MS ESI-MS	12.7	Antiviral activity against HIV-1	Mori et al., 2005
Eucheuma serra *Galaxaura marginata*	Lectins ESA GMA	Phosphate buffer Ethanol	–	–	Antibacterial activity against marine *vibrios*	Liao et al., 2003

Pheophyceae						
Undaria pinnatifida	UPGP	Distilled water Ethanol	SDS-PAGE FT-IR	~10	Anti-Alzheimer's and anti-inflammatory activities	Rafiquzzaman et al., 2015
Fucus serratus *Fucus vesiculosus*	GPs	EAE HCl	SDS-PAGE		Antioxidant	Surendraraj et al., 2015
Saccharina japonica	SJGP	Distilled water Ethanol HPLC	SDS-PAGE FT-IR	~10	Probiotic properties Antioxidant and DNA protection	Kim et al., 2015 Kim et al., 2012
Laminaria japonica	LJGP	Distilled water Ethanol $(NH_4)_2SO_4$	SDS-PAGE	~10	antiproliferative effects on HT-29 colon cancer cells	Go et al., 2010
Hizikia fusiformis	HFGP	Distilled water Ethanol $(NH_4)_2SO_4$	SDS-PAGE	–	Chemoprotective effects against acetaminophen-induced liver injury	Hwang et al., 2008

–, data no reported; *EAE*, Enzyme-Assisted Extraction.

Barre, Simplicien, Benoist, Van Damme, and Rougé (2019), have reviewed their structure and potential biomedical applications extensively.

1.2.4 Cell wall-attached proteins

Arabinogalactan proteins (AGPs) belong to the hydroxyproline-rich glycoproteins (HRGP), they are highly glycosylated proteins reported in the cell wall of few species of seaweeds. AGP are implicated in terrestrial plant developmental process but their roles in seaweed has been yet poorly known (Popper et al., 2011). The genome sequence of *Ectocarpus siliculosus* encodes AGP protein backbone motifs shown that differs considerably from land plants in a gene context, while AGP on the cell surfaces of the brown seaweed embryos *F. serratus*, was confirm by positively stained with β-Glucosyl and β-Galactosyl Yariv reactive (Hervé et al., 2016). In AGP, the carbohydrates represent between 90 and 98% (*w/w*), composed mainly of arabinose, rhamnose and glucuronic acid residues (Tan et al., 2012), while the protein backbone typically contains repeating motifs rich in hydroxyproline/proline, alanine and serine/threonine (Gorres & Raines, 2010). The presence of HRGP was confirmed by immunolabeling and by the β-Glc Yariv reactive in the green seaweeds *Codium vermilara* (Fernández, Ciancia, Miravalles, & Estevez, 2010) and *C. fragile*, in which an arabinose-rich fraction showed a furanosic α-arabinosyl structure (Estevez, Fernández, Kasulin, Dupree, & Ciancia, 2009).

1.2.5 Phycobiliproteins

Phycobiliproteins (PBPs) are the main light-harvesting pigments in red seaweeds and the only water-soluble algal pigments (Fig. 1A and B), representing up to 20% dw. PBPs are grouped into four classes: phycoerythrin (PE), phycocyanin (PC), phycoerythrocyanins (PEC), and allophycocyanin (APC). PE is the main pigment, and it is divided into R- for Rhodophyta (R-PE) and B- for Bangiales (B-PE). Biliproteins are ensemble of α and β subunits, and only R-PE possess an extra γ subunit (Dumay, Morançais, Munier, Le Guillard, & Fleurence, 2014). Isolation of PE has been reported in many species, for example *Gelidium pusillum* (Mittal, Sharma, & Raghavarao, 2019), *Grateloupia turuturu* (Denis, Massé, Fleurence, & Jaouen, 2009; Le Guillard et al., 2016), *R. pseudopalmata* (Pliego-Cortés et al., 2017).

1.2.6 Mycosporine-like amino acids

Mycosporine-like amino acids (MAAs) are small-sized (<400 Da) secondary metabolites with strong absorption of UVR, including A and

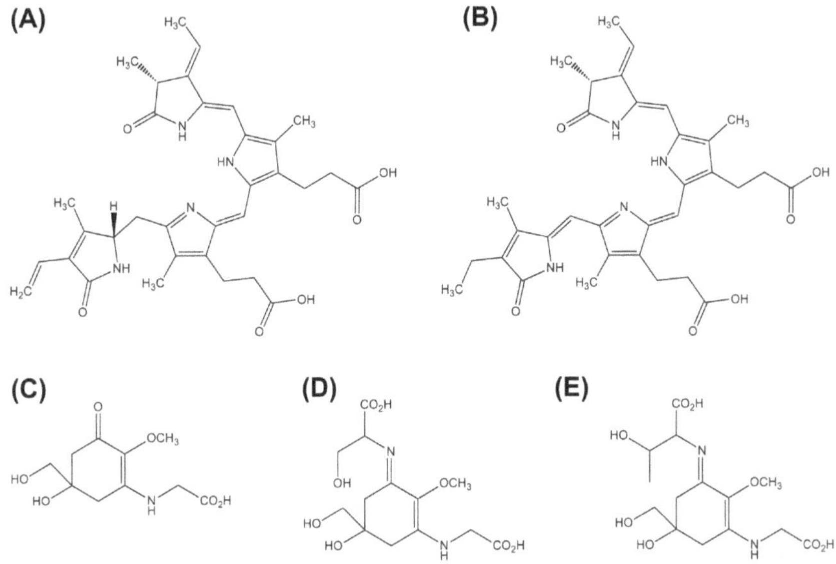

Fig. 1 Chemical structures of (A) phycoerythrobiline, (B) phycocyanobiline (based on ChemSpider ID 16741434), and the mycosporine-like amino acids: (C) mycosporine-glycine, (D) shinorine, and (E) porphyra-334.

B (292–365 nm), colorless, and stable in changing conditions of light, temperature, and pH (Fig. 1C to E). Their main functions are UVR-protection and antioxidant activities (Carreto & Carignan, 2011). More than 20 MAAs have been identified in seaweeds, and their capacity to synthesize a large variety of MAAs facilitates their adaptation to stressful environments (Pliego-Cortés et al., 2019; Wada et al., 2015). A full description of MAAs is given in Chapter 8 by Bedoux et al.

2. Challenge: variability and availability of biomass sources

2.1 Wild harvested versus cultivated

According to literature, high levels of variability in protein content have been reported due to seasons, environmental factors, or harvesting locations (Table 3). This represents an important challenge, since protein quantity and quality cannot be the same through the year. Generally higher amounts of proteins are reported in winter, such as in *Gracilaria tikvahiae* (Madden, Mitra, Ruby, & Schwarz, 2012) collected in winter (~30% dw) compared to spring (~17% dw) or in *P. palmata* (Galland-Irmouli

Table 3 Protein content, extraction and quantification of selected seaweeds species according to the season, source and geographical location.

Species	Source	Location	Sampling	Protein content % Dry weight	Protein extraction and quantification	References
Rhodophyta						
Rhodymenia pseudopalmata	Wild IMTA	Yucatan, Mexico	July	9.4 ± 0.4 18.7 ± 0.4	Freeze-dried BCA Lowry	Pliego-Cortés et al., 2019
Gracilaria conferta	IMTA	Haifa, Israel	March–August	17.7 ± 1.52	Dried (60 °C)	Ashkenazi et al., 2019
Hypnea musciformis	IMTA Controls	Herzliya, Israel		25.32 ± 1.25 ~5 to 10	Kjeldahl N x 6.25	
Chondracanthus chamissoi	Wild	Coquimbo, Chile	January–March	45.2 ± 4.2	Dried (60 °C) EAE, Lowry	Vásquez et al., 2019
Chondrus crispus	Wild	North-Central coast, Portugal	April–July October–November	19.5 ± 0.16 19.1 ± 0.33	Dried (52 °C), Kjeldahl N x 6.25	Vieira et al., 2018
Gracilaria sp.	Wild	Portugal	April–July October – November	24.7 ± 0.24 24.4 ± 0.24	Dried (52 °C), Kjeldahl N x 6.25	Vieira et al., 2018
Osmundea pinnatifida	Wild	Portugal	April–July October – November	24.3 ± 0.73 22.8 ± 0.33	Dried (52 °C), Kjeldahl N x 6.25	Vieira et al., 2018
Porphyra spp.	Wild	Portugal	April–July October – November	27.4 ± 0.08 28.2 ± 0.16	Dried (52 °C), Kjeldahl N x 6.25	Vieira et al., 2018

R. pseudopalmata	Cultivated	Yucatan, Mexico	30 psu	10.5 ± 1.42	Freeze-dried	Pliego-Cortés et al., 2017
			40 psu	4.9 ± 0.43	BCA Lowry	
Palmaria palmata	Wild	Brittany, France	Winter–spring	21.9 ± 3.5	Freeze-dried	Galland-Irmouli et al., 1999
			Summer– early autumn	11.9 ± 2.0	Kjeldahl N x 6.25	
Chlorophyta						
Ulva ohnoi	Commercial cultivated	Queensland, Australia	October – November		Fresh, milli-Q water, HCl, ultrafiltration, EAE, MAE. Sum of all amino acids	Magnusson et al., 2019
			Unprocessed	22.2 ± 0.4		
			Protein-enriched-I	~41%		
			Protein-enriched-II	~42%		
			Protein-enriched-III	~40%		
			Protein-enriched-IV	39.5 ± 1.9		
Ulva spp.	Wild	Portugal	April–July	20.5 ± 0.49	Dried (52 °C), Kjeldahl N x 6.25	Vieira et al., 2018
			October – November	23.3 ± 0.01		
Pheophyceae						
Macrocystis pyrifera	Wild	Coquimbo, Chile	January–March	61.6 ± 4.7	Dried (60 °C) EAE, Lowry	Vásquez et al., 2019
Ascophyllum nodosum	Wild	Portugal	April–July	6.90 ± 0.16	Dried (52 °C), Kjeldahl N x 6.25	Vieira et al., 2018
			October – November	9.40 ± 0.08		

(*Continued*)

Table 3 Protein content, extraction and quantification of selected seaweeds species according to the season, source and geographical location.—cont'd

Species	Source	Location	Sampling	Protein content % Dry weight	Protein extraction and quantification	References
Fucus spiralis	Wild	Portugal	April–July October – November	11.8 ± 0.16 11.7 ± 0.24	Dried (52 °C), Kjeldahl N x 6.25	Vieira et al., 2018
Saccorhiza polyschides	Wild	Portugal	April–July October – November	12.4 ± 0.01 11.8 ± 0.16	Dried (52 °C), Kjeldahl N x 6.25	Vieira et al., 2018
Undaria pinnatifida	Wild	Portugal	April–July October – November	16.5 ± 0.08 19.5 ± 0.21	Dried (52 °C), Kjeldahl N x 6.25	Vieira et al., 2018
Sargassum horneri	Wild	Fukuoka, Japan	Feb. Immature Feb. Male/Female May Male May Female	12.8 ± 1 10.7/10.4 ± 0 5.6 ± 0.2 8.4 ± 0.1	Freeze-dried Kjeldahl N x 6.25	Murakami et al., 2011

N x 6.25, Nitrogen-to-Protein conversion factor of 6.25; *psu*, practical salinity units.

et al., 1999), during winter–spring (21.9% dw) *vs.* summer–early autumn (11.9% dw). The brown algae *Ascophyllum nodosum, U. pinnatifida* and the green *Ulva* sp., analyzed during October–November exhibited higher protein than in April–June. The protein quality was not affected since total amino acids, EAA and EAA/NEAA ratios remain unchanged (Vieira et al., 2018). *Sargassum horneri* also showed significant differences in protein content between seasons and thalli maturity. Immature thalli collected in February showed the highest value (128 mg g^{-1}), while in May, male and female thalli showed lower content (56 and 84 mg g^{-1}), being early spring the best season to harvest the alga for human consumption, when its growth and maturity are complete (Murakami et al., 2011). Concerning the harvesting localization, the total content of AA and the ratio EAA/NEAA in *Porphyra* sp. collected in Portugal (38.8 mg AA 100 mg^{-1} dw) (Vieira et al., 2018) was higher than those harvested in Chile (17.2 mg AA 100 mg^{-1} dw) (Astorga-España et al., 2016). According to Angell et al. (2016b), a meta-analysis including 382 species, the mean protein content was lower in tropical seaweeds (14.5% dw), compared to temperate and polar (17.6% dw) by using the Nx6.25 factor. However, polar seaweed showed lower content (8.14% dw) when the total amino acid determination was used. The authors indicated the difficulty to compare the protein content between geographic regions since differences were related to the methods of protein determination. In contrast, the cultivated seaweeds showed higher protein contents in the meta-analysis, regardless of the method of protein determination.

The cultivation of seaweed could face the challenge of variability and availability of biomass, cultivation usually enhance the content of protein and also decrease the pressure of harvesting large quantities of biomass from wild populations. For instance, in Europe more than 97% of seaweed production is harvested from the wild in Norway, France and Ireland (Grote, 2019). The same *scenario* is reported in Latin America, 96% harvested from natural beds in Mexico, Peru, Ecuador (Alemañ, Robledo, & Hayashi, 2019; Vázquez-Delfín, Freile-Pelegrín, Pliego-Cortés, & Robledo, 2019). Eco-intensification of seaweed cultivation can be improved through integrated multitrophic aquaculture (IMTA). Further, biomass could be maximized using biorefinery strategies. Under an IMTA approach, thalli from *Rhodymenia pesudopalmata* showed an increase of total protein content after 3 days of cultivation (18.7% dw) in fishpond effluents rich in ammonium (NH_4^+), than thalli from wild populations (9.4% dw), and a fivefold increase in the MAAs total content as well (Pliego-Cortés et al., 2019). The cultivation of *P. palmata* under different cultivation systems and nutrient treatments has been addressed, and proved to be a good candidate for IMTA because their high affinity for NH_4^+, which resulted in an increase of 20% of protein content in thalli fed with NH_4^+ (Grote, 2019). The IMTA systems does not just allow the production of algal biomass but also reduces negative impact on ecosystems because the capacity of seaweeds to absorb the inorganic nutrients. Recently, high protein content from *U. lactuca* (37.4% dw) and *Gracilaria conferta* (29.4% dw) produced in IMTA were used to fed sea urchins as a sole feed compared with a formulated diet. *U. lactuca* enhanced somatic and gonad growth, and better protein assimilation efficiency (81%) compare to formulated diet (59%), which was attributed to the relatively high protein and EEA (44%) content. *Ulva* produced in IMTA compared to wild harvested, was more productive and had lower biochemical composition variability. The use of *Ulva* seems to be cost effective and a more sustainable method to produce echinoderms (Shpigel, Shauli, Odintsov, Ashkenazi, & Ben-Ezra, 2018). The cultivation of *R. pseudopalmata* under different conditions of light and salinity increased the content of EAA up to 43% (Pliego-Cortés et al., 2017). Furthermore, other plausible alternative to face this challenge is to obtain complementary products to seaweed protein and seaweed derived proteins in the same biomass using a biorefinery approach.

Using a multi-step biorefinery process, the quality and yield of the protein content from commercially produced *Ulva ohnoi* was enhanced, by producing four fractions (PEB-I to PEB-IV) and one protein isolate (PI). The protein increased from 22% dw in unprocessed biomass to 39% in PEB-IV and 45% in the PI. Authors projected an annual production (t dw ha^{-1} $year^{-1}$) of 29 t of PEB-I, or 3.2 t of PI, and 24 t of salt, and 4.3 t of ulvan (Magnusson et al., 2019). In *U. lactuca*, a protein-enriched fraction containing 343 g protein kg^{-1} dw was achieved, and its content in EEA and *in vitro* digestibility (90%), led the authors to propose it as a protein source for monogastric animals, but also as an effective source to produce acetone, butanol, ethanol and 1,2-propanediol by fermentation (Bikker et al., 2016). A sustainable production of biomolecules from macroalgae could strongly contribute to marine bio-economy, since the recovery of proteins (11% dw) from *U. lactuca* with a high digestibility (85.8%) and good amino acid content shows their suitability for use as food supplements. Other products such as mineral salts (14%), fatty acids (1.5%), ulvan (19.9% dw) and cellulose (10.35% dw) were coproduced (Gajaria et al., 2017).

3. Challenge: protein extraction and purification

The raw biomass from seaweed after harvesting must be preserved by drying or freezing, or used in fresh as soon as possible to avoid protein degradation. The effect of drying methods (freeze-, vacuum-, solar-, and convective-drying) has been established in *Ulva* sp. The convective-drying showed the higher content of crude protein (20% dw) with the lower EAA content (41%), the highest EAA content (56%) was obtained in vacuum-drying (Uribe et al., 2018). While, the processes of freeze-dried, oven-dried, and −20°C-frozen on *Saccharina latissima* resulted in significantly higher protein yield than the −80°C-frozen, sun-dried and ensiled processes. The sun-dried had the highest EAA content and freeze-dried the lowest (51.7 vs. 40.2 mg 100 g^{-1} protein); however, a high degree of protein degradation was observed in sun-drying and −20 °C-freezing methods. These processes showed that post-harvest preservation must be carefully chosen according to the final use of the protein (Abdollahi et al., 2019). Afterward, seaweeds need to be processed for use as a high quality protein source.

The development of seaweeds as a protein source challenges the complex nature of the cell wall. The presence of polysaccharide-bound cell wall mucilage including anionic or neutral polysaccharides, and polyphenols reduces protein extractability and requires additional adapted steps for fractionation and purification. Studies in model systems using carrageenan associated to whey protein, or ι- κ-carrageenan or dextran sulfate with bovine serum albumin, have proven that polysaccharides induced strong electrostatic interactions (Weinbreck, Nieuwenhuijse, Robijn, & de Kruif, 2004). Whereas, polyphenols may form reversible hydrogen bonds with proteins or oxidize to quinones, to bind irreversibly to proteins (Vilg & Undeland, 2017). Thus, the combination of extraction methods and purification techniques are necessary to improve protein yield and purity, or digestibility. This challenge becomes greater on an industrial scale, because require methods with low-time, -cost and -energy consumption (i.e. environmentally friendly).

3.1 Conventional extractions

There are various liquid systems which allowed the protein extraction, such as distilled water, buffers, acid or alkaline solution (Fleurence, Le Coeur, Mabeau, Maurice, & Landrein, 1995; Galland-Irmouli et al., 1999), urea (Contreras et al., 2008), lysis solutions (Kim et al., 2010; Wijesekara et al.,

2017), and phenol-based extraction systems (Nagai, Yotsukura, Ikegami, Kimura, & Morimoto, 2008; Wong, Tan, Nawi, & AbuBakar, 2006). Combining with physical methods, as osmotic shock, freeze/thawing or grinding, the extraction in some seaweeds have been enhanced (Table 3). Extraction with alkaline solubilization, isoelectric precipitation and frozen/thawing showed higher protein concentration in *Porphyra umbilicalis* (71%), *S. latissima* (51%), and *U. lactuca* (40%) compared to water extraction (Harrysson et al., 2018). Precipitation of solubilized proteins at pH 2 yielded 34.5% of proteins, and osmoshocking with 60 vol of water gave 59% yield in *S. latissima* (Vilg & Undeland, 2017). Glycoproteins (GP) and lectins are commonly extracted using water or buffers, followed by ethanolic, acidic or ammonium sulfate precipitation (Table 2), such as the GP from *Laminaria japonica* (Go, Hwang, & Nam, 2010). While, protein hydrolysis using 6 M HCl in either a liquid or vapor phase, allows high recovery rate for a majority of amino acids, but tryptophan is destroyed, and methionine and cysteine are also sensible, and therefore underestimated (Kazir et al., 2019). Alternate methods must be used, as alkaline hydrolysis using 4 N LiOH (Dawczynski, Schubert, & Jahreis, 2007). The content of free amino acids (amino acids not embedded into proteins) was extracted by ultrasound-assisted perchloric acid (Vieira et al., 2018).

3.2 Assisted extractions and alternative methods

Enzyme-Assisted Extraction (EAE) has been applied by using enzymes to assist and to extract proteins, and hydrolysates from seaweeds. The use of commercially enzymatic preparations is frequently used with limited practices using specific enzyme on seaweeds, such as κ-/ι-carragenase or β-agarase (Denis, Morançais, Gaudin, & Fleurence, 2009; Fleurence, Massiani, Guyader, & Mabeau, 1995). In a recent study, protein extraction from *Chondracanthus chamissoi* and *Macrocystis pyrifera* was enhanced using a cellulose cocktail (Cellic CTec3, Novozymes), producing 452 and 616 mg g^{-1} dw of protein. This cocktail contained multiple hydrolytic activities (cellulase, hemicellulase and β-glucosidase) for carrageenan, agar, alginate and cellulose. Furthermore, did not exhibit proteolytic activity (Vásquez, Martínez, & Bernal, 2019). On the contrary, the subtilisin protease (Novozymes), which raised an 8.9% dw protein extraction in *S. chordalis*, allowed the cleavage by subtilases or serine proteases by breaking down the cell wall of seaweeds through a selective degradation of structural proteins, a degradation of serine glycoproteins of the cell wall might be possible (Burlot, Bedoux, & Bourgougnon, 2016). The EAE of *Ulva armoricana*, *Sargassum muticum*,

S. chordalis and *Pyropia columbina* yielded bioactive extracts and hydrolysates (Cian, Garzón, Ancona, Guerrero, & Drago, 2015; Hardouin et al., 2013, 2016; Puspita et al., 2017). Xilanase enzyme allowed to a significant increase from 0.11 to 1.15 g kg^{-1} dw (dry ground algae) in R-PE from *P. palmata* (Dumay, Clément, Morançais, & Fleurence, 2013). These studies have shown that the EAE is free from application of organic solvents, leads to higher extraction rate of bioactive hydrolysates in reduced extraction times and costs. Different digestion enzymes have been used to release bioactive peptides from the parent proteins, chymotrypsin, trypsin and pepsin being the most commonly used.

Protein extraction using Ultrasound-Assisted Extraction (UAE) during 2 h followed by ion exchange purification yielded a protein content of 70% in *Ulva sp.* and 86% in *Gracilaria* sp. This procedure was up-scalable and suitable to obtain a 'food-grade' product (Kazir et al., 2019). The effects of UAE in the brown seaweed *A. nodosum*, showed that an amplitude of 68.4 μm, using acid and alkali treatments yielded 43% and 57% of protein recovery, respectively. The effect of UAE can be attributed to bubble cavitation, facilitating the degradation of biological matrices (Kadam, Tiwari, & O'Donnell, 2013). A recovery of 71% and 74% of the initial material into the soluble phase rich in proteins, carbohydrates, amino acids and salts, was obtained by using EAE or UAE in *G. turuturu*. However, when these processes were combined, up to 91% of solubilized material was archived due to a synergistic effect (Le Guillard et al., 2016). Extraction of R-PE has been effective by combining these two processes in *G. turuturu* (Le Guillard et al., 2015). In *G. pusillum*, UAE followed by maceration and buffers, allowed 77%–93% of recovery for R-PE and R-PC (Mittal, Tavanandi, Mantri, & Raghavarao, 2017). Even a very high purity of R-PE with a removal of up to 95% of total sugars was achieved (Mittal et al., 2019). Iodinated proteins in *Porphyra* sp. raised extraction yields close to 100% using UAE with pancreatin hydrolysis (Romarís-Hortas, Bermejo-Barrera, & Moreda-Piñeiro, 2013).

The Pulsed Electric Field (PEF) extraction uses electric potential generated from an electric field using high voltage (kV) and different duration time in the range of microseconds to milliseconds, to disrupt cell membranes (Okolie, Akanbi, Mason, Udenigwe, & Aryee, 2019). PEF treatments in *U. lactuca* showed that the higher content of protein (15% dw) was achieved at an electric field strength of 7.5 kV cm^{-1}, with two pulses of 0.05 ms, and a specific energy input of 6.6 kW h kg_{prot}^{-1} (Postma et al., 2018). While 50 pulses of 50 kV, applied at a 70.3 mm electrode gap, resulted in an

~sevenfold increase in the total protein in *Ulva* sp. compared to osmotic shock (Robin et al., 2018). The PEF extracted juice from *U. lactuca* showed higher protein content than the control (59 vs. 23 μg mL^{-1}). The developed process has almost no thermal effects on the produced proteins (Polikovsky et al., 2016).

Membrane filtration is based on the molecular size of the compounds, which can be separated by a semi-permeable membrane to obtain different fractions, namely microfiltration, ultrafiltration, nanofiltration and reverse osmosis. For example, ultrafiltration allowed to isolate proteins between 1 and 200 kDa (Bleakley & Hayes, 2017). An ultrafiltration unit at industrial scale, employing polyethersulfone 25—30 kDa membrane showed the recovery of 100% of R-PE, 32.9% of other proteins and 64.6% of sugars from *G. turuturu* (Denis, Massé, Fleurence, & Jaouen, 2009).

Other procedures could be applied to enhance protein extraction, like Microwave Assisted Extraction (MAE), mainly used for carbohydrate extractions. This method is based on the transfer of energy to the solution, as microwaves induce the vibration of water molecules in the matrixes, disrupting the hydrogen bonds and allowing migration of dissolved ions (Kadam et al., 2013). The quality and yield of the protein in *U. ohnoi* was improved through EAE, MAE and ultrafiltration extractions (Magnusson et al., 2019). Pressurized Liquid Extraction (PLE) also known as Accelerated Solvent Extraction (ASE), combines temperature (50—200 °C) and pressure (35—200 bar) to increase mass transfer rate and reduce the viscosity and surface tension of the solvents, allowing to keep it in a liquid state (Grosso, Valentão, Ferreres, & Andrade, 2015). Under a biorefinery approach applying ASE to *U. lactuca, S. latissima* and *P. umbilicalis,* the protein content was lower than in the crude biomass, probably because proteins were removed during phlorotannins extraction (Harrysson et al., 2018). However, ASE technique allowed to concentrated ash and glutamate, which can be used as salt replacers and/or umami enhancers.

Purification of extracted protein also represents a challenge, especially for novel proteins, because their physico-chemical characteristics are often unknown. Indeed, the selection of methods depends on the nature and final application, and the scale of production. Single or combined methods could be used, such as anion-exchange chromatography, hydrophobic interactions, or gel permeation, sodium dodecylsulfate polyacrylamide gel electrophoresis SDS-PAGE (1-D or 2-D), and lectin or dyes affinity, for example, the Yariv reagent to bind glycoproteins.

4. Challenge: protein characterization

Characterization and/or identification of isolated proteins is possible by direct comparison with standards molecules, or with literature data. However, when compounds are unknown and standard are not available, comprehensive and systematic techniques are required.

4.1 SDS-PAGE

SDS-PAGE is performed to identify the molecular weight of dominant protein sub-units bands. The Tris-Glycine SDS-PAGE stained with silver nitrate allowed to identify that the protein profile of *Ulva* sp. varies with seasons, showing sharper protein bands during October and November than September and March (Wijesekara et al., 2017). SDS method was effective to recognize five proteins bands in *Himantalia elongata* with molecular weights of71, 53, 43, 36 and 27 kD (Garcia-Vaquero, Lopez-Alonso, & Hayes, 2017). R-PE purified in protein extracted from *Furcelaria lumbricalis* analyzed by denaturing SDS-PAGE, allowed to identify a band near the 25 kDa region (Saluri, Kaldmäe, & Tuvikene, 2019). The periodic acid-Schiff (PAS) staining after SDS-PAGE was used to detect glycoproteins in *L. japonica* (Go et al., 2010).

4.2 Chromatography

High Performance Liquid Chromatograph (HPLC) is widely used for separation and purification of many compounds. R-PE was over 60% recovered and identified by combining HPLC coupled to photodiode array detector (PAD), and Size Exclusion Chromatography (HPSEC) in *F. lumbricalis* (Saluri et al., 2019). HPSEC reveled molecular weights from 2.6 to 3.8 kDa, in protein extracted from *A. nodosum* (Kadam et al., 2013). HPSEP and SDS-PAGE highlighted protein degradation, based on the apparent molecular weight profiles from *S. latissima* after different postharvest stabilization methods (Abdollahi et al., 2019). The analyses of amino acids through gas chromatography (GC) or LC requires a derivatization step, treated using Ortho Phtalaldehyde (OPA) or Fluorenylmethoxy chloroformate (FMOC). These reaction allow the separation of amino acids on reversed phase and their detection using UV or fluorescence detectors (Fernández-Segovia, Lerma-García, Fuentes, & Barat, 2018; Vieira et al., 2018). Recently, Kazir et al. (2019) presented the possibility to analyze the amino acid profile without derivatization using anion exchange (HPAEC-PAD).

4.3 Spectrometry

Fourier Transform Infrared (FTIR) spectroscopy provides information about the structural composition of proteins, especially for the secondary structural composition. The FRIT-ATR (Attenuated Total Reflection) spectrum of protein extracts from *M. pyrifera* and *C. chamissoi*, showed bands at 3281 cm^{-1} and 3274 cm^{-1} may corresponds to amide of the protein polypeptide skeleton because the N–H vibrations; bands at 1637 cm^{-1} and 1544 cm^{-1} could be identified as C=O vibrations of the peptide bond of proteins. While bands at 1220 cm^{-1} and 1243 cm^{-1} may corresponds to S=O vibrations from sulfated polysaccharide (fucoidan and carrageenan), and could indicate the presence of polysaccharides co-precipitated with proteins (Vásquez et al., 2019). In protein concentrate from *Kappaphycus alvarezzi*, bands at 704 cm^{-1} revealing N–H bending, and band at 616 cm^{-1} could be related to phosphate group (Suresh Kumar et al., 2014). An accurate protein identification and characterization can be done through mass spectrometry (MS), since supports mass determination, purity, mass fingerprint, sequence analysis, post-translational modifications analysis and protein-protein interactions. Electrospray ionization (ESI) and matrix-assisted laser desorption ionizations/time-of-flight (MALDI-TOF) have appeared to be important tools. MALDI-TOF/MS was used to confirm the presence of protein part in fractions rich in glycoproteins obtained from *Ulva* sp. (Wijesekara et al., 2017). The combination of RP-HPLC with inductively coupled plasma mass (ICP-MS) allowed the succeed determination of iodinate proteins such as mono-iodotyrosine (MIT) and di-iodotyrosine (DIT), with limits of detection of 1.1 and 4.3 ng g^{-1}, respectively, in the edible seaweeds *P. palmata*, *P. umbilicalis*, *Ulva rigida*, *U. pinnatifida* and *Laminaria ochroleuca* (Romarís-Hortas et al., 2013). The bioactive peptides in *P. palmata* and *S. chordalis* were identified by ESI-MS/MS online nanoscale capillary LC (RP-nano-LC), the method also allowed to identify the peptides origin (Bondu et al., 2015). The fibrinolytic enzyme codiase was determined by MS and SDS-PAGE in *C. fragile* (Choi et al., 2013).

5. Biological activities

5.1 Antioxidant activity

Seaweeds have been developed efficient and complex defense systems of antioxidants to protect themselves against oxidation. The antioxidant activity evaluated by the Oxygen Radical Absorbance Capacity (ORAC) of extract rich in proteins obtained by EAE, was higher in *M. pyrifera* than *C. chamissoi* (83 vs. 35 μmol Trolox Equivalent (TE) g^{-1} protein) (Vásquez et al., 2019). While the extracts containing 20% protein from *Ulva* sp. and *Gracilaria* sp., showed 10–20 times higher antioxidant capacity than β-Lactoglobulin (β-Lg), bovine serum albumin (BSA), and potato protein isolates, evaluated by the ferric reducing antioxidant power (FRAP) assay and ORAC. The activity of the protein was attributed to cysteine, methionine, tyrosine, phenylalanine, and tryptophan (Robin et al., 2018), and to the presence of 0.7%–1.0% dw of polyphenols (Kazir et al., 2019). These amino acids, including histidine were also reported in *R. pseudopalmata* cultivated under stress conditions of light and salinities. These results suggested that they are involved in antioxidant protection (Pliego-Cortés et al., 2017). Peptides <1 kDa from *S. chordalis* exhibited higher antioxidant activities (2,2-diphenyl-1-picrylhydrazyl (DPPH), FRAP and ORAC), than the parent proteins (Bondu et al., 2015). Due to high bioactive of peptides, they are the most commonly occurring antioxidant substances in food (Admassu, Gasmalla, Yang, & Zhao, 2018). Antioxidant activity of seaweed proteins and their types of proteins have been previously reported (Admassu et al., 2018; Harnedy & FitzGerald, 2011; Pangestuti & Kim, 2015; Wada et al., 2015).

5.2 Antivirus activity

The high-mannose specific lectin and its recombinants from *K. alvarezii* showed a potent anti-HIV activity. This lectin has strong affinity to bind to the viral envelope glycoprotein gp120 (Hirayama, Shibata, Imamura, Sakaguchi, & Hori, 2016). Very low concentrations (EC_{50} 0.043 and 0.63 nM) of lectin Griffithsin GRFT, isolated from the red alga *Griffithsia* sp. were needed to inhibit the cytopathic effects of laboratory strains and clinical primary isolates of HIV-1 on T-lymphoblastic cells, and macrophage-tropic (M-tropic) strains of HIV-1. GRFT also aborted cell-to-cell fusion and transmission of HIV-1 infection. The lectin produced with success by expression of a corresponding DNA sequence in *Escherichia coli*, being suitable for large-scale production and provide an effective and economical prophylaxis strategy for HIV infection (Mori et al., 2005). Since then, recombinant GRFT has been produced in diverse organisms, such as *Nicotiana benthamiana* plant and/or *E. coli*, which showed a strong *in vitro* activity against Hepatitis C Virus (HCV) (Meuleman et al., 2011). GRFT also showed good bioavailability after subcutaneous injection, with a significant *in vivo* efficacy in reducing HCV viral titers in a mouse model system with engrafted human hepatocytes (Takebe et al., 2013). The recombinant GRFT combined with carrageenan obtained from Gelymar (Puerto Montt, Chile), showed higher activity against herpes simplex virus 2 (HSV-2) and human papillomavirus (HPV) compared to placebo (Levendosky et al., 2015).

A wide range of antiviral spectrum against influenza A and B virus has been obtained from lectins. ESA-2 from the red seaweed *Echeuma serra*, showed the highest activity at EC_{50} 0.8 nM (Sato et al., 2015). HRL40 from the green algae *H. renschii*, inhibited the infection of influenza virus A strain A/H3N2/Udorn/72 (Mu et al., 2017). While, KAA-2 from *K. alvarezii*, effectively inactivated swine origin H1N1 influenza virus. Lectins inhibits the virus entry into the cells by binding to viral envelop glycoproteins (Sato, Morimoto, Hirayama, & Hori, 2011).

5.3 Antimicrobial activity

Antibacterial peptides (>10 kDa) isolated from *S. longicruris* exhibited activity against the bacterium *Staphylococcus aureus*. The identified peptides came from precursors such as a protein similar to ubiquitin, histones, a hypothetical leukine rich repeat protein and a ribosomal protein, which might be associated with the innate immune defenses of this brown seaweed (Beaulieu, Bondu, Doiron, Rioux, & Turgeon, 2015). Bacterial inhibition obtained from *H. elongata* extracts, was almost 100% against *Salmonella abony* and *Listeria monocytogenes*, using protein and carbohydrates model food systems. A bacteriostatic effect was observed preventing further growth of the bacteria (Cox, Hamilton Turley, Rajauria, Abu-Ghannam, & Jaiswal, 2014). Likewise growth reductions of *S. aureus*, *S. epidermidis*, and *Pseudomonas aeruginosa*, was caused by the lectins, BSL from *Bryothamnion seaforthii*, and HML from *Hypnea musciformis* (Vasconcelos et al., 2014). This later red seaweed also showed antifungal activity on human pathogenic yeasts *Candida albicans* and *C. guilliermondii* using a protein fraction rich in lectin (Cordeiro, Gomes, Carvalho, & Melo, 2006). While, extracts from *E. serra* and *Pterocladia capillacea*, rich in lectins, inhibited markedly the growth rate of *Vibrio vulnificus*. Futher, the antibacterial activity from *P. capillacea* was improved when algae was grown under higher irradiance, severe nutrient stress and temperature of 20 °C (Liao, Lin, Shieh, Jeng, & Huang, 2003).

5.4 Anti-inflammatory activity

Seaweed proteins are potential anti-inflammatory candidates. The *Pyropia yezoensis* peptide (PYP1) showed a potent inhibition activity on the release of pro-inflammatory cytokines, and induced the proliferation of intestinal epithelial cells IEC-6 (Lee et al., 2015). In a recent study, the synthesized peptide Cyclophilin, from *P. yezoensis* (pyCyp), activated the epidermal growth factor receptor (EGFR) signaling pathway, which is involved in the regulation of cell growth, proliferation, and survival of IEC-6 cells (Jung, Choi, Lee, Choi, & Nam, 2019). The anti-inflammatory activity of UPGP, a purified glycoprotein from *U. pinnatifida*, was evident since inhibited COX-1 and COX-2 enzymes (IC_{50} 53 and 193 $\mu g\, mL^{-1}$) in the cyclooxygenase assay, whereas inhibition of nitric oxide (NO) production reached 94.2% in lipopolysaccharide (LPS) activated RAW 264.7 macrophages (Rafiquzzaman et al., 2015). The protein hydrolysate from *Porphyra columbina* inhibits the production of pro-inflammatory cytokines by macrophages under LPS stimulation (Cian, López-Posadas, Drago, Sánchez de Medina, & Martínez-Augustin, 2012). A potentially valuable tool for studies the complex event of inflammation was observed using the lectin CcL, from *Caulerpa cupressoides* var. *lycopodium,* which revealed that the anti-inflammatory effect involves the inhibition of cytokines secretion (IL-1b, TNF-a, IL-6), and COX-2 in dextran or histamine-induced rat paw edema model (de Queiroz et al., 2015). Similarly, CcL showed a potential anti-inflammatory effects in the temporomandibular joint disorders (da Conceição Rivanor et al., 2014).

5.5 Anti-hypertensive activity

Cardiovascular diseases are affliction that affect human increasingly, one of the most at risk is hypertension, which can be regulated by the action of two key enzymes, namely Angiotensin Converting Enzyme (ACE) and Renin Converting Enzymes (Admassu et al., 2018). Many studies show the activity of seaweed-derived proteins in the downregulation of these enzymes. An extract of *M. pyrifera* containing 1.72 μg protein mL^{-1}, showed an ACE inhibition of 38.8% (Vásquez et al., 2019). The protein hydrolysates from *P. palmata,* showed that ACE inhibitory activity was independent from the time of harvesting (Harnedy, Soler-Vila, Edwards, & FitzGerald, 2014). Nine ACE-inhibitory peptides derived from phycobiliproteins were identified, and the oligopeptide LRY (Leu-Arg-Tyr) demonstrated particularly high inhibitory activities, with an IC_{50} of 0.044 μM (Furuta, Miyabe, Yasui, Kinoshita, & Kishimura, 2016). The activity of ACE-inhibitory peptide Pro-Ala-Phe-Gly extracted from *E. clathrata,* showed an IC_{50} value of 35.9 μM. Moreover, their work confirmed that this product was resistant to gastrointestinal protease, making it a potential pharmaceutical or nutraceutical product (Pan et al., 2016). Most recently, two peptides obtained from *Caulerpa lentillifera*, the oligo-peptides FP-5 (Phe-Asp-Gly-Ile-Pro) and AA-7 (Ala-Ile-Asp-Pro-Val-Arg-Ala), exhibits IC_{50} of 58.9 and 65.8 μM respectively (Joel, Sutopo, Prajitno, Su, & Hsu, 2018). Phycoerythrins, obtained from red seaweeds, also showed ACE inhibitory activity (Furuta et al., 2016).

5.6 Anti-cancer activity

The anti-cancer activity from glycoproteins (GP) have been reported in *C. decorticatum*, which revealed a significant inhibition of cell growth in breast, cervical and lung cancer cells, the cell membrane damages was induced by releasing the LDH enzyme producing apoptosis (Senthilkumar & Jayanthi, 2016). The *U. pinnatifida* glycoprotein UPGP, showed hypoglycaemic activity against yeast and rat intestinal α-glucosidase (Rafiquzzaman et al., 2015), while in *Capsosiphon fulvescens,* Cf-GP induces apoptosis signaling in gastric cancer (AGS) cells, through the inhibition of β-catenin, Wnt-1 and transcription factors (Kim, Kim, & Nam, 2012) and downregulated the expression of growth-related proteins, especially integrin (Kim, Kim, & Nam, 2013a). Cf-GP could be used for the development of functional foods and therapeutic agents, since this glycoprotein also inhibit cell invasion and proliferation (Kim, Kim, & Nam, 2013b). Another glycoprotein, LJGP, from the brown alga *L. japonica*, inhibited the proliferation of several cancer cell lines (AGS, HepG2 and HT-29), HT-29 were the most sensitive, reaching 60% inhibition, and apoptosis of colon cancer cells were observed (Go et al., 2010). R-PE induced apoptosis of AGS cancer cells by arresting the SGC-7901 cell, due to the expression of the CDC25A protein, and reduced the formation of Cyclin-CDK complex (Tan et al., 2016). R-PE showed *in vitro* activity against A549 and HepG2 cancer cell lines, the cell shrinkage, membrane blebbing, and nuclear DNA fragmentation were observed as characteristic apoptotic features (Senthilkumar et al., 2013). The use of R-PE subunits (α, β, γ) were an attractive option for improving the selectivity of photodynamic therapy (PDT) in mouse cancer tumor cell S180 and human liver carcinoma cell SMC 7721 (Bei, Guang-Ce, Chen-Kui, & Zhen-gang, 2002). The mixture of two *S. filiformis* lectin (SfL) isoforms (SfL-1 and SfL-2) inhibited 50% of viability in human breast adenocarcinoma (MCF-7) cells, and 34% of inhibition in Human Dermal Fibroblasts (HDF), and also could potentially exert antitumor activity because the ability to bind to high-mannose oligosaccharides present in the MCF-7 cells (Chaves et al., 2018). The lectin *E. serra* agglutinin (ESA), induced apoptotic cancer cell death, such as in human osteosarcoma cells (OST) and murine osteosarcoma cell line (LM8), by decreasing cell viability of 54% and 41%, respectively (Hayashi et al., 2012). The inhibition of human colon cancer (Colo201) cell growth and diminishing tumor growth implanted in nude mice by treated with ESA in Span 80 vesicles as drug delivery system was observed (Omokawa et al., 2010).

5.7 Others biological activities

The anti-Alzheimer activity was reported for the glycoprotein UPGP purified from *U. pinnatifida*, which effectively inhibits the enzymatic activity of acetylcholinesterase (AChE), butarylcholinesterase (BChE) and β—secretase (BACE-I) enzymes involved in the Alzheimer's disease, with IC_{50} values of 63.5, 99.0, and 73.3 $\mu g\, mL^{-1}$, respectively (Rafiquzzaman et al., 2015). Protection against dexamethasone (DEX)-induced myotube atrophy has been reported from *P. yezoensis* peptides PYP15 (Lee, Choi, Choi, & Nam, 2019), and crude protein PYCP (Lee, Choi, Choi, & Nam, 2018). The former inhibited the insulin-like growth factor-I receptor (IGF-IR) and the Akt-mTORC1 signaling pathway, and the latter increased the body weight, calf thickness, and muscle weight in mice. The bioactivity against neurodegenerative diseases associated with age have been studied. The oral administration (15 $mg\, kg^{-1}\, day^{-1}$) of hydrophilic green algae *C. fulvescens* glycoproteins (Cf-hGP), during four weeks, promoted the activation of the synaptosomal BDNF-ERK1/2 signaling in the dorsal hippocampus due to the inhibition of the endoplasmic reticulum (ER) stress-induced cognitive dysfunction (Oh & Nam, 2019). Similarly, microinjections (1 $\mu g\, kg^{-1}$, 0.54 nmol) of phycoerythrin-derived peptide of *P. yezoensis* (PYP) attenuated the glucose-regulated protein 78 (GRP78) expression in rat prefrontal cortex caused by perfluorooctane sulfonate (PFOS, 10 mg/kg) a stable fluorosurfactant, which causes ERs in the brain (Oh, Kim, & Nam, 2018). While the supplementation of *Hizikia fusiformis* glycoprotein in fish diet improved the immunity in juvenile fish olive flounder, because the interleukin IL-2 and IL-6 levels increased significantly in fish fed with glycoprotein compared to those in the control (Choi, Kim, Han, Nam, & Lee, 2014). An antidepressant-like effect, was reported from *S. filiformis* lectin (SfL), the results showed no sychostimulant and anxiolytic-like effects in mice treated with intravenous injections (Abreu et al., 2018). Lectin also has the potential to be used as a bio-pesticide, the *Gracilaria ornata* lectin significantly affected the development of *Callosobruchus maculatus* larvae, a common pest of stored cowpea seeds (Leite et al., 2005).

6. Concluding remarks

Seaweed have been identified as excellent sources for protein and seaweed-derived proteins since their quality, bioactivities, and availability; IMTA practices following by biorefinery strategies are potential alternatives to face the protein variability and availability from wild harvesting. Seaweed proteins and seaweed-derived proteins are yet to be explored and they are promising candidates to incorporate in the food, pharmaceutical and cosmeceutical products. Moreover, they are an excellent material for the animal feed industry. The molecular structures of these proteins need to be elucidated, the global advances in high-resolution instruments and bioinformatics support these analyses for reliable results, such as the increased attention focused on metabolomics and proteomics.

Acknowledgments

First author, Hugo Pliego-Cortés, gratefully acknowledges the Région Bretagne, France (SAD-Glycoalg-2017) for funding support; the research leading to these results has received funding from the People Programme (Marie Curie Actions) of the European Union's Seventh Framework Programme (FP7/2007–2013) under REA grant agreement n. PCOFUND-GA-2013-609102, through the PRESTIGE programme coordinated by Campus France. ECOS Nord Mexico M14A03. HPC acknowledges to SECTEI, Ciudad de Mexico for the scholarship (SECTEI/130/2019). Isuru Wijesekara acknowledges National Research Council, Sri Lanka (NRC Grant: 19-095).

References

Abdel-fattah, A. F., & Sary, H. H. (1987). Glycoproteins from *Ulva lactuca*. *Phytochemistry, 26*(5), 1447–1448. https://doi.org/10.1016/S0031-9422(00)81831-2.

Abdollahi, M., Axelsson, J., Carlsson, N.-G., Nylund, G. M., Albers, E., & Undeland, I. (2019). Effect of stabilization method and freeze/thaw-aided precipitation on structural and functional properties of proteins recovered from brown seaweed (*Saccharina latissima*). *Food Hydrocolloids, 96*, 140–150. https://doi.org/10.1016/J.FOODHYD.2019.05.007.

Abreu, T. M., Monteiro, V. S., Martins, A. B. S., Teles, F. B., da Conceição Rivanor, R. L., Mota, É. F., et al. (2018). Involvement of the dopaminergic system in the antidepressant-like effect of the lectin isolated from the red marine alga *Solieria filiformis* in mice. *International Journal of Biological Macromolecules, 111*, 534–541. https://doi.org/10.1016/J.IJBIOMAC.2017.12.132.

Admassu, H., Gasmalla, M. A. A., Yang, R., & Zhao, W. (2018). Bioactive peptides derived from seaweed protein and their health benefits: Antihypertensive, antioxidant, and antidiabetic properties. *Journal of Food Science, 83*(1), 6–16. https://doi.org/10.1111/1750-3841.14011.

Alemañ, A. E., Robledo, D., & Hayashi, L. (2019). Development of seaweed cultivation in Latin America: Current trends and future prospects. *Phycologia, 58*(5), 462–471. https://doi.org/10.1080/00318884.2019.1640996.

Angell, A. R., Angell, S. F., de Nys, R., & Paul, N. A. (2016a). Seaweed as a protein source for mono-gastric livestock. *Trends in Food Science & Technology, 54*, 74–84. https://doi.org/10.1016/J.TIFS.2016.05.014.

Angell, A. R., Mata, L., de Nys, R., & Paul, N. A. (2016b). The protein content of seaweeds: A universal nitrogen-to-protein conversion factor of five. *Journal of Applied Phycology, 28*(1), 511–524. https://doi.org/10.1007/s10811-015-0650-1.

Ashkenazi, D. Y., Israel, A., & Abelson, A. (2019). A novel two-stage seaweed integrated multi-trophic aquaculture. *Reviews in Aquaculture, 11*(1), 246–262. https://doi.org/10.1111/raq.12238.

Astorga-España, M. S., Rodríguez-Galdón, B., Rodríguez-Rodríguez, E. M., & Díaz-Romero, C. (2016). Amino acid content in seaweeds from the Magellan Straits (Chile). *Journal of Food Composition and Analysis, 53*, 77–84. https://doi.org/10.1016/j.jfca.2016.09.004.

Barbarino, E., & Lourenço, S. O. (2005). An evaluation of methods for extraction and quantification of protein from marine macro- and microalgae. *Journal of Applied Phycology, 17*(5), 447–460. https://doi.org/10.1007/s10811-005-1641-4.

Barre, A., Simplicien, M., Benoist, H., Van Damme, E. J. M., & Rougé, P. (2019). Mannose-specific lectins from marine algae: Diverse structural Scaffolds associated to common virucidal and anti-cancer properties. *Marine Drugs, 17*(8), 440. https://doi.org/10.3390/md17080440.

Beaulieu, L., Bondu, S., Doiron, K., Rioux, L.-E., & Turgeon, S. L. (2015). Characterization of antibacterial activity from protein hydrolysates of the macroalga *Saccharina longicruris* and identification of peptides implied in bioactivity. *Journal of Functional Foods, 17*, 685–697. https://doi.org/10.1016/J.JFF.2015.06.026.

Bei, H., Guang-Ce, W., Chen-Kui, Z., & Zhen-gang, L. (2002). The experimental research of R-phycoerythrin subunits on cancer treatment: A new photosensitizer in PDT. *Cancer Biotherapy and Radiopharmaceuticals, 17*(1), 35–42. https://doi.org/10.1089/10849780252824055.

Bikker, P., van Krimpen, M. M., van Wikselaar, P., Houweling-Tan, B., Scaccia, N., van Hal, J. W., et al. (2016). Biorefinery of the green seaweed *Ulva lactuca* to produce animal feed, chemicals and biofuels. *Journal of Applied Phycology, 28*(6), 3511–3525. https://doi.org/10.1007/s10811-016-0842-3.

Bleakley, S., & Hayes, M. (2017). Algal proteins: Extraction, application, and challenges concerning production. *Foods, 6*(5), 33. https://doi.org/10.3390/foods6050033.

Bondu, S., Bonnet, C., Gaubert, J., Deslandes, É., Turgeon, S. L., & Beaulieu, L. (2015). Bioassay-guided fractionation approach for determination of protein precursors of proteolytic bioactive metabolites from macroalgae. *Journal of Applied Phycology, 27*(5), 2059–2074. https://doi.org/10.1007/s10811-014-0425-0.

Burlot, A.-S., Bedoux, G., & Bourgougnon, N. (2016). Response surface methodology for enzyme-assisted extraction of water- soluble antiviral compounds from the proliferative macroalga *Solieria chordalis*. *Enzyme Engineering, 05*(02), 1–8. https://doi.org/10.4172/2329-6674.1000148.

Carreto, J. I., & Carignan, M. O. (2011). Mycosporine-like amino acids: Relevant secondary metabolites. Chemical and ecological aspects. *Marine Drugs, 9*(3), 387–446. https://doi.org/10.3390/md9030387.

Chaves, R. P., da Silva, S. R., Nascimento Neto, L. G., Carneiro, R. F., da Silva, A. L. C., Sampaio, A. H., et al. (2018). Structural characterization of two isolectins from the marine red alga *Solieria filiformis* (Kützing) P.W. Gabrielson and their anticancer effect on MCF-7 breast cancer cells. *International Journal of Biological Macromolecules, 107*, 1320–1329. https://doi.org/10.1016/J.IJBIOMAC.2017.09.116.

Choi, Y. H., Kim, K.-W., Han, H.-S., Nam, T. J., & Lee, B.-J. (2014). Dietary *Hizikia fusiformis* glycoprotein-induced IGF-I and IGFBP-3 associated to somatic growth, polyunsaturated fatty acid metabolism, and immunity in juvenile olive flounder *Paralichthys olivaceus*. *Comparative Biochemistry and Physiology Part A: Molecular & Integrative Physiology, 167*, 1–6. https://doi.org/10.1016/J.CBPA.2013.09.011.

Choi, J.-H., Sapkota, K., Park, S.-E., Kim, S., & Kim, S.-J. (2013). Thrombolytic, anticoagulant and antiplatelet activities of codiase, a bi-functional fibrinolytic enzyme from *Codium fragile*. *Biochimie, 95*(6), 1266–1277. https://doi.org/10.1016/J.BIOCHI.2013.01.023.

Cian, R. E., Garzón, A. G., Ancona, D. B., Guerrero, L. C., & Drago, S. R. (2015). Hydrolyzates from *Pyropia columbina* seaweed have antiplatelet aggregation, antioxidant and ACE I inhibitory peptides which maintain bioactivity after simulated gastrointestinal digestion. *Lebensmittel-Wissenschaft und -Technologie- Food Science and Technology, 64*(2), 881–888. https://doi.org/10.1016/j.lwt.2015.06.043.

Cian, R. E., López-Posadas, R., Drago, S. R., Sánchez de Medina, F., & Martínez-Augustin, O. (2012). A *Porphyra columbina* hydrolysate upregulates IL-10 production in rat macrophages and lymphocytes through an NF-κB, and p38 and JNK dependent mechanism. *Food Chemistry, 134*(4), 1982–1990. https://doi.org/10.1016/J.FOODCHEM.2012.03.134.

da Conceição Rivanor, R. L., Chaves, H. V., do Val, D. R., de Freitas, A. R., Lemos, J. C., Rodrigues, J. A. G., et al. (2014). A lectin from the green seaweed *Caulerpa cupressoides* reduces mechanical hyper-nociception and inflammation in the rat temporomandibular joint during zymosan-induced arthritis. *International Immunopharmacology, 21*(1), 34–43. https://doi.org/10.1016/J.INTIMP.2014.04.009.

Contreras, L., Ritter, A., Dennett, G., Boehmwald, F., Guitton, N., Pineau, C., et al. (2008). Two-dimensional gel electrophoresis analysis of brown algal protein extracts. *Journal of Phycology, 44*(5), 1315–1321. https://doi.org/10.1111/j.1529-8817.2008.00575.x.

Cordeiro, R. A., Gomes, V. M., Carvalho, A. F. U., & Melo, V. M. M. (2006). Effect of proteins from the red seaweed *Hypnea musciformis* (Wulfen) Lamouroux on the growth of human pathogen yeasts. *Brazilian Archives of Biology and Technology, 49*(6), 915–921. https://doi.org/10.1590/S1516-89132006000700008.

Cox, S., Hamilton Turley, G., Rajauria, G., Abu-Ghannam, N., & Jaiswal, A. K. (2014). Antioxidant potential and antimicrobial efficacy of seaweed (*Himanthalia elongata*) extract in model food systems. *Journal of Applied Phycology, 26*(4), 1823–1831. https://doi.org/10.1007/s10811-013-0215-0.

Dawczynski, C., Schubert, R., & Jahreis, G. (2007). Amino acids, fatty acids, and dietary fibre in edible seaweed products. *Food Chemistry, 103*(3), 891–899. https://doi.org/10.1016/j.foodchem.2006.09.041.

Denis, C., Massé, A., Fleurence, J., & Jaouen, P. (2009a). Concentration and pre-purification with ultrafiltration of a R-phycoerythrin solution extracted from macro-algae *Grateloupia turuturu*: Process definition and up-scaling. *Separation and Purification Technology, 69*(1), 37–42. https://doi.org/10.1016/J.SEPPUR.2009.06.017.

Denis, C., Morançais, M., Gaudin, P., & Fleurence, J. (2009b). Effect of enzymatic digestion on thallus degradation and extraction of hydrosoluble compounds from *Grateloupia turuturu*. *Botanica Marina, 52*(3), 262–267. https://doi.org/10.1515/BOT.2009.035.

Dumay, J., Clément, N., Morançais, M., & Fleurence, J. (2013). Optimization of hydrolysis conditions of *Palmaria palmata* to enhance R-phycoerythrin extraction. *Bioresource Technology, 131*, 21–27. https://doi.org/10.1016/J.BIORTECH.2012.12.146.

Dumay, J., Morançais, M., Munier, M., Le Guillard, C., & Fleurence, J. (2014). Phycoerythrins: Valuable proteinic pigments in red seaweeds. *Advances in Botanical Research, 71*, 321–343. https://doi.org/10.1016/B978-0-12-408062-1.00011-1.

Estevez, J. M., Fernández, P. V., Kasulin, L., Dupree, P., & Ciancia, M. (2009). Chemical and in situ characterization of macromolecular components of the cell walls from the green seaweed *Codium fragile*. *Glycobiology, 19*(3), 212–228. https://doi.org/10.1093/glycob/cwn101.

Fernández-Segovia, I., Lerma-García, M. J., Fuentes, A., & Barat, J. M. (2018). Characterization of Spanish powdered seaweeds: Composition, antioxidant capacity and technological properties. *Food Research International, 111*(January), 212–219. https://doi.org/10.1016/j.foodres.2018.05.037.

Fernández, P. V., Ciancia, M., Miravalles, A. B., & Estevez, J. M. (2010). Cell-wall polymer mapping in the coenocytic macroalga *Codium vermilara* (bryopsidales, chlorophyta). *Journal of Phycology, 46*(3), 456–465. https://doi.org/10.1111/j.1529-8817.2010.00821.x.

Fleurence, J., Le Coeur, C., Mabeau, S., Maurice, M., & Landrein, A. (1995a). Comparison of different extractive procedures for proteins from the edible seaweeds *Ulva rigida* and *Ulva rotundata*. *Journal of Applied Phycology, 7*(6), 577–582. https://doi.org/10.1007/BF00003945.

Fleurence, J., Massiani, L., Guyader, O., & Mabeau, S. (1995b). Use of enzymatic cell wall degradation for improvement of protein extraction from *Chondrus crispus, Gracilaria verrucosa* and *Palmaria palmata*. *Journal of Applied Phycology, 7*(4), 393–397. https://doi.org/10.1007/BF00003796.

Fleurence, J., Morançais, M., & Dumay, J. (2018). Seaweed proteins. *Proteins in Food Processing*, 245–262. https://doi.org/10.1016/B978-0-08-100722-8.00010-3.

Furuta, T., Miyabe, Y., Yasui, H., Kinoshita, Y., & Kishimura, H. (2016). Angiotensin I converting enzyme inhibitory peptides derived from phycobiliproteins of dulse *Palmaria palmata*. *Marine Drugs, 14*(2). https://doi.org/10.3390/md14020032.

Gajaria, T. K., Suthar, P., Baghel, R. S., Balar, N. B., Sharnagat, P., Mantri, V. A., et al. (2017). Integration of protein extraction with a stream of byproducts from marine macroalgae: A model forms the basis for marine bioeconomy. *Bioresource Technology, 243*, 867–873. https://doi.org/10.1016/J.BIORTECH.2017.06.149.

Galland-Irmouli, A.-V., Fleurence, J., Lamghari, R., Luçon, M., Rouxel, C., Barbaroux, O., et al. (1999). Nutritional value of proteins from edible seaweed *Palmaria palmata* (dulse). *The Journal of Nutritional Biochemistry, 10*(6), 353–359. https://doi.org/10.1016/S0955-2863(99)00014-5.

Garcia-Vaquero, M., Lopez-Alonso, M., & Hayes, M. (2017). Assessment of the functional properties of protein extracted from the brown seaweed *Himanthalia elongata* (Linnaeus) S. F. Gray. *Food Research International, 99*, 971–978. https://doi.org/10.1016/J.FOODRES.2016.06.023.

Go, H., Hwang, H.-J., & Nam, T.-J. (2010). A glycoprotein from *Laminaria japonica* induces apoptosis in HT-29 colon cancer cells. *Toxicology In Vitro, 24*(6), 1546–1553. https://doi.org/10.1016/J.TIV.2010.06.018.

Gorres, K. L., & Raines, R. T. (2010). Prolyl 4-hydroxylase. *Critical Reviews in Biochemistry and Molecular Biology, 45*(2), 106–124. https://doi.org/10.3109/10409231003627991.

Gressler, V., Yokoya, N. S., Fujii, M. T., Colepicolo, P., Filho, J. M., Torres, R. P., & Pinto, E. (2010). Lipid, fatty acid, protein, amino acid and ash contents in four Brazilian red algae species. *Food Chemistry, 120*(2), 585–590. https://doi.org/10.1016/j.foodchem.2009.10.028.

Grosso, C., Valentão, P., Ferreres, F., & Andrade, P. (2015). Alternative and efficient extraction methods for marine-derived compounds. *Marine Drugs, 13*(5), 3182–3230. https://doi.org/10.3390/md13053182.

Grote, B. (2019). Recent developments in aquaculture of *Palmaria palmata* (Linnaeus) (Weber & Mohr 1805): Cultivation and uses. *Reviews in Aquaculture, 11*(1), 25–41. https://doi.org/10.1111/raq.12224.

Hardouin, K., Bedoux, G., Burlot, A.-S., Donnay-Moreno, C., Bergé, J.-P., Nyvall-Collén, P., et al. (2016). Enzyme-assisted extraction (EAE) for the production of antiviral and antioxidant extracts from the green seaweed *Ulva armoricana* (Ulvales, Ulvophyceae). *Algal Research, 16*, 233–239. https://doi.org/10.1016/j.algal.2016.03.013.

Hardouin, K., Burlot, A.-S., Umami, A., Tanniou, A., Stiger-Pouvreau, V., Widowati, I., et al. (2013). Biochemical and antiviral activities of enzymatic hydrolysates from different invasive French seaweeds. *Journal of Applied Phycology, 26*(2), 1029–1042. https://doi.org/10.1007/s10811-013-0201-6.

Harnedy, P. A., & FitzGerald, R. J. (2011). Bioactive proteins, peptides, and amino acids from macroalgae. *Journal of Phycology, 47*(2), 218–232. https://doi.org/10.1111/j.1529-8817.2011.00969.x.

Harnedy, P. A., Soler-Vila, A., Edwards, M. D., & FitzGerald, R. J. (2014). The effect of time and origin of harvest on the in vitro biological activity of *Palmaria palmata* protein hydrolysates. *Food Research International, 62*, 746–752. https://doi.org/10.1016/J.FOODRES.2014.04.035.

Harrysson, H., Hayes, M., Eimer, F., Carlsson, N.-G., Toth, G. B., & Undeland, I. (2018). Production of protein extracts from Swedish red, green, and brown seaweeds, *Porphyra umbilicalis* Kützing, *Ulva lactuca* Linnaeus, and *Saccharina latissima* (Linnaeus) J. V. Lamouroux using three different methods. *Journal of Applied Phycology, 30*(6), 3565–3580. https://doi.org/10.1007/s10811-018-1481-7.

Hayashi, K., Walde, P., Miyazaki, T., Sakayama, K., Nakamura, A., Kameda, K., et al. (2012). Active targeting to osteosarcoma cells and apoptotic cell death induction by the novel lectin *Eucheuma serra* agglutinin isolated from a marine red alga. *Journal of Drug Delivery, 2012*, 842785. https://doi.org/10.1155/2012/842785.

Hervé, C., Siméon, A., Jam, M., Cassin, A., Johnson, K. L., Salmeán, A. A., et al. (2016). Arabinogalactan proteins have deep roots in eukaryotes: Identification of genes and epitopes in brown algae and their role in *Fucus serratus* embryo development. *New Phytologist, 209*(4), 1428–1441. https://doi.org/10.1111/nph.13786.

Hirayama, M., Shibata, H., Imamura, K., Sakaguchi, T., & Hori, K. (2016). High-mannose specific lectin and its recombinants from a Carrageenophyta *Kappaphycus alvarezii* represent a potent anti-HIV activity through high-affinity binding to the viral envelope glycoprotein gp120. *Marine Biotechnology, 18*(1), 144–160. https://doi.org/10.1007/s10126-015-9677-1.

Hwang, H. J., Kim, I. H., & Nam, T. J. (2008). Effect of a glycoprotein from *Hizikia fusiformis* on acetaminophen-induced liver injury. *Food and Chemical Toxicology, 46*(11), 3475–3481. https://doi.org/10.1016/J.FCT.2008.08.032.

Ingram, G. A. (1985). Lectins and lectin-like molecules in lower plants — I. Marine algae. *Developmental & Comparative Immunology, 9*(1), 1–10. https://doi.org/10.1016/0145-305X(85)90054-0.

Joel, C. H., Sutopo, C. C. Y., Prajitno, A., Su, J. H., & Hsu, J. L. (2018). Screening of angiotensin-I converting enzyme inhibitory peptides derived from *caulerpa lentillifera*. *Molecules, 23*(11), 1–15. https://doi.org/10.3390/molecules23113005.

Jung, J.-H., Choi, J.-W., Lee, M.-K., Choi, Y.-H., & Nam, T.-J. (2019). Effect of Cyclophilin from *Pyropia yezoensis* on the proliferation of intestinal epithelial cells by epidermal growth factor receptor/ras signaling pathway. *Marine Drugs, 17*(5). https://doi.org/10.3390/md17050297.

Kadam, S. U., Tiwari, B. K., & O'Donnell, C. P. (2013). Application of novel extraction technologies for bioactives from marine algae. *Journal of Agricultural and Food Chemistry, 61*(20), 4667–4675. https://doi.org/10.1021/jf400819p.

Kazir, M., Abuhassira, Y., Robin, A., Nahor, O., Luo, J., Israel, A., et al. (2019). Extraction of proteins from two marine macroalgae, *Ulva* sp. and *Gracilaria* sp., for food application, and evaluating digestibility, amino acid composition and antioxidant properties of the protein concentrates. *Food Hydrocolloids, 87*(February 2018), 194–203. https://doi.org/10.1016/j.foodhyd.2018.07.047.

Kim, E.-Y., Kim, D.-G., Kim, Y.-R., Hwang, H.-J., Nam, T.-J., & Kong, I.-S. (2010). An improved method of protein isolation and proteome analysis with *Saccharina japonica* (Laminariales) incubated under different pH conditions. *Journal of Applied Phycology, 23*(1), 123–130. https://doi.org/10.1007/s10811-010-9550-6.

Kim, E.-Y., Rafiquzzaman, S. M., Lee, J. M., Noh, G., Jo, G., Lee, J. H., & Kong, I. S. (2015). Structural features of glycoprotein purified from *Saccharina japonica* and its effects on the selected probiotic properties of *Lactobacillus plantarum* in Caco-2 cell. *Journal of Applied Phycology, 27*(2), 965–973. https://doi.org/10.1007/s10811-014-0390-7.

Kim, Y.-M., Kim, I.-H., & Nam, T.-J. (2012). Induction of apoptosis signaling by glycoprotein of *Capsosiphon fulvescens* in human gastric cancer (AGS) cells. *Nutrition and Cancer, 64*(5), 761–769. https://doi.org/10.1080/01635581.2012.683228.

Kim, Y.-M., Kim, I., & Nam, T.-J. (2013a). *Capsosiphon fulvescens* glycoprotein reduces AGS gastric cancer cell migration by downregulating transforming growth factor-β1 and integrin expression. *International Journal of Oncology, 43*(4), 1059–1065. https://doi.org/10.3892/ijo.2013.2055.

Kim, Y.-M., Kim, I., & Nam, T.-J. (2013b). Inhibition of AGS human gastric cancer cell invasion and proliferation by *Capsosiphon fulvescens* glycoprotein. *Molecular Medicine Reports, 8*(1), 11–16. https://doi.org/10.3892/mmr.2013.1492.

Laohakunjit, N., Selamassakul, O., & Kerdchoechuen, O. (2014). Seafood-like flavour obtained from the enzymatic hydrolysis of the protein by-products of seaweed (*Gracilaria* sp.). *Food Chemistry, 158*, 162–170. https://doi.org/10.1016/J.FOODCHEM.2014.02.101.

Le Guillard, C., Bergé, J.-P., Donnay-Moreno, C., Bruzac, S., Ragon, J.-Y., Baron, R., et al. (2016). Soft liquefaction of the red seaweed *Grateloupia turuturu* Yamada by ultrasound-assisted enzymatic hydrolysis process. *Journal of Applied Phycology, 28*(4), 2575–2585. https://doi.org/10.1007/s10811-015-0788-x.

Le Guillard, C., Dumay, J., Donnay-Moreno, C., Bruzac, S., Ragon, J. Y., Fleurence, J., et al. (2015). Ultrasound-assisted extraction of R-phycoerythrin from *Grateloupia turuturu* with and without enzyme addition. *Algal Research, 12*, 522–528. https://doi.org/10.1016/j.algal.2015.11.002.

Lee, M.-K., Choi, J.-W., Choi, Y. H., & Nam, T.-J. (2018). *Pyropia yezoensis* protein supplementation prevents dexamethasone-induced muscle atrophy in C57BL/6 mice. *Marine Drugs, 16*(9). https://doi.org/10.3390/md16090328.

Lee, M.-K., Choi, J.-W., Choi, Y. H., & Nam, T.-J. (2019). Protective effect of *Pyropia yezoensis* peptide on dexamethasone-induced myotube Atrophy in C2C12 myotubes. *Marine Drugs, 17*(5), 284. https://doi.org/10.3390/md17050284.

Lee, M.-K., Kim, I.-H., Choi, Y.-H., Choi, J.-W., Kim, Y.-M., & Nam, T.-J. (2015). The proliferative effects of *Pyropia yezoensis* peptide on IEC-6 cells are mediated through the epidermal growth factor receptor signaling pathway. *International Journal of Molecular Medicine, 35*(4), 909–914. https://doi.org/10.3892/ijmm.2015.2111.

Leite, Y. F. M. M., Silva, L. M. C. M., Amorim, R. C. das N., Freire, E. A., de Melo Jorge, D. M., Grangeiro, T. B., et al. (2005). Purification of a lectin from the marine red alga *Gracilaria ornata* and its effect on the development of the cowpea weevil *Callosobruchus maculatus* (Coleoptera: Bruchidae). *Biochimica et Biophysica Acta (BBA) - General Subjects, 1724*(1–2), 137–145. https://doi.org/10.1016/J.BBAGEN.2005.03.017.

Levendosky, K., Mizenina, O., Martinelli, E., Jean-Pierre, N., Kizima, L., Rodriguez, A., et al. (2015). Griffithsin and Carrageenan combination to target herpes simplex virus 2 and human papillomavirus. *Antimicrobial Agents and Chemotherapy, 59*(12), 7290–7298. https://doi.org/10.1128/AAC.01816-15.

Liao, W.-R., Lin, J.-Y., Shieh, W.-Y., Jeng, W.-L., & Huang, R. (2003). Antibiotic activity of lectins from marine algae against marine vibrios. *Journal of Industrial Microbiology and Biotechnology, 30*(7), 433–439. https://doi.org/10.1007/s10295-003-0068-7.

Madden, M., Mitra, M., Ruby, D., & Schwarz, J. (2012). Seasonality of selected nutritional constituents of edible delmarva seaweeds. *Journal of Phycology, 48*(5), 1289–1298. https://doi.org/10.1111/j.1529-8817.2012.01207.x.

Magnusson, M., Glasson, C. R. K., Vucko, M. J., Angell, A., Neoh, T. L., & de Nys, R. (2019). Enrichment processes for the production of high-protein feed from the green seaweed *Ulva ohnoi*. *Algal Research, 41*, 101555. https://doi.org/10.1016/J.ALGAL.2019.101555.

Marsham, S., Scott, G. W., & Tobin, M. L. (2007). Comparison of nutritive chemistry of a range of temperate seaweeds. *Food Chemistry, 100*(4), 1331–1336. https://doi.org/10.1016/J.FOODCHEM.2005.11.029.

Matsubara, K., Hori, K., Matsuura, Y., & Miyazawa, K. (1999). A fibrinolytic enzyme from a marine green alga. *Codium latum*. *Phytochemistry, 52*(6), 993–999. https://doi.org/10.1016/S0031-9422(99)00356-8.

Meuleman, P., Albecka, A., Belouzard, S., Vercauteren, K., Verhoye, L., Wychowski, C., et al. (2011). Griffithsin has antiviral activity against hepatitis C virus. *Antimicrobial Agents and Chemotherapy, 55*(11), 5159–5167. https://doi.org/10.1128/AAC.00633-11.

Mittal, R., Sharma, R., & Raghavarao, K. (2019). Aqueous two-phase extraction of R-Phycoerythrin from marine macro-algae, *Gelidium pusillum*. *Bioresource Technology, 280*, 277–286. https://doi.org/10.1016/J.BIORTECH.2019.02.044.

Mittal, R., Tavanandi, H. A., Mantri, V. A., & Raghavarao, K. S. M. S. (2017). Ultrasound assisted methods for enhanced extraction of phycobiliproteins from marine macro-algae, *Gelidium pusillum* (Rhodophyta). *Ultrasonics Sonochemistry, 38*, 92–103. https://doi.org/10.1016/j.ultsonch.2017.02.030.

Mori, T., O'Keefe, B. R., Sowder, R. C., Bringans, S., Gardella, R., Berg, S., et al. (2005). Isolation and characterization of griffithsin, a novel HIV-inactivating protein, from the red alga *Griffithsia* sp. *Journal of Biological Chemistry, 280*(10), 9345–9353. https://doi.org/10.1074/jbc.M411122200.

Mu, J., Hirayama, M., Sato, Y., Morimoto, K., & Hori, K. (2017). A novel high-mannose specific lectin from the green alga *Halimeda renschii* exhibits a potent anti-influenza virus activity through high-affinity binding to the viral hemagglutinin. *Marine Drugs, 15*(8), 255. https://doi.org/10.3390/md15080255.

Murakami, K., Yamaguchi, Y., Noda, K., Fujii, T., Shinohara, N., Ushirokawa, T., et al. (2011). Seasonal variation in the chemical composition of a marine brown alga, *Sargassum horneri* (Turner) C. Agardh. *Journal of Food Composition and Analysis, 24*(2), 231–236. https://doi.org/10.1016/J.JFCA.2010.08.004.

Nagai, K., Yotsukura, N., Ikegami, H., Kimura, H., & Morimoto, K. (2008). Protein extraction for 2-DE from the lamina of *Ecklonia kurome* (laminariales): Recalcitrant tissue containing high levels of viscous polysaccharides. *Electrophoresis, 29*(3), 672–681. https://doi.org/10.1002/elps.200700461.

Oh, J., Kim, E.-Y., & Nam, T.-J. (2018). Phycoerythrin peptide from *Pyropia yezoensis* alleviates endoplasmic reticulum stress caused by perfluorooctane sulfonate-induced calcium dysregulation. *Marine Drugs, 16*(2), 44. https://doi.org/10.3390/md16020044.

Oh, J., & Nam, T.-J. (2019). Hydrophilic glycoproteins of an edible green alga *Capsosiphon fulvescens* prevent aging-induced spatial memory impairment by suppressing GSK-3β-mediated ER stress in dorsal Hippocampus. *Marine Drugs, 17*(3), 168. https://doi.org/10.3390/md17030168.

Okolie, C. L., Akanbi, T. O., Mason, B., Udenigwe, C. C., & Aryee, A. N. A. (2019). Influence of conventional and recent extraction technologies on physicochemical properties of bioactive macromolecules from natural sources: A review. *Food Research International, 116*, 827–839. https://doi.org/10.1016/J.FOODRES.2018.09.018.

Omokawa, Y., Miyazaki, T., Walde, P., Akiyama, K., Sugahara, T., Masuda, S., et al. (2010). In vitro and in vivo anti-tumor effects of novel Span 80 vesicles containing immobilized

Eucheuma serra agglutinin. *International Journal of Pharmaceutics, 389*(1–2), 157–167. https://doi.org/10.1016/J.IJPHARM.2010.01.033.
Paiva, L., Lima, E., Patarra, R. F., Neto, A. I., & Baptista, J. (2014). Edible Azorean macroalgae as source of rich nutrients with impact on human health. *Food Chemistry, 164*, 128–135. https://doi.org/10.1016/J.FOODCHEM.2014.04.119.
Pangestuti, R., & Kim, S.-K. (2015). Seaweed proteins, peptides, and amino acids. *Seaweed Sustainability*, 125–140. https://doi.org/10.1016/B978-0-12-418697-2.00006-4.
Pan, S., Wang, S., Jing, L., & Yao, D. (2016). Purification and characterisation of a novel angiotensin-I converting enzyme (ACE)-inhibitory peptide derived from the enzymatic hydrolysate of *Enteromorpha clathrata* protein. *Food Chemistry, 211*, 423–430. https://doi.org/10.1016/j.foodchem.2016.05.087.
Peumans, W. J., & Van Damme, E. J. (1995). Lectins as plant defense proteins. *Plant Physiology, 109*(2), 347–352. https://doi.org/10.1104/pp.109.2.347.
Pliego-Cortés, H., Bedoux, G., Boulho, R., Taupin, L., Freile-Pelegrín, Y., Bourgougnon, N., et al. (2019). Stress tolerance and photoadaptation to solar radiation in *Rhodymenia pseudopalmata* (Rhodophyta) through mycosporine-like amino acids, phenolic compounds, and pigments in an Integrated Multi-Trophic Aquaculture system. *Algal Research, 41*, 101542. https://doi.org/10.1016/J.ALGAL.2019.101542.
Pliego-Cortés, H., Caamal-Fuentes, E., Montero-Muñoz, J., Freile-Pelegrín, Y., & Robledo, D. (2017). Growth, biochemical and antioxidant content of *Rhodymenia pseudopalmata* (Rhodymeniales, Rhodophyta) cultivated under salinity and irradiance treatments. *Journal of Applied Phycology, 29*(5), 2595–2603. https://doi.org/10.1007/s10811-017-1085-7.
Polikovsky, M., Fernand, F., Sack, M., Frey, W., Müller, G., & Golberg, A. (2016). Towards marine biorefineries: Selective proteins extractions from marine macroalgae *Ulva* with pulsed electric fields. *Innovative Food Science & Emerging Technologies, 37*, 194–200. https://doi.org/10.1016/J.IFSET.2016.03.013.
Popper, Z. A., Michel, G., Hervé, C., Domozych, D. S., Willats, W. G. T., Tuohy, M. G., et al. (2011). Evolution and diversity of plant cell walls: From algae to flowering plants. *Annual Review of Plant Biology, 62*(1), 567–590. https://doi.org/10.1146/annurev-arplant-042110-103809.
Postma, P. R., Cerezo-Chinarro, O., Akkerman, R. J., Olivieri, G., Wijffels, R. H., Brandenburg, W. A., et al. (2018). Biorefinery of the macroalgae *Ulva lactuca*: Extraction of proteins and carbohydrates by mild disintegration. *Journal of Applied Phycology, 30*(2), 1281–1293. https://doi.org/10.1007/s10811-017-1319-8.
Puspita, M., Déniel, M., Widowati, I., Radjasa, O. K., Douzenel, P., Marty, C., et al. (2017). Total phenolic content and biological activities of enzymatic extracts from *Sargassum muticum* (Yendo) Fensholt. *Journal of Applied Phycology, 29*(5), 2521–2537. https://doi.org/10.1007/s10811-017-1086-6.
de Queiroz, I. N. L., Quinderé, A. L. G., Rodrigues, J. A. G., de Sousa Oliveira Vanderlei, E., Ribeiro, N. A., da Conceição Rivanor, R. L., et al. (2015). Dual effects of a lectin from the green seaweed *Caulerpa cupressoides* var. *lycopodium* on inflammatory mediators in classical models of inflammation. *Inflammation Research, 64*(12), 971–982. https://doi.org/10.1007/s00011-015-0880-3.
Rafiquzzaman, S. M., Kim, E. Y., Lee, J. M., Mohibbullah, M., Alam, M. B., Soo Moon, I., et al. (2015). Anti-Alzheimers and anti-inflammatory activities of a glycoprotein purified from the edible brown alga *Undaria pinnatifida*. *Food Research International,* 77, 118–124. https://doi.org/10.1016/J.FOODRES.2015.08.021.
Robin, A., Kazir, M., Sack, M., Israel, A., Frey, W., Mueller, G., et al. (2018). Functional protein concentrates extracted from the green marine macroalga Ulva sp., by high voltage pulsed electric fields and mechanical press. *ACS Sustainable Chemistry & Engineering, 6*(11), 13696–13705. https://doi.org/10.1021/acssuschemeng.8b01089.

Romarís-Hortas, V., Bermejo-Barrera, P., & Moreda-Piñeiro, A. (2013). Ultrasound-assisted enzymatic hydrolysis for iodinated amino acid extraction from edible seaweed before reversed-phase high performance liquid chromatography—inductively coupled plasma-mass spectrometry. *Journal of Chromatography A, 1309*, 33—40. https://doi.org/10.1016/J.CHROMA.2013.08.022.
Saluri, M., Kaldmäe, M., & Tuvikene, R. (2019). Extraction and quantification of phycobiliproteins from the red alga *Furcellaria lumbricalis*. *Algal Research, 37*, 115—123. https://doi.org/10.1016/J.ALGAL.2018.11.013.
Sato, Y., Morimoto, K., Hirayama, M., & Hori, K. (2011). High mannose-specific lectin (KAA-2) from the red alga *Kappaphycus alvarezii* potently inhibits influenza virus infection in a strain-independent manner. *Biochemical and Biophysical Research Communications, 405*(2), 291—296. https://doi.org/10.1016/J.BBRC.2011.01.031.
Sato, Y., Morimoto, K., Kubo, T., Sakaguchi, T., Nishizono, A., Hirayama, M., et al. (2015). Entry inhibition of influenza viruses with high mannose binding lectin ESA-2 from the red alga *Eucheuma serra* through the recognition of viral hemagglutinin. *Marine Drugs, 13*(6), 3454—3465. https://doi.org/10.3390/md13063454.
Senthilkumar, D., & Jayanthi, S. (2016). Partial characterization and anticancer activities of purified glycoprotein extracted from green seaweed *Codium decorticatum*. *Journal of Functional Foods, 25*, 323—332. https://doi.org/10.1016/J.JFF.2016.06.010.
Senthilkumar, N., Kurinjimalar, C., Thangam, R., Suresh, V., Kavitha, G., Gunasekaran, P., et al. (2013). Further studies and biological activities of macromolecular protein R-Phycoerythrin from *Portieria hornemannii*. *International Journal of Biological Macromolecules, 62*, 107—116. https://doi.org/10.1016/J.IJBIOMAC.2013.08.004.
Shpigel, M., Shauli, L., Odintsov, V., Ashkenazi, N., & Ben-Ezra, D. (2018). *Ulva lactuca* biofilter from a land-based integrated multi trophic aquaculture (IMTA) system as a sole food source for the tropical sea urchin *Tripneustes gratilla elatensis*. *Aquaculture, 496*, 221—231. https://doi.org/10.1016/J.AQUACULTURE.2018.06.038.
Singh, R. S., & Walia, A. K. (2018). Lectins from red algae and their biomedical potential. *Journal of Applied Phycology, 30*(3), 1833—1858. https://doi.org/10.1007/s10811-017-1338-5.
Stiger-Pouvreau, V., Bourgougnon, N., & Deslandes, E. (2016). Carbohydrates from seaweeds. *Seaweed in Health and Disease Prevention*, 223—274. https://doi.org/10.1016/B978-0-12-802772-1.00008-7.
Suetsuna, K., Maekawa, K., & Chen, J. R. (2004). Antihypertensive effects of *Undaria pinnatifida* (wakame) peptide on blood pressure in spontaneously hypertensive rats. *Journal of Nutritional Biochemistry*. https://doi.org/10.1016/j.jnutbio.2003.11.004.
Surendraraj, A., Habeebullah, S. F. K., & Jacobsen, C. (2015). *Isolation of glycoproteins from brown algae*. Retrieved from: https://orbit.dtu.dk/en/publications/isolation-of-glycoproteins-from-brown-algae(0f503087-a93d-413e-819e-19a701912562).html.
Suresh Kumar, K., Ganesan, K., Selvaraj, K., & Subba Rao, P. V. (2014). Studies on the functional properties of protein concentrate of *Kappaphycus alvarezii* (Doty) Doty — an edible seaweed. *Food Chemistry, 153*, 353—360. https://doi.org/10.1016/J.FOODCHEM.2013.12.058.
Takebe, Y., Saucedo, C. J., Lund, G., Uenishi, R., Hase, S., Tsuchiura, T., et al. (2013). Antiviral lectins from red and blue-green algae show potent in vitro and in vivo activity against hepatitis C virus. *PLoS One, 8*(5), e64449. https://doi.org/10.1371/journal.pone.0064449.
Tan, H., Gao, S., Zhuang, Y., Dong, Y., Guan, W., Zhang, K., et al. (2016). R-phycoerythrin induces SGC-7901 apoptosis by arresting cell cycle at S phase. *Marine Drugs, 14*(9), 166. https://doi.org/10.3390/md14090166.
Tan, L., Showalter, A. M., Egelund, J., Hernandez-Sanchez, A., Doblin, M. S., & Bacic, A. (2012). Arabinogalactan-proteins and the research challenges for these enigmatic plant

cell surface proteoglycans. *Frontiers of Plant Science, 3*, 140. https://doi.org/10.3389/fpls.2012.00140.

Thangam, R., Senthilkumar, D., Suresh, V., Sathuvan, M., Sivasubramanian, S., Pazhanichamy, K., et al. (2014). Induction of ROS-Dependent Mitochondria-Mediated Intrinsic Apoptosis in MDA-MB-231 Cells by Glycoprotein from *Codium decorticatum*. *Journal of Agricultural and Food Chemistry, 62*(15), 3410–3421. https://doi.org/10.1021/jf405329e.

Tee, M. Z., Yong, Y. S., Rodrigues, K. F., & Yong, W. T. L. (2015). Growth rate analysis and protein identification of *Kappaphycus alvarezii* (Rhodophyta, Gigartinales) under pH induced stress culture. *Aquaculture Reports, 2*, 112–116. https://doi.org/10.1016/J.AQREP.2015.09.001.

Ünlü, E. S., Ünüvar, Ö. C., & Aydın, M. (2019). Identification of alternative oxidase encoding genes in *Caulerpa cylindracea* by de novo RNA-Seq assembly analysis. *Marine Genomics, 46*, 41–48. https://doi.org/10.1016/J.MARGEN.2019.03.004.

Uribe, E., Vega-Gálvez, A., García, V., Pastén, A., López, J., & Goñi, G. (2018). Effect of different drying methods on phytochemical content and amino acid and fatty acid profiles of the green seaweed, *Ulva* spp. *Journal of Applied Phycology*. https://doi.org/10.1007/s10811-018-1686-9.

Vasconcelos, M. A., Arruda, F. V. S., Carneiro, V. A., Silva, H. C., Nascimento, K. S., Sampaio, A. H., et al. (2014). Effect of algae and plant lectins on planktonic growth and biofilm formation in clinically relevant bacteria and yeasts. *BioMed Research International, 2014*, 365272. https://doi.org/10.1155/2014/365272.

Vásquez, V., Martínez, R., & Bernal, C. (2019). Enzyme-assisted extraction of proteins from the seaweeds *Macrocystis pyrifera* and *Chondracanthus chamissoi*: Characterization of the extracts and their bioactive potential. *Journal of Applied Phycology, 31*(3), 1999–2010. https://doi.org/10.1007/s10811-018-1712-y.

Vázquez-Delfín, E., Freile-Pelegrín, Y., Pliego-Cortés, H., & Robledo, D. (2019). Seaweed resources of Mexico: Current knowledge and future perspectives. *Botanica Marina, 0*(0). https://doi.org/10.1515/bot-2018-0070.

Vieira, E. F., Soares, C., Machado, S., Correia, M., Ramalhosa, M. J., Oliva-teles, M. T., et al. (2018). Seaweeds from the Portuguese coast as a source of proteinaceous material: Total and free amino acid composition profile. *Food Chemistry, 269*(April), 264–275. https://doi.org/10.1016/j.foodchem.2018.06.145.

Vilg, J. V., & Undeland, I. (2017). pH-driven solubilization and isoelectric precipitation of proteins from the brown seaweed *Saccharina latissima*—effects of osmotic shock, water volume and temperature. *Journal of Applied Phycology, 29*(1), 585–593. https://doi.org/10.1007/s10811-016-0957-6.

Wada, N., Sakamoto, T., Matsugo, S., Wada, N., Sakamoto, T., & Matsugo, S. (2015). Mycosporine-like amino acids and their derivatives as natural antioxidants. *Antioxidants, 4*(3), 603–646. https://doi.org/10.3390/antiox4030603.

Weinbreck, F., Nieuwenhuijse, H., Robijn, G., & de Kruif, C. (2004). Complexation of whey proteins with carrageenan. *Journal of Agricultural and Food Chemistry, 52*(11), 3550–3555. https://doi.org/10.1021/JF034969T.

Wijesekara, I., & Kim, S.-K. (2015). Application of marine algae derived nutraceuticals in the food industry. In *Marine algae extracts* (pp. 627–638). https://doi.org/10.1002/9783527679577.ch35.

Wijesekara, I., Lang, M., Marty, C., Gemin, M.-P., Boulho, R., Douzenel, P., et al. (2017). Different extraction procedures and analysis of protein from *Ulva* sp. in Brittany, France. *Journal of Applied Phycology, 29*(5), 2503–2511. https://doi.org/10.1007/s10811-017-1239-7.

Wong, P.-F., Tan, L.-J., Nawi, H., & AbuBakar, S. (2006). Proteomics of the red alga, *Gracilaria changii* (gracilariales, rhodophyta). *Journal of Phycology, 42*(1), 113–120. https://doi.org/10.1111/j.1529-8817.2006.00182.x.

Yang, D., Wang, J., Bao, Y., & An, L. (2003). Purification and characterization of *Ulva pertusa* Kjellm alkaline phosphatase. *Preparative Biochemistry & Biotechnology, 33*(2), 113–123. https://doi.org/10.1081/PB-120021436.

Yoshiie, T., Maeda, M., Kimura, M., Hama, Y., Uchida, M., & Kimura, Y. (2012). Structural Features of N -Glycans of Seaweed Glycoproteins: Predominant Occurrence of High-Mannose Type N-Glycans in Marine Plants. *Bioscience, Biotechnology, and Biochemistry, 76*(10), 1996–1998. https://doi.org/10.1271/bbb.120463.

CHAPTER TEN

Defence on surface: macroalgae and their surface-associated microbiome

Shareen A. Abdul Malik[a,*], Gilles Bedoux[a], Jose Q. Garcia Maldonado[b], Yolanda Freile-Pelegrín[b], Daniel Robledo[b] and Nathalie Bourgougnon[a]

[a]Laboratoire de Biotechnologie et Chimie Marines, EA3884, UBS, IUEM, F-56000 Vannes, France
[b]CINVESTAV Marine Resources Department, Mérida, Yucatan, Mexico
*Corresponding author: E-mail: shareen.a-abdul-malik-arockiasamy@univ-ubs.fr

Contents

Advances in Botanical Research, Volume 95
ISSN 0065-2296
https://doi.org/10.1016/bs.abr.2019.11.009

Abstract

Macroalgae surfaces are considered as potentially suitable substrata present underwater for the dynamic epibiotic microbiome. They secrete various organic substances as nutrients for the multiplication of bacteria and the formation of microbial biofilms. The microbial biofilm itself acts as a biological defence mechanism to prevent the establishment of pathogenic microbes as well as to protect the surface of the seaweed from the macrofoulers. In most cases, the presence and diversity of these organisms are host-specific. The macroalgal host and their associated microbiont are often referred to as a 'Holobiont'.

The surface-associated microbiome of macroalgae plays a crucial role in the growth, morphogenesis and defence in the normal favourable environmental conditions of the macroalgae (Symbiotism). Whereas, when there is a change i.e. reversed, unfavourable conditions, the microbiome reverses its symbiotic nature and affects growth and degrades the macroalgae by causing different diseases in different species. In parallel, marine macroalgae employ different defence mechanisms, a physical defence such as continuous shedding of the outer layer of cells, and the mucilaginous covering and chemical defence such as the production of secondary metabolites.

The following chapter focuses on the macroalgal surface-associated microbiome and its defence system with a certain number of relevant examples from related literature on this subject.

1. Introduction

Macroalgae, being important habitat formers and primary producers in the marine ecosystem, provide potential settling substrate and protective environments for the early life and development of many invertebrates. Attached to their substrate, marine seaweeds are linked to their environment and confronted with various abiotic and biotic stresses. They are affected by wave action, light radiation, desiccation, and variations in salinity, temperature, nutrient supply, herbivore predation, epibiosis, microfouling and human activity. In the marine environment, competition for space and nutrients is intense. Each species lives in this complex environment according to its own physiological parameters, and its ability to resist changes in the state of the environment, particularly, during low tides or in reduced space. Due to the extraordinary complexity of the marine environment, seaweeds establish close associations with other species to ensure their survival and competitiveness for space.

Macroalgal surfaces represent a major physiological and highly functional interface with the environment. Such an interface is always threatened by fouling, and thus macroalgae need to develop many defence strategies to protect their surface from a vast number of consequences (Wahl, Goecke, Labes, Dobretsov, & Weinberger, 2012). *Biofilms on solid surfaces* are omnipresent and therefore macroalgae are of no exception to this. Biofilm formation is an interactive process affected by local hydrodynamics of the fluid environment, physicochemical properties of the surface and behavioural response of bacterial colonizers (Geesey, 2001). In biofilms, the self-produced matrix is mainly composed of polysaccharides, proteins and eDNA synthesized by the associated bacteria, diatoms, fungi, and other microorganisms (Garg, Jain, & Bhosle, 2009).

Host-microbe interactions play a crucial role in marine ecosystems, in a symbiotic community where the largest partner, based on its size, is considered as a host, e.g. macroalgae (Dittami et al., 2019). The host and all of its associated microbiota, which can be acknowledged as all the microorganisms which are present or associated in a particular environment is now considered as a *Holobiont* (da Gama, Plouguerné, & Pereira, 2014; Duarte et al., 2018; Egan et al., 2013; Hollants, Leliaert, De Clerck, & Willems, 2013; ; Wahl et al., 2012). The term 'holobiont' was coined by Lynn Margulis which referred to *symbiotic associations* that last for a significant part of an organism's lifetime (Cavalier-Smith, 1992). As it includes all the associated microbiota, their interactions may be harmful, beneficial or of no consequence, which may be coevolved or opportunistic, competitive or cooperative (Skillings, 2016). The combined genetic information of the associated microbiota is known as a microbiome (Dittami et al., 2019; Skillings, 2016) These microbiomes are functionally connected with their eukaryotic hosts and they are highly diverse. Specific functional genes of the associated microbiome are now regarded as secondary genomes (Singh & Reddy, 2016). *Any disturbances to the microbiome* or even *uncontrolled microbial colonization* could be *associated with stress and disease of the host organism* (Beattie, Lachnit, Dinsdale, Thomas, & Steinberg, 2018; Lachnit, Blümel, Imhoff, & Wahl, 2009; Morris et al., 2016; Rohde, Hiebenthal, Wahl, Karez, & Bischof, 2008).

Lacking a cell-based adaptive immune response, macroalgae have two different defence capabilities: constitutive defence under normal circumstances, such as the *production of metabolites by macroalgae* or its *bacterial symbionts* and *induced defence activated by any tissue damage* (Egan, Fernandes, Kumar, Gardiner, & Thomas, 2014; Weinberger, 2007). Surface-associated microbiome proved a wide range of bioactive compounds with a broad spectrum of activities, such as antibacterial, antifungal, antiviral, antiplasmodial, nematicidal, anti-inflammatory, anti-cancer and antiangiogenic activities (Singh,

Baghel, Reddy, & Jha, 2015). Here we consider the defence provided by the macroalgae themselves through the production of secondary metabolites as 'Chemical Defence' and the defence provided by the associated microbiota as 'Microbial/Biological Defence', however, it is now known that many of the bioactive compounds previously attributed to seaweeds were in fact produced or metabolized by their associated microbiota, the microbiome (da Gama et al., 2014; Egan et al., 2013; Hollants et al., 2013; Wahl et al., 2012). In this association, a highly and dynamic complex set of chemical, physical and biological interactions regulates the relationships between the *basibiont and its epibionts* (Singh, Kumari, & Reddy, 2014).

2. Seaweed surface-microbe interactions – biofilm development

Biofilms, a consortia of surface-attached microbial cells dispersed in a self-secreted extracellular polymeric matrix, constitute the principal form of microbial growth and a widespread survival strategy among microorganisms (Costerton, Geesey, & Cheng, 1978; Donlan, 2002). In the marine environment, surfaces are rapidly colonized by marine organisms in a process known as biofouling (Callow & Callow, 2002). The adsorption of dissolved macromolecules by the substratum is a preliminary step before bacterial colonization. The *algal cell wall structure* is a selective factor in the settlement of macromolecules and microorganisms. Bacteria are typically the primary colonizers of submerged surfaces. Marine macroalgae are known to produce secondary metabolites to defend themselves from the colonization of their exposed surfaces. In fact, *these metabolites and pioneering bacterial species* contribute to the chemical and physicochemical conditions responsible for a secondary recruitment of other microorganisms, but also sponges, macroalgae and invertebrates (Dang & Lovell, 2000). In addition, multi-layered Extracellular Polymeric Substances (EPSs), which include polysaccharides and also proteins, nucleic acids and lipids, form a base of biofilm structural organization (Brian-Jaisson et al., 2016; Lage & Graça, 2016).

2.1 Seaweed surface: as a substratum

A highly specific association exists between bacterial communities and marine macroalgae. The physiological and biochemical properties of the macroalgae surface predetermine the composition of the adhering microbial communities (Goecke, Labes, Wiese, & Imhoff, 2010). Macroalgae surfaces provide a rich habitat and release large amounts of organic carbon and oxygen into the surrounding environment, providing nutrients for microorganisms, triggering chemotactic behaviour of bacteria or inducing microbial colonization and reproduction (Goecke et al., 2010; Steinberg, De Nys, & Kjelleberg, 2002).

2.1.1 Cell wall structure

The cell wall structure of macroalgae is mainly composed of mannans, xylans, cellulose, hemicellulose and sulfated or/and anionic polysaccharides. The cell wall matrix polysaccharides are specific: Chlorophyta (green seaweeds) contain sulphated xyloarabinogalactans, glucuronoxylorhamnans, glucuronoxylorhamnogalactans and ulvans. Phaeophyceae (brown seaweeds) produce alginate and fucose-containing sulfated polysaccharides, whereas Rhodophyta (red seaweeds) contain neutral and sulfated polysaccharides such as agars and carrageenans (Guedes, Amaro, Sousa-Pinto, & Malcata, 2019; Percival, 1979). Within the algal lineages, these polysaccharides are highly diversified in terms of their degree of sulfation, esterification, molecular weight and conformation of sugar residue (Kloareg & Quatrano, 1988; Mabeau & Kloareg, 1987; Popper et al., 2011). Apart from the above mentioned structural polysaccharides, the red, green and, brown macroalgae also contain storage polysaccharides such as floridean starch, and, laminarin and mannitol, respectively (Stiger-Pouvreau, Bourgougnon, & Deslandes, 2016).

2.1.2 Surface topographical features

The composition and function of surface consortia are significantly influenced by the chemically mediated interactions and communication between the associated microorganisms and the hosts (Egan, Thomas, & Kjelleberg, 2008). The surface topographical features include a mucilage and sheath demonstrating a gelatinous covering of the macroalgae (Chapman et al., 2014) and it influences the overall epibiosis (Sullivan & Regan, 2011). The microtopography of the interface is a selective factor in the settlement of macromolecules and microorganisms. (Chapman et al., 2014; Sullivan & Regan, 2011). Therefore, the physical and chemical properties of the substratum play a key role in the attachment and establishment of the initial bacterial colonization. Surface roughness is often speculated to have a deterring role either in attaching or promoting the easy release of fouling organisms (Callow & Callow, 2011). In a study on 'bioinspired synthetic macroalgae' the surface topography of the brown algae, *Saccharina latissima* and *Fucus guiryi*, was replicated on synthetic elastomer and evaluated for its antifouling nature. Together with the chemical defence, the surface topography efficiently reduced the fouling by 40%, where it has been proved that the topography and chemistry work synergistically (Chapman et al., 2014).

2.2 Macroalgae defence

Macroalgae control epiphytic microbial colonization by several processes.

2.2.1 Removal of surface layers

Under pressure from epibionts, some algal thalli can decompose, generating a kind of desquamation of tissues on the surface. This phenomenon has been observed in several species of brown algae *Ascophyllum nodosum, Sargassum muticum*, some Corallinaceae and *Dilsea carnosa.* In the summer where light intensity and temperature are at extremes, physical processes involve regular shedding of biofilm surface cell layers. For example, the Gigartinales *Chondracanthus chamissoi* whitens and decomposes. During this period, many epiphytes and herbivores were observed. Over time, the alga, which is more vulnerable, produces defence molecules.

2.2.2 Production of Reactive Oxygen Species

Many seaweed species are known to produce bioactive compounds through their specialized metabolism. The production of ROS or oxidative burst forms an important component of the stress produced in marine organisms. This production is triggered by the contact of the cells with products resulting from the degradation of the algae wall - oligosaccharides (for example oligoguluronates in *Laminaria*; oligo-agars in *Gracilaria*) (da Gama et al., 2014). When a macroalga is colonized by bacteria or herbivores, it receives a signal that will enable it to activate its defence systems. These signals of aggression or Elicitors are defined as chemical compounds that can induce physiological changes in the target organism. In marine macroalgae, they are most often oligosaccharides resulting from the degradation of the algae walls. The presence of Polysaccharides (xyloglucan, laminarin), peptides or proteins or lipid derivatives have also been reported (Potin, Bouarab, Salaün, Pohnert, & Kloareg, 2002).

2.2.3 Antimicrobial compounds

Exuded algal bioactive compounds on the surface have antimicrobial properties, which protect the seaweed's surface from grazers, fouling organisms and bacterial pathogens. In different macroalgae species, the defence molecules rise to the surface as an antifouling. They can also be stored in different types of specialized structures, for example, gland cells of *Delisea pulchra*, 'corps en cerise' in *Laurencia* species. Physodes in brown seaweeds present in the thallus cortex are known for their antifouling compound delivery systems (Dahms & Dobretsov, 2017; Dworjanyn, Wright, Paul, de Nys, & Steinberg, 2006; Harizani, Ioannou, & Roussis, 2016). In green algae, the identified compounds mainly belong to di- and sesquiterpenes. They have been relatively little studied (Fig. 1). The diversity and richness of Rhodophyta in terms of secondary metabolites are the most important of all

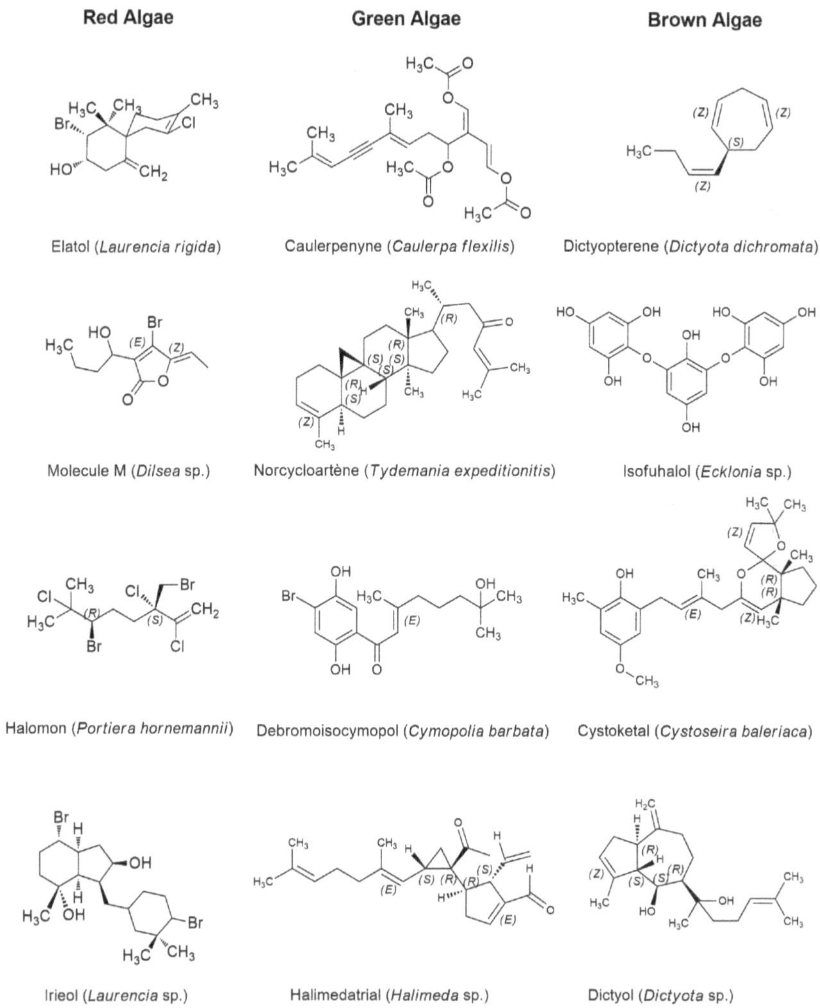

Fig. 1 Bioactive compounds produced by the red, green and brown macroalgae.

macroalgae with more than 1600 compounds identified. Red macroalgae mainly produce terpenoids and nonterpenoids such as acetogenins. Acetogenins are fatty acid derivatives and are usually halogenated, particularly C-15 acetogenins (Bawakid et al., 2017). Red algae are also characterized by the production of halogenated compounds containing bromine and chlorine. For example, the red alga *D. pulchra* produces an analogous molecule of bacterial N-Acyl-Homoserine Lactones (AHLs) known to inhibit the signal pathways in the Gram-positive and Gram-negative bacteria *Bacillus subtilis* (Ren, Sims, & Wood, 2002), *Escherichia coli* (Ren, Sims, & Wood, 2001) and *Pseudomonas aeruginosa* (Hentzer et al., 2002) these lead to selective colonization of bacteria on thalli (Steinberg & de Nys, 2002).

Another mechanism involved in red algae defence reactions against pathogen attack is the oxidation (enzymatic or non-enzymatic) of fatty acids that results in the production of oxylipins. Sugars (floridoside) from photosynthesis are involved in the defence mechanisms. In brown algae, more than 1200 molecules have been characterized, including terpenes, galactoglycerolipids, fatty acids, phlorotannins and small C11 acetogenins. For example, in *Dictyota ciliolata*, pachydictyol A, dictyol B acetate, dictyodial, and sterols were shown to be present in higher concentrations in older, less palatable tissues than in apical meristem (Cronin & Hay, 1996). Therefore, the different allocation and concentration of the chemical compounds may lead to different microbial communities at the different parts of macroalgae. There is little data about the authenticity and ecological significance of algal chemical defence compounds having verified their production, presence and/or release at ecologically realistic concentrations (Kubanek et al., 2003; Lane et al., 2009; Maximilien et al., 1998; Nyadong et al., 2009; Nylund et al., 2010). In 2010 Lachnit et al. demonstrated that available surface chemical compounds on *Fucus vesiculosus* have different main effects on bacterial colonizers; one group of compounds reduced the overall abundance of epiphytic bacteria non-discriminatively, whereas the other displayed selective and attractant features for bacteria. Consequently, the community patterns and overall densities of epiphytic bacteria are likely to have resulted from some sort of equilibrium between these opposite effects. These authors propose that the effect size will vary over the course of the year due to environmental changes, thus resulting in algal-specific but temporally variable patterns of epibiosis in accordance with previous observations. The results of this study suggest that unidentified compounds and fucoxanthin may be significant drivers of the interaction between the alga and colonizing bacteria. DiMethylSulphoPropionate (DMSP)and proline from the surface of *F. vesiculosus* are active against the attachment of five different bacterial strains isolated from the algae (Saha et al., 2012). Further dynamics downstream of the initial colonization by bacteria, such as microorganism– microorganism interactions are expected to add another effective layer that results in spatial heterogeneity of epiphytic bacterial biofilms.

3. Microbiome on the surface

3.1 Bacteria

Descriptive studies of bacteria isolated from the surface of macroalgae were reported as early as 1875 (Johansen, Nielsen, & Sjoholm, 1999). The interest in bacterial populations living in association with macroalgae has increased during recent decades (Goecke et al., 2010). In 2010, Goecke et al. found 107 studies on bacterial communities associated to a total of 148 macroalgae (36 Chlorophyta, 46 Phaeophyceae, 55 Rhodophyta, 12 undetermined algae) within the last 40 years. Bacteria often have a protective role through the production of compounds into the water that prevents the installation of biofilm on the surface. The primary biofilm of bacteria serves as cues for the further settlement of unicellular and multicellular eukaryotes. These cues can be either positive or negative (Rao et al., 2007). The presence of both attractants and antibacterial molecules will determine the promoter or inhibitory effect on biofilm establishment, predation deterrence, and germination of an epiphyte spore or metamorphosis of an epibiont invertebrate.

Earlier studies have mainly focused on elucidation of the roles of the epiphytic bacterial communities in the ecophysiology of the host macroalga. However, mutualistic interactions have become a topic of current interest. It is evident from recent studies that a fraction of epiphytic bacterial communities can be categorized as "core microbial species", suggesting an obligate association. However, the precise composition of microbiomes and their functional partnership with hosts are not fully understood (Singh & Reddy, 2016). Bacterial communities associated with marine macroalgae differ strongly from those of their surrounding seawater, sediment or substrate (Aires, Serrão, & Engelen, 2016; Bengtsson, Sjøtun, & Øvreås, 2010; Bengtsson & Øvreås, 2010). For example, the surface microbiome of *F. vesiculosus* was analysed comparatively by amplicon sequencing, using seawater as reference, and it was found that the majority of microbial OTUs (Operational Taxonomical Units) were assigned to *Alphaproteobacteria* (64%–68%)

and their share in the seawater bacterial community was even higher (84%) (Parrot et al., 2019). Bacterial communities associated to marine macroalgae are not fixed and can change temporally and spatially across seasons, life-spans, life stages and tissue types by biotic and abiotic factors (Aires et al., 2016; Mancuso, D'Hondt, Willems, Airoldi, & De Clerck, 2016; Michelou, Caporaso, Knight, & Palumbi, 2013; Staufenberger, Thiel, Wiese, & Imhoff, 2008). For example, the epibacterial diversity of *Laminaria hyperborea* (*Phaeophyta*) increases with the age/successive colonization of the kelp surface (Bengtsson, Sjøtun, Lanzén, & Øvreås, 2012). Bacterial density and community composition follow the brown seaweed seasonal growth cycle. As most of the biofilm seems to consist of bacteria utilizing carbon produced by the host (Bengtsson & Øvreås, 2010), microbiome dynamics are probably strongly linked to seasonal changes in the kelp metabolome and seawater temperature (Bengtsson et al., 2010). Table 1 illustrates the list of bacteria isolated from the major '3' phyla, red, green and brown, of macroalgae.

Table 1 List of bacteria isolated from the '3'major phyla of macroalgae.

Macroalgae	Bacteria isolated	References
Red		
Polysiphonia stricta	*Pseudoalteromonas arctica* *Shewanella basaltis*	(Nasrolahi et al., 2012)
Gracilariopsis heteroclada	*Vibrio parahaemolyticus*; *V. alginolyticus*	(Martinez & Padilla, 2016)
Gracilaria blodgetti	*Agarilytica rhodophyticola*	(Ling, Xia, Liu, Chen, & Du, 2017)
Laurenciae papillosa	*Bacillus amyloliquefaciens*	(Chakraborty et al., 2017b)
Ahnfeltia tobuchiensis	*Polaribacter staleyi*	(Nedashkovskaya et al., 2018a)
Tichocarpus crinitus	*Aquimarina algiphila*	(Nedashkovskaya et al., 2018b)
Iridaea cordata	*Streptomyces pratensis*	(Alvarado, Huang, Wang, Garrido, & Leiva, 2018)
Gracilaria corticate	*Bacillus* sp.; *Pseudomonas stutzeri*; *Vibrio owensii*	(Karthick & Mohanraju, 2018)
Mastophora rosea	*Vibrio* sp.	(Karthick & Mohanraju, 2018)
Gracilaria edulis	*Corynebacterium* sp. *Bacillus megaterium* *Klebsiella oxytoca* *Bacillus pasteurii*	(Suvega & Arunkumar, 2019)

Table 1 List of bacteria isolated from the '3'major phyla of macroalgae.—cont'd

Macroalgae	Bacteria isolated	References
	Bacillus cereus *Aeromonas* sp. *Lysinibacillus xylanilyticus* *Lactobacillus casei* *Aeromonas hydrophila*	
Undetermined	*Alteromonas macleodii*	(Naval & Chandra, 2019)
Kappaphycus alvarezii	*Bacillus* sp., *Vibrio* sp.	(Mohammed Riyaz et al. 2019)
Green		
Ulva australis	*Pseudoalteromonas tunicate* *Phaeobacter* sp.	(Rao et al. 2007)
Ulva sp.	*Kordia ulvae*	(Qi, Huang, Lai, Li, & Shao, 2016)
Ulva fenestrata	*Olleya algicola*	(Nedashkovskaya, Kim, Zhukova, & Mikhailov, 2017)
Ulva lactuca	*Bacillus* sp. *Shewanella oneidensis* *Shewanella* sp. *Pseudoalteromonas* sp. *Pseudoalteromonas rubra* *Vibrio* sp. *Bacillus cereus* *Pseudomonas pseudoalcaligenes* *Alcanivorax dieselolei*	(Ramya et al., 2017) (Karthick & Mohanraju, 2018)
Monostroma hariotii	*S. pratensis; S. brevispora; Pseudonocardia adelaidensis; Agrococcus baldri*	(Alvarado et al. 2018)
Brown		
Fucus vesiculosus	*Pseudoalteromonas mariniglutinosa* *Pseudoalteromonas tunicata* *Shewanella baltica* *Bacillus foraminis*	(Nasrolahi et al. 2012)
Fucus serratus	*Ulvibacter litoralis, Photobacterium halotolerans*	(Nasrolahi et al. 2012)
Ascophyllum nodosum	*Zobellia* sp.; *Maribacter thermophilus; Maribacter forsetii;* *Cellulophaga baltica; Cellulophaga geojensis;*	(Martin et al. 2015)

(*Continued*)

Table 1 List of bacteria isolated from the '3'major phyla of macroalgae.—cont'd

Macroalgae	Bacteria isolated	References
	Algibacter miyuki; Vibrio sp.; *Shewanella* sp.; *Paraglaciecola* sp.; *Colwellia* sp.; *Pseudoalteromonas* sp.; *Marinomonas* sp.; *Cobetia* sp.;	
Padina tetrastromatica	*Halolactibacillus alkaliphilus*	(Suresh, Renugadevi, Brammavidhya, Iyapparaj, & Anantharaman, 2015)
Splachnidium rugosum	*Vibrio cyclitrophicus; Vibrio splendidus; Vibrio comitans; Vibrio celticus; Pseudoalteromonas carrageenovora; Pseudoalteromonas espejiana; Alteromonas stellipolaris; Psychromonas arctica; Pseudomonas poae; Pseudomonas fluorescens; Neptumonas naphthovorans; Cobetia amphilecti; Shewanella* sp.; *Sphingomonas* sp.; *Sulfitobacter* sp.; *Polaribacter* sp.; *Bacillus* sp.; *Staphylococcus saprophyticus.*	(Albakosh, Naidoo, Kirby, & Bauer, 2016)
Padina pavonica	*Paracoccus* sp.; *Devosia* sp.; *Vibrio* sp.; *Pseudomonas* sp.; *Pseudomonas putida*; *Brevibacterium iodinum; Staphylococcus* sp.; *Bacillus pumilus*; *Pseudoalteromonas* sp.; *Acinetobacter* sp.; *Planomicrobium glaciei; Erwinia bilingiae.*	(Ismail et al. 2016)
Anthophycus longifolius	*Bacillus subtilis*	(Chakraborty et al., 2017a)
Sargassum myriocystum	*Bacillus subtilis*	(Chakraborty et al. 2017a)
S. polycystum *S. duplicatum* *S. echinocarpum*		(Susilowati, Sabdono, & Widowati, 2015)

Table 1 List of bacteria isolated from the '3'major phyla of macroalgae.—cont'd

Macroalgae	Bacteria isolated	References
Ecklonia cava	*Flavivirga eckloniae; F. aquimarina*	(Lee, Kang, Choe, & Seong, 2017a)
Sargassum fulvellum	*Winogradskyella flava* *Dokdonia lutea*	(Lee, Kang, Shin, & Seong, 2017b) (Choi, Kang, Lee, & Seong, 2017)
Adenocystis utricularis	*Arthrobacter flavus; Staphylococcus haemolyticus; Micrococcus luteus; Tessaracoccus flavescens; Pseudarthrobacter oxydans*	(Alvarado et al. 2018)
Fucus serratus L.	*Cellulophaga baltica* sp. *Cellulophaga fucicola* sp.	(Johansen, Nielsen, & Sjraholm, 2018)
Turbinaria ornate	*Exiguobacterium profundum*	(Karthick & Mohanraju, 2018)
Pelvitia canaliculata	*Kocuria marina*	(Uzair et al. 2018)
Laminaria japonica	*Algibacter alginolytica; Alginococcus marinus*	Sun et al., 2016 (Ying et al., 2019)
Undertermined	*Gilvimarinus polysaccharolyticus*	(Cheng et al. 2015)
Undertermined	*Marinirhabdus citrea*	(Yang, Oh, Seo, Lee, & Kwon, 2018)

Macroalgal growth and morphogenesis are particularly shown to depend on the associated bacterial communities which produce plant growth-promoting substances, quorum sensing signal molecules and several bioactive compounds (Singh & Reddy, 2014). Bacterial biofilms can have a repellent effect, for example, nonviable bacteria tended to repel cyprids when compared to the surfaces without a biofilm, which ultimately prevent heavy fouling on the surface (Nasrolahi, Stratil, Jacob, & Wahl, 2012). Bacterial communities belonging to the phyla *Proteobacteria* and *Firmicutes* are generally the most abundant epiphytic bacteria associated with macroalgal host (Singh & Reddy, 2016). The '71' heterotrophic bacteria isolated from the green alga *Ulva rigida* belong to the four phyla *Proteobacteria, Bacteroidetes, Firmicutes* and *Actinobacteria* (Ismail et al., 2018). *Marinomonas* sp. and *Bacillus* spp. were observed to have the morphogenesis-inducing ability toward its host macroalgae *Ulva fasciata* (Singh et al., 2011). An in-depth study on the 16s rRNA gene libraries of the bacterial community on the surface of *Ulva australis* and the surrounding water determines the uniqueness of the association. Both the surface and seawater libraries were dominated by *Alphaproteobacteria* and *Bacteroidetes* families, but they were observed to be clearly distinct from each other through clustering into OTUs (Burke, Thomas, Lewis, Steinberg, & Kjelleberg, 2011). *Pseudoalteromonas tunicata* and *Phaeobacter sp.* in association with the green alga *U. australis* inhibits the common fouling organisms by producing extracellular compounds even at low densities, 10^2–10^3 cells cm^{-2} and 10^3–10^4 cells cm^{-2}, respectively (Rao et al., 2007).

The *Roseobacter* species from the surface of a green alga *Ulva mutabilis* exhibits a specific chemotactic affinity to the alga and chemically communicates with the alga as well as with another associated bacterium *Cytophaga* sp. forming a symbiotic tripartite community (Spoerner, Wichard, Bachhuber, Stratmann, & Oertel, 2012). Additionally, some bacterial species produce regulatory compounds resembling cytokinin (from *Roseobacter, Sulfitobacter*, and *Halomonas*) and auxin (from *Cytophaga*) that assist in the differentiation of *U. mutabilis* (Spoerner et al., 2012). The epibiotic bacteria associated with the brown alga *F. vesiculosus* tended to repel the cyprids benefitting the host by preventing heavy fouling on its surface (Nasrolahi et al., 2012). Specific Functionally Active Bacterial Species (FABS) will serve to improve the health and performance of the associated host by suppressing the growth of pathogens (Singh & Reddy, 2016). Bondoso et al. (2017) hypothesized that specific association of *Planctomycetes* and their macroalgal hosts, such as the brown alga *Fucus spiralis*, the green alga *Ulva* sp. and the red alga *Chondrus crispus*, is likely to be determined by the nutrients provided by the host and the set of sulfatases inherent to the associated *Planctomycetes* species.

In red alga, *Asparagopsis armata*-associated bacteria revealed to be a potential source of compounds with antitumor and antibacterial activity. The extract of *Shewanella* sp. ASP 26 showed increased antimicrobial activity against *Bacillus subtilis* and *Staphylococcus aureus* (Horta et al., 2019). Ma, Liu, Yu, Li, and Cao (2009) isolated 192 bacterial strains from the surfaces of seaweeds, of which '63' isolates were shown to inhibit the settlement of algal spores and '62′ isolates inhibited larval settlement under laboratory conditions. About '25' heterotrophic algae-associated bacteria from the surfaces of *U. rigida* show antibacterial activity whereas the free-living bacteria from the surrounding seawater did not show such activity (Ismail et al., 2018). Sixty of the '180' strains isolated from different marine macroalgae exhibited antibiotic activity contrary to a series of fouling bacteria and therefore have the potential to control the microbial population on the seaweed's surface either by inhibiting their growth or by influencing the tactic behaviour of potentially competing bacteria (Boyd, Adams, & Burgess, 1999).

A broad range of antibacterial activity contra important food pathogens such as *Vibrio parahaemolyticus*, *Vibrio vulnificus* and *Aeromonas hydrophilia* was observed in *Bacillus subtilis* associated with the brown alga *Anthophycus longifolius* (Chakraborty, Thilakan, Raola, & Joy, 2017). Two polyketides from *Bacillus amyloliquefaciens*, a heterotrophic bacteria associated with the edible red alga, *Laurencia papillosa*, 3-(octahydro-9-isopropyl-2H-benzo[h]chromen-4-yl)-2-methylpropyl benzoate and methyl 8-(2- (benzoyloxy)-ethyl)-hexahydro-4-((E)-pent-2-enyl)-2H-chromene-6-carboxylate exhibits an activity contra to human opportunistic food pathogens (Chakraborty, Thilakan, & Raola, 2017b). Members of the genus *Pseudoalteromonas*, *Bacillus*, *Vibrio* and *Shewanella* isolated from the surface of the green alga, *Ulva lactuca*, showed an antialgal activity on the pennate diatom, *Cylindrotheca fusiformis* (Kumar, Rao, Thomas, Kjelleberg, & Egan, 2011). Fig. 2 illustrates bioactive metabolites isolated from the macroalgal associated bacteria.

3.2 Fungi

The fungal population on the surface of seaweed mainly consists of parasites, saprobes or asymptomatic fungal endosymbionts where the ascomycetes and anamorphic fungi were dominant. Suryanarayanan et al. (2010) observed that among the different red, green and brown macroalgae studied, the brown algae supported a higher diversity of fungal endophytes whereas the green algae had a low diversity with dense colonization. In different seaweeds, members of genera *Acremonium*, *Alternaria*, *Arthrinium*, *Aspergillus*, *Cladosporium*, *Fusarium*, *Geomyces*, *Penicillium* and *Phoma* were the most common fungal endosymbionts (reviewed by Singh et al., 2014). More recently, fungi of the Eurotiomycetes class were visualised for the first time on the surface of the brown alga *F. vesiculosus*, by CARD-FISH imaging (Parrot et al., 2019).

Aspergillus terreus isolated from *Halimeda macroloba, Gracilaria edulis* and *U. lactuca* produced metabolites which were active against bacteria, algae and fungi (Suryanarayanan et al., 2010). However, in contrast to the well-studied bacteria-algae surface interactions, there is only a little evidence on fungi and their role on macroalgal surfaces (Egan et al., 2013). Most of the studies on macroalgal associated fungi dealt with the endophytic fungi and they are well detailed by Habbu, Warad, Shastri, Madagundi, and Kulkarni (2016) together with their antimicrobial activities. The filamentous fungal community associated with the healthy and the decaying brown alga, *Fucus serratus*, is observed to be different. Such a change in the community structure could

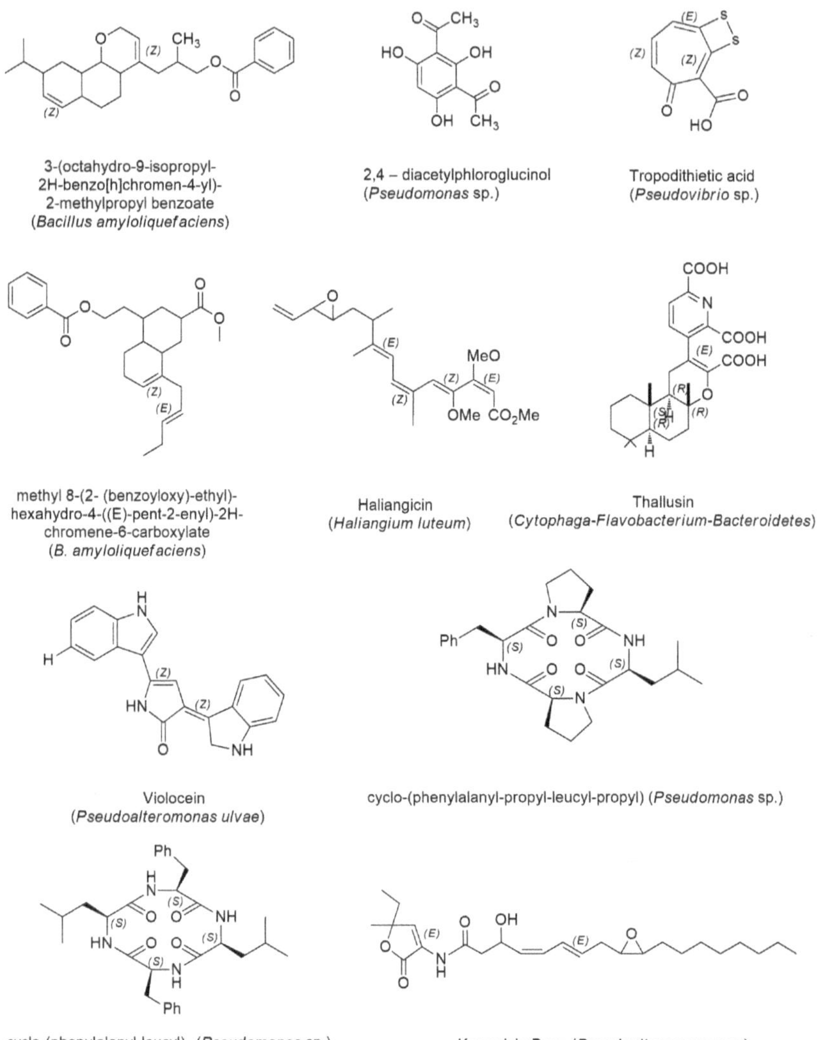

Fig. 2 Bioactive metabolites from the macroalgal associated bacteria.

be related to the release of nutrients resulting from tissue breakdown (Zuccaro et al., 2008). A total of '75' fungi were isolated from marine macroalgae of Antarctica and despite the fact that these fungi were isolated from living surfaces most of the isolates are probably saprobic decomposers, which may have an important role in the ecosystem for organic matter recycling (Loque et al., 2010). Table 2 illustrates the list of fungi isolated from the '3' major phyla of macroalgae, such as red, green and brown.

Table 2 List of fungi isolated from the '3'major phyla of macroalgae.

Macroalgae	Fungi isolated	References
Red		
Eucheuma sp.	*Aspergillus flavus, A. niger*	(Vala, Anand, Bhatt, & Joshi, 2004)
Palmaria decipiens	*Metschnikowia australis, Cryptococcus carnescens*	(Loque et al. 2010)
Grateloupia lithophila	*Cladosporium* sp., *Chaetomium* sp., *Nigrospora* sp.,	(Murali, 2011)
Halymenia sp.	*Emericella nidulans, Chaetomium* sp., *Emericella nidulans,*	(Murali, 2011)
Portieria hornemonii	*Cladosporium* sp., *Emericella nidulans, Phomopsis* sp.	(Murali, 2011)
Jania adherens	*Nigrospora* sp.	(Murali, 2011)
Bostrychia radicans	*Phomopsis longicolla*	(Erbert et al. 2012)
Green		
Halimeda macroloba	*Aspergillus* sp., *Aspergillus terreus, Penicillium* sp.	(Murali, 2011)
Caulerpa racemosa	*Aspergillus niger*	(Murali, 2011)
Caulerpa scalpelliformis	*Aspergillus niger, Aspergillus terreus, Paecilomyces* sp.	(Murali, 2011)
Ulva lactuca	*Aspergillus terreus, Chaetomium* sp., *Cladosporium* sp.,	(Murali, 2011)
Chondrus ocellatus	*Paraconiothyrium* sp.	(Suzuki et al. 2019)
Brown		
Fucus serratus	*Lindra, Lulworthia, Engyodontium, Sigmoidea marina, Acremonium fuci.*	(Zuccaro et al. 2008)
Desmarestia anceps	*Geomyces pannorum, M. australis*	(Loque et al. 2010)
Adenocystis utricularis	*Antarctomyces psychrotrophicus, G. pannorum, Oidiodendron* sp, *Penicillium* sp, *Phaeosphaeria herpotrichoides, M. australis, Rhodotorula mucilaginosa, Rhodotorula mucilaginosa*	(Loque et al. 2010)
Sargassum ilicifolium	*Aspergillus terreus*	(Murali, 2011)
Turbinaria sp.	*Alternaria* sp., *Aspergillus* sp., *Colletotrichum* sp, *Curvularia tuberculate, Drechslera* sp., *Fusarium* sp., *Phaeotrichoconis* sp., *Paecilomyces* sp., *Nigrospora* sp.,	(Murali, 2011)

(*Continued*)

Table 2 List of fungi isolated from the '3'major phyla of macroalgae.—cont'd

Macroalgae	Fungi isolated	References
Stoechospermum marginatum	*Alternaria* sp.	(Murali, 2011)
Sargassum wightii	*Alternaria* sp., *Aspergillus* sp., *Cladosporium* sp., *Curvularia* sp., *Emericella nidulans, Trichoderma* sp., *Pestalotiopsis* sp. *Cladosporium cladosporioides*	(Murali, 2011) (Hulikere et al. 2017)
Enteromorpha flexuosa	*Penicillium daleae*	(Iswarya & Ramesh, 2019)

Seaweed associated-fungi, *Aspergillus* spp. and *Penicillium* were common colonizers and also serve as a potential source of pharmacologically active compounds (Singh et al., 2015). *Aspergillus terreus* from the tissues of *Caulerpa scalpelliformis, C. sertularioides, G. edulis, Sargassum ilicifolium* and *U. lactuca* produced metabolites which have insecticidal activity (Suryanarayanan et al., 2010). *Xylaria* sp. isolated from red alga *Bostrychia tenella* synthesize cytochalasin D, a known antibiotic and antitumor compound (de Felício et al., 2015) whereas *Aspergillus niger* EN-13, isolated from brown alga, *Colpomenia sinuosa,* produces a sphingolipid and their corresponding glycosphingolipid named Asperamides A and B, where compound A exhibits antifungal activity against *Candida albicans* (Zhang et al., 2007) (Fig. 3).

3.3 Microalgae

Table 3 illustrates the list of microalgae isolated from different macroalgal species. The epiphytic microalgal community is mostly dominated by

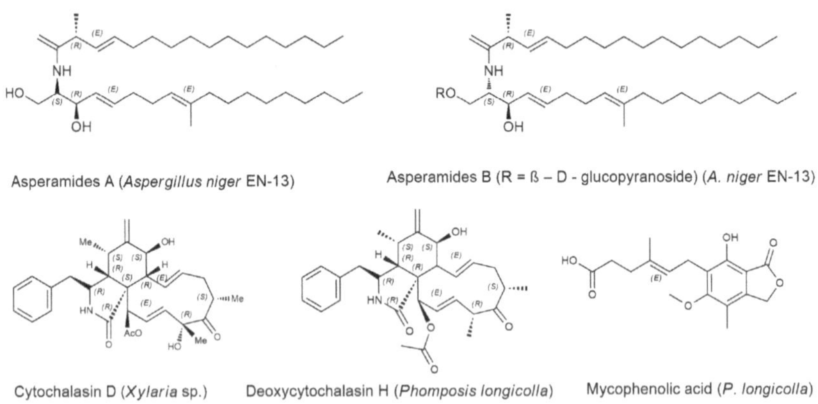

Fig. 3 Bioactive metabolites from the macroalgal associated fungi.

Table 3 List of microalgae isolated from the macroalgae.

Macroalgae	Microalgae isolated	References
Chaetomorpha linum	*Leptocylindrus danicus; Navicula distans; Nitzschia hungarica; Thalassionema frauenfeldi*	(Al-Harbi, 2017)
Enteromorpha intestinalis	*Chaetoceros fragile; Gyrosigma fasciola; L. danicus; N. distans; N. transitans; Pleurosigma angulatum; P. normanii*	(Al-Harbi, 2017)
Ulva lactuca	*Licmophora flabellata; L. abbreviata; P. normanii; T. frauenfeldii*	(Al-Harbi, 2017)
Laminaria sp.	*L. danicus; L. flabellata; Navicula ramosissma;*	(Al-Harbi, 2017)
Padina fraseri	*Cylindrotheca closterium; Navicula* sp.; *Nitzschia* sp.; *Prorocentrum lima; Cocconeis lineatus*	(Al-Harbi, 2017)
Sargassum muticum	*N. distans; N. socialis; N. vanhoeffeni*	(Al-Harbi, 2017)
Turbinaria ornata	*Leptocylindrus minimus; L. danicus; Bacillaria paxillifer; N. hungarica; T. frauenfeldii*	(Al-Harbi, 2017)
Ecklonia maxima	*Amphora; Gomphoseptatum; Navicula, Rhoicosphenia*	(Mayombo, Majewska, & Smit, 2019)
Laminaria pallida	*Amphora; Gomphoseptatum; Navicula, Rhoicosphenia*	(Mayombo et al. 2019)

benthic diatoms and a few centric species possessing an attached mode of life (Lage & Graça, 2016). As photosynthetic eukaryotes, diatoms can also serve as food for heterotrophic bacteria through specific interactions with growing diatoms as well as colonizing the sinking diatom and decomposing their organic matter. The interactions between diatoms and bacteria are generally confined and invariably, it is observed with *Proteobacteria* and *Bacteroidetes* phyla (Amin, Parker, & Armbrust, 2012). Diatoms release Transparent Exopolymer Particles (TEPs) which are often colonized by bacteria and in response, bacteria produce exopolysaccharides probably to initiate attachment (Amin et al., 2012; Fukao, Kimoto, & Kotani, 2010; Schafer, Abbas, Witte, & Muyzer, 2002). The scanning electron micrographs of the red alga

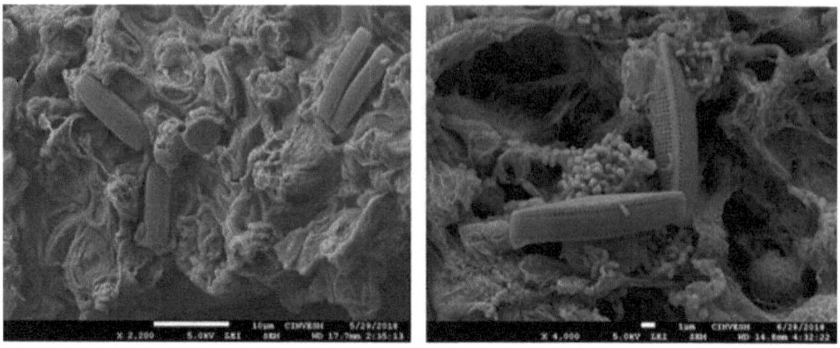

Fig. 4 Diatoms embedded on the extracellular matrix of *H. floresii* (left) and associated with the macroalgal surface bacteria (right).

Halymenia floresii depict the association of different bacterial phylotypes with the surface attached diatoms (Fig. 4). The abundance and taxonomic composition of the diatoms on the macroalgae are highly affected by the architecture of the macroalgal thallus and its surface characteristics (Totti et al., 2009).

3.4 Virus

As ubiquitous components of marine holobionts, viruses are now emerging as pathogens or potential pathogens in various marine organisms including macroalgae (Beattie et al., 2018; Lachnit, Thomas, & Steinberg, 2016). Viruses can have both positive and negative impacts on the holobiont which either may spread diseases among hosts or may be a part of the holobiont immune system (Ying et al., 2018). Lachnit et al. (2016) reported the viruses associated with a red macroalga, *D. pulchra*, identifying diverse morphotypes of virus-like particles from icosahedral to bacilliform to coiled pleomorphic as well as bacteriophages. The virome sequencing revealed the presence of a diverse group of dsRNA viruses affiliated to the genus *Totivirus*, known to infect plant pathogenic fungi. Furthermore, a ssRNA virus belonging to the *Picornavirales* order with a close phylogenetic relationship to a pathogenic virus infecting marine diatoms was also identified in association with *D. pulchra*. Viruses have a large potential effect by not only infecting the host but also other microorganisms, like bacteriophages which regulate the bacterial community composition, thus adding a complexity to the holobiont structure (Gachon, Sime-Ngando, Strittmatter, Chambouvet, & Kim, 2010; Ying et al., 2018). Table 4 includes the viruses isolated from different macroalgae.

Table 4 List of viruses isolated from the macroalgae.

Macroalgae	Viruses	References
Delisea pulchra	*Totivirus*; *Picornavirales*	(Lachnit et al. 2016)
Chondrus crispus	*Totivirus-like entities*	(Rousvoal, Bouyer, López-Cristoffanini, Boyen, & Collén, 2016)
Ectocarpus siliculosus	*Ectocarpus* DNA virus	(Lanka, Klein, Ramsperger, Müller, & Knippers, 1993)
Heterosigma akashiwo	*Phycodnaviridae*	(Nagasaki, Shirai, Tomaru, Nishida, & Pietrokovski, 2005)
Ecklonia radiata	*Siphoviridae*; *Myoviridae*; *Inoviridae*; *Podoviridae*	(Beattie et al. 2018)
Laminaria sp. and *Saccharina* sp.	*Phaeovirus*	(McKeown et al. 2017)

4. Role of microbiome: microbial interactions and their hosts response

The microbiome performs a 'protective' role with epibiotic bacteria present on surfaces that release chemicals that prevent biofouling by other organisms (Armstrong, Yan, Boyd, Wright, & Burgess, 2001). Similarly, macroalgal growth and development are shown to depend on associated microorganisms, particularly on bacterial communities (Singh & Reddy, 2014; Wichard et al., 2015). With such an obligate association, a fraction of epiphytic bacterial communities present on the surface of different individuals of the macroalgae can be regarded as 'core microbial species' or 'core community' (Singh & Reddy, 2016; Burke et al., 2011). The interactions between the macroalgae and the microbiota are extremely diverse (van der Loos, Eriksson, & Falcão Salles, 2019). As the microorganisms in the diverse marine environment are concentrated on microscale patches, their closeness, proximity suggests many potential cell-cell communications (Amin et al., 2012; Azam & Malfatti, 2007). The interactions of macroalgae and bacteria are mainly for nutrient exchange, signal transduction and gene transfer (Kouzuma & Watanabe, 2015).

4.1 Quorum sensing

Quorum Sensing (QS), an essential language of bacteria, is a cell density-dependent communication system that regulates a diverse array of physiological activities such as symbiosis, virulence, competence, conjugation, antibiotic production, motility, sporulation, and biofilm formation. This phenomenon relies on the accumulation of signal molecules in the surrounding environment up to threshold concentrations, at which the target genes are activated (Romero, Martin-Cuadrado, Roca-Rivada, Cabello, & Otero, 2011; Williams, Winzer, Chan, & Cámara, 2007). These QS signals, otherwise known as Auto Inducers (AIs), broadly fall into three categories. Acyl Homoserine Lactones (AHLs) are the major group of auto inducer signals in Gram-negative bacteria (Di Cagno, De Angelis, Calasso, & Gobbetti, 2011) (Fig. 5); Auto Inducer Peptides (AIPs) are the major auto inducers in Gram-positive bacteria (Thoendel, Kavanaugh, Flack, & Horswill, 2011; Zhang & Li, 2016), whereas, regardless of Gram-positive

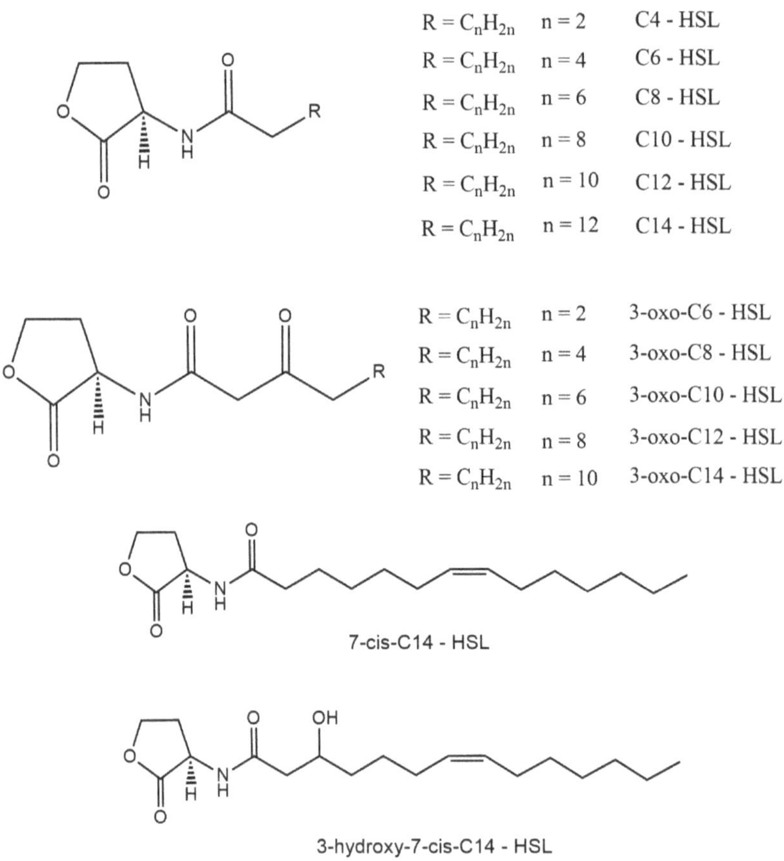

Fig. 5 Different homoserine lactones with differing chain length produced by the bacteria.

or Gram-negative bacteria, Auto inducer 2 (AI 2) known as universal language, is used for intra- and inter-species communication (Lowery, Dickerson, & Janda, 2008). Besides, there have been cyclic dipeptides, bradyoxetin, AI -3 in *E. coli* and diffusible signal factors, documented in different QS systems (Gonzalez & Keshavan, 2006).

Apart from defending the attached basibiont the surface-associated bacteria also assist the host in growth and morphogenesis. AHLs, N-acyl homoserine lactones, produced by the macroalgae-associated bacteria facilitates the settlement of zoospores in *Ulva spp.* (Joint et al., 2002; Singh et al., 2015; Williams, 2007). The bacterial isolates, *Shewanella algae, P. aeruginosa, Photobacterium sp., P. lutimaris, Vibrio gallicus, Vibrio fluvialis, V. parahaemolyticus*, associated with green macroalgae *Ulva* (*U. fasciata* and *U. lactuca*) and red macroalgae *Gracilaria* (*G. corticata* and *Gracilaria dura*), induced carpospore liberation from cyctocarps of *G. dura* by producing different AHLs (Singh et al., 2015). The AHLs produced by *Acrochaetium* sp. control spore release in the red alga, *G. chilensis* whereas its production was inhibited in the brown alga, *C. sinuosa* (Kanagasabhapathy, Yamazaki, Ishida, Sasaki, & Nagata, 2009; Weinberger et al., 2007). The motile zoospores of *Enteromorpha* exploit the diffusible signal molecules produced by bacterial biofilms for temporary settlement on the surface (Joint et al., 2002).

QS-mediated interactions widely influence algal-bacterial symbiotic relationships, which in turn determine community organization, population structure, and ecosystem functioning (Zhou, Lyu, Richlen, Anderson, & Cai, 2016). In *U. mutabilis* the chemical communication signals between the associated bacteria, *Roseobacter* and *Cytophaga*, up-regulate the growth and thallus morphogenesis (Spoerner et al., 2012). A chemical elicitor protein, probably a quorum signal N-acyl homoserine lactone produced by a gram-positive probiotic *Lysinibacillus xylanilyticus* associated with a red alga *G. edulis* promotes the growth of the host (Suvega & Arunkumar, 2019).

4.2 Quorum quenching

Quorum Quenching (QQ), is defined as any interference on the QS communication system either by enzymatic degradation of the AHLs, or by production of inhibitors or antagonists of signal reception. QQ strategy plays an important role in preventing microbial disease and affecting beneficial bacteria (Chen, Gao, Chen, Yu, & Li, 2013). Many algae use defence strategies to protect themselves from the potentially opportunistic pathogens associated with them by altering their QS communication systems i.e. quenching their signal molecules.

The first QS inhibitory activity was reported by the halogenated furanones from the red alga *D. pulchra* which exhibited antibiofilm activity against *Bacillus subtilis, P. aeruginosa* and *E. coli,* and these compounds were released at the surface at concentrations capable of inhibition (Givskov et al., 1996; Ren et al., 2002, 2001; Steinberg & de Nys, 2002). In particular the brominated furanones significantly inhibited the settlement of cyprid larvae of *Balanus amphitrite* (de Nys et al., 1995). A number of halogenated furanones from this alga antagonize the bacterial AHLs and interfere with their associated biological activity, such as competitively inhibiting swarming motility (Givskov et al., 1996). A mixture of relatively less lipophilic compounds, floridoside, betonicine and isethionic acid, isolated from the red alga *Ahnfeltiopsis flabelliformis* was observed to inhibit the exogenous QS signaling molecule, N-Octanoyl-DL-Homoserine Lactone (OHL), in the quorum-sensing inhibition assay (Kim, Kim, Seo, & Park, 2007). The methanolic extract of the red alga, *Asparagopsis taxiformis,* showed quorum sensing inhibitory activity against the reporter strain, *Chromobacterium violaceum* CV026 (Jha, Kavita, Westphal, Hartmann, & Schmitt-Kopplin, 2013) whereas the polar extracts of the green alga *U. fasciata* and *Codium* sp. exhibited QS inhibitory activity on *C. violaceum* CV017 (Batista, Carvalho, Costa, Coutinho, & Dobretsov, 2014). In the brown alga *Laminaria digitata,* QQ activity was demonstrated by the inactivation of AHLs based on the enzymatic production of the Oxidized Bromine HOBr which reacts specifically with 3-oxo-acyl HSLs (Borchardt et al., 2001; Romero et al., 2011). The polar and non-polar extracts of the different brown algae, *Sargassum* spp. *S. vulgare, S. furcatum,* showed anti-quorum sensing activity on the reporter strain, *C. violaceum* CV017 (Batista et al., 2014; Carvalho, Batista, Dobretsov, & Coutinho, 2017; Schwartz, Dobretsov, Rohde, & Schupp, 2017).

By interfering with the AHL signaling of *Vibrio anguillarum*, the settlement of *Enteromorpha,* an important fouling organism, can be prevented (Joint et al., 2002). In aquaculture systems, several algal species were used for controlling pathogenic bacteria by disrupting the quorum sensing communication between them, for example, *Vibrio harveyi* (Natrah, Bossier, Sorgeloos, Yusoff, & Defoirdt, 2014; Ramanan, Kim, Cho, Oh, & Kim, 2016). The halogenated furanones of *D. pulchra* protect both shrimp and fish from vibriosis by interfering with their AHL signaling system (Rasch et al., 2004; Torres et al., 2019). A QS antagonist, (5Z)-4-bromo-5-(bromomethylene)-3-butyl-2(5H)-furanone (furanone) from the *D. pulchra* was found to inhibit the formation of the siderophore produced by *Pseudomonas putida* and *P. aeruginosa* (Ren, Zuo, & Wood, 2005). QQ is a promising approach for very effective next-generation antibacterial drugs which block the QS-mediated pathogenic infection by interfering with the bacterial communication (Chen et al., 2013).

4.3 Positive effects: symbiotism

There are many beneficial functions for the macroalgal host, for example, normal growth, morphological development, spore germination and nitrogen fixation activity. This is despite the low average microbial density (10^5 to 10^6 cells per gram of seawater), the closeness, proximity and the heterogeneous distribution that suggest their interaction is across multiple spatial scales (Amin et al., 2012). Generally, the associated microorganisms were assumed to use the readily available range of organic carbon sources produced by the macroalgae in the form of complex diversified polysaccharides (Martin, Portetelle, Michel, & Vandenbol, 2014). The microbial metabolites produced by these symbiotic microorganisms help in chemical defences of the macroalgae and its biological interactions with foulers and pathogens (Chakraborty et al., 2017a; Wahl et al., 2012). In such a microhabitat, the macroalgae supply bacteria with oxygen and fixed carbon, and the bacteria supply CO_2, minerals, vitamins and growth factors. The microbial community on the surface of the macroalgae was mainly formed by bacteria but also included different eukaryotes in a ratio of 640:4:1 of bacteria:diatoms: flagellates (Lage & Graça, 2016; Railkin, 2003).

Epiphytic bacteria are known to maintain the health of the host. Several algal species need specific vitamins for their growth and bacteria may be partly responsible for the production of these substances (Villarreal-Gómez, Soria-Mercado, Guerra-Rivas, & Ayala-Sánchez, 2010). Conversely, bacterial colonization could be especially rapid if the surface is a potential source of nutrients, such as polysaccharides in exudates of seaweeds that may serve as a source of carbon for heterotrophic bacteria living on the surface. For example, Kelp exudates could shape bacterial community composition, and create communities that are kelp-specific rather than randomly assembled from the surrounding seawater (Collen & Davison, 2001; Michelou et al., 2013). Bacteria may degrade algal polysaccharides such as for fucoidan (Nedashkovskaia et al., 2002) or alginates (Brown, Preston, & Ingram, 1991; Preston, Romeo, & Bromley, 1986).

An overarching clade of bacteria such as *Proteobacteria* and *Bacteroidetes* are more likely to be associated with the green alga and furthermore, they are functionally equipped to be associated with them (Ramanan et al., 2015,

2016). The temperate green alga *U. lactuca* relies on the defense provided by the microbial community on its surface to regulate the fouling by colonizers (Kumar et al., 2011) and '73' different bacterial strains were isolated from the surfaces of *U. lactuca*, *U. pertusa* and *Enteromorpha* sp. (Ma et al., 2009). In the green alga, *U. rigida* '71' bacterial phylotypes were isolated from its surface which belongs to four phyla: *Proteobacteria*, *Bacteroidetes*, *Firmicutes* and *Actinobacteria* (Ismail et al., 2018). Different studies showed that induction of cell divisions, differentiation, wall formation, and normal morphogenesis in green algae belonging to Ulvophyceae depend on bacterial compounds (Grueneberg, Engelen, Costa, & Wichard, 2016; Matsuo, Imagawa, Nishizawa, & Shizuri, 2005; Wichard, 2015). *Pseudoalteromonas ulvae* and *P. tunicata* isolated from the surface of the green seaweed *U. lactuca* have shown the ability to reduce the adhesion of colonizing organisms and inhibit the germination of spores of this alga. This result was also observed in the red algae *Polysiphonia* sp. in the presence of *P. tunicata. Ulva* spores can be attracted by the emission of QS molecules from biofilms dominated by *Pseudoalteromonas*. In other species, spore germination may be inhibited by the presence of *P. tunicata. U. australis* is colonized by a very poorly diversified microflora, dominated by *Roseobacter gallaeciensis*, which facilitates the selective establishment of the bacterium *P. tunicata*. The two bacteria use different strategies. *R. gallaeciensis* invades and disintegrates the biofilms of competing bacteria, with the exception of *P. tunicata. P. tunicata* takes advantage of its "immunity" to proliferate in the face of competition that it cannot manage alone and then forms spatially separated microcolonies. These bacterial behaviors on the surface of *U. australis* allow the algae to have an inexpensive metabolic control of the bacterial biofilm.

The red alga *Agarophyton vermiculophyllum* chemically 'gardens' and recruits protective epibacterial strains in order to strengthen its disease resistance against potentially pathogenic microbes (Saha & Weinberger, 2019). Twenty one phenotypically different bacterial strains were isolated from the red alga *A. armata* which belong to seven different genera: *Vibrio*, *Staphylococcus*, *Bacillus*, *Cobetia*, *Photobacterium*, *Shewanella* and *Alteromonadaceae* (Horta et al., 2019). Ma et al. (2009) isolated '88' bacterial strains from the surfaces of the red algae: *Chondrus ocellatus*, *Dumontia simplex*, *Grateloupia filicina*, *Halymenia sinensis* and *Porphyra* sp.

A study by Ji, Zhang, Wang, and Zhao (2017) suggested a complementary and mutualistic relationship between the female gametophyte of the Brown seaweed *Saccharina japonica* and its microbiome, in which bacteria seem to benefit from seaweed polysaccharides, and the kelp profits from enhanced growth and nutrient uptake by bacterial bioactive compounds such as vitamins and hormones. Analysis of gene functions within this kelp epimicrobiome was mainly symbiosis-associated, indicating that selective pressures shape these microbiomes to sustain a mutual benefit for both kelp and bacteria (Ji et al., 2017). *F. vesiculosus* L. is recognized as an important habitat-forming macroalga and could be able to produce secondary metabolites that directly act on the invertebrate larvae to prevent fouling (Kersen, Kotta, Bučas, Kolesova, & Deķere, 2011; Lachnit, Wahl, & Harder, 2010). Several species of marine bacteria have been isolated from brown algae such as the spot-wounded fronds of *Laminaria japonica* (Sawabe et al., 2000), the surface of *F. serratus* (Johansen et al., 1999; Lachnit et al., 2009) and from the rotten thallus of *Fucus evanescens* (Ivanova et al., 2004), *Undaria pinnatifida* (Lee, Jung, & Lee, 2006), *Padina arborescens*, *Padina pavonica*, *P. tetrastromatica* (Ismail et al., 2016), *Macrocystis pyrifera* (Michelou et al., 2013), *Ecklonia radiata* (Qiu et al., 2019), *Taonia otamaria* (Othmani, Briand, Ayé, Molmeret, & Culioli, 2016). Around '18' different bacteria were isolated from the surfaces of two different brown algae, *Sargassum thunbergii* and *Sargassum* sp. (Ma et al., 2009). Such a complex microbiome on the surface of macroalgae is regarded as an indirect defence mechanism which may strengthen the direct antifouling defence mechanism of the host against any macrofoulers (Nasrolahi et al., 2012).

4.4 Negative effects: pathogenism

Any disturbance and disruption of the mutualistic association between host and microbiota are regarded as an 'Holobiont break-up' which can be caused by environmental stress (Ying et al., 2018). Such disturbances are widely caused by spatial and temporal effects. The spatial and temporal shifts of seaweed associated microbiota are induced by the changing environmental conditions and seaweed physiology. Such shifts may result in potential consequences provoked by the structural, functional and behavioral changes in the host posing a challenge for its fitness (Serebryakova, Aires, Viard, Serrão, & Engelen, 2018). This shifts the symbiotic relationship of the microbiota to a potential pathogenic relationship with its associated hosts. The interplay between temperature, the algal chemical defence mechanism and bacterial virulence demonstrates a complex environmental change impact on an ecosystem (Case et al., 2011). Though the epibacterial community structures on the surface of macroalgae are host-specific, they are highly prone to spatial and temporal variation (Wu, Liu, Zhang, & Xiao, 2014). A shift in the microbiome from a stable state to a disturbed state where the colonizers may become opportunist pathogens when the host system is stressed by abiotic or biotic challenges is referred to as 'dysbiosis' (Minich et al., 2018).

Pathogenic bacteria can cause severe degradation of algal host cells or even lead to seaweed mortality, causing major financial losses for seaweed mariculture every year (Correa, Flores, & Sanchez, 1993; Goecke et al., 2010; Vairappan et al., 2008). Also, biofouling forms a permanent threat to macroalgae as bacterial biofilms increase the hydrodynamic drag on their host and enhance the attachment of other fouling organisms and grazers. Biofilms may also compete for nutrients, inhibit gaseous exchange, or block light essential for photosynthesis.

Microbial colonization can lead to the degradation of algal tissue, decreased photosynthesis and gas and nutrient exchange and ultimately disease. Marine macroalgal diseases have been recognised for a long time, but identifying the causative agents and distinguishing them from other opportunistic bacteria is still an obstacle for describing the virulence mechanism (Egan et al., 2013). Macroalgae are under constant colonization pressure from the millions of microorganisms within the surrounding seawater, some of which are potential pathogens. The primary biofilm undergoes a transition to a much more complex community which eventually ends in fouling on the surface (Bhadury & Wright, 2004). All such biofouling communities vary spatially and temporally (Fitridge, Dempster, Guenther, & de Nys, 2012). Due to the effects of environmental change on host organisms and their pathogens, temporal-mediated diseases can have severe consequences for marine macroalgal ecosystems (Campbell, Harder, Nielsen, Kjelleberg, & Steinberg, 2011).

4.4.1 Spatial effects

The close spatial proximity of surface-associated microorganisms on macroalgae operates and controls specific intercellular interactions (Egan et al., 2008). The spatial distribution and the abundance of the epiphytes are largely determined by several abiotic and biotic factors which ultimately leads to uncontrolled growth (Kersen et al., 2011; Kotta, Paalme, Püss, Herkül, & Kotta, 2008; Kotta & Witman, 2009; Nylund & Pavia, 2005). Spatial variability of the epiphytic communities varies both on the large and small scale (Fitridge et al., 2012). The protective effects of the microbiome on the surface were affected by severe settlement pressure and stressful temperatures as observed in *F. vesiculosus* (Nasrolahi et al., 2012), in which the composition and dominance were significantly influenced by wave exposure and frond segment (Kersen et al., 2011). Also, reduced light conditions due to the uncontrolled colonization of the surface have detrimental effects on the growth of the macroalgae (Rohde et al., 2008). The occurrence and richness of the epiphytes were observed to be specific and different, according to the different parts of the host macroalgae (Kersen et al., 2011). The spatial distribution of epibiotic organisms is determined by their ability to tolerate desiccation during low tides (Molina-Montenegro et al., 2005). Multiple opportunistic pathogens, *Alteromonas sp.* BL110, *Aquimarina sp.* AD1 and BL5 and *Agarivorans sp* BL7 rather than the previously identified *Nautella* R11, cause bleaching disease in a temperate red alga *D. pulchra* with loss of its photosynthetic pigments followed by tissue necrosis and death. Multiple opportunistic pathogens are involved in bleaching disease in *D. pulchra* (Campbell et al., 2011; Case et al., 2011; Fernandes et al., 2011; Gonzalez & Keshavan, 2006; Kumar, Zozaya-Valdes, Kjelleberg, Thomas, & Egan, 2016).

4.4.2 Temporal effects

The temporal variation of the macroalgal associated microbiome may be affected by biotic factors such as algal growth, exudation of macroalgal metabolites, the composition of the microbial community in the surrounding seawater and abiotic factors like temperature, wave action, light conditions and nutrient levels in seawater (Bengtsson et al., 2010). Different seasons affect the temporal variations of macroalgal surface-associated bacterial communities (Singh et al., 2015). As a result of climate change, environmental stress can render macroalgae susceptible to the presence of opportunistic pathogens which are otherwise commensal to them, but can still take advantage of the weakened host and this results in disease outbreak (Egan et al., 2014). As most of the biofilm seems to consist of bacteria utilizing carbon produced by the host (Bengtsson & Øvreås, 2010), microbiome dynamics are probably strongly linked to seasonal changes in the kelp metabolome and seawater temperature (Bengtsson et al., 2010).

The temporal shifts of bacterial communities associated with the brown alga *S. muticum* were reflected in a significant abundance of unidentified *Rhodobacterales* and *Laktonella* in September—March and the substantial prevalence of unidentified *Pirellulales* in the summer. Such changes were related to the temporally changing algal exudates (Serebryakova et al., 2018). In another brown alga *L. hyperborea*, the substratum and seawater temperature manifested a variation in their associated microbiome where *Planctomycetes* and *Alphaproteobacteria* were frequently detected throughout the year and

Verrucomicrobia, *Cyanobacteria*, *Gammaproteobacteria*, *Betaproteobacteria*, and *Bacteroidetes* were more sporadically detected (Bengtsson et al., 2010). The kelp microbiome was most influenced by an elevated temperature exhibiting a reduction in the dominant kelp-associated Alteromondales, which negatively affects its growth (Minich et al., 2018). Any temporal variations in the primary biofilm will cause major temporal changes in the marine invertebrate populations, particularly, seasonality (Fitridge et al., 2012).

5. Future perspectives

Increased physical disturbances and stress, resulting from changing environmental conditions, are known to affect the macroalgae-associated microbial composition (directly or via its physiological responses) and cause its structural, functional or behavioral changes. In the marine environment, changing climatic conditions such as ocean warming and acidification can have direct physiological effects on the marine-habitat forming macroalgal species, which sequentially, can have negative impacts on the associated microbial communities. The microbiome of the brown alga, *E. radiata*, was affected by warming, followed by acidification which resulted in disease-like symptoms in the host (Qiu et al., 2019). The characteristic 'core' microbial community was replaced by taxonomically diverse microbial communities following the host's stress (Marzinelli et al., 2015). Case et al. (2011) showed that temperate marine macroalga, *D. pulchra,* has become more susceptible to the temperature-enhanced bacterial virulence of *Ruegeria* sp. R11 as a result of high-water temperatures or increased solar radiation. This also affects the production or delivery of furanones potentially leading to *in situ* bleaching (Case et al., 2011). Ice-ice disease in the red alga *Eucheuma/Kappaphycus* is still a major problem in this seaweed cultivation where the changes in extreme environmental conditions make the alga susceptible to the infection of *Vibrio* sp. (Irmawati & Sudirjo, 2017).

Bioactive metabolites of macroalgae surface-associated bacteria open up a new approach as a novel source of therapeutic molecules. For instance, *A. armata*-associated bacteria revealed to be a potential source of compounds with antitumor and antibacterial activity (Horta et al., 2019). Marine bacteria can be a vast source of new antimicrobial molecules and antibiofilm molecules with a wide range of applications in human health and Industry (Desriac et al., 2018). In the future, beneficial bacteria identified from healthy seaweed holobionts could be applied to diseased plantlets in order to suppress the growth of detrimental ones and/or to prevent disease outbreaks in aquaculture settings. In addition to bacteria, these macroalgae frequently host endophytic fungi that may have protective functions for the algae (Dittami et al., 2019). Thus, metabolites of these bacteria will provide a new possibility of disease management in aquaculture systems.

The surface-associated microbiome of macroalgae is phylogenetically diverse and ecologically important. Thus, assessing the microbial functions and interactions has emerged as a critical need in macroalgal research. Massive sequencing of the 16S rRNA gene and using a metagenomic approach have been the main tools to explore these aspects. However, the complete understanding of the relationships between host and microorganisms, as well as the dynamics of the symbiotic community, probably requires the generation of comprehensive insights with another novel integrative "omics" approach, such as single-cell sequencing, metagenome binning, GeoChip, RNA-Seq, etc. To characterise the interactions of organisms with their environment, environmental metabolomics shall be further explored in this area (Bundy, Davey, & Viant, 2009). Bioactive compounds from macroalgae and its associated microbiome may provide candidates with a high-efficiency and have broad implications in disease control. Concentration and quantity are the two main challenges to be addressed here. Different biotechnological tools should be developed for the isolation and identification of these compounds even at lesser concentrations, which again are the biggest challenges in this field.

References

Aires, T., Serrão, E. A., & Engelen, A. H. (2016). Host and environmental specificity in bacterial communities associated to two highly invasive marine species (genus *Asparagopsis*). *Frontiers in Microbiology, 7*, 559.

Al-Harbi, S. M. (2017). Epiphytic microalgal species composition and dynamics on host green seaweeds (Ulvaphyceae) on the northern coast of Jeddah, Saudi Arabia. *Environment and Ecology Research, 5*, 212–219.

Albakosh, M. A., Naidoo, R. K., Kirby, B., & Bauer, R. (2016). Identification of epiphytic bacterial communities associated with the brown alga *Splachnidium rugosum*. *Journal of Applied Phycology, 28*, 1891–1901.

Alvarado, P., Huang, Y., Wang, J., Garrido, I., & Leiva, S. (2018). Phylogeny and bioactivity of epiphytic gram-positive bacteria isolated from three co-occurring antarctic macroalgae. *Antonie Van Leeuwenhoek, 111*, 1543–1555.

Amin, S. A., Parker, M. S., & Armbrust, E. V. (2012). Interactions between diatoms and bacteria. *Microbiology and Molecular Biology Reviews, 76*, 667–684.

Armstrong, E., Yan, L., Boyd, K. G., Wright, P. C., & Burgess, J. G. (2001). The symbiotic role of marine microbes on living surfaces. *Hydrobiologia, 461*, 37–40.

Azam, F., & Malfatti, F. (2007). Microbial structuring of marine ecosystems. *Nature Reviews Microbiology, 5*, 782–791.

Batista, D., Carvalho, A. P., Costa, R., Coutinho, R., & Dobretsov, S. (2014). Extracts of macroalgae from the Brazilian coast inhibit bacterial quorum sensing. *Botanica Marina, 57*, 441–447.

Bawakid, N. O., Alarif, W. M., Alburae, N. A., Alorfi, H. S., Al-Footy, K. O., Al-Lihaibi, S. S., et al. (2017). Isolaurenidificin and bromlaurenidificin, two new C15-acetogenins from the red alga *Laurencia obtusa*. *Molecules, 22*, E807.

Beattie, D. T., Lachnit, T., Dinsdale, E. A., Thomas, T., & Steinberg, P. D. (2018). Novel ssDNA viruses detected in the virome of bleached, habitat-forming kelp *Ecklonia radiata*. *Frontiers in Marine Science, 4*, 441.

Bengtsson, M. M., & Øvreås, L. (2010). Planctomycetes dominate biofilms on surfaces of the kelp *Laminaria hyperborea*. *BMC Microbiology, 10*, 261.

Bengtsson, M. M., Sjøtun, K., Lanzén, A., & Øvreås, L. (2012). Bacterial diversity in relation to secondary production and succession on surfaces of the kelp *Laminaria hyperborea*. *The ISME Journal, 6*, 2188–2198.

Bengtsson, M., Sjøtun, K., & Øvreås, L. (2010). Seasonal dynamics of bacterial biofilms on the kelp *Laminaria hyperborea*. *Aquatic Microbial Ecology, 60*, 71–83.

Bhadury, P., & Wright, P. (2004). Exploitation of marine algae: Biogenic compounds for potential antifouling applications. *Planta, 219*, 561–578.

Bondoso, J., Godoy-Vitorino, F., Balagué, V., Gasol, J. M., Harder, J., & Lage, O. M. (2017). Epiphytic *Planctomycetes* communities associated with three main groups of macroalgae. *FEMS Microbiology Ecology, 93*, fiw255.

Borchardt, S. A., Allain, E. J., Michels, J. J., Stearns, G. W., Kelly, R. F., & McCoy, W. F. (2001). Reaction of acylated homoserine lactone bacterial signaling molecules with oxidized halogen antimicrobials. *Applied and Environmental Microbiology, 67*, 3174–3179.

Boyd, K. G., Adams, D. R., & Burgess, J. G. (1999). Antibacterial and repellent activities of marine bacteria associated with algal surfaces. *Biofouling, 14*, 227–236.

Brian-Jaisson, F., Molmeret, M., Fahs, A., Guentas-Dombrowsky, L., Culioli, G., Blache, Y., et al. (2016). Characterization and anti-biofilm activity of extracellular polymeric substances produced by the marine biofilm-forming bacterium *Pseudoalteromonas ulvae* strain TC14. *Biofouling, 32*, 547–560.

Brown, B. J., Preston, J. F., & Ingram, L. O. (1991). Cloning of alginate lyase gene (alxM) and expression in *Escherichia coli*. *Applied and Environmental Microbiology, 57*, 1870–1872.

Bundy, J. G., Davey, M. P., & Viant, M. R. (2009). Environmental metabolomics: A critical review and future perspectives. *Metabolomics, 5*, 3–21.

Burke, C., Thomas, T., Lewis, M., Steinberg, P., & Kjelleberg, S. (2011). Composition, uniqueness and variability of the epiphytic bacterial community of the green alga *Ulva australis*. *The ISME Journal, 5*, 590–600.

Callow, M. E., & Callow, J. E. (2002). Marine biofouling: A sticky problem. *Biologist, 49*, 10–14.

Callow, J. A., & Callow, M. E. (2011). Trends in the development of environmentally friendly fouling-resistant marine coatings. *Nature Communications, 2*, 244.

Campbell, A. H., Harder, T., Nielsen, S., Kjelleberg, S., & Steinberg, P. D. (2011). Climate change and disease: Bleaching of a chemically defended seaweed. *Global Change Biology, 17*, 2958–2970.

Carvalho, A. P., Batista, D., Dobretsov, S., & Coutinho, R. (2017). Extracts of seaweeds as potential inhibitors of quorum sensing and bacterial growth. *Journal of Applied Phycology, 29*, 789–797.

Case, R. J., Longford, S. R., Campbell, A. H., Low, A., Tujula, N., Steinberg, P. D., et al. (2011). Temperature induced bacterial virulence and bleaching disease in a chemically defended marine macroalga. *Environmental Microbiology, 13*, 529–537.

Cavalier-Smith, T. (1992). Symbiosis as a source of evolutionary innovation: Speciation and morphogenesis. In L. Margulis, & R. Fester (Eds.), *Trends in ecology & evolution* (Vol. 7). MIT Press.

Chakraborty, K., Thilakan, B., Chakraborty, R. D., Raola, V. K., & Joy, M. (2017a). O-heterocyclic derivatives with antibacterial properties from marine bacterium *Bacillus subtilis* associated with seaweed, *Sargassum myriocystum*. *Applied Microbiology and Biotechnology, 101*, 569–583.

Chakraborty, K., Thilakan, B., & Raola, V. K. (2017b). Antimicrobial polyketide furanoterpenoids from seaweed-associated heterotrophic bacterium *Bacillus subtilis* MTCC 10403. *Phytochemistry, 142*, 112–125.

Chakraborty, K., Thilakan, B., Raola, V. K., & Joy, M. (2017). Antibacterial polyketides from *Bacillus amyloliquefaciens* associated with edible red seaweed *Laurenciae papillosa*. *Food Chemistry, 218*, 427–434.

Chapman, J., Hellio, C., Sullivan, T., Brown, R., Russell, S., Kiterringham, E., et al. (2014). Bioinspired synthetic macroalgae: Examples from nature for antifouling applications. *International Biodeterioration & Biodegradation, 86*, 6–13.

Chen, F., Gao, Y., Chen, X., Yu, Z., & Li, X. (2013). Quorum quenching enzymes and their application in degrading signal molecules to block quorum sensing-dependent infection. *International Journal of Molecular Sciences, 14*, 17477–17500.

Cheng, H., Zhang, S., Huo, Y.-Y., Jiang, X.-W., Zhang, X.-Q., Pan, J., et al. (2015). *Gilvimarinus polysaccharolyticus* sp. nov., an agar-digesting bacterium isolated from seaweed, and emended description of the genus *Gilvimarinus*. *International Journal of Systematic and Evolutionary Microbiology, 65*, 562–569.

Choi, S., Kang, J. W., Lee, J. H., & Seong, C. N. (2017). *Dokdonia lutea* sp. nov., isolated from *Sargassum fulvellum* seaweed. *International Journal of Systematic and Evolutionary Microbiology, 67*, 4482–4486.

Collen, J., & Davison, I. R. (2001). Seasonality and thermal acclimation of reactive oxygen metabolism in *Fucus vesiculosus* (Phaeophyceae). *Journal of Phycology, 37*, 474–481.

Correa, J. A., Flores, V., & Sanchez, P. (1993). Deformative disease in *Iridaea laminarioides* (Rhodophyta): Gall development associated with an endophytic cyanobacterium. *Journal of Phycology, 29*, 853–860.

Costerton, J. W., Geesey, G. G., & Cheng, K.-J. (1978). How bacteria stick. *Scientific American, 238*, 86–95.

Cronin, G., & Hay, M. E. (1996). Within-plant variation in seaweed palatability and chemical defenses: Optimal defense theory versus the growth-differentiation balance hypothesis. *Oecologia, 105*, 361–368.

Dahms, H., & Dobretsov, S. (2017). Antifouling compounds from marine macroalgae. *Marine Drugs, 15*, E265.

Dang, H., & Lovell, C. R. (2000). Bacterial primary colonization and early succession on surfaces in marine waters as determined by Amplified rRNA gene restriction analysis and sequence analysis of 16S rRNA genes. *Applied and Environmental Microbiology, 66*, 467–475.

Desriac, F., Rodrigues, S., Doghri, I., Sablé, S., Lanneluc, I., Fleury, Y., et al. (2018). Antimicrobial and antibiofilm molecules produced by marine bacteria. In *Blue biotechnology* (pp. 789–819). Weinheim, Germany: Wiley-VCH Verlag GmbH & Co. KGaA.

Di Cagno, R., De Angelis, M., Calasso, M., & Gobbetti, M. (2011). Proteomics of the bacterial cross-talk by quorum sensing. *Journal of Proteomics, 74*, 19–34.

Dittami, S. M., Arboleda, E., Auguet, J.-C., Bigalke, A., Briand, E., Cardenas, P., et al. (2019). A community perspective on the concept of marine holobionts: State-of-the-art, challenges, and future directions. *PeerJ Preprints, 7*, e27519v3.

Donlan, R. M. (2002). Biofilms: Microbial life on surfaces. *Emerging Infectious Diseases, 8*, 881–890. @BULLET.

Duarte, C., Navarro, J. M., Quij, P. A., Loncon, D., Torres, R., Manríquez, P. H., et al. (2018). The energetic physiology of juvenile mussels, *Mytilus chilensis* (Hupe): The prevalent role of salinity under current and predicted pCO_2 scenarios. *Environmental Pollution, 242*, 156–163.

Dworjanyn, S. A., Wright, J. T., Paul, N. A., de Nys, R., & Steinberg, P. D. (2006). Cost of chemical defence in the red alga *Delisea pulchra*. *Oikos, 113*, 13–22.

Egan, S., Fernandes, N. D., Kumar, V., Gardiner, M., & Thomas, T. (2014). Bacterial pathogens, virulence mechanism and host defence in marine macroalgae. *Environmental Microbiology, 16*, 925–938.

Egan, S., Harder, T., Burke, C., Steinberg, P., Kjelleberg, S., & Thomas, T. (2013). The seaweed holobiont: Understanding seaweed–bacteria interactions. *FEMS Microbiology Reviews, 37*, 462–476.

Egan, S., Thomas, T., & Kjelleberg, S. (2008). Unlocking the diversity and biotechnological potential of marine surface associated microbial communities. *Current Opinion in Microbiology, 11*, 219–225.

Erbert, C., Lopes, A. A., Yokoya, N. S., Furtado, N. A. J. C., Conti, R., Pupo, M. T., et al. (2012). Antibacterial compound from the endophytic fungus *Phomopsis longicolla* isolated from the tropical red seaweed *Bostrychia radicans*. *Botanica Marina, 55*, 435–440.

de Felício, R., Pavão, G. B., de Oliveira, A. L. L., Erbert, C., Conti, R., Pupo, M. T., et al. (2015). Antibacterial, antifungal and cytotoxic activities exhibited by endophytic fungi from the Brazilian marine red alga *Bostrychia tenella* (Ceramiales). *Revista Brasileira de Farmacognosia, 25*, 641–650.

Fernandes, N., Case, R. J., Longford, S. R., Seyedsayamdost, M. R., Steinberg, P. D., Kjelleberg, S., et al. (2011). Genomes and virulence factors of novel bacterial pathogens causing bleaching disease in the marine red alga *Delisea pulchra*. *PLoS One, 6*, e27387.

Fitridge, I., Dempster, T., Guenther, J., & de Nys, R. (2012). The impact and control of biofouling in marine aquaculture: A review. *Biofouling, 28*, 649–669.

Fukao, T., Kimoto, K., & Kotani, Y. (2010). Production of transparent exopolymer particles by four diatom species. *Fisheries Science, 76*, 755–760.

Gachon, C. M. M., Sime-Ngando, T., Strittmatter, M., Chambouvet, A., & Kim, G. H. (2010). Algal diseases: Spotlight on a black box. *Trends in Plant Science, 15*, 633–640.

da Gama, B. A. P., Plouguerné, E., & Pereira, R. C. (2014). The antifouling defence mechanisms of marine macroalgae. In N. Bourgougnon (Ed.), *Advances in botanical research* (Vol. 71, pp. 413–440). Elsevier.

Garg, A., Jain, A., & Bhosle, N. B. (2009). Chemical characterization of a marine conditioning film. *International Biodeterioration & Biodegradation, 63*, 7–11.

Geesey, G. G. (2001). Bacterial behavior at surfaces. *Current Opinion in Microbiology, 4*, 296–300.

Givskov, M., de Nys, R., Manefield, M., Gram, L., Maximilien, R., Eberl, L., et al. (1996). Eukaryotic interference with homoserine lactone-mediated prokaryotic signalling. *Journal of Bacteriology, 178*, 6618–6622.

Goecke, F., Labes, A., Wiese, J., & Imhoff, J. (2010). Chemical interactions between marine macroalgae and bacteria. *Marine Ecology Progress Series, 409*, 267–299.

Gonzalez, J. E., & Keshavan, N. D. (2006). Messing with bacterial quorum sensing. *Microbiology and Molecular Biology Reviews, 70*, 859–875.

Grueneberg, J., Engelen, A. H., Costa, R., & Wichard, T. (2016). Macroalgal morphogenesis induced by waterborne compounds and bacteria in coastal seawater. Coastal seawater. *PLoS One, 11*, 146307.

Guedes, A. C., Amaro, H. M., Sousa-Pinto, I., & Malcata, F. X. (2019). Algal spent biomass—A pool of applications. In A. Pandey, J.-S. Chang, C. R. Soccol, D.-J. Lee, & Y. Chisti (Eds.), *Biofuels from algae (II)* (pp. 397–433). Elsevier.

Habbu, P., Warad, V., Shastri, R., Madagundi, S., & Kulkarni, V. H. (2016). Antimicrobial metabolites from marine microorganisms. *Chinese Journal of Natural Medicines, 14*, 101–116.

Harizani, M., Ioannou, E., & Roussis, V. (2016). The laurencia paradox: An endless source of chemodiversity. *Progress in the Chemistry of Organic Natural Products, 102*, 91–252.

Hentzer, M., Riedel, K., Rasmussen, T. B., Heydorn, A., Andersen, J. B., Parsek, M. R., et al. (2002). Inhibition of quorum sensing in *Pseudomonas aeruginosa* biofilm bacteria by a halogenated furanone compound. *Microbiology, 148*, 87–102.

Hollants, J., Leliaert, F., De Clerck, O., & Willems, A. (2013). What we can learn from sushi: A review on seaweed-bacterial associations. *FEMS Microbiology Ecology, 83*, 1–16.

Horta, A., Alves, C., Pinteus, S., Lopes, C., Fino, N., Silva, J., et al. (2019). Identification of *Asparagopsis armata* -associated bacteria and characterization of their bioactive potential. *MicrobiologyOpen, 29*, e00824.

Hulikere, M., Joshi, C. G., Danagoudar, A., Poyya, J., Kudva, A. K., & BL, D. (2017). Biogenic synthesis of gold nanoparticles by marine endophytic fungus-*Cladosporium cladosporioides* isolated from seaweed and evaluation of their antioxidant and antimicrobial properties. *Process Biochemistry, 63*, 137–144.

Irmawati, Y., & Sudirjo, F. (2017). IOP conference series: Earth and environmental science infection *Vibrio* sp. bacteria on *Kappaphycus* seaweed varieties Brown and green infection *Vibrio* sp. bacteria on *Kappaphycus* seaweed varieties Brown and green. *IOP Conference Series: Earth and Environmental Science, 89*, 012016.

Ismail, A., Ktari, L., Ahmed, M., Bolhuis, H., Boudabbous, A., Stal, L. J., et al. (2016). Antimicrobial activities of bacteria associated with the Brown alga *Padina pavonica. Frontiers in Microbiology, 7*, 1072.

Ismail, A., Ktari, L., Ahmed, M., Bolhuis, H., Bouhaouala-Zahar, B., Stal, L. J., et al. (2018). Heterotrophic bacteria associated with the green alga *Ulva rigida*: Identification and antimicrobial potential. *Journal of Applied Phycology, 30*, 2883–2899.

Iswarya, B., & Ramesh, K. V. (2019). Promising antimicrobial compounds from an endophytic fungus Penicillium daleae EF 4 isolated from the seaweed *Enteromorpha flexuosa* Linn. *Research & Reviews: A Journal of Microbiology & Virology, 9*, 26–39.

Ivanova, E. P., Alexeeva, Y. A., Zhukova, N. V., Gorshkova, N. M., Buljan, V., Nicolau, D. V., et al. (2004). *Bacillus algicola* sp. nov., a novel filamentous organism isolated from Brown alga *Fucus evanescens. Systematic & Applied Microbiology, 27*, 301–307.

Jha, B., Kavita, K., Westphal, J., Hartmann, A., & Schmitt-Kopplin, P. (2013). Quorum sensing inhibition by *Asparagopsis taxiformis*, a marine macro alga: Separation of the compound that interrupts bacterial communication. *Marine Drugs, 11*, 253–265.

Ji, P., Zhang, Y., Wang, J., & Zhao, F. (2017). MetaSort untangles metagenome assembly by reducing microbial community complexity. *Nature Communications, 8*, 14306.

Johansen, J. E., Nielsen, P., & Sjoholm, C. (1999). Description of *Cellulophaga baltica* gen. nov., sp. nov. and *Cellulophaga fucicola* gen. nov., sp. nov. and reclassification of [Cytophaga] lytica to *Cellulophaga lytica* gen. nov., comb. nov. *International Journal of Systematic Bacteriology, 49*, 1231–1240.

Johansen, J. E., Nielsen, P., & Sjraholm, C. (2018). Description of *Cellulophaga baltica* gen. nov., sp. nov. and *Cellulophaga fucicola* gen. nov., sp. nov. and reclassification of [Cytophaga] Iytica to *Cellulophaga lytica* gen. nov., comb. nov. *International Journal of Systematic Bacteriology, 2346*, 1–23.

Joint, I., Tait, K., Callow, M. E., Callow, J. A., Milton, D., Williams, P., et al. (2002). Cell-to-Cell communication across the prokaryote-eukaryote boundary. *Science, 298*, 1207.

Kanagasabhapathy, M., Yamazaki, G., Ishida, A., Sasaki, H., & Nagata, S. (2009). Presence of quorum-sensing inhibitor-like compounds from bacteria isolated from the brown alga *Colpomenia sinuosa. Letters in Applied Microbiology, 49*, 573–579.

Karthick, P., & Mohanraju, R. (2018). Antimicrobial potential of epiphytic bacteria associated with seaweeds of little Andaman, India. *Frontiers in Microbiology, 9*, 611.

Kersen, P., Kotta, J., Bučas, M., Kolesova, N., & Deķere, Z. (2011). Epiphytes and associated fauna on the brown alga *Fucus vesiculosus* in the Baltic and the North Seas in relation to different abiotic and biotic variables. *Marine Ecology, 32*, 87–95.

Kim, J. S., Kim, Y. H., Seo, Y. W., & Park, S. (2007). Quorum sensing inhibitors from the red alga,*Ahnfeltiopsis flabelliformis*. *Biotechnology and Bioprocess Engineering, 12*, 308–311.

Kloareg, B., & Quatrano, R. S. (1988). Structure of the cell walls of marine algae and ecophysiological functions of the matrix polysaccharides. *Oceanography and Marine Biology an Annual Review, 26*, 259–315.

Kotta, J., Paalme, T., Püss, T., Herkül, K., & Kotta, I. (2008). Contribution of scale-dependent environmental variability on the biomass patterns of drift algae and associated invertebrates in the Gulf of Riga, northern Baltic Sea. *Journal of Marine Systems, 74*, S116–S123.

Kotta, J., & Witman, J. D. (2009). Regional-scale patterns. In M. Wahl (Ed.), *Marine hard bottom communities* (pp. 89–99). Berlin, Heidelberg: Springer.

Kouzuma, A., & Watanabe, K. (2015). Exploring the potential of algae/bacteria interactions. *Current Opinion in Biotechnology, 33*, 125–129.

Kubanek, J., Jensen, P. R., Keifer, P. A., Sullards, M. C., Collins, D. O., & Fenical, W. (2003). Seaweed resistance to microbial attack: A targeted chemical defense against marine fungi. *Proceedings of the National Academy of Sciences of the United States of America, 100*, 6916–6921.

Kumar, V., Rao, D., Thomas, T., Kjelleberg, S., & Egan, S. (2011). Antidiatom and antibacterial activity of epiphytic bacteria isolated from *Ulva lactuca* in tropical waters. *World Journal of Microbiology and Biotechnology, 27*, 1543–1549.

Kumar, V., Zozaya-Valdes, E., Kjelleberg, S., Thomas, T., & Egan, S. (2016). Multiple opportunistic pathogens can cause a bleaching disease in the red seaweed *Delisea pulchra*. *Environmental Microbiology, 18*, 3962–3975.

Lachnit, T., Blümel, M., Imhoff, J. F., & Wahl, M. (2009). Specific epibacterial communities on macroalgae: Phylogeny matters more than habitat. *Aquatic Biology, 5*, 181–186.

Lachnit, T., Thomas, T., & Steinberg, P. (2016). Expanding our understanding of the seaweed holobiont: RNA viruses of the red alga *Delisea pulchra*. *Frontiers in Microbiology, 6*, 1489.

Lachnit, T., Wahl, M., & Harder, T. (2010). Isolated thallus-associated compounds from the macroalga *Fucus vesiculosus* mediate bacterial surface colonization in the field similar to that on the natural alga. *Biofouling, 26*, 247–255.

Lage, O. M., & Graça, A. P. (2016). Biofilms: An extra coat on macroalgae. In N.T., & D. Dhanasekaran (Eds.), *Algae - organisms for imminent biotechnology*. InTech.

Lane, A. L., Nyadong, L., Galhena, A. S., Shearer, T. L., Stout, E. P., Parry, R. M., et al. (2009). Desorption electrospray ionization mass spectrometry reveals surface-mediated antifungal chemical defense of a tropical seaweed. *Proceedings of the National Academy of Sciences of the United States of America, 106*, 7314–7319.

Lanka, S. T. J., Klein, M., Ramsperger, U., Müller, D. G., & Knippers, R. (1993). Genome structure of a virus infecting the marine Brown alga *Ectocarpus siliculosus*. *Virology, 193*, 802–811.

Lee, Y. K., Jung, H. J., & Lee, H. K. (2006). Marine bacteria associated with the Korean brown alga, *Undaria pinnatifida*. *Journal of Microbiology, 44*, 694–698.

Lee, J. H., Kang, J. W., Choe, H. N., & Seong, C. N. (2017a). *Flavivirga eckloniae* sp. nov. and *Flavivirga aquimarina* sp. nov., isolated from seaweed *Ecklonia cava*. *International Journal of Systematic and Evolutionary Microbiology, 67*, 3089–3094.

Lee, J. H., Kang, J. W., Shin, S. B., & Seong, C. N. (2017b). *Winogradskyella flava* sp. nov., isolated from the brown alga, *Sargassum fulvellum*. *International Journal of Systematic and Evolutionary Microbiology, 67*, 3540–3546.

Ling, S.-K., Xia, J., Liu, Y., Chen, G.-J., & Du, Z.-J. (2017). *Agarilytica rhodophyticola* gen. nov., sp. nov., isolated from *Gracilaria blodgettii*. *International Journal of Systematic and Evolutionary Microbiology, 67*, 3778–3783.

van der Loos, L. M., Eriksson, B. K., & Falcão Salles, J. (2019). The macroalgal holobiont in a changing sea. *Trends in Microbiology, 27*, 635–650.
Loque, C. P., Medeiros, A. O., Pellizzari, F. M., Oliveira, E. C., Rosa, C. A., & Rosa, L. H. (2010). Fungal community associated with marine macroalgae from Antarctica. *Polar Biology, 33*, 641–648.
Lowery, C. A., Dickerson, T. J., & Janda, K. D. (2008). Interspecies and interkingdom communication mediated by bacterial quorum sensing. *Chemical Society Reviews, 37*, 1337.
Mabeau, S., & Kloareg, B. (1987). Isolation and analysis of the cell walls of Brown algae: *Fucus spiralis, F. Ceranoides, F. Vesiculosus, F. Serratus, Bifurcaria bifurcata* and *Laminaria digitata. Journal of Experimental Botany, 38*, 1573–1580.
Ma, Y., Liu, P., Yu, S., Li, D., & Cao, S. (2009). Inhibition of common fouling organisms in mariculture by epiphytic bacteria from the surfaces of seaweeds and invertebrates. *Acta Ecologica Sinica, 29*, 222–226.
Mancuso, F. P., D'Hondt, S., Willems, A., Airoldi, L., & De Clerck, O. (2016). Diversity and temporal dynamics of the epiphytic bacterial communities associated with the canopy-forming seaweed cystoseira compressa (Esper) Gerloff and Nizamuddin. *Frontiers in Microbiology, 7*, 476.
Martin, M., Barbeyron, T., Martin, R., Portetelle, D., Michel, G., & Vandenbol, M. (2015). The cultivable surface microbiota of the Brown alga *Ascophyllum nodosum* is enriched in macroalgal-polysaccharide-degrading bacteria. *Frontiers in Microbiology, 6*, 1487.
Martinez, J. N., & Padilla, P. I. P. (2016). Isolation and characterization of agar-digesting *Vibrio* species from the rotten thallus of *Gracilariopsis heteroclada* Zhang et Xia. *Marine Environmental Research, 119*, 156–160.
Martin, M., Portetelle, D., Michel, G., & Vandenbol, M. (2014). Microorganisms living on macroalgae: Diversity, interactions, and biotechnological applications. *Applied Microbiology and Biotechnology, 98*, 2917–2935.
Marzinelli, E. M., Campbell, A. H., Zozaya Valdes, E., Vergés, A., Nielsen, S., Wernberg, T., et al. (2015). Continental-scale variation in seaweed host-associated bacterial communities is a function of host condition, not geography. *Environmental Microbiology, 17*, 4078–4088.
Matsuo, Y., Imagawa, H., Nishizawa, M., & Shizuri, Y. (2005). Isolation of an algal morphogenesis inducer from a marine bacterium. *Science, 307*, 1598.
Maximilien, R., de Nys, R., Holmström, C., Gram, L., Givskov, M., Crass, K., et al. (1998). Chemical mediation of bacterial surface colonisation by secondary metabolites from the red alga *Delisea pulchra. Aquatic Microbial Ecology, 15*, 233–246.
Mayombo, N. A. S., Majewska, R., & Smit, A. J. (2019). Diatoms associated with two South African kelp species: *Ecklonia maxima* and *Laminaria pallida. African Journal of Marine Science, 41*, 221–229.
McKeown, D. A., Stevens, K., Peters, A. F., Bond, P., Harper, G. M., Brownlee, C., et al. (2017). Phaeoviruses discovered in kelp (Laminariales). *The ISME Journal, 11*, 2869–2873.
Michelou, V. K., Caporaso, J. G., Knight, R., & Palumbi, S. R. (2013). The ecology of microbial communities associated with *Macrocystis pyrifera. PLoS One, 8*, e67480.
Minich, J. J., Morris, M. M., Brown, M., Doane, M., Edwards, M. S., Michael, T. P., et al. (2018). Elevated temperature drives kelp microbiome dysbiosis, while elevated carbon dioxide induces water microbiome disruption. *PLoS One, 13*, e0192772.
Mohammed Riyaz, S. U., Nalini, S., Kavitha, G., Arul, S., Sutha, D., & Inbakandan, & D. (2019). Characterization and identification of isolated bacteria from ice-ice disease infected seaweed *Kappaphycus alvarezii. Indian Journal of Geo Marine Sciences, 48*, 1286–1290.

Molina-Montenegro, M., Muñoz, A., Badano, E., Morales, B., Fuentes, K., & Cavieres, L. (2005). Positive associations between macroalgal species in a rocky intertidal zone and their effects on the physiological performance of Ulva lactuca. *Marine Ecology Progress Series, 292*, 173–180.

Morris, M. M., Haggerty, J. M., Papudeshi, B. N., Vega, A. A., Edwards, M. S., & Dinsdale, E. A. (2016). Nearshore Pelagic Microbial Community Abundance Affects Recruitment Success of Giant Kelp, Macrocystis pyrifera. *Frontiers in Microbiology, 7*, 1800.

Murali, T. (2011). L-asparaginase from marine derived fungal endophytes of seaweeds. *Mycosphere*, 147–155.

Nagasaki, K., Shirai, Y., Tomaru, Y., Nishida, K., & Pietrokovski, S. (2005). Algal viruses with distinct intraspecies host specificities include identical intein elements. *Applied and Environmental Microbiology, 71*, 3599–3607.

Nasrolahi, A., Stratil, S. B., Jacob, K. J., & Wahl, M. (2012). A protective coat of microorganisms on macroalgae: Inhibitory effects of bacterial biofilms and epibiotic microbial assemblages on barnacle attachment. *FEMS Microbiology Ecology, 81*, 583–595.

Natrah, F. M. I., Bossier, P., Sorgeloos, P., Yusoff, F. M., & Defoirdt, T. (2014). Significance of microalgal-bacterial interactions for aquaculture. *Reviews in Aquaculture, 6*, 48–61.

Naval, P., & Chandra, T. S. (2019). Characterization of membrane vesicles secreted by seaweed associated bacterium *Alteromonas macleodii* KS62. *Biochemical and Biophysical Research Communications, 514*, 422–427.

Nedashkovskaia, O. I., Ivanova, E. P., Bakunina, I. I., Svetashev, V. I., Zviagintseva, T. N., & Mikhaĭlov, V. V. (2002). Characterization of marine bacteria *Pseudoalteromonas citrea* degradating fucoidan. *Mikrobiolohichnyi Zhurnal, 64*, 3–10.

Nedashkovskaya, O. I., Kim, S.-G., Balabanova, L. A., Zhukova, N. V., Bakunina, I. Y., & Mikhailov, V. V. (2018a). *Polaribacter staleyi* sp. nov., a polysaccharide-degrading marine bacterium isolated from the red alga *Ahnfeltia tobuchiensis*. *International Journal of Systematic and Evolutionary Microbiology, 68*, 623–629.

Nedashkovskaya, O. I., Kim, S.-G., Stenkova, A. M., Kukhlevskiy, A. D., Zhukova, N. V., & Mikhailov, V. V. (2018b). *Aquimarina algiphila* sp. nov., a chitin degrading bacterium isolated from the red alga *Tichocarpus crinitus*. *International Journal of Systematic and Evolutionary Microbiology, 68*, 892–898.

Nedashkovskaya, O. I., Kim, S.-G., Zhukova, N. V., & Mikhailov, V. V. (2017). *Olleya algicola* sp. nov., a marine bacterium isolated from the green alga *Ulva fenestrata*. *International Journal of Systematic and Evolutionary Microbiology, 67*, 2205–2210.

Nyadong, L., Hohenstein, E. G., Galhena, A., Lane, A. L., Kubanek, J., Sherrill, C. D., et al. (2009). Reactive desorption electrospray ionization mass spectrometry (DESI-MS) of natural products of a marine alga. *Analytical and Bioanalytical Chemistry, 394*, 245–254.

Nylund, G., & Pavia, H. (2005). Chemical versus mechanical inhibition of fouling in the red alga *Dilsea carnosa*. *Marine Ecology Progress Series, 299*, 111–121.

Nylund, G. M., Persson, F., Lindegarth, M., Cervin, G., Hermansson, M., & Pavia, H. (2010). The red alga Bonnemaisonia asparagoides regulates epiphytic bacterial abundance and community composition by chemical defence. *FEMS Microbiology Ecology, 71*, 84–93.

de Nys, R., Steinberg, P. D., Willemsen, P., Dworjanyn, S. A., Gabelish, C. L., & King, R. J. (1995). Broad spectrum effects of secondary metabolites from the red alga *Delisea pulchra* in antifouling assays. *Biofouling, 8*, 259–271.

Othmani, A., Briand, J.-F., Ayé, M., Molmeret, M., & Culioli, G. (2016). Surface metabolites of the brown alga *Taonia atomaria* have the ability to regulate epibiosis. *Biofouling, 32*, 801–813.

Parrot, D., Blümel, M., Utermann, C., Chianese, G., Krause, S., Kovalev, A., et al. (2019). Mapping the surface microbiome and metabolome of Brown seaweed *Fucus vesiculosus* by amplicon sequencing, integrated metabolomics and imaging techniques. *Scientific Reports, 9*, 1061.

Percival, E. (1979). The polysaccharides of green, red and brown seaweeds: Their basic structure, biosynthesis and function. *British Phycological Journal, 14*, 103–117.

Popper, Z. A., Michel, G., Hervé, C., Domozych, D. S., Willats, W. G. T., Tuohy, M. G., et al. (2011). Evolution and diversity of plant cell walls: From algae to flowering plants. *Annual Review of Plant Biology, 62*, 567–590.

Potin, P., Bouarab, K., Salaün, J.-P., Pohnert, G., & Kloareg, B. (2002). Biotic interactions of marine algae. *Current Opinion in Plant Biology, 5*, 308–317.

Preston, J. F., Romeo, T., & Bromley, J. C. (1986). Selective alginate degradation by marine bacteria associated with the algal genus*Sargassum*. *Journal of Industrial Microbiology, 1*, 235–244.

Qi, F., Huang, Z., Lai, Q., Li, D., & Shao, Z. (2016). *Kordia ulvae* sp. nov., a bacterium isolated from the surface of green marine algae *Ulva* sp. *International Journal of Systematic and Evolutionary Microbiology, 66*, 2623–2628.

Qiu, Z., Coleman, M. A., Provost, E., Campbell, A. H., Kelaher, B. P., Dalton, S. J., et al. (2019). Future climate change is predicted to affect the microbiome and condition of habitat-forming kelp. *Proceedings of the Royal Society B: Biological Sciences, 286*, 20181887.

Railkin, A. I. (2003). Marine biofouling: Colonization processes and defenses. In *Marine biofouling: Colonization processes and defenses* (pp. 1–317). CRC Press.

Ramanan, R., Kang, Z., Kim, B.-H., Cho, D.-H., Jin, L., Oh, H.-M., et al. (2015). Phycosphere bacterial diversity in green algae reveals an apparent similarity across habitats. *Algal Research, 8*, 140–144.

Ramanan, R., Kim, B.-H., Cho, D.-H., Oh, H.-M., & Kim, H.-S. (2016). Algae–bacteria interactions: Evolution, ecology and emerging applications. *Biotechnology Advances, 34*, 14–29.

Ramya, R., Sangeetha Devi, R., Manikandan, A., & Rajesh Kannan, V. (2017). Standardization of biopolymer production from seaweed associative bacteria. *International Journal of Biological Macromolecules, 102*, 550–564.

Rao, D., Webb, J. S., Holmström, C., Case, R., Low, A., Steinberg, P., et al. (2007). Low densities of epiphytic bacteria from the marine alga *Ulva australis* inhibit settlement of fouling organisms. *Applied and Environmental Microbiology, 73*, 7844–7852.

Rasch, M., Buch, C., Austin, B., Slierendrecht, W. J., Ekmann, K. S., Larsen, J. L., et al. (2004). An inhibitor of bacterial quorum sensing reduces mortalities caused by vibriosis in rainbow trout (*Oncorhynchus mykiss*, walbaum). *Systematic & Applied Microbiology, 27*, 350–359.

Ren, D., Sims, J. J., & Wood, T. K. (2001). Inhibition of biofilm formation and swarming of *Escherichia coli* by (5Z)-4-bromo-5-(bromomethylene)-3-butyl-2(5H)-furanone. *Environmental Microbiology, 3*, 731–736.

Ren, D., Sims, J. J., & Wood, T. K. (2002). Inhibition of biofilm formation and swarming of *Bacillus subtilis* by (5Z)-4-bromo-5-(bromomethylene)-3-butyl-2(5H)-furanone. *Letters in Applied Microbiology, 34*, 293–299.

Ren, D., Zuo, R., & Wood, T. K. (2005). Quorum-sensing antagonist (5Z)-4-bromo-5-(bromomethylene)-3-butyl-2(5H)-furanone influences siderophore biosynthesis in *Pseudomonas putida* and *Pseudomonas aeruginosa*. *Applied Microbiology and Biotechnology, 66*, 689–695.

Rohde, S., Hiebenthal, C., Wahl, M., Karez, R., & Bischof, K. (2008). Decreased depth distribution of *Fucus vesiculosus* (Phaeophyceae) in the western baltic: Effects of light deficiency and epibionts on growth and photosynthesis. *E*uropean Journal of Phycology, *43*, 143–150.

Romero, M., Martin-Cuadrado, A.-B., Roca-Rivada, A., Cabello, A. M., & Otero, A. (2011). Quorum quenching in cultivable bacteria from dense marine coastal microbial communities. *FEMS Microbiology Ecology, 75*, 205–217.

Rousvoal, S., Bouyer, B., López-Cristoffanini, C., Boyen, C., & Collén, J. (2016). Mutant swarms of a totivirus-like entities are present in the red macroalga *Chondrus crispus* and have been partially transferred to the nuclear genome. *Journal of Phycology, 52*, 493–504.

Saha, M., Rempt, M., Gebser, B., Grueneberg, J., Pohnert, G., & Weinberger, F. (2012). Dimethylsulphopropionate (DMSP) and proline from the surface of the brown alga *Fucus vesiculosus* inhibit bacterial attachment. *Biofouling, 28*, 593–604.

Saha, M., & Weinberger, F. (2019). Microbial "gardening" by a seaweed holobiont: Surface metabolites attract protective and deter pathogenic epibacterial settlement. *Journal of Ecology*, 1365–2745.

Sawabe, T., Tanaka, R., Iqbal, M. M., Tajima, K., Ezura, Y., Ivanova, E. P., et al. (2000). Assignment of *Alteromonas elyakovii* KMM 162(T) and five strains isolated from spot-wounded fronds of Laminaria japonica to *Pseudoalteromonas elyakovii* comb. nov. and the extended description of the species. *International Journal of Systematic and Evolutionary Microbiology, 50*, 265–271.

Schäfer, H., Abbas, B., Witte, H., & Muyzer, G. (2002). Genetic diversity of "satellite" bacteria present in cultures of marine diatoms. *FEMS Microbiology Ecology, 42*, 25–35.

Schwartz, N., Dobretsov, S., Rohde, S., & Schupp, P. J. (2017). Comparison of antifouling properties of native and invasive *Sargassum* (Fucales, Phaeophyceae) species. *European Journal of Phycology, 52*, 116–131.

Serebryakova, A., Aires, T., Viard, F., Serrão, E. A., & Engelen, A. H. (2018). Summer shifts of bacterial communities associated with the invasive brown seaweed Sargassum muticum are location and tissue dependent. *PLoS One, 13*, e0206734.

Singh, R. P., Baghel, R. S., Reddy, C. R. K., & Jha, B. (2015). Effect of quorum sensing signals produced by seaweed-associated bacteria on carpospore liberation from *Gracilaria dura*. *Frontiers of Plant Science, 6*, 117.

Singh, R. P., Kumari, P., & Reddy, C. R. K. (2014). Antimicrobial compounds from seaweeds-associated bacteria and fungi. *Applied Microbiology and Biotechnology, 99*, 1571–1586.

Singh, R. P., & Reddy, C. R. K. (2014). Seaweed-microbial interactions: Key functions of seaweed-associated bacteria. *FEMS Microbiology Ecology, 88*, 213–230.

Singh, R. P., & Reddy, C. R. K. (2016). Unraveling the functions of the macroalgal microbiome. *Frontiers in Microbiology, 6*, 1–8.

Singh, R. P., Shukla, M. K., Mishra, A., Kumari, P., Reddy, C. R. K., & Jha, B. (2011). Isolation and characterization of exopolysaccharides from seaweed associated bacteria *Bacillus licheniformis*. *Carbohydrate Polymers, 84*, 1019–1026.

Skillings, D. (2016). Holobionts and the ecology of organisms: Multi-species communities or integrated individuals? *Biology and Philosophy, 31*, 875–892.

Spoerner, M., Wichard, T., Bachhuber, T., Stratmann, J., & Oertel, W. (2012). Growth and thallus morphogenesis of *Ulva mutabilis* (Chlorophyta) depends on A combination of two bacterial species excreting regulatory factors. *Journal of Phycology, 48*, 1433–1447.

Staufenberger, T., Thiel, V., Wiese, J., & Imhoff, J. F. (2008). Phylogenetic analysis of bacteria associated with *Laminaria saccharina*. *FEMS Microbiology Ecology, 64*, 65–77.

Steinberg, P. D., & de Nys, R. (2002). Chemical mediation of colonization of seaweed surfaces. *Journal of Phycology, 38*, 621–629.

Steinberg, P. D., De Nys, R., & Kjelleberg, S. (2002). Chemical cues for surface colonization. *Journal of Chemical Ecology, 28*, 1935–1951.

Stiger-Pouvreau, V., Bourgougnon, N., & Deslandes, E. (2016). Carbohydrates from seaweeds. In J. Fleurence, & I. Levine (Eds.), *Seaweed in health and disease prevention* (pp. 223–274). Elsevier Inc. Academic Press.

Sullivan, T., & Regan, F. (2011). The characterization, replication and testing of dermal denticles of *Scyliorhinus canicula* for physical mechanisms of biofouling prevention. *Bioinspiration & Biomimetics, 6*, 046001.

Sun, C., Fu, G. Y., Zhang, C. Y., Hu, J., Xu, L., Wang, R. J., ... Wu, M. (2016). Isolation and complete genome sequence of Algibacter alginolytica sp. nov., a novel seaweed-degrading Bacteroidetes bacterium with diverse putative polysaccharide utilization loci. *Applied and Environmental Microbiology, 82*, 2975–2987.

Suresh, M., Renugadevi, B., Brammavidhya, S., Iyapparaj, P., & Anantharaman, P. (2015). Antibacterial activity of red pigment produced by *Halolactibacillus alkaliphilus* MSRD1—an isolate from seaweed. *Applied Biochemistry and Biotechnology, 176*, 185–195.

Suryanarayanan, T. S., Venkatachalam, A., Thirunavukkarasu, N., Ravishankar, J. P., Doble, M., & Geetha, V. (2010). Internal mycobiota of marine macroalgae from the Tamilnadu coast: Distribution, diversity and biotechnological potential. *Botanica Marina, 53*, 457–468.

Susilowati, R., Sabdono, A., & Widowati, I. (2015). Isolation and characterization of bacteria associated with Brown algae *Sargassum* spp. from panjang Island and their antibacterial activities. *Procedia Environmental Sciences, 23*, 240–246.

Suvega, T., & Arunkumar, K. (2019). Probiotic bacteria promote the growth of associating host(red seaweed, *Gracilaria edulis*) also synthesize antibacterial protein. *Biocatalysis and Agricultural Biotechnology, 19*, 101136.

Suzuki, T., Ariefta, N. R., Koseki, T., Furuno, H., Kwon, E., Momma, H., et al. (2019). New polyketides, paralactonic acids A–E produced by *Paraconiothyrium* sp. SW-B-1, an endophytic fungus associated with a seaweed, *Chondrus ocellatus* Holmes. *Fitoterapia, 132*, 75–81.

Thoendel, M., Kavanaugh, J. S., Flack, C. E., & Horswill, A. R. (2011). Peptide signaling in the staphylococci. *Chemical Reviews, 111*, 117–151.

Torres, M., Dessaux, Y., Llamas, I., Torres, M., Dessaux, Y., & Llamas, I. (2019). Saline environments as a source of potential quorum sensing disruptors to control bacterial infections: A review. *Marine Drugs, 17*, 191.

Totti, C., Poulin, M., Romagnoli, T., Perrone, C., Pennesi, C., & De Stefano, M. (2009). Epiphytic diatom communities on intertidal seaweeds from Iceland. *Polar Biology, 32*, 1681–1691.

Uzair, B., Menaa, F., Khan, B. A., Mohammad, F. V., Ahmad, V. U., Djeribi, R., et al. (2018). Isolation, purification, structural elucidation and antimicrobial activities of kocumarin, a novel antibiotic isolated from actinobacterium *Kocuria marina* CMG S2 associated with the brown seaweed *Pelvetia canaliculata*. *Microbiological Research, 206*, 186–197.

Vairappan, C. S., Chung, C. S., Hurtado, A. Q., Soya, F. E., Lhonneur, G. B., & Critchley, A. (2008). Distribution and symptoms of epiphyte infection in major carrageenophyte-producing farms. *Journal of Applied Phycology, 20*, 477–483.

Vala, A. K., Anand, N., Bhatt, P. N., & Joshi, H. V. (2004). Tolerance and accumulation of hexavalent chromium by two seaweed associated aspergilli. *Marine Pollution Bulletin, 48*, 983–985.

Villarreal-Gómez, L. J., Soria-Mercado, I. E., Guerra-Rivas, G., & Ayala-Sánchez, N. E. (2010). Antibacterial and anticancer activity of seaweeds and bacteria associated with their surface. *Revista de Biología Marina y Oceanografía, 45*, 267–275.

Wahl, M., Goecke, F., Labes, A., Dobretsov, S., & Weinberger, F. (2012). The second skin: Ecological role of epibiotic biofilms on marine organisms. *Frontiers in Microbiology, 3*, 292.

Weinberger, F. (2007). Pathogen-induced defense and Innate immunity in macroalgae. *The Biological Bulletin, 213*, 290–302.

Weinberger, F., Beltran, J., Correa, J. A., Lion, U., Pohnert, G., Kumar, N., et al. (2007). Spore release in *Acrochaetium* sp. (Rhodophyta) is bacterially controlled. *Journal of Phycology, 43*, 235–241.

Wichard, T. (2015). Exploring bacteria-induced growth and morphogenesis in the green macroalga order *Ulvales* (Chlorophyta). *Frontiers of Plant Science, 6*, 86.

Wichard, T., Charrier, B., Mineur, F., Bothwell, J. H., Clerck, O. De, & Coates, J. C. (2015). The green seaweed *Ulva*: A model system to study morphogenesis. *Frontiers of Plant Science, 6*, 72.
Williams, P. (2007). Quorum sensing, communication and cross-kingdom signalling in the bacterial world. *Microbiology, 153*, 3923–3938.
Williams, P., Winzer, K., Chan, W. C., & Cámara, M. (2007). Look who's talking: Communication and quorum sensing in the bacterial world. *Philosophical Transactions of the Royal Society B: Biological Sciences, 362*, 1119–1134.
Wu, H., Liu, M., Zhang, W., & Xiao, T. (2014). Phylogenetic analysis of epibacterial communities on the surfaces of four red macroalgae. *Journal of Ocean University of China, 13*, 1025–1032.
Yang, S.-H., Oh, J. H., Seo, H.-S., Lee, J.-H., & Kwon, K. K. (2018). *Marinirhabdus citrea* sp. nov., a marine bacterium isolated from a seaweed. *International Journal of Systematic and Evolutionary Microbiology, 68*, 547–551.
Ying, C., Chang, M.-J., Hu, C.-H., Chang, Y.-T., Chao, W.-L., Yeh, S.-L., et al. (2018). The effects of marine farm-scale sequentially integrated multi-trophic aquaculture systems on microbial community composition, prevalence of sulfonamide-resistant bacteria and sulfonamide resistance gene *sul*1. *The Science of the Total Environment, 643*, 681–691.
Ying, J.-J., Zhang, S.-L., Huang, C.-Y., Xu, L., Zhao, Z., Wu, M., et al. (2019). *Algicoccus marinus* gen. nov. sp. nov., a marine bacterium isolated from the surface of brown seaweed *Laminaria japonica*. *Archives of Microbiology, 201*, 943–950.
Zhang, W., & Li, C. (2016). Exploiting quorum sensing interfering strategies in gram-negative bacteria for the enhancement of environmental applications. *Frontiers in Microbiology, 6*, 1535.
Zhang, Y., Wang, S., Li, X.-M., Cui, C.-M., Feng, C., & Wang, B.-G. (2007). New sphingolipids with a previously unreported 9-methyl-C20-sphingosine moiety from a marine algous endophytic fungus *Aspergillus Niger* EN-13. *Lipids, 42*, 759–764.
Zhou, J., Lyu, Y., Richlen, M. L., Anderson, D. M., & Cai, Z. (2016). Quorum sensing is a language of chemical signals and plays an ecological role in algal-bacterial interactions. *Critical Reviews in Plant Sciences, 35*, 81–105.
Zuccaro, A., Schoch, C. L., Spatafora, J. W., Kohlmeyer, J., Draeger, S., & Mitchell, J. I. (2008). Detection and identification of fungi intimately associated with the brown seaweed *Fucus serratus*. *Applied and Environmental Microbiology, 74*, 931–941.

CHAPTER ELEVEN

Green and integrated processing approaches for the recovery of high-value compounds from brown seaweeds

Achinta Bordoloi and Neill Goosen*
Department of Process Engineering, Stellenbosch University, Stellenbosch, South Africa
*Corresponding author: E-mail: njgoosen@sun.ac.za

Contents

Abstract

The rapidly increasing interest in utilizing seaweeds as sustainable raw material from which multiple high-value bioactive compounds can be recovered, has necessitated the development of new processing methods to achieve this. Brown seaweeds contain a range of bioactive compounds that are of commercial interest, including hydrocolloid and non-hydrocolloid polysaccharides, antioxidant polyphenols and unique pigments. Traditional polysaccharide extraction from brown seaweeds has relied on methods that require harsh chemical processing and sometimes also large amounts of energy and

Advances in Botanical Research, Volume 95
ISSN 0065-2296
https://doi.org/10.1016/bs.abr.2019.11.011

organic solvents. Commercial operation requires high recovery of the targeted compounds on the one hand, but simultaneously also the preservation of biological activity of the products on the other, an objective that is difficult to achieve using conventional methods. In order to address this shortcoming and to move toward more environmentally friendly processing in general, several greener extraction methods are being developed which have lower environmental footprint and sometimes also greater process efficiency, including enzymatic, sub and supercritical fluid, ultrasonic and microwave extraction techniques. This chapter explores these extraction techniques within the context of particularly brown seaweeds and provides an overview of potential advantages compared to conventional technologies. The work further elucidates how certain physiological characteristics of brown seaweeds can impact the processing thereof and looks toward possible future developments within the field of brown seaweed processing.

1. Introduction

Seaweeds have been highlighted as a potentially sustainable biomass feedstock for production and recovery of energy and a unique range of products for feed, food, nutraceutical and pharmaceutical industries (Holdt & Kraan, 2011; Song, Duc Pham, Seon, & Chul Woo, 2015). Specifically brown seaweeds (Phaedophyta) harbor a range of high value compounds such as lamarinin, fucoidans, alginates and phlorotannins, which could potentially have widespread applications in the food, cosmetic and the pharmaceutical industry (Barbosa, Lopes, Andrade, & Valentão, 2019; Deniaud-Bouët, Hardouin, Potin, Kloareg, & Hervé, 2017; Pádua, Rocha, Gargiulo, & Ramos, 2015). The fact that seaweed cultivation does not compete with food crops for land use and fresh water resources is one of the biggest advantages, and it also does not require fertilizer application. Brown seaweed can reach high growth rates of up to ~13 kg dry weight $m^{-2}.y^{-1}$, which is significantly higher than some terrestrial crops like sugarcane at ~10 kg dry weight $m^{-2}.y^{-1}$ (Leu & Boussiba, 2014), which is a contributing factor to numerous efforts focused on commercializing brown seaweed cultivation.

Development of sustainable and high efficiency processing methods is a requirement to drive value-addition within the globally expanding seaweed industry (FAO, 2018). Conventional processing methods often employ harsh processing conditions, use organic solvents, or require extensive downstream processing for product separation and purification, which can lead to generation of hazardous wastes and create a significant carbon footprint. Alternative and so-called 'greener' processing methods are being developed and research has shown that these can consistently yield better process efficacy, shorter processing times and result in enhanced bioactive properties of the products. These novel extraction methods range from enzyme assisted extraction (EAE), ultrasound assisted extraction (UAE), microwave assisted extraction (MAE), pressurized liquid extraction (PLE), sub-critical water extraction (SWE) and super-critical fluid extraction (SFE).

This chapter reviews various emerging extraction techniques as being applied particularly to brown seaweeds and compares these to conventional methods. It further explores the coupling of various techniques in terms of increasing extraction efficiency and the benefits of considering the influence of different extraction processes on the functional characteristics of the product. The review was also extended to explore the physiological characteristics of brown seaweed over different seasons that could impact its processing and can lead to an integrated process development through employing a biorefinery approach.

2. Green extraction techniques for valorizing macroalgae

The extraction route for recovering products from seaweeds can be generalized to some extent, and is divided into pre-treatment of the raw material, followed by extraction in single or multiple stages, separation of valuable compounds from residues, and finally downstream recovery and concentration of the final products. Fig. 1 is a depiction of the general route that can be expected in a seaweed processing operation.

2.1 Macroalgae pre-treatments

In general, seaweed biomass is washed post-harvesting to remove impurities like sand and epiphytes and can be further dried in cases where immediate further processing is not feasible. Industrially, drying can be achieved through sun drying or using heated drying equipment (although this comes at a significant energy expenditure); freeze drying is reserved almost solely for laboratory investigations due to the high cost of this technique. A further commonly employed pre-treatment is size reduction of the material in order to enable more efficient mass transfer during compound extraction. Different milling/grinding techniques are used, and samples are classified into different sizes depending on the further processing requirements, which serves to enlarge the surface to volume ratio to improve extraction efficiency

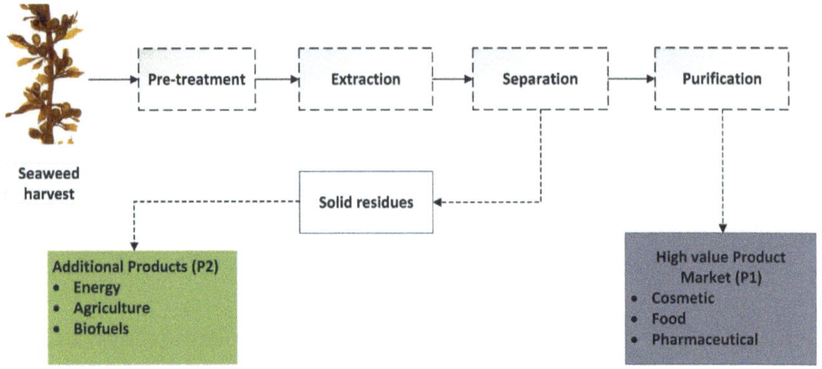

Fig. 1 A general schematic of the work-flow under a biorefinery approach showing different processing stages from harvesting to final recovery of products.

(Hahn, Lang, Ulber, & Muffler, 2012; Imbs, Ermakova, Malyarenko, Isakov, & Zvyagintseva, 2016). Proprietary cell burst and micronization techniques are increasingly being implemented for size reduction within the industry. While these routines are followed often, because of the nature of the algal biomass with water content ranging from 80% to 85%, drying can incur significant expenses upfront from an energy perspective. Developing extraction techniques which allows using size reduced wet samples would inevitably boost the workflow and can be either species-specific or end-product driven. However, storage is critical to account for seasonality and maintaining inputs through the macroalgal supply chain while maintaining bioactive functionalities of high value products within the raw materials at the same time. Drying and storage is currently a requirement as it increases the shelf life of the harvested fractions.

Based on the target end products, the samples are also pre-treated to avoid cross-contamination and removal of unwanted compounds from the sample. Mostly organic solvents are used for removing lipids (defatting), proteins and phenols (Hahn et al., 2012). Table 1 outlines different methods of pre-treatments used in various studies. Recently, alternative pre-treatment techniques have been favorably used to improve cell membrane permeability such as supercritical-CO_2 to defat algal biomass prior to extraction (Quitain, Kai, Sasaki, & Goto, 2013a), hydrothermal puffing, etc. Brown seaweed *Cystoseira trinodis* pre-treatment by fungal fermentation lead to reduced molecular weights of the fucoidan and alginate fraction with significantly improved antioxidant capacity (Hifney, Fawzy, Abdel-Gawad, & Gomaa, 2018).

Table 1 Different type of algal biomass pre-treatment methods employed prior to extraction of compounds.

Pre-treatment	Condition/source/coupled technique	Application	Reference
Dilute HCl	0.1 M HCl, 2 h, 60°C, solvent to algae ratio of 5 (mL g^{-1})	Used prior to fucoidan extraction by disrupting h—bonds between polysaccharides	(Anastyuk et al., 2017)
	0.1 M HCl, pH 2, 65°C, 3 h, 3 times, solvent to algae ratio of 10 (mL g^{-1})	For alginate extraction it helps remove unwanted impurities (polyphenols, proteins) while converting insoluble alginate salts to alginic acids for further extraction as sodium alginate	(Rostami, Tabarsa, You, & Rezaei, 2017)
Ethanol	85% ethanol, solvent to algae ratio of 10 (mL g^{-1}), stirring (2000 rpm), 24 h, room temperature	Use to selectively extract polyphenols, pigments and other low molecular weight compounds prior to subsequent extraction methods	(Alboofetileh et al., 2019)
Formaldehyde	2% (w/v) formaldehyde solution overnight, dried for 48 h, 40°C	Used to remove phenolics from the cell wall matrix and improve product purity	(Mohammed et al., 2018)
Enzymatic	a)5% solids content, alginate lyase (0.05% w/w), 37°C, 2 h, 125 rpm agitation	Helps in degradation of alginate in the cell wall matrix to improve extraction yield	(Billakanti, Catchpole, Fenton, Mitchell, & MacKenzie, 2013)
	b) Viscozyme intensified with microwave, 30 min, 50°C, pH 4.5	Coupling strategy was used to test pre-biotic potentials of different molecular weight fractions	(Charoensiddhi, Conlon, Vuaran, Franco, & Zhang, 2017)
Supercritical-CO_2	300 bar, 50°C, 4 h	Employed as defatting technique prior to polysaccharide extraction to enhance bioactivity of target polysaccharides	(Saravana et al., 2018)
Ultrasound assisted (UA)	750 W capacity and 20 kHz frequency for 60 min pre-treatment	UA pre-treatment was followed by alkali-based extraction of proteins	(Kadam, Álvarez, Tiwari, & O'Donnell, 2017)

2.2 Novel extraction techniques

The bottleneck in harnessing the potential of the seaweed compound repository lies in improving extraction efficiency to get viable yields, while also preserving the bioactive compounds functionalities. The macroalgal cell wall structure, with interlinked polysaccharide fractions, provides a very robust physical barrier which needs to be overcome in order to access various high value compounds that occur as part of the cell wall structure, or are found internally within the cells. The macroalgal matrix has a unique structural integrity and conventional extraction has relied on harsh treatments (high temperatures and high- and low pH, with or without the use of organic solvents) to overcome this barrier. However, these harsh extraction conditions and use of organic solvents raises safety issues for end use in food, pharma and cosmetics, in addition to environmental concerns. Downstream separation and purifications steps might further lead to production of waste that raise disposal concerns.

Novel and greener processes have started making headway recently, which have showed promise through improved yields and shorter extraction times, with lower environmental footprints and energy demands compared to conventional processes (Bowyer, Van Altena, & Scarlett, 2018; Flórez-Fernández, Torres, González-Muñoz, & Domínguez, 2019; Kadam, Tiwari, & O'Donnell, 2013; Nadar, Rao, & Rathod, 2018; Saravana et al., 2018). Although development of these novel processes require significant investment in terms of initial process development, piloting and scale up, long term economic prospects seem promising (Charoensiddhi et al., 2018).

2.2.1 Enzyme assisted extraction

Enzyme assisted extraction is gaining significant attention over conventional solvent based extraction techniques in terms of offering better extraction efficacy, specificity in targeting desired end products, reduced processing time and an environmentally friendly process devoid of the need for harsh solvents (Nadar et al., 2018; Wijesinghe & Jeon, 2012b). Macroalgae, specifically the brown seaweed species harbor a range of unique biomolecules such as phlorotannins, fucoxanthins, alginates and the sulfated polysaccharide fractions with illustrated bioactivities which are enmeshed within the structural makeup of the cell wall (Holdt & Kraan, 2011). Breaking down these structural barriers is pivotal to release the entrapped compounds and EAE is a promising technique in selectively breaking down the cellular barriers under mild conditions (Charoensiddhi, Conlon, Franco, & Zhang, 2017).

EAE offers important advantages within integrated processing schemes. Extraction conditions enable the recovery of safe products suitable for use in nutraceutical, cosmetic and pharmaceutical industries. Furthermore, EAE products have been shown to possess enhanced bioactivities in various studies, which have been linked to the preservation of the structural integrity of target compounds (Casas, Conde, Domínguez, & Moure, 2018; Lee et al., 2012; Puspita et al., 2017; Rodrigues et al., 2015). EAE furthermore offers process scalability and good selectivity toward individual target compounds (Jaswir, Amid, Alam, Asiyanbi-H, & Ramli, 2013), and both individual enzymes or enzyme cocktails have proved valuable for targeting specific components of the macroalgae (Puspita et al., 2017).

Despite the promises of EAE, one of the main obstacles to this processing method in seaweeds is the fact that very few industrially available enzymes have been specifically tailored toward substrates found only in seaweed. Research on enzymatic seaweed processing to date has employed enzymes developed for use in terrestrial feedstocks due to lack of alternatives, and in some cases results have been variable. In the case of employing cellulases for release of fermentable sugars like glucose, results have been favorable and high glucose yields are achievable (Borines, De Leon, & Cuello, 2013; Noraini, Ong, Badrul, & Chong, 2014), although sometimes only after extended hydrolysis times. Other attempts to degrade seaweed polysaccharides using carbohydrases have been less successful, where a number of carbohydrases exhibited almost no ability to significantly solubilize seaweed biomass (Hardouin et al., 2016). The implication of this lack of industrial enzymes specifically tailored toward seaweed processing is that the terrestrial biomass-oriented enzymes available may exhibit slower enzyme kinetics and poorer substrate specificity, and that for some substrates found in seaweeds, no industrial enzymes are available to employ during processing. Thus, tapping into the potential of novel enzymes from marine sources more specific to the macroalgal chemistry can boost the EAE process significantly in terms of increased extraction efficiency, with likely improvement in the techno-economics of EAE-based seaweed processing operations.

Table 2 compiles the various enzymes and associated extracted product and its application from different brown seaweed species. EAE has been reported to deliver yields in the range of 20%–50% on a dry algal weight basis at enzyme dosing of 0.5%–5% (w/v), where target products included polyphenols, alginates, sulfated polysaccharides and proteins. Comparatively, the

Table 2 Applications of EAE for bioactive compounds from various brown macroalgae species.

Species	Enzyme	Conditions	Compounds	Application	Reference
Ecklonia radiate	Viscozyme® L, Celluclast® 1.5 L, Ultraflo® L, Alacalase® 2.4 L FG, Neutrase® 0.8 L and Flavourzyme® 1000 L	Enzymatic hydrolysis at 50°C for 24 h at pH 4.5–8 with buffers and water	Carbohydrates	Food industry	(Charoensiddhi et al., 2017)
Sargassum muticum (Yendo)	Neutrase, alcalase, ultraflo, amyloglucosidase (AMG), shearzyme, termamyl, viscozyme, and celluclast	3 h at 50°C followed by deactivation for 15 min at 90°C	Phenolic compounds	Antioxidant, antimicrobial, cytotoxicity activity, tyrosinase inhibition activity	(Puspita et al., 2017)
Nizamuddinia zanardinii	Alcalase, flavourzyme, celluclast, and viscozyme	5% (v/v), 50°C, 24 h at optimal pH of each enzyme	Sulfphated polysaccharides	Anticancer and immune-enhancing activities	(Alboofetileh, Rezaei, & Tabarsa, 2018)
Undaria pinnatifida	Alginate lyase	Enzyme pre-processing (0.05% w/w) followed by extraction using dimethyl ether (DME) with ethanol as a co-solvent. 37°C and pH 6.5	Fucoxanthin and lipids	Therapeutics	(Billakanti et al., 2013)

Ecklonia cava	Celluclast	24 h at 50°C followed by deactivation for 10 min at 100°C	Fucoidans	Anti-inflammatory activity	(Lee et al., 2012)
Sargassum muticum	Proteases (endo-peptidase, endo-protease) and carbohydrases (cellulase, xylanase, β-glucanase, arabanase)	50°C for 5 h	Crude hydrolysate	Anti-viral activity	(Hardouin et al., 2014)
Sargassum muticum	Celluclast, viscozyme, rapidase, amylase alcalase, protamex	3 h of extraction using 2%–5% (wt%) enzyme loads	Crude hydrolysate	Antioxidant, cytotoxicity activity	(Casas et al., 2018)
Sargassum muticum	Alcalase, flavourzyme, cellulase, viscozyme L	24 h at 50°C and deactivated by heating the sample at 90–100°C for 10 min.	Phenolics, sulfphated polysaccharides	Hydroxyl-radical scavenging activity, prebiotic activity	(Rodrigues et al., 2015)

fractions with different molecular weights from the native molecules, have often shown enhanced bioactivities, including antioxidant, anti-viral, anti-cancer and anti-inflammatory activities.

2.2.2 Microwave assisted extraction

Microwave assisted extraction operates in the electromagnetic waves spectrum with frequencies ranging between 300 MHz and 300 GHz (Routray & Orsat, 2012). These waves are non-ionizing and the energy translates into heat which dissipates causing structural changes within the material matrix (Yuan & Macquarrie, 2015). The dielectric properties of the fluid within the matrix can be a deciding parameter for extraction efficiency. Absorption of the waves by the sample initiates two different phenomena, dipole rotation and ionic conduction which are influenced by temperature (Garcia-Vaquero, Rajauria, O'Doherty, & Sweeney, 2017). Ionic conduction happens due to the movement of the charged ions under the influence of the electrostatic field created by the microwaves and the resulting friction owing to these movements within the sample space creates heat. The other heat generating mechanism from the microwaves happens due to the tendency of the dipolar fractions in the sample to align according to the alternating electric field generated by the microwaves and collides among themselves and generates heat (Zhang, Yang, & Wang, 2011).

MAE has gained attention as a good alternative to solvent extraction techniques for extracting bioactive compounds from various natural matrices including macroalgae. MAE offers several advantages in terms of using less solvent, shorter processing times and increased extraction yields (Yuan et al., 2018). It enables deeper penetration of solvents into the biomass matrix which aid in better mass transfer and dissolution of solutes. Its industrial scale applicability has been recommended which does involve an initial capital expenditure, but more investigation is required in terms of operational logistics for large-scale operations (Leone, Tamborrino, Romaniello, Zagaria, & Sabella, 2014). In addition, due to the generation of heat during the process it could also hamper extraction of heat labile compounds and possibly contribute toward structural alteration of the compounds (Ren et al., 2017). A significant portion of polyphenols, up to 40% are cell-wall bound and thus allows better accessibility via MAE compared to solvent extraction. MAE increased polyphenols yields in the *Carpophyllum flexuosum* species by up to 70% compared to solvent extraction (Magnusson et al., 2017). Table 3 compiles the various MAE isolated products and its illustrated application from different brown seaweed species. Fucoidans and

Table 3 Applications of MAE for bioactive compounds from various brown macroalgae species.

Species	Conditions	Compounds	Application	Reference
Fucus vesiculosus	MAE at 120 psi, 1 min, solvent to algae ratio of 25 (mL g^{-1})	Fucoidans	Not characterized	(Rodriguez-Jasso, Mussatto, Pastrana, Aguilar, & Teixeira, 2011)
Ascophyllum nodosum	120°C for 15 min extraction, up to 300W	Fucoidans	Anti-oxidant activity	(Yuan & Macquarrie, 2015)
Undaria pinnatifida	Extraction time 1–3 min, microwave power 600 W, temperature 110–200°C	Fucoidans	Low-molecular-weight components (activity not evaluated)	(Quitain et al., 2013a)
Undaria pinnatifida and *Sargassum fusiforme*	Extraction time 10–30 min, microwave power 300–500 W, temperature 30–70°C,	Phytol Fucosterol	Not characterized	(Xiao, Yuan, & Li, 2013)
Saccharina japonica Aresch	Extraction time 5–25 min, microwave power 400–600 W, temperature 60°C	Phlorotannin	Inhibited HepG2 cancer cells	(He et al., 2013)
Ascophyllum nodosum, Laminaria japonica, Lessonia trabeculate and *Lessonia nigrecens*	Irradiation (2.45 GHz) for 15 min (5 min climbing and 10 min holding) at 110 °C. Agitation: 300 rpm	Phlorotannin	Antioxidant activities and inhibitory effects on α-amylase, α-glucosidase, pancreatic lipase and tyrosinase	(Yuan et al., 2018)

(Continued)

Table 3 Applications of MAE for bioactive compounds from various brown macroalgae species.—cont'd

Species	Conditions	Compounds	Application	Reference
Sargassum thunbergii	Extraction time 23 min, microwave power 547 W, extraction temperature 80°C, and solvent to algae ratio of 27 (mL g^{-1})	Polysaccharides	Antioxidant and hypoglycemic activities	(Ren et al., 2017)
Ten different species from Australia and New Zealand were studied	Biomass:solvent ratio 1:30, 160°C, 3 min, max power 850 W	Phenolic compounds	Antioxidant capacity	(Magnusson et al., 2017)
Sargassum vestitum	1200 W, frequency 2450 MHz extracted in 70% ethanol at solvent to algae ratio 50 (mL g^{-1})	Phenolic compounds	Antioxidant capacity	(Bowyer et al., 2018)

polyphenolic compounds from a number of different brown seaweed species have been isolated, and a range of bioactivities have been demonstrated, but MAE research can benefit from further functional characterization of the isolated products.

2.2.3 Ultrasound assisted extraction

Ultrasound assisted extraction has shown potential as a viable extraction technique for the bioactive compounds in the macroalgal cellular matrix (Kadam, Tiwari, Smyth, & O'Donnell, 2015). It works based on the propagation of high frequency sound waves (>20 kHz) through the medium in compression and rarefaction waves which leads to the formation of cavitation bubbles. Vapor from the liquid medium gets entrapped within the bubble which expands until the bubble collapses generating mechanical energy leading to the algal biomass matrix disruption through microturbulence (Vinatoru, Mason, & Calinescu, 2017). Liquid-solid suspensions (e.g. algal biomass) promotes the formation of asymmetrical bubbles which upon implosion forms micro-jets causing surface erosion, local thermal events and algal cell wall breakdown, thus enhancing extraction efficiency (Chemat et al., 2017; Shirsath, Sonawane, & Gogate, 2012). Additionally, it promotes mass transfer and micro-mixing within the matrix. A range of parameters influences the cavitation formation and its resulting physical impact such as frequency and intensity of the waves, viscosity and surface tension of the solvent along with operating temperature and pressure (Tiwari, 2015).

UAE is a favourable technique offering a range of benefits compared to conventional methods: it is an organic solvent-free process, has high extraction efficiency, it is a relatively fast extraction method, operates at mild temperatures, requires low maintenance of equipment, offers the possibility of direct scale-up to industrial scale, and it has been shown to be cost effective in a number of processing industries (Flórez-Fernández, López-García, González-Muñoz, Vilariño, & Domínguez, 2017; Garcia-Vaquero et al., 2017; Hmelkov, Zvyagintseva, Shevchenko, Rasin, & Ermakova, 2018; Peshkovsky, Peshkovsky, & Bystryak, 2013; Roselló-Soto et al., 2015). Dang et al. (2017) reported a 142.6% increase in total phenolics content using UAE for 1 h compared to conventional solvent extraction with ethanol for 12 h with higher antioxidant potential. Low temperature operation is favourable for extraction of thermo labile compounds, while the clean operational conditions that can be achieved with UAE open avenues in product applications within the food and pharma industries. However, UAE does require significant process energy input, and initial equipment requires significant starting capital for operation at industrial scale.

Table 4 compiles the various UAE studies and associated extracted product and its application from different brown seaweed species. A range of fucoidans, polyphenols and laminarins has been isolated successfully at varying operating parameters, also with ethanol as a co-solvent primarily focusing on polyphenol extractions. An interesting fucoidan with previously unreported regular 1 → 3; 1 → 4-α-L-fucan structure, with the fucose residues sulfated at C2 and acetylated at C3, was isolated from a brown alga *Fucus evanescens* by ultrasound extraction, and was shown to possess anti-cancer activity (Hmelkov et al., 2018). Illustrated bioactivity of the isolated fractions achieved in a short time and at lower operating temperature with UAE compared to conventional processes, is one of its core advantages which needs to be translated in an industrial setting. Thus, while further research is warranted, UAE can be a promising methodology which can serve as a pre-treatment technique or can be used either as stand-alone extraction technique or in conjunction with other methods for developing an integrated extraction process.

2.2.4 Sub-critical water

Sub-critical water extraction is gaining traction owing to its zero toxic waste production, as water is employed as extraction solvent. SWE operates in the sub-critical region of water between 100 and 374 °C and pressures between 0.1 and 20 MPa to keep the water in liquid state, and allows manipulation of the properties of the water due to the temperature dependence of properties like polarity, viscosity, density and surface tension. Classical solvent extraction of non-polar compounds from macroalgae uses organic solvents such as hexane, methanol, chloroform, ethanol etc. Some of these solvents can be toxic and requires considerable downstream processing for both product and solvent recovery, and in order to adhere to the food and pharmaceutical industry regulations for safe end-product application. SWE in contrast offers a safe, green and high efficacy extraction technique for product use in therapeutics, food and cosmetic industry (Carr, Mammucari, & Foster, 2011).

A key operating principle of SWE is the ability to manipulate the properties of the extraction solvent and thereby to target specific classes of compounds. Water in ambient state is a polar solvent with a dielectric constant of 80 at 25°C which can be modulated in subcritical state and reduce to 40 at 200°C, thus enhancing solubilities of hydrophobic organic compounds (Reddy et al., 2014). The reduced dielectric constant of water at elevated

Table 4 Applications of UAE for bioactive compounds from various brown macroalgae species.

Species	Operational conditions	Compounds	Applications	Reference
Ascophyllum nodosum	20 kHz between ultrasound amplitude (22.8–114 μm) for 10 min	Total phenolics, fucose, and uronic acid	Bioactive properties	(Kadam et al., 2015)
Ascophyllum nodosum	20 kHz and 700 W, amplitude levels of 22.8 and 68.4 μm for 10 min	Proteins and amino acids	Bioactive properties	(Kadam et al., 2017)
Ascophyllum nodosum, Fucus vesiculosus and *Bifurcaria bifurcata*	Operated at room temperature for 30 min with water/ethanol (50:50, v:v) as The extraction solvent	Crude extract	Antioxidant potential	(Agregán et al., 2018)
Undaria pinnatifida	20 kHz/100–300 W/40–60°C for 30 min solvent to algae ratio of 30 (mL g^{-1}) with 410 rpm agitation at pH 11	Pigments: carotenoid and chlorophyll	Not evaluated in this study	(Zhu et al., 2017)
Sargassum muticum	24 kHz/400 W/20–60°C for 5–25 min	Post hydrolysis of crude fucoidan derived from hydrothermal treatment	Enhanced antiradical and cytotoxic properties	(Flórez-Fernández, González-Muñoz, & Domínguez, 2017)
Sargassum muticum	150 W and 40 kHz at room temperature (25°C) for 5–30 min	Crude fucoidan and phlorotannins	Antioxidant and anti-tumour activity	(Flórez-Fernández, López-García et al., 2017)

(Continued)

Table 4 Applications of UAE for bioactive compounds from various brown macroalgae species.—cont'd

Species	Operational conditions	Compounds	Applications	Reference
Sargassum muticum	400 W, 24 kHz ay 60% amplitude for 5 min and off periods of 25 min	Phenolics and carbohydrates	Antioxidant activity, cytotoxic activity	(Casas et al., 2018)
Sargassum muticum	400 W, 50/60 Hz for 60 min (sonicate for 10 min and pause for 2 min) at 50°C	Sulfated polysaccharides and phenolics	Antioxidant and prebiotic potential	(Rodrigues et al., 2015)
Sargassum muticum	Solvent to algae ratio of 20 (mL g^{-1}), 40 Hz/150W, 5–30 min and 25°C and amplitude of 40%–100%	Alginates	Cytotoxicity	(Flórez-Fernández, Torres, González et al., 2019)
Fucus evanescens	35 kHz/150 W at room temperature for 5–30 min	Fucoidans	Anticancer activity	(Hmelkov et al., 2018)
Hormosira banksii	50 Hz/150–250 W, 20–60 min and 30–50°C. 70% ethanol (v/v) with a solvent to algae ratio of 50 (mL g^{-1})	Phenolic content	Antioxidant activity	(Dang et al., 2017)
Laminaria digitata	20 kHz/500 W, 10–30 min and 40–80°C and amplitude of 40–100%	Fucose, glucans	Antioxidant activity	(Garcia-Vaquero, Rajauria, Tiwari, Sweeney, & O'Doherty, 2018)

temperature in the sub-critical range is closer to organic solvents like ethanol (27) and methanol (33) (IAPWS, 1997). SWE entails considerably higher yields and shorter extraction times for a range of moderately non-polar compounds (Herrero, Cifuentes, & Ibañez, 2006). The SWE techniques follows a combination of reaction kinetics from desorption of the target compounds from the matrix, diffusion of the solvent into the matrix and finally dissolution of the compound into the solvent (Carr et al., 2011). In addition to the compound chemistry and biological matrix types, altered solvent properties, including surface tension, viscosity, density and polarity, plays a vital role in the overall SWE process efficiency. Surface tension of water significantly reduces with increasing temperature which entails enhanced wetting of the matrix and facilitates mass transfer. There is a 50% reduction in surface tension of water from 25 to 200°C (Plaza & Turner, 2015). Also, both density of water and viscosity reduces at elevated temperature which further enhances mass transfer and diffusion of analytes out of the matrix to the solvent phase. Density of water decreases from 0.997 g/mL at 25°C to 0.579 g/mL at 350°C (NIST, 2014). Cumulatively, all these parameter changes happen in parallel which contributes toward higher yields and lower extraction times with greater selectivity than most conventional techniques, all of which can be achieved in a clean and solvent-free extraction process. In addition to the use of water as a green solvent, the luxury of using wet algal biomass without the requirement of drying the biomass can cut down energy expenses up to 2–8 times (Reddy et al., 2014; Suganya, Varman, Masjuki, & Renganathan, 2016). A significant disadvantage of SWE is the higher equipment costs associated with the high pressure and high temperature operating regime, which necessitates high initial capital outlay, but also higher maintenance costs along with strict safety measures.

Table 5 compiles the various SWE studies and associated extracted product and its application from different brown seaweed species. A review of the recent investigations with SWE clearly puts it as a promising method for isolating a number of high-value products including fucoidans, alginates and polyphenols. The combination of temperature and pressure entails significantly higher overall yields compared to conventional as well as some novel methods with ~80% yields achieved in the 180–220°C/7–50 bar operating range (Flórez-Fernández, Torres, González et al., 2019; Saravana, Choi, Park, Woo, & Chun, 2016). The attributed bioactivity of the isolated products showed higher antioxidant activities along with antibacterial and anti-cancer activity. A number of studies also explored using catalysts like NaOH, acetic acid, formic acid, ionic liquids (ILs), and deep

Table 5 Applications of SWE for bioactive compounds from various brown macroalgae species.

Species	Operating conditions	Compounds	Applications	Reference
Saccharina japonica	Addition of 0.1% NaOH 100–180°C 100–300 rpm 20–80 bar solvent to algae ratio of 10–25 (mL g^{-1}), 10–20 min	Fucoidans, mannitol, amino acids	Antioxidant and anti-proliferation activity	(Saravana et al., 2018)
Saccharina japonica	1% acetic acid-assisted SWE 200–280°C 150 rpm 13–60 bar 28–42 min	Crude hydrolysate	Antibacterial	(Meillisa, Siahaan, Park, Woo, & Chun, 2013)
Saccharina japonica	1% formic acid-assisted SWE 180–260°C 140 rpm 15–65 bar	Monosaccharides and bioactive compounds	Antioxidant activity	(Meillisa, Woo, & Chun, 2015)
Saccharina japonica	180–420°C 150 rpm 13–520 bar for 5 min	Mannitol, amino acids	Fermentation industry	(Saravana et al., 2016)
Saccharina japonica	Ionic liquid-assisted SWE 100–250°C 200 rpm 50 bar 5 min	Phenolics	Antioxidant activity	(Vo Dinh, Saravana, Woo, & Chun, 2018)

Saccharina japonica	SWE with DES Optimized condition 150 °C, 19.85 bar, 70% water content and solvent to algae ratio of 38.8 (mL g^{-1})	Fucoidans and alginates	Antioxidant activity	(Saravana, Cho, Woo, & Chun, 2018)
Laminaria saccharina	Optimized condition, solvent to algae ratio of 10 (mL g^{-1}), 350°C 15 min	Biocrude	Bio-oil, biorefinery	(Anastasakis & Ross, 2011)
Nizamuddinia zanardinni	150 °C for 29 min	Fucoidans	Antibacterial and antiviral activity	(Alboofetileh, Ritta et al., 2019)
Laminaria ochroleuca	Non-isothermal autohydrolysis from temperature 120–220°C	Bioactive and gelling extracts	Antioxidant activity, viscoelastic properties, antitumor activity	(Flórez-Fernández, Torres, González-Muñoz et al., 2019)

eutectic solvents (DESs) to increase yields, some of which showed promising results; however, further investigation is needed with the use of some additives. Overall, SWE holds immense potential for scalability in industrial scale and research should be directed toward facilitating scaled-up operations.

2.2.5 Super-critical fluid extraction

Super-critical fluid extraction operates above the critical points of temperature and pressure of the fluid while utilizing the modulated physico-chemical properties of the fluid to extract target compounds from a range of natural biomass matrices (Herrero et al., 2006). In the super-critical state, the density of the supercritical fluid is akin to a liquid, while the viscosity is in the intermediate range of liquid and gas. The relatively low viscosity and the associated increased diffusivity facilitates better mass transfer of the solutes from the matrix into the solvent (McHugh and Krukonis, 1994).

CO_2 is the most widely used fluid for Sc-FE and is increasingly being used for bioactive compounds extraction from plant materials, food by-products, macroalgae and microalgae (Esquivel-Hernández et al., 2017). Its critical points of temperature (30°C) and pressure (7.38 MPa) are favourable for extracting heat labile compounds. The use of CO_2 makes the process clean, non-toxic and environmentally safe which adheres to the safe regulatory requirements for solvents. Bioactive compounds extracted using Sc-FE has found application in the food and pharmaceutical industry (Herrero et al., 2006; Pérez-López et al., 2014). However, Sc-CO_2 is generally a non-polar solvent which is not ideal for extracting polar compounds. Addition of certain modifiers have been proven useful to considerably modulate the solvating capability of extracting polar natural compounds (Goto, Kanda, Wahyudiono & Machmudah, 2015). Table 6 compiles the various Sc-FE studies and associated extracted product and its application from different brown seaweed species. Sc-CO_2 extraction studies with brown seaweeds have been primarily focused on extracting fucoxanthin as a stand-alone method as well as with co-solvents. Significantly improved fucoxanthin yields have been reported with attributed bifunctionalities ranging from antioxidant activity, plant biostimulants and anti-hypersensitive activity. In addition, Sc-CO_2 have been used to extract both fatty acids and phenolic compounds with proposed application in the food and cosmetic industry. In a nutshell, further research is needed for its scale-up applicability given the environmentally benign and safe nature of the solvent.

Table 6 Applications of Sc-CO_2 for bioactive compounds from various brown macroalgae species.

Species	Conditions	Compounds	Application	Reference
Fucus vesiculosus	Raw material 9 wt% (air-dry raw material), temperature 60°C, pressure 304 bar, extraction time 60 min	Fatty acids, polyphenols	Bactericidal, fungicidal, and immunostimulating activities	(Bogolitsyn et al., 2017)
Saccharina japonica	45–55°C, pressures 200–300 bar, and co-solvent flow rates 0.50–2.00 (% of CO_2, w/w)	Fucoxanthin, phlorotannin	Antioxidant	(Saravana et al., 2017)
Sargassum hemiphyllum	40–50°C, 24.1–379 bar, 60 min	ω-3 fatty acids	Functional food supplements	(Cheung, Leung, & Ang, 1998)
Sargassum muticum	45°C, 290 bar with super-critical fluid flow of 10 kg h^{-1}	Glycolipids	Functional food	(Terme et al., 2017)
Undaria pinnatifida	40–70°C; 400 bar for 150 min	Fucoxanthin	Food ingredients	(Goto et al., 2015)
Undaria pinnatifida	40°C and 400 bar, extraction time of 180 min	Fucoxanthin	Food and pharmaceutical additive	(Quitain, Kai, Sasaki, & Goto, 2013b)
Ascophyllum nodosum	500 bar; 50°C; load mass 257 g; CO_2 passed through a bed of algal biomass 23.8 kg	Plant biostimulants	Agriculture and horticulture	(Michalak et al., 2016)

(Continued)

Table 6 Applications of Sc-CO_2 for bioactive compounds from various brown macroalgae species.—cont'd

Species	Conditions	Compounds	Application	Reference
Sargassum muticum	100 bar, 50°C	Fatty acids, phenolics and fucoxanthin	Anti-browning activity, inhibition of lipogenesis and cytotoxicity effects	(Conde, Moure, & Domínguez, 2015)
Saccharina japonica and *Sargassum horneri*	45°C and pressure of 250 bar. The flow rate of CO_2 27 g min^{-1} and ethanol as co-solvent	Fucoxanthin in oil	Antihypertensive and antioxidant activity	(Sivagnanam et al., 2015)

2.2.6 Other novel green solvents

Developing green solvents has got a great impetus in the quest for replacing the toxic organic solvents employed in various industrial processes. Deep eutectic solvents (DESs) are mixtures formed by combinations of Brønsted Lewis bases and acids with different types of anionic and cationic groups imparting specific tunable properties (Saravana et al., 2018). These mixtures have lower melting points compared to any of its individual components, partly owing to intermolecular hydrogen bonds (Dai, van Spronsen, Witkamp, Verpoorte, & Choi, 2013). They are biodegradable, easy to handle and have decreased toxicity (Ruesgas-Ramón, Figueroa-Espinoza, & Durand, 2017). Ionic liquids (ILs) are another category of multimolecular solvents comprising of an organic cation with a coordinating inorganic or organic anion (Han & Row, 2010). In addition to their tunable properties, ILs have the advantage of possessing low vapor pressure, are non-flammable and are stable at higher temperature (Ruesgas-Ramón et al., 2017). However, despite these advantages the significant economic challenge of high solvent costs remains an obstacle for process commercialization.

3. Coupling of techniques for enhanced extractions of bioactive compounds

Recent research is driven toward development of eco-friendly and cost-effective extraction techniques which leads to high yields and shorter processing times, while recovering bioactive compounds with preserved or enhanced functionalities. Each extraction technique has its own advantages and disadvantages in terms of operational logistics, capital costs and downstream processing requirements of the desired end-products (Table 7). Also, the comparison of different methods clearly illustrates that there is a variation across methods in terms of process efficacy as well as compound selectivity. A good extraction technique is a function of multiple variables ranging from the species type, target compounds and the desired application of the products. Products for food and cosmetic industry needs adherence to normally strict regulations, while pharmaceutical applications need to follow extremely stringent standards. Greener extraction techniques hold a distinct advantage in comparison to conventional solid-liquid extractions which often employ toxic organic solvents. Coupling of extraction methodologies offers the advantages of circumventing some pitfalls of any single method which can lead to better yields and product recovery and lead to the development of more efficient and profitable processes.

Table 7 Comparison of the advantages and disadvantages of some novel green extraction methodologies currently in focus.

Extraction technique	Advantages	Disadvantages	Reference
Enzyme-assisted extraction	Green technology, mild operating conditions, water can be used, enhanced bioactive properties of extracts	Extended hydrolysis time, enzyme specificity and activity, expensive leading to high cost for industrial scale-up	(Nadar et al., 2018; Rhein-Knudsen, Ale, & Meyer, 2015; Wijesinghe & Jeon, 2012b)
Microwave-assisted extraction	Shorter processing time, no harsh co-solvents involved, environment friendly, low cost, both open and closed vessel operation is possible at ambient temperature	Structural modification to the compound structure may alter/ or limit bioactive properties, recovery of non-polar compound- High extraction temperature	(Joana Gil-Chávez et al., 2013; Ren et al., 2017; Yuan et al., 2018; Yuan & Macquarrie, 2015)
Ultrasound-assisted extraction	Non-thermal extraction technique, low solvent requirement, high extraction yields, greater level of automation, Affordable and reproducible method	Economic feasibility for scale up, Non-linearity of the process, scale-up operation needs to be validated, high energy inputs	(Chemat et al., 2017; Kadam et al., 2015; Roselló-Soto et al., 2015; Tiwari, 2015; Vinatoru et al., 2017)
Sub-critical water extraction	Green and sustainable process, tuneable properties of water under sub-critical conditions, higher yields, enhanced mass transfer and lower extraction time, relatively simple equipment set-up	High pressure and temperature operation, high initial capital costs, May not be suitable of thermolabile compounds	(Meillisa et al., 2013; Plaza, Amigo-Benavent, del Castillo, Ibáñez, & Herrero, 2010; Saravana, Cho, et al., 2018; Saravana, Tilasun, et al., 2018)

Super-critical fluid extraction	Non-toxic and solvent free, mild extraction parameters, highly selective	High power consumption, high operating pressure, expensive set-up, laborious sample processing	(Bogolitsyn et al., 2017; Conde et al., 2015; Michalak, Chojnacka, & Saeid, 2017)
Deep eutectic solvents	Biodegradable, easy to handle and have lower environmental footprints then organic solvents, non-volatile, non-flammable, wide range of selectivity for compounds,	High densities and viscosities of DES, recovery of target compounds can be difficult due to low vapor pressure, re-cyclability and industrial applicability yet to be evaluated	(Dai et al., 2013; Ruesgas-Ramón et al., 2017; Saravana et al., 2018)
Ionic liquids	Low vapor pressure, non-flammable and stability at higher temperature	Expensive, water stability, high toxicity, issues in purifying, poor biodegradability	(Han & Row, 2010; Ruesgas-Ramón et al., 2017)

Deep eutectic solvent combined with subcritical water extraction was used to improve the extraction of seaweed polysaccharides (fucoidans and alginates) from *Saccharina japonica* (Saravana et al., 2018). A combined EAE-pressurized liquid extraction process was studied by Sánchez-Camargo et al. (2016) to extract phlorotannins from the seaweed *Sargassum muticum* for antioxidant potential, but suggested further research with more substrate-specific enzymes is required to selectively recover the target compounds. Ionic liquid assisted SWE has been shown to enhance the extraction of phenolic compounds from the brown macroalgae *Saccharina japonica* with good antioxidant activity (Vo Dinh et al., 2018). Casas et al. (2018) illustrated EAE coupled with ultrasound extraction to be effective in selectively enhancing the phenolic extraction with good antioxidant activity in addition to improved yields in *S. muticum*. The extracts also had additional bioactive properties in terms of cytotoxic activity against murine melanoma B16F10 cells and human liposarcoma SW872 cells. Recently an EDTA based extraction of fucoidans from *Laminaria japonica* resulted in high yield with antioxidant and anti-cancer activity (Zhao, Xu, & Xu, 2018). Supercritical CO_2 (Sc-CO_2) defatted *S. japonica* samples were coupled with SWE to extract fucoidans with good antioxidant and anti-proliferative potential suggesting potential industrial applications (Saravana et al., 2018).

There is an intricate link between the extracted product, structural characteristics and its bioactivity to the extraction techniques used (Okolie, Akanbi, Mason, Udenigwe, & Aryee, 2018; Zhu et al., 2016). Linking extraction techniques to the products and its characteristics and decoupling various facets can help gain significant insight in integrating processes toward maximizing the valorized macroalgal biomass potential. Alginates extracted from the species *Colpomenia peregrina* through enzymatic hydrolysis showed lowest protein (<0.7%) and polyphenol (<5 mg gallic acid/g sample) contents in comparison to extractions using water and acid (Rostami et al., 2017). Enzyme assisted methodologies have been suggested to be less effective for polyphenol recovery but much more effective for proteins, uronic acids and sulfate groups (Hardouin et al., 2014; Puspita et al., 2017). EAE, while being effective, also ranks as one of the safest extraction methods for food grade products end-uses (Nadar et al., 2018; Rodrigues et al., 2015). Polyphenol recovery from *S. muticum* through EAE in combination with ultrasound significantly improved in selectively recovering polyphenols than enzyme alone extractions (Casas et al., 2018). Choice of enzyme is particularly critical owing to its specificities to target compounds (Jaswir et al., 2013). In comparison to a range of carbohydrase, and protease and hot water extractions from *Nizamuddinia zanardinii*, Alcalase was the most effective in extracting fucoidans with higher molecular weight and sulfur content with enhanced anticancer and immune-enhancing activities (Alboofetileh et al., 2018). Quitain et al. (2013a) illustrated the advantages of using microwave aided extraction in comparison to conventional heating to generate low-molecular-weight fucoidan components of about 5–30 kDa, from wakame de-oiled using Sc-CO_2. Charoensiddhi et al. (2017) coupled ethanol pre-treatment of *Ecklonia radiata* to extract polyphenols and outlined a sequential process by combining enzyme aided (Viscozyme) with microwave-intensified extraction to generate different molecular weight polysaccharide with good prebiotic potential. Other techniques like SWE often results in structural modifications and leads to neoformation of compounds with potent antioxidant capacity. Further research is warranted pertaining to characterizing these compounds and assessing their application as nutraceuticals and functional food components. Thus, with further research novel techniques can cater toward a sustainable process development.

4. Potential high-value seaweed products

There is a broad range of potentially valuable compounds found in brown seaweeds, including polyphenols and other metabolites, polysaccharides, proteins, lipids, minerals and a host of other minor constituents. Various of these compounds are valued for their nutritional value or potential positive impacts on health via favourable biological activities. Table 8 provides a summary of the bioactivities characterized in various brown seaweeds and to which types of compounds these activities were linked. Seaweeds are often the subject of bioprospecting investigations in further efforts to discover novel compounds. Two of the classes of compounds which have potentially high-value applications and which are unique to seaweeds, are the different polysaccharides and polyphenols found in these organisms.

4.1 Polysaccharides

Fucoidan, laminarin and alginate constitutes the three major valuable polysaccharide fractions in the brown seaweeds, and are increasingly finding applications in the food, cosmetic and the pharmaceutical industries, mainly owing to their functional and bioactive characteristics.

Table 8 Compilation of some routinely investigated brown seaweed species and their illustrated bioactivities.

Brown seaweed species	Compounds	Application	Reference
Fucus vesiculosus	Fucoidans, fatty acids, polyphenols	Anticancer and anti-angiogenic potential, bactericidal and immunostimulating activities	(Bogolitsyn et al., 2017; Oliveira et al., 2019)
Laminaria digitata	Fucose, glucans	Antioxidant activities	(Garcia-Vaquero et al., 2018)
Laminaria ochroleuca	Fucoidans, phlorotannins, proteins, alginates	Antioxidant, antitumor	(Flórez-Fernández, Torres, González-Muñoz et al., 2019)
Sargassum muticum	Phlorotannins, phenolics, sulphated polysaccharides, alginates, glycolipids, fatty acids	Antioxidant, hydroxyl-radical scavenging activity, prebiotic activity	(Conde et al., 2015; Flórez-Fernández et al., 2019b; Pérez-Larrán et al., 2019; Rodrigues et al., 2015; Sánchez-Camargo et al., 2016; Terme et al., 2017)
Halidrys siliquosa	Phlorotannins	Sunscreen, antioxidant, and bactericide	(Le Lann et al., 2016)
Laminaria japonica	Fucoidans, phlorotannin	Antioxidant and antitumor activity	(Yuan et al., 2018; Zhao et al., 2018)
Himanthalia elongata (Linnaeus)	Protein	Food industry	(Garcia-Vaquero, Lopez-Alonso, & Hayes, 2017)
Sargassum turbinarioides	Alginates	Food and non- food application	(Fenoradosoa et al., 2010)
Ecklonia radiata	Polysaccharides and phlorotannins	Dietary supplements with pre-biotic potential	(Charoensiddhi et al., 2017)

Himanthalia elongata	Phenolic compounds	Antioxidants	(Rajauria, Foley, & Abu-Ghannam, 2016)
Nizamuddinia zanardinii	Fucoidans	Antibacterial and antiviral activity	(Alboofetileh et al., 2019)
Durvillaea potatorum	Fucoidans, alginates and laminarin	Different molecular weight distribution for various uses	(Abraham, Su, Puri, Raston, & Zhang, 2019)
Fucus evanescens	Fucoidans	Anticancer activity	(Hmelkov et al., 2018)
Silvetia compressa and *Ecklonia arborea*	Dietary fiber, total polyphenol	Food and nutraceutical industries	(Tapia-Salazar et al., 2019)
Padina pavonica, *Dictyopteris membranaceae* and *Cystoseira compressa*	Alginates	Gastroprotective activity	(Ammar et al., 2018)
Colpomenia peregrina	Alginates	Immunomodulatory and antioxidant properties	(Rostami et al., 2017)
Coccophora langsdorfii	Fucoidans	Anticancer activity	(Imbs et al., 2016)
Sargassum hemiphyllum	Fucoidans	Neuroprotective effects	(Huang, Kuo, & Chen, 2017)
Cystoseira trinodis	Fucoidans, alginates	Antioxidants	(Hifney et al., 2018)
Cystoseira abies-marina	Phlorotannin	Biological properties	(Sánchez-Camargo et al., 2016)

Laminarins are unique internal storage polysaccharides usually found within the cell vacuoles comprising of linear β-1,3 D-glucose units along with β-1,6-linkages which influences the water solubility of the compounds (Rioux, Turgeon, & Beaulieu, 2010). They are low molecular weight fractions of ~ 5 kDa and can account for up to 35% DW of the biomass, depending on the season, species and growth conditions (Kadam, Tiwari, & O'Donnell, 2015). Structural variations are dependent on the degree of polymerization (25) and the ratio of (1,3)- and (1,6)-glycosidic bonds. Laminarins have demonstrated good bioactive properties such as antibacterial, antioxidative and anticoagulant potential (Zhang & Row, 2015). Complex laminarin structures are stabilized by inter-chained hydrogen bonds thus rendering them resistant to hydrolysis in the upper gastrointestinal tract and are therefore classified as dietary fibers (O'Sullivan et al., 2010). The basic structural units of laminarin are shown in Fig. 2.

Fucoidans are sulfated polysaccharides comprising of (1,2) or (1,3) -linked α-L-fucopyranose residues along with other monomers such as galactose, mannose, xylose, glucose and glucuronic acid (Ale, Mikkelsen, & Meyer, 2011). These hetropolysaccharides are mostly found in the cell walls and the intercellular spaces (Mizuno et al., 2009). They are generally large macromolecules, with molecular weight ranges from 100 to 1600 kDa and are mostly soluble in water and dilute acids (Zhang & Row, 2015). The functional and bioactivity of these polysaccharides are influenced by their molecular weight, composition of the monosaccharide units, sulfate content and the position of the sulfate ester group and the extraction technique used for isolation of the compound (Hahn et al., 2012). Seasonal variations in terms of quantity and structure of fucoidans for brown macroalgae species have also been found (Fletcher, Biller, Ross, & Adams, 2017). Fucoidans have generated significant interest in recent times due to their broad spectrum of applications in cosmetic, food and

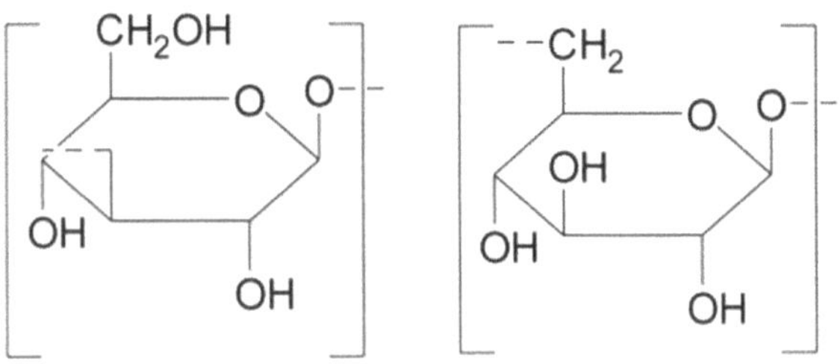

Fig. 2 Basic chemical units of laminarin, made up of β-(1,3) and β-(1,6) linked glucose (O'Sullivan et al., 2010) (Used under open access license CC BY 3.0).

the pharmaceutical industry (Fitton, Stringer, & Karpiniec, 2015). It has shown good bioactivities such as antiviral, anti-inflammatory (Lee et al., 2012), immunomodulatory, antithrombotic, anticoagulant (Wijesinghe & Jeon, 2012a), antioxidative (Hifney, Fawzy, Abdel-Gawad, & Gomaa, 2016; Yuan & Macquarrie, 2015; Zhao et al., 2018), antitumor (Wu et al., 2016), antibacterial, and anticancer activity (Ale, Maruyama, Tamauchi, Mikkelsen, & Meyer, 2011). The structure of fucoidan is shown in Fig. 3.

Alginates are linear unbranched polymers comprising of D-mannuronic (M) and L-guluronic (G) acid with β(1,4)-linkages. The monomeric unit arrangement sequence varies from homopolymeric blocks (MM and GG) to heteropolymeric blocks (MG or GM). Their molecular weight ranges from 500 to 1000 kDa. Alginate forms gels in presence of divalent cations (e.g. Ba^{2+}, Ca^{2+}, Mg^{2+}), and can stabilize emulsions which is a critical property finding widespread application in the food, textile and the cosmetic industry (Borazjani, Tabarsa, You, & Rezaei, 2017; Flórez-Fernández et al., 2019a; O'Sullivan et al., 2010). The gel properties in terms of strength and viscosity of the alginate solution is influenced by the concentration and the length of the polymer chains, and the ratio of G:M monomers (Lee & Mooney, 2012). Higher guluronic acid concentrations imparts greater strength to the gels while mannuronic acid and heterogenous blocks (MG or GM) forms weak but more flexible gels (Costa et al., 2018). Alginates also finds application in the biomedical field in wound dressing, cell immobilization and is a viable tool for drug delivery due to favourable properties like biocompatibility, biodegradability and non-toxicity (Lee & Mooney, 2012). The different building blocks of alginate are shown in Fig. 4.

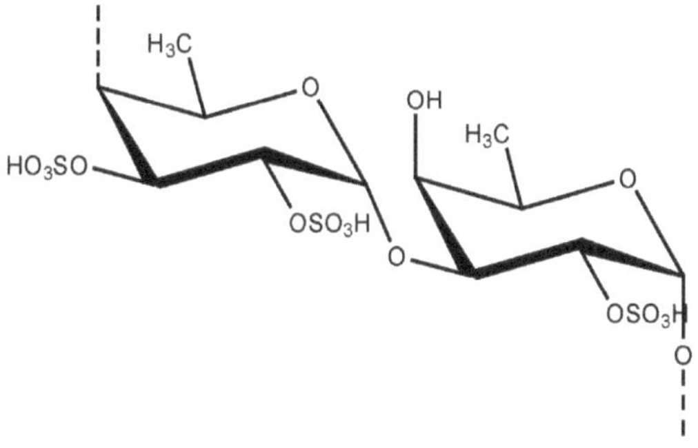

Fig. 3 Chemical structure of fucoidan (Thomas & Kim, 2013) (Used under open access license CC BY 3.0).

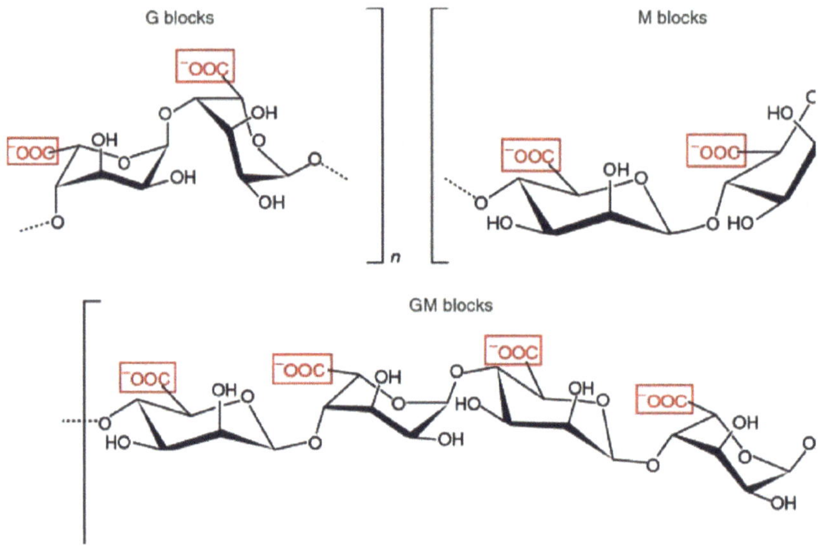

Fig. 4 Brown seaweed alginates showing units of (A) poly-D-glucoronic acid (G blocks), (B) poly-D-mannuronic acid (M blocks), and (C) alternate D-glucoronic and D-mannuronic acid (GM block) (Vera, Castro, Gonzalez, & Moenne, 2011) (Image reproduced under open access license CC BY 3.0).

4.2 Polyphenols

Polyphenols in macroalgae comprise of phenolic acids, flavonoids, tannins and phlorotannins. Brown seaweed polyphenols are known as phlorotannins which are formed by polymerization of phloroglucinol units and are unique to these species. Phlorotannins can be classified based on their monomer linkage of phloroglucinol units which are known as fucols, phlorethols, fucophlorethols and eckols (Isaza Martínez & Torres Castañeda, 2013). These hydrophilic compounds have molecular weights ranging between 126 Da and 650 kDa and its content can vary from 1 to 14% between species (Kadam et al., 2013). They reside in special vesicles (physodes), form an intricate linkage with alginates in the cell wall structure and impart protection against microbes and other environmental stressors (Pádua et al., 2015). Barbosa et al. (2019) have comprehensively reviewed the anti-inflammatory and anti-allergic potential of phlorotannins isolated from brown seaweeds. Polyphenols content in brown algae varies as a function of both taxonomy and geography (Magnusson et al., 2017) and can exhibit different bioactive and functional properties (Agregán et al., 2018; Li, Wijesekara, Li, & Kim, 2011; Puspita et al., 2017). The structure of some phlorotannins are shown in Fig. 5.

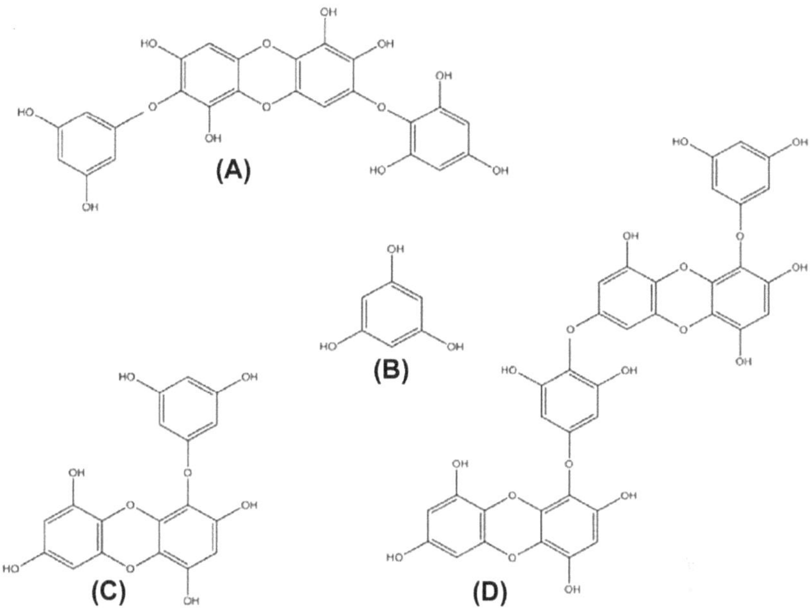

Fig. 5 Chemical structures of some phlorotannins: (A) diphlorethohydroxycarmalol; (B) phloroglucinol; (C) eckol; (D) dieckol (Thomas & Kim, 2013). *(Image reproduced under open access licence CC BY 3.0).*

5. Impact of physiological characteristics and seasonality on processing

The processing of seaweeds cannot be developed independently from the organisms physiology and seasonal variation in composition. Careful consideration of these factors during processing operations could contribute to improved scheduling of harvesting, and more optimal resource utilization. Seaweed growth and productivity are intricately tied to environmental and seasonally dependent factors like solar irradiance, nutrient availability, temperature, salinity and water movement. Seasonal variability thus plays a key role in seaweed physiological characteristics, as most of these factors change with seasons. Chemical composition of seaweeds has also been linked to the seasonal variability which culminates through a range of abiotic factors and nutrient synthesis (Marinho-Soriano, Fonseca, Carneiro, & Moreira, 2006).

Seasonal variations are furthermore different for the various compound classes found in brown seaweeds. Investigations of seasonality of fucoidan in three brown macroalgae species revealed significant difference in the quantity and structure of fucoidans across species and seasons (Fletcher et al., 2017), with the highest fucoidan content being reported for the autumn months. Metabolomic and bioactivity profile study of *Fucus vesiculosus* as a function of seasonal variation revealed significant changes in compound concentrations, where phlorotannin content was highest in the summer months, while lipids and carotenoids increased over the winter months (Heavisides et al., 2018). Nutritional composition of the *Sargassum oligocystem* species was reported to be most favourable during the summer season, while highest anti-proliferative activity was found from samples from the monsoon months (Praiboon, Palakas, Noiraksa, & Miyashita, 2018). Significant seasonal variation in the total phenolics content and antioxidant capacity for the two brown macroalgae *Macrocystis pyrifera* and *Lessonia spicata* was found, with the higher concentration of phenolics reported during winter (Beratto-Ramos, Castillo-Felices, Troncoso-Leon, Agurto-Muñoz, & Agurto-Muñoz, 2019).

6. Seaweed biorefinery

Substantial improvements in resource utilization, environmental impact, and in some instances, processing profitability can be achieved through employing the biorefinery approach. The biorefinery concept involves integration of various sequential processing steps which enables maximum utilization of the seaweed biomass, by producing multiple products and minimizing residues. The focus is to extract maximum value through biomass fractionation and to produce a range of products that find application in different industries, including relatively higher value markets like cosmetics, food and pharmaceuticals, and in bulk industries like energy and platform chemicals (Milledge, Nielsen, & Bailey, 2016). A key aspect within a biorefinery approach is to preserve the structural integrity and thus bioactive properties of the higher-value components derived from seaweeds, which requires selection of appropriate extraction and purification steps.

Green processing methods that minimize or eliminate waste production, decrease energy requirements and avoid the use of carbon-intensive organic solvents may lead to substantial reduction of environmental impacts, and may also offer improved overall process efficiency (Balboa, Moure, & Domínguez, 2015). There are substantial advantages of using seaweed biomass for biorefineries compared to terrestrial biomass, including no competition for agricultural land use and zero agricultural inputs like fresh water and fertilizer. Development and optimization of green seaweed processing methods and implementing these within a traditionally carbon-intensive processing sector can substantially improve sustainability (Ibañez & Cifuentes, 2013). However, the commercial success of seaweed based biorefineries will be dependent on the financial viability, which in turn will be determined by the specific biomass processed, processing efficiency and the market worth of the product spectrum followed by the associated logistics of running a large scale industrial facility (Jung, Lim, Kim, & Park, 2013; Suganya et al., 2016; Trivedi, Gupta, Reddy, & Jha, 2013).

Despite these potential advantages, empirical evidence on the economics associated with seaweed biorefineries is extremely limited, as is the important associated components such as quantitative environmental footprint data, sustainability and scale-up feasibility of newer and so-called green processing techniques. Balboa et al. (2015) valorized *S. muticum* under a biorefinery concept to recover six different products and favored the integration of greener techniques. A cascading biorefinery process targeting ulvans along with seaweed salts and proteins were defined by Glasson, Sims, Carnachan, de Nys, and Magnusson (2017) investigating eight biorefinery processes. Flórez-Fernández, Torres, González-Muñoz et al. (2019) proposed a biorefinery approach by using autohydrolysis of *S. muticum* biomass enabling simultaneous extraction of alginate, fucoidan and phlorotannin fractions. Charoensiddhi et al. (2018) did a techno-economic case study for a simulated industrial scale production of functional food from *Ecklonia radiata* and demonstrated economic feasibility with a best-case scenario showing a payback time and net present value (NPV) of 1.1 years and US$ 89.43 m respectively.

7. Conclusion and future perspectives

Seaweed biomass as a renewable feedstock has gained significant attention in the quest to move away from a fossil fuel-based economy. Macroalgal biomass can play an important role in providing food and energy products in future. Efforts toward commercial cultivation of brown seaweeds are making substantial progress, and brown seaweed biomass can be particularly suited to build processing operations around due to high yields and the range of unique compounds that it contains, with widespread application in the food, cosmetic, pharmaceutical and agricultural industries.

The economic success of future seaweed processing enterprises is likely to depend on how successfully such enterprises can move away from environmentally unsustainable processing operations. For this reason, the development of so-called green processing methods that are less energy and resource intensive, produce less waste and avoid utilization of high carbon footprint organic solvents are particularly desirable. Substantial progress has been made toward adapting technologies developed for terrestrial biomass processing for seaweed processing, and individual technologies are being optimized mostly at the proof-of concept and laboratory demonstration stages. The concept of integrated seaweed biorefineries needs to be demonstrated beyond just modeling investigations in order to decrease technical risk for investors, and to demonstrate the technical viability of integrated approaches.

A major shortcoming in especially enzyme-assisted seaweed processing is the lack of commercially available enzymes that have been tailored to seaweed biomass. While it is acknowledged that enzymatic processing is relatively mild and environmentally friendly, currently available commercial enzyme preparations are invariably tailored to terrestrial biomass and therefore the seaweed cellular matrix thus presents a unique challenge to efficient processing. In brown seaweeds, the gelling properties of alginate can lead to mass transfer limitations during extraction, as well as difficulties in downstream processing. There is therefore a distinct need for enzymes which are more specific to substrates found in seaweed, which could improve extraction efficiency during seaweed processing and thus lead to improved economic performance of such operations.

Substantial future research is required to further sustainability and economic performance of processing of the various brown seaweeds around the world. The integration and coupling of different extraction and downstream processing methods need to be optimized, and the impact of processing steps needs to describe impacts on bioactivity-structure relationship of target compounds. Furthermore, taking into account physiological characteristics and seasonality in the composition of the raw material is important to ensure appropriate scheduling of harvesting and processing operations. Pertaining to the biorefinery concept, there is a scarcity of data regarding the downstream processing stages post-extraction and their associated efficiencies, which increases the uncertainty around technical feasibility and techno-economic analysis of current seaweed biorefinery research.

Overall, brown seaweed biomass has been highlighted as being a potentially sustainable biomass feedstock for the future. If processing technologies can be optimized and implemented successfully, the processing of brown seaweed can contribute significantly toward the development of a strong bio-based economy.

References

Abraham, R. E., Su, P., Puri, M., Raston, C. L., & Zhang, W. (2019). Optimisation of biorefinery production of alginate, fucoidan and laminarin from brown seaweed *Durvillaea potatorum*. *Algal Research, 38*, 101389.

Agregán, R., Munekata, P., Franco, D., Carballo, J., Barba, F., & Lorenzo, J. (2018). Antioxidant potential of extracts obtained from macro- (*Ascophyllum nodosum, Fucus vesiculosus* and *Bifurcaria bifurcata*) and micro-algae (*Chlorella vulgaris* and *Spirulina platensis*) assisted by ultrasound. *Medicines, 5*(2), 33.

Alboofetileh, M., Rezaei, M., & Tabarsa, M. (2018). Enzyme-assisted extraction of *Nizamuddinia zanardinii* for the recovery of sulfated polysaccharides with anticancer and immune-enhancing activities. *Journal of Applied Phycology, 31*(2), 1391–1402.

Alboofetileh, M., Rezaei, M., Tabarsa, M., Rittà, M., Donalisio, M., Mariatti, F., et al. (2019). Effect of different non-conventional extraction methods on the antibacterial and antiviral activity of fucoidans extracted from *Nizamuddinia zanardinii*. *International Journal of Biological Macromolecules, 124*, 131–137.

Alboofetileh, M., Rezaei, M., Tabarsa, M., You, S., Mariatti, F., & Cravotto, G. (2019). Subcritical water extraction as an efficient technique to isolate biologically-active fucoidans from *Nizamuddinia zanardinii*. *International Journal of Biological Macromolecules, 128*, 244–253.

Ale, M. T., Maruyama, H., Tamauchi, H., Mikkelsen, J. D., & Meyer, A. S. (2011). Fucoidan from *Sargassum sp.* and *Fucus vesiculosus* reduces cell viability of lung carcinoma and melanoma cells in vitro and activates natural killer cells in mice in vivo. *International Journal of Biological Macromolecules, 49*(3), 331–336.

Ale, M. T., Mikkelsen, J. D., & Meyer, A. S. (2011). Important determinants for fucoidan bioactivity: A critical review of structure-function relations and extraction methods for fucose-containing sulfated polysaccharides from brown seaweeds. *Marine Drugs, 9*(10), 2106.

Ammar, H. H., Lajili, S., Sakly, N., Cherif, D., Rihouey, C., Le Cerf, D., et al. (2018). Influence of the uronic acid composition on the gastroprotective activity of alginates from three different genus of Tunisian brown algae. *Food Chemistry, 239*, 165–171.

Anastasakis, K., & Ross, A. B. (2011). Hydrothermal liquefaction of the brown macro-alga *Laminaria Saccharina:* Effect of reaction conditions on product distribution and composition. *Bioresource Technology, 102*(7), 4876–4883.

Anastyuk, S. D., Shevchenko, N. M., Usoltseva, R. V., Silchenko, A. S., Zadorozhny, P. A., Dmitrenok, P. S., et al. (2017). Structural features and anticancer activity in vitro of fucoidan derivatives from brown alga *Saccharina cichorioides*. *Carbohydrate Polymers, 157*, 1503–1510.

Balboa, E., Moure, A., & Domínguez, H. (2015). Valorization of Sargassum muticum biomass according to the biorefinery concept. *Marine Drugs, 13*(6), 3745.

Barbosa, M., Lopes, G., Andrade, P. B., & Valentão, P. (2019). Bioprospecting of brown seaweeds for biotechnological applications: Phlorotannin actions in inflammation and allergy network. *Trends in Food Science & Technology, 86*, 153–171.

Beratto-Ramos, A., Castillo-Felices, R. P., Troncoso-Leon, N. A., Agurto-Muñoz, A., & Agurto-Muñoz, C. (2019). Selection criteria for high-value biomass: Seasonal and morphological variation of polyphenolic content and antioxidant capacity in two brown macroalgae. *Journal of Applied Phycology, 31*(1), 653–664.

Billakanti, J. M., Catchpole, O. J., Fenton, T. A., Mitchell, K. A., & MacKenzie, A. D. (2013). Enzyme-assisted extraction of fucoxanthin and lipids containing polyunsaturated fatty acids from *Undaria pinnatifida* using dimethyl ether and ethanol. *Process Biochemistry, 48*(12), 1999–2008.

Bogolitsyn, K. G., Kaplitsin, P. A., Dobrodeeva, L. K., Druzhinina, A. S., Ovchinnikov, D. V., Parshina, A. E., et al. (2017). Fatty acid composition and biological activity of supercritical extracts from arctic brown algae *Fucus vesiculosus*. *Russian Journal of Physical Chemistry B, 11*(7), 1144–1152.

Borazjani, N. J., Tabarsa, M., You, S., & Rezaei, M. (2017). Effects of extraction methods on molecular characteristics, antioxidant properties and immunomodulation of alginates from *Sargassum angustifolium*. *International Journal of Biological Macromolecules, 101*, 703–711.

Borines, M. G., De Leon, R. L., & Cuello, J. L. (2013). Bioethanol production from the macroalgae *Sargassum* spp. *Bioresource Technology, 138*, 22–29.

Bowyer, M. C., Thanh, T Dang, Van Altena, I. A., & Scarlett, C. J. (2018). Optimum conditions of microwave-assisted extraction for phenolic compounds and antioxidant capacity of the brown alga *Sargassum vestitum*. *Separation Science and Technology, 53*(11), 1711–1723.

Carr, A. G., Mammucari, R., & Foster, N. R. (2011). A review of subcritical water as a solvent and its utilisation for the processing of hydrophobic organic compounds. *Chemical Engineering Journal, 172*(1), 1–17.

Casas, M. P., Conde, E., Domínguez, H., & Moure, A. (2018). Ecofriendly extraction of bioactive fractions from *Sargassum muticum*. *Process Biochemistry, 79*, 166–173.

Charoensiddhi, S., Conlon, M. A., Franco, C. M. M., & Zhang, W. (2017). The development of seaweed-derived bioactive compounds for use as prebiotics and nutraceuticals using enzyme technologies. *Trends in Food Science & Technology, 70*, 20–33.

Charoensiddhi, S., Conlon, M. A., Vuaran, M. S., Franco, C. M. M., & Zhang, W. (2017). Polysaccharide and phlorotannin-enriched extracts of the brown seaweed *Ecklonia radiata* influence human gut microbiota and fermentation in vitro. *Journal of Applied Phycology, 29*(5), 2407–2416.

Charoensiddhi, S., Lorbeer, A. J., Franco, C. M. M., Su, P., Conlon, M. A., & Zhang, W. (2018). Process and economic feasibility for the production of functional food from the brown alga *Ecklonia radiata*. *Algal Research, 29*, 80–91.

Chemat, F., Rombaut, N., Sicaire, A.-G., Meullemiestre, A., Fabiano-Tixier, A.-S., & Abert-Vian, M. (2017). Ultrasound assisted extraction of food and natural products. Mechanisms, techniques, combinations, protocols and applications. A review. *Ultrasonics Sonochemistry, 34*, 540–560.

Cheung, P. C. K., Leung, A. Y. H., & Ang, P. O. (1998). Comparison of supercritical carbon dioxide and Soxhlet extraction of lipids from a brown seaweed, *Sargassum hemiphyllum* (turn.) C. Ag. *Journal of Agricultural and Food Chemistry, 46*(10), 4228–4232.

Conde, E., Moure, A., & Domínguez, H. (2015). Supercritical CO_2 extraction of fatty acids, phenolics and fucoxanthin from freeze-dried *Sargassum muticum*. *Journal of Applied Phycology, 27*(2), 957–964.

Costa, M. J., Marques, A. M., Pastrana, L. M., Teixeira, J. A., Sillankorva, S. M., & Cerqueira, M. A. (2018). Physicochemical properties of alginate-based films: Effect of ionic crosslinking and mannuronic and guluronic acid ratio. *Food Hydrocolloids, 81*, 442–448.

Dai, Y., van Spronsen, J., Witkamp, G.-J., Verpoorte, R., & Choi, Y. H. (2013). Natural deep eutectic solvents as new potential media for green technology. *Analytica Chimica Acta, 766*, 61–68.

Dang, T. T., Van Vuong, Q., Schreider, M. J., Bowyer, M. C., Van Altena, I. A., & Scarlett, C. J. (2017). Optimisation of ultrasound-assisted extraction conditions for phenolic content and antioxidant activities of the alga *Hormosira banksii* using response surface methodology. *Journal of Applied Phycology, 29*(6), 3161–3173.

Deniaud-Bouët, E., Hardouin, K., Potin, P., Kloareg, B., & Hervé, C. (2017). A review about brown algal cell walls and fucose-containing sulfated polysaccharides: Cell wall context, biomedical properties and key research challenges. *Carbohydrate Polymers, 175*, 395–408.

Esquivel-Hernández, D. A., Ibarra-Garza, I. P., Rodríguez-Rodríguez, J., Cuéllar-Bermúdez, S. P., Rostro-Alanis, M. J., Alemán-Nava, G. S., et al. (2017). Green extraction technologies for high-value metabolites from algae: A review. *Biofuels, Bioproducts and Biorefining, 11*(1), 215–231.

FAO. (2018). *The state of world fisheries and aquaculture 2018 – meeting the sustainable development goals*. Rome, Italy: Food and Agricultural Organization of the United Nations.

Fenoradosoa, T. A., Ali, G., Delattre, C., Laroche, C., Petit, E., Wadouachi, A., et al. (2010). Extraction and characterization of an alginate from the brown seaweed *Sargassum turbinarioides* Grunow. *Journal of Applied Phycology, 22*(2), 131–137.

Fitton, J. H., Stringer, D. N., & Karpiniec, S. S. (2015). Therapies from fucoidan: An update. *Marine Drugs, 13*(9), 5920.

Fletcher, H. R., Biller, P., Ross, A. B., & Adams, J. M. M. (2017). The seasonal variation of fucoidan within three species of brown macroalgae. *Algal Research, 22*, 79–86.

Flórez-Fernández, N., Domínguez, H., & Torres, M. D. (2019a). Advances in the biorefinery of *Sargassum muticum*: Valorisation of the alginate fractions. *Industrial Crops and Products, 138*, 111483.

Flórez-Fernández, N., Domínguez, H., & Torres, M. D. (2019b). A green approach for alginate extraction from *Sargassum muticum* brown seaweed using ultrasound-assisted technique. *International Journal of Biological Macromolecules, 124*, 451–459.

Flórez-Fernández, N., González-Muñoz, M. J., & Domínguez, H. (2017). Feasibility of post-hydrolysis processing of hydrothermal extracts from *Sargassum muticum*. *Algal Research, 27*, 73–81.

Flórez-Fernández, N., López-García, M., González-Muñoz, M. J., Vilariño, J. M. L., & Domínguez, H. (2017). Ultrasound-assisted extraction of fucoidan from *Sargassum muticum*. *Journal of Applied Phycology, 29*(3), 1553–1561.

Flórez-Fernández, N., Torres, M. D., González-Muñoz, M. J., & Domínguez, H. (2019). Recovery of bioactive and gelling extracts from edible brown seaweed *Laminaria ochroleuca* by non-isothermal autohydrolysis. *Food Chemistry, 277*, 353–361.

Garcia-Vaquero, M., Lopez-Alonso, M., & Hayes, M. (2017). Assessment of the functional properties of protein extracted from the brown seaweed *Himanthalia elongata* (Linnaeus) S. F. Gray. *Food Research International, 99*, 971–978.

Garcia-Vaquero, M., Rajauria, G., O'Doherty, J. V., & Sweeney, T. (2017). Polysaccharides from macroalgae: Recent advances, innovative technologies and challenges in extraction and purification. *Food Research International, 99*, 1011–1020.

Garcia-Vaquero, M., Rajauria, G., Tiwari, B., Sweeney, T., & O'Doherty, J. (2018). Extraction and yield optimisation of fucose, glucans and associated antioxidant activities from *Laminaria digitata* by applying response surface methodology to high intensity ultrasound-assisted extraction. *Marine Drugs, 16*(8), 257.

Glasson, C. R. K., Sims, I. M., Carnachan, S. M., de Nys, R., & Magnusson, M. (2017). A cascading biorefinery process targeting sulfated polysaccharides (ulvan) from *Ulva ohnoi*. *Algal Research, 27*, 383–391.

Goto, M., Kanda, H., Wahyudiono, & Machmudah, S. (2015). Extraction of carotenoids and lipids from algae by supercritical CO_2 and subcritical dimethyl ether. *The Journal of Supercritical Fluids, 96*, 245–251.

Hahn, T., Lang, S., Ulber, R., & Muffler, K. (2012). Novel procedures for the extraction of fucoidan from brown algae. *Process Biochemistry, 47*(12), 1691–1698.

Han, D., & Row, K. H. (2010). Recent applications of ionic liquids in separation technology. *Molecules, 15*(4), 2405.
Hardouin, K., Bedoux, G., Burlot, A.-S., Donnay-Moreno, C., Bergé, J.-P., Nyvall-Collén, P., et al. (2016). Enzyme-assisted extraction (EAE) for the production of antiviral and antioxidant extracts from the green seaweed *Ulva armoricana* (Ulvales, Ulvophyceae). *Algal Research, 16*, 233–239.
Hardouin, K., Burlot, A. S., Umami, A., Tanniou, A., Stiger-Pouvreau, V., Widowati, I., et al. (2014). Biochemical and antiviral activities of enzymatic hydrolysates from different invasive French seaweeds. *Journal of Applied Phycology, 26*(2), 1029–1042.
Heavisides, E., Rouger, C., Reichel, A. F., Ulrich, C., Wenzel-Storjohann, A., Sebens, S., et al. (2018). Seasonal variations in the metabolome and bioactivity profile of *Fucus vesiculosus* extracted by an optimised, pressurised liquid extraction protocol. *Marine Drugs, 16*(12), 503.
He, Z., Chen, Y., Chen, Y., Liu, H., Yuan, G., Fan, Y., et al. (2013). Optimization of the microwave-assisted extraction of phlorotannins from *Saccharina japonica* Aresch and evaluation of the inhibitory effects of phlorotannin-containing extracts on $HepG_2$ cancer cells. *Chinese Journal of Oceanology and Limnology, 31*(5), 1045–1054.
Herrero, M., Cifuentes, A., & Ibañez, E. (2006). Sub- and supercritical fluid extraction of functional ingredients from different natural sources: Plants, food-by-products, algae and microalgae: A review. *Food Chemistry, 98*(1), 136–148.
Hifney, A. F., Fawzy, M. A., Abdel-Gawad, K. M., & Gomaa, M. (2016). Industrial optimization of fucoidan extraction from *Sargassum sp.* and its potential antioxidant and emulsifying activities. *Food Hydrocolloids, 54*, 77–88.
Hifney, A. F., Fawzy, M. A., Abdel-Gawad, K. M., & Gomaa, M. (2018). Upgrading the antioxidant properties of fucoidan and alginate from *Cystoseira trinodis* by fungal fermentation or enzymatic pretreatment of the seaweed biomass. *Food Chemistry, 269*, 387–395.
Hmelkov, A. B., Zvyagintseva, T. N., Shevchenko, N. M., Rasin, A. B., & Ermakova, S. P. (2018). Ultrasound-assisted extraction of polysaccharides from brown alga *Fucus evanescens*. Structure and biological activity of the new fucoidan fractions. *Journal of Applied Phycology, 30*(3), 2039–2046.
Holdt, S. L., & Kraan, S. (2011). Bioactive compounds in seaweed: Functional food applications and legislation. *Journal of Applied Phycology, 23*(3), 543–597.
Huang, C.-Y., Kuo, C.-H., & Chen, P.-W. (2017). Compressional-puffing pretreatment enhances neuroprotective effects of fucoidans from the brown seaweed *Sargassum hemiphyllum* on 6-hydroxydopamine-induced apoptosis in SH-SY5Y cells. *Molecules, 23*(1), 78.
IAPWS. (1997). *The International Association for the Properties of Water and Steam*.
Ibañez, E., & Cifuentes, A. (2013). Benefits of using algae as natural sources of functional ingredients. *Journal of the Science of Food and Agriculture, 93*(4), 703–709.
Imbs, T. I., Ermakova, S. P., Malyarenko, O. S., Isakov, V. V., & Zvyagintseva, T. N. (2016). Structural elucidation of polysaccharide fractions from the brown alga *Coccophora langsdorfii* and in vitro investigation of their anticancer activity. *Carbohydrate Polymers, 135*, 162–168.
Isaza Martínez, J. H., & Torres Castañeda, H. G. (2013). Preparation and chromatographic analysis of phlorotannins. *Journal of Chromatographic Science, 51*(8), 825–838.
Jaswir, I., Ademola, M. H., Amid, A., Alam, Z., Asiyanbi-H, T. T., & Ramli, N. (2013). Enzymatic hydrolysis of plants and algae for extraction of bioactive compounds. *Food Reviews International, 29*(4), 352–370.
Joana Gil-Chávez, G., Villa, J. A., Fernando Ayala-Zavala, J., Basilio Heredia, J., Sepulveda, D., Yahia, E. M., et al. (2013). Technologies for extraction and production of bioactive compounds to be used as nutraceuticals and food ingredients: An overview. *Comprehensive Reviews in Food Science and Food Safety, 12*(1), 5–23.

Jung, K. A., Lim, S.-R., Kim, Y., & Park, J. M. (2013). Potentials of macroalgae as feedstocks for biorefinery. *Bioresource Technology, 135*, 182–190.
Kadam, S. U., Álvarez, C., Tiwari, B. K., & O'Donnell, C. P. (2017). Extraction and characterization of protein from Irish brown seaweed *Ascophyllum nodosum*. *Food Research International, 99*, 1021–1027.
Kadam, S. U., Tiwari, B. K., & O'Donnell, C. P. (2015). Extraction, structure and biofunctional activities of laminarin from brown algae. *International Journal of Food Science and Technology, 50*(1), 24–31.
Kadam, S. U., Tiwari, B. K., & O'Donnell, C. P. (2013). Application of novel extraction technologies for bioactives from marine algae. *Journal of Agricultural and Food Chemistry, 61*(20), 4667–4675.
Kadam, S. U., Tiwari, B. K., Smyth, T. J., & O'Donnell, C. P. (2015). Optimization of ultrasound assisted extraction of bioactive components from brown seaweed *Ascophyllum nodosum* using response surface methodology. *Ultrasonics Sonochemistry, 23*, 308–316.
Le Lann, K., Surget, G., Couteau, C., Coiffard, L., Cérantola, S., Gaillard, F., et al. (2016). Sunscreen, antioxidant, and bactericide capacities of phlorotannins from the brown macroalga *Halidrys siliquosa*. *Journal of Applied Phycology, 28*(6), 3547–3559.
Lee, K. Y., & Mooney, D. J. (2012). Alginate: Properties and biomedical applications. *Progress in Polymer Science, 37*(1), 106–126.
Lee, S.-H., Ko, C.-I., Ahn, G., You, S., Kim, J.-S., Heu, M. S., et al. (2012). Molecular characteristics and anti-inflammatory activity of the fucoidan extracted from *Ecklonia cava*. *Carbohydrate Polymers, 89*(2), 599–606.
Leone, A., Tamborrino, A., Romaniello, R., Zagaria, R., & Sabella, E. (2014). Specification and implementation of a continuous microwave-assisted system for paste malaxation in an olive oil extraction plant. *Biosystems Engineering, 125*, 24–35.
Leu, S., & Boussiba, S. (2014). Advances in the production of high-value products by microalgae. *Industrial Biotechnology, 10*(3), 169–183.
Li, Y.-X., Wijesekara, I., Li, Y., & Kim, S.-K. (2011). Phlorotannins as bioactive agents from brown algae. *Process Biochemistry, 46*(12), 2219–2224.
Magnusson, M., Yuen, A. K. L., Zhang, R., Wright, J. T., Taylor, R. B., Maschmeyer, T., et al. (2017). A comparative assessment of microwave assisted (MAE) and conventional solid-liquid (SLE) techniques for the extraction of phloroglucinol from brown seaweed. *Algal Research, 23*, 28–36.
Marinho-Soriano, E., Fonseca, P. C., Carneiro, M. A. A., & Moreira, W. S. C. (2006). Seasonal variation in the chemical composition of two tropical seaweeds. *Bioresource Technology, 97*(18), 2402–2406.
McHugh, M. A., & Krukonis, V. J. (1994). *Supercritical fluid extraction: Principles and practice*. Boston, MA: Butterworth-Heinemann.
Meillisa, A., Siahaan, E. A., Park, J.-N., Woo, H.-C., & Chun, B.-S. (2013). Effect of subcritical water hydrolysate in the brown seaweed *Saccharina japonica* as a potential antibacterial agent on food-borne pathogens. *Journal of Applied Phycology, 25*(3), 763–769.
Meillisa, A., Woo, H.-C., & Chun, B.-S. (2015). Production of monosaccharides and bio-active compounds derived from marine polysaccharides using subcritical water hydrolysis. *Food Chemistry, 171*, 70–77.
Michalak, I., Chojnacka, K., Dmytryk, A., Wilk, R., Gramza, M., & Rój, E. (2016). Evaluation of supercritical extracts of algae as biostimulants of plant growth in field trials. *Frontiers of Plant Science, 7*, 1591.
Michalak, I., Chojnacka, K., & Saeid, A. (2017). Plant growth biostimulants, dietary feed supplements and cosmetics formulated with supercritical CO_2 algal extracts. *Molecules, 22*(1), 66.

Milledge, J. J., Nielsen, B. V., & Bailey, D. (2016). High-value products from macroalgae: The potential uses of the invasive brown seaweed, *Sargassum muticum*. *Reviews in Environmental Science and Biotechnology, 15*(1), 67–88.

Mizuno, M., Nishitani, Y., Tanoue, T., Matoba, Y., Ojima, T., Hashimoto, T., et al. (2009). Quantification and localization of fucoidan in *Laminaria japonica* using a novel antibody. *Bioscience, Biotechnology, and Biochemistry, 73*(2), 335–338.

Mohammed, A., Bissoon, R., Bajnath, E., Mohammed, K., Lee, T., Bissram, M., et al. (2018). Multistage extraction and purification of waste *Sargassum natans* to produce sodium alginate: An optimization approach. *Carbohydrate Polymers, 198*, 109–118.

Nadar, S. S., Rao, P., & Rathod, V. K. (2018). Enzyme assisted extraction of biomolecules as an approach to novel extraction technology: A review. *Food Research International, 108*, 309–330.

NIST. (2014). *NIST data tables*.

Noraini, M. Y., Ong, H. C., Badrul, M. J., & Chong, W. T. (2014). A review on potential enzymatic reaction for biofuel production from algae. *Renewable and Sustainable Energy Reviews, 39*, 24–34.

Okolie, C. L., Akanbi, T. O., Mason, B., Udenigwe, C. C., & Aryee, A. N. A. (2018). Influence of conventional and recent extraction technologies on physicochemical properties of bioactive macromolecules from natural sources: A review. *Food Research International, 116*, 827–839.

Oliveira, C., Granja, S., Neves, N. M., Reis, R. L., Baltazar, F., Silva, T. H., et al. (2019). Fucoidan from *Fucus vesiculosus* inhibits new blood vessel formation and breast tumor growth in vivo. *Carbohydrate Polymers, 223*, 115034.

O'Sullivan, L., Murphy, B., McLoughlin, P., Duggan, P., Lawlor, P. G., Hughes, H., et al. (2010). Prebiotics from marine macroalgae for human and animal health applications. *Marine Drugs, 8*(7).

Pádua, D., Rocha, E., Gargiulo, D., & Ramos, A. A. (2015). Bioactive compounds from brown seaweeds: Phloroglucinol, fucoxanthin and fucoidan as promising therapeutic agents against breast cancer. *Phytochemistry Letters, 14*, 91–98.

Pérez-Larrán, P., Torres, M. D., Flórez-Fernández, N., Balboa, E. M., Moure, A., & Domínguez, H. (2019). Green technologies for cascade extraction of *Sargassum muticum* bioactives. *Journal of Applied Phycology, 31*(4), 2481–2495.

Pérez-López, P., Balboa, E. M., González-García, S., Domínguez, H., Feijoo, G., & Moreira, M. T. (2014). Comparative environmental assessment of valorization strategies of the invasive macroalgae *Sargassum muticum*. *Bioresource Technology, 161*, 137–148.

Peshkovsky, A. S., Peshkovsky, S. L., & Bystryak, S. (2013). Scalable high-power ultrasonic technology for the production of translucent nanoemulsions. *Chemical Engineering and Processing: Process Intensification, 69*, 77–82.

Plaza, M., Amigo-Benavent, M., del Castillo, M. D., Ibáñez, E., & Herrero, M. (2010). Facts about the formation of new antioxidants in natural samples after subcritical water extraction. *Food Research International, 43*(10), 2341–2348.

Plaza, M., & Turner, C. (2015). Pressurized hot water extraction of bioactives. *TrAC Trends in Analytical Chemistry, 71*, 39–54.

Praiboon, J., Palakas, S., Noiraksa, T., & Miyashita, K. (2018). Seasonal variation in nutritional composition and anti-proliferative activity of brown seaweed, *Sargassum oligocystum*. *Journal of Applied Phycology, 30*(1), 101–111.

Puspita, M., Déniel, M., Widowati, I., Radjasa, O. K., Douzenel, P., Marty, C., et al. (2017). Total phenolic content and biological activities of enzymatic extracts from *Sargassum muticum* (Yendo) Fensholt. *Journal of Applied Phycology, 29*(5), 2521–2537.

Quitain, A. T., Kai, T., Sasaki, M., & Goto, M. (2013a). Microwave–hydrothermal extraction and degradation of fucoidan from supercritical carbon dioxide deoiled *Undaria pinnatifida*. *Industrial & Engineering Chemistry Research, 52*(23), 7940–7946.

Quitain, A. T., Kai, T., Sasaki, M., & Goto, M. (2013b). Supercritical carbon dioxide extraction of fucoxanthin from *Undaria pinnatifida*. *Journal of Agricultural and Food Chemistry, 61*(24), 5792–5797.

Rajauria, G., Foley, B., & Abu-Ghannam, N. (2016). Identification and characterization of phenolic antioxidant compounds from brown Irish seaweed *Himanthalia elongata* using LC-DAD–ESI-MS/MS. *Innovative Food Science & Emerging Technologies, 37*, 261–268.

Reddy, H. K., Muppaneni, T., Sun, Y., Li, Y., Ponnusamy, S., Patil, P. D., et al. (2014). Subcritical water extraction of lipids from wet algae for biodiesel production. *Fuel, 133*, 73–81.

Ren, B., Chen, C., Li, C., Fu, X., You, L., & Liu, R. H. (2017). Optimization of microwave-assisted extraction of *Sargassum thunbergii* polysaccharides and its antioxidant and hypoglycemic activities. *Carbohydrate Polymers, 173*, 192–201.

Rhein-Knudsen, N., Ale, M., & Meyer, A. (2015). Seaweed hydrocolloid production: An update on enzyme assisted extraction and modification technologies. *Marine Drugs, 13*(6), 3340.

Rioux, L.-E., Turgeon, S. L., & Beaulieu, M. (2010). Structural characterization of laminaran and galactofucan extracted from the brown seaweed *Saccharina longicruris*. *Phytochemistry, 71*(13), 1586–1595.

Rodrigues, D., Sousa, S., Silva, A., Amorim, M., Pereira, L., Rocha-Santos, T. A., et al. (2015). Impact of enzyme- and ultrasound-assisted extraction methods on biological properties of red, brown, and green seaweeds from the central west coast of Portugal. *Journal of Agricultural and Food Chemistry, 63*(12), 3177–3188.

Rodriguez-Jasso, R. M., Mussatto, S. I., Pastrana, L., Aguilar, C. N., & Teixeira, J. A. (2011). Microwave-assisted extraction of sulfated polysaccharides (fucoidan) from brown seaweed. *Carbohydrate Polymers, 86*(3), 1137–1144.

Roselló-Soto, E., Galanakis, C. M., Brnčić, M., Orlien, V., Trujillo, F. J., Mawson, R., et al. (2015). Clean recovery of antioxidant compounds from plant foods, by-products and algae assisted by ultrasounds processing. Modeling approaches to optimize processing conditions. *Trends in Food Science & Technology, 42*(2), 134–149.

Rostami, Z., Tabarsa, M., You, S., & Rezaei, M. (2017). Relationship between molecular weights and biological properties of alginates extracted under different methods from *Colpomenia peregrina*. *Process Biochemistry, 58*, 289–297.

Routray, W., & Orsat, V. (2012). Microwave-assisted extraction of flavonoids: A review. *Food and Bioprocess Technology, 5*(2), 409–424.

Ruesgas-Ramón, M., Figueroa-Espinoza, M. C., & Durand, E. (2017). Application of deep eutectic solvents (DES) for phenolic compounds extraction: Overview, challenges, and opportunities. *Journal of Agricultural and Food Chemistry, 65*(18), 3591–3601.

Sánchez-Camargo, A. P., Montero, L., Cifuentes, A., Herrero, M., & Ibáñez, E. (2016). Application of Hansen solubility approach for the subcritical and supercritical selective extraction of phlorotannins from *Cystoseira abies-marina*. *RSC Advances, 6*(97), 94884–94895.

Sánchez-Camargo, A. P., Montero, L., Stiger-Pouvreau, V., Tanniou, A., Cifuentes, A., Herrero, M., et al. (2016). Considerations on the use of enzyme-assisted extraction in combination with pressurized liquids to recover bioactive compounds from algae. *Food Chemistry, 192*, 67–74.

Saravana, P. S., Choi, J. H., Park, Y. B., Woo, H. C., & Chun, B. S. (2016). Evaluation of the chemical composition of brown seaweed (*Saccharina japonica*) hydrolysate by pressurized hot water extraction. *Algal Research, 13*, 246–254.

Saravana, P. S., Cho, Y.-N., Woo, H.-C., & Chun, B.-S. (2018). Green and efficient extraction of polysaccharides from brown seaweed by adding deep eutectic solvent in subcritical water hydrolysis. *Journal of Cleaner Production, 198*, 1474–1484.

Saravana, P. S., Getachew, A. T., Cho, Y.-J., Choi, J. H., Park, Y. B., Woo, H. C., et al. (2017). Influence of co-solvents on fucoxanthin and phlorotannin recovery from brown seaweed using supercritical CO_2. *The Journal of Supercritical Fluids, 120*, 295–303.

Saravana, P. S., Tilahun, A., Gerenew, C., Tri, V. D., Kim, N. H., Kim, G.-D., et al. (2018). Subcritical water extraction of fucoidan from *Saccharina japonica*: Optimization, characterization and biological studies. *Journal of Applied Phycology, 30*(1), 579–590.

Shirsath, S. R., Sonawane, S. H., & Gogate, P. R. (2012). Intensification of extraction of natural products using ultrasonic irradiations—a review of current status. *Chemical Engineering and Processing: Process Intensification, 53*, 10–23.

Sivagnanam, S. P., Yin, S., Choi, J. H., Park, Y. B., Woo, H. C., & Chun, B. S. (2015). Biological properties of fucoxanthin in oil recovered from two brown seaweeds using supercritical CO_2 extraction. *Marine Drugs, 13*(6), 3422–3442 (Special Issue -Marine Functional Food).

Song, M., Duc Pham, H., Seon, J., & Chul Woo, H. (2015). Marine brown algae: A conundrum answer for sustainable biofuels production. *Renewable and Sustainable Energy Reviews, 50*, 782–792.

Suganya, T., Varman, M., Masjuki, H. H., & Renganathan, S. (2016). Macroalgae and microalgae as a potential source for commercial applications along with biofuels production: A biorefinery approach. *Renewable and Sustainable Energy Reviews, 55*, 909–941.

Tapia-Salazar, M., Arévalo-Rivera, I. G., Maldonado-Muñiz, M., Garcia-Amezquita, L. E., Nieto-López, M. G., Ricque-Marie, D., et al. (2019). The dietary fiber profile, total polyphenol content, functionality of *Silvetia compressa* and *Ecklonia arborea*, and modifications induced by high hydrostatic pressure treatments. *Food and Bioprocess Technology, 12*(3), 512–523.

Terme, N., Boulho, R., Kendel, M., Kucma, J.-P., Wielgosz-Collin, G., Bourgougnon, N., et al. (2017). Selective extraction of lipid classes from *Solieria chordalis* and *Sargassum muticum* using supercritical carbon dioxide and conventional solid–liquid methods. *Journal of Applied Phycology, 29*(5), 2513–2519.

Thomas, N. V., & Kim, S.-K. (2013). Beneficial effects of marine algal compounds in cosmeceuticals. *Marine Drugs, 11*(1), 146–164.

Tiwari, B. K. (2015). Ultrasound: A clean, green extraction technology. *TrAC Trends in Analytical Chemistry, 71*, 100–109.

Trivedi, N., Gupta, V., Reddy, C. R. K., & Jha, B. (2013). Enzymatic hydrolysis and production of bioethanol from common macrophytic green alga *Ulva fasciata Delile*. *Bioresource Technology, 150*, 106–112.

Vera, J., Castro, J., Gonzalez, A., & Moenne, A. (2011). Seaweed polysaccharides and derived oligosaccharides stimulate defense responses and protection against pathogens in plants. *Marine Drugs, 9*(12).

Vinatoru, M., Mason, T. J., & Calinescu, I. (2017). Ultrasonically assisted extraction (UAE) and microwave assisted extraction (MAE) of functional compounds from plant materials. *TrAC Trends in Analytical Chemistry, 97*, 159–178.

Vo Dinh, T., Saravana, P. S., Woo, H. C., & Chun, B. S. (2018). Ionic liquid-assisted subcritical water enhances the extraction of phenolics from brown seaweed and its antioxidant activity. *Separation and Purification Technology, 196*, 287–299.

Wijesinghe, W. A. J. P., & Jeon, Y.-J. (2012a). Biological activities and potential industrial applications of fucose rich sulfated polysaccharides and fucoidans isolated from brown seaweeds: A review. *Carbohydrate Polymers, 88*(1), 13–20.

Wijesinghe, W. A. J. P., & Jeon, Y.-J. (2012b). Enzyme-assistant extraction (EAE) of bioactive components: A useful approach for recovery of industrially important metabolites from seaweeds: A review. *Fitoterapia, 83*(1), 6–12.

Wu, L., Sun, J., Su, X., Yu, Q., Yu, Q., & Zhang, P. (2016). A review about the development of fucoidan in antitumor activity: Progress and challenges. *Carbohydrate Polymers, 154*, 96–111.

Xiao, X.-H., Yuan, Z.-Q., & Li, G.-K. (2013). Preparation of phytosterols and phytol from edible marine algae by microwave-assisted extraction and high-speed counter-current chromatography. *Separation and Purification Technology, 104*, 284–289.

Yuan, Y., & Macquarrie, D. (2015). Microwave assisted extraction of sulfated polysaccharides (fucoidan) from *Ascophyllum nodosum* and its antioxidant activity. *Carbohydrate Polymers, 129*, 101–107.

Yuan, Y., Zhang, J., Fan, J., Clark, J., Shen, P., Li, Y., et al. (2018). Microwave assisted extraction of phenolic compounds from four economic brown macroalgae species and evaluation of their antioxidant activities and inhibitory effects on α-amylase, α-glucosidase, pancreatic lipase and tyrosinase. *Food Research International, 113*, 288–297.

Zhang, H., & Row, K. H. (2015). Extraction and separation of polysaccharides from *Laminaria japonica* by size-exclusion chromatography. *Journal of Chromatographic Science, 53*(4), 498–502.

Zhang, H.-F., Yang, X.-H., & Wang, Y. (2011). Microwave assisted extraction of secondary metabolites from plants: Current status and future directions. *Trends in Food Science & Technology, 22*(12), 672–688.

Zhao, D., Xu, J., & Xu, X. (2018). Bioactivity of fucoidan extracted from *Laminaria japonica* using a novel procedure with high yield. *Food Chemistry, 245*, 911–918.

Zhu, Z., Wu, Q., Di, X., Li, S., Barba, F. J., Koubaa, M., et al. (2017). Multistage recovery process of seaweed pigments: Investigation of ultrasound assisted extraction and ultra-filtration performances. *Food and Bioproducts Processing, 104*, 40–47.

Zhu, Z-y, Dong, F., Liu, X., Lv, Q., YingYang, Liu, F., et al. (2016). Effects of extraction methods on the yield, chemical structure and anti-tumor activity of polysaccharides from *Cordyceps gunnii* mycelia. *Carbohydrate Polymers, 140*, 461–471.

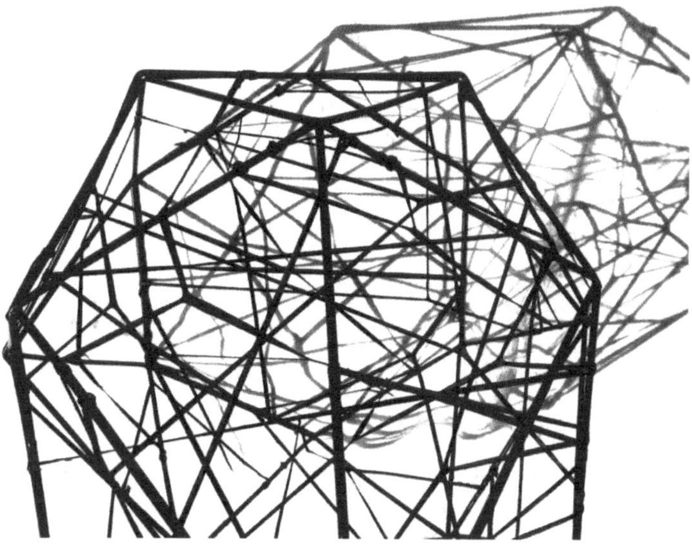

Sculptural dried kelp, ©Violaine Buet, 2016.*(Photo credit Violaine Buet).*

Whilst the poem (art, architecture) frees the earth from its self identity by opening it into new worlds, science dissects the world in order to find out the earth. And so comes about, in this indefinitely resumed controversy, (...) movement and constant renewal, the reality of human milieus. (Berque, 2019c, p. 70).

CHAPTER TWELVE

From seaweed to *seaweeders*[**]

Violaine Buet[*,a,b]
Studio in Auray, Bretagne, France
*Corresponding author: E-mail: contact@violainebuet.com

Contents

[a] Independant, seaweed designer.
[b] With the participation of Julia Lohmann.
[**] Over our correspondences in our group "the Department of seaweed" we slowly started greeting each other this way: Dear Seaweeders.

Advances in Botanical Research, Volume 95
ISSN 0065-2296
https://doi.org/10.1016/bs.abr.2019.12.003

Abstract

At the threshold of design, craft and research, seaweed inspire the construction of a multidisciplinary know-how; a conversation between macro-algae and textile techniques, between nature and manufacture. The identity of these promising, biodegradable materials -magnificent kelp- raise a series of questions and reflections about our relationship to nature, and about our hopes and expectations from materials and objects. This eye-opening concept of reconciling to the impermanent rhythm of living beings invites us to draw on new ways of conceiving our relationship to our milieu and question where and when the action of our hands can re-align to our thoughts.

It all started with a desire; the wish to dialog with a natural, living material. My choice was intuitive: seaweed. This ancestral marine plant has been part of the landscape of my Breton childhood and it was about to open a vast field of research, triggering connections within the keystones of my own relationship to the world: creativity (the tool), seashore (the ground), living with nature (the direction). Together they combined to conceive the Milieu of a newborn practice: Seaweed Design.

At the threshold of design, craft and research, I am developing, accompanied by a precious network, a multidisciplinary know-how; a dialog conversation between macro-algae and textile techniques, between nature and manufacture. The identity of these promising, biodegradable materials - magnificent kelp-raise a series of questions and reflections about our relationship to nature, and about our hopes and expectations from materials and objects. This eye-opening concept of reconciling to the impermanent rhythm of living beings invites us to draw on new ways of conceiving our relationship to our milieu and question where and when the action of our hands can re-align to our thoughts. "*Separating ourselves from nature has profoundly rendered us into waifs. To reconnect with the simple idea that we are viscerally a part of it, (...) to re-establish this is a blending, with the joyous tidings; we are not outsiders.*" (Reeves, 2017).

Colored, woven laminaria, ©Violaine Buet, 2017. *Photo credit: Pierre-Yves Dinasquet.*

Genesis

In this era of consumption frenzy, I wondered about our attachment to objects, materials. What good are sustainable energy thoughts and discussions when all we are producing is unsustainable waste? Within this equation of material and emotional obsolescence, where attachments are quick to disappear, a challenge was raised for designers; to introduce sustainability while inspiring a lasting meaning for the future. To abstain one's self from the idea of an immediate object, which is no longer of interest as soon as it has served a rapid purpose, but to look toward objects and materials that can still speak to us. "*(…) products must mutually evolve alongside users, sustaining value by revealing their true beauty only through the slow passing of time (…), modern products lack character; they are too smart and precise, removing all possible surprise, mystery and, perhaps above all, charm from the process of engagement.*" (Chapman, 2005). Reconnecting with emotions as an entry to durability; for its capacity of renewal and generating the impulse of life.

The following extract is from a text written by the French Writer Alain Damasio; "*(…) So, to be truly alive is to increase our capacity to be affected, within our spectrum or our amplitude to be touched, changed, moved; To grasp a sensation, contemplate, occupy a moment or a book: These are the noble connections*".[1] That was exactly it; the desire to design noble connections; not anything, anywhere, at any cadence … an approach that would host a whole; *the dynamic coupling of the Living and its milieu.*[2] I wanted to find the perspectives and the value in this movement itself, transforming *an environment (universal and virtual) into a milieu (existing and singular).*[3]

As a contemporary designer I seriously questioned making new objects unless they were not to become "things"; born out of a milieu, and not an addition to an already *junk space* (Koolhaas, 2002, pp. 175—190) resulting from "*the loss of landmarks that would hold together beings and the things of a milieu*" (Berque, 2018a, p. 15). Set apart from their milieu,"*objects could become absolute*" (Berque, 2013). I was dreaming of objects and materials that would take us -thanks to an attention to a concrescence[4] in the design process-beyond their physical envelope, until what they have triggered has been caught in the weaves of our lives, and their stories survive their materiality. I have here the image of the water-soluble membrane used in embroidery, whose vocation is to disappear once the structure exists. I was conscious it was about writing a story, with the inevitable presence of the time factor; the time it takes for the living, the hand, and the meaning, to get in tune with one another. The time it takes for the value to rise from the heart of a system, without being snatched by the external regular decoy; exploitation, and financial gain.

1 Damasio, A. Boomerang. France Inter, May 13th, 2019. 32 min.

2 "Mediance: the structural moment of human existence (…)that is the dynamic coupling of Being* and its milieu"*; corresponds to (..) the reciprocal adequacy of the animal (or more generally the living) and its milieu. Berque, A. (2018). *Glossaire de mésologie, definitions of mediance* (p. 24). Bastia: Eoliennes.

3 "Trajection: *To-and-fro motion of reality* between the two theoretical poles of the subjective and the objective: reality belongs neither to the sole realm of the object*, nor to the sole realm of the subject*; belonging to the trajection of both terms, it is trajective (…)*". Berque, A. (2018). *Glossaire de mésologie, definitions of trajection* (p. 41). Bastia: Eoliennes.

4 Co-arousal.

From this reflection, and the multiple possible paths that opened, I followed this one: the exploration of a natural and living material, and above all a material that makes me feel alive, Seaweed. To me, seaweed is a powerful material because it is a time traveler; coming from so far and taking us ways ahead. As an isolated designer I can't embrace a spectrum of such amplitude unless I grow on a philosophical grounding and a collaborative affiliation, to invent a practice in this direction.

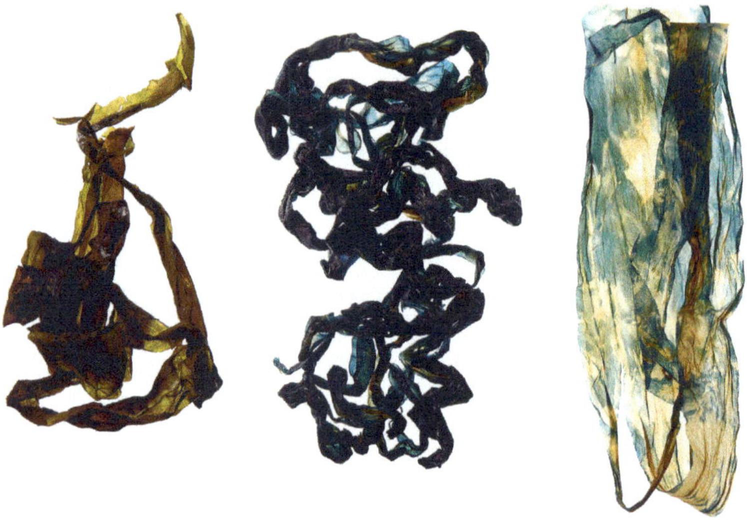

Processing Sacchoriza polyshides, ©Violaine Buet, 2016. *Photo credit: Pierre-Yves Dinasquet.*

1. Lineage

Seaweed comes from so far —1,3 billion years old for the brown lineage, my best ally - working with it connects me with the origin of life and with the quest for meaning that accompanies any new foundation.

1.1 Mesology

Developing a new craft with such a wild material is an audacious yet uncertain path. Manual work alone is not enough, as the need to align its process with that of thought is essential. In this unpredictable environment I had chosen, I was looking for an unwavering horizon. Where is there a better balance than when the earth meets the sky? In a *summetria*, the lost signification of symmetry "*which is a common measure between the sky and the earth (...) nature and human*" (Berque, 2019a, p. 19). Away from a dualism separating human subjects from objective orders promoted by science, I was interested to cultivate the relationship between a microcosmic intervention and a macrocosmic belonging. This relationship itself projected me into the work; the mysterious strength and wisdom of this ancient marine plant which connects me somehow with the origin of life nourishes system I am developing toward a future.

In search for a holistic system that overcomes the specialization of knowledge I came across the field of Mesology, as defined by Augustin Berque,[5] its eminent explorer, as " *the study of the milieus and in particular of the ecumene, the eco-techno-symbolic relationship of humankind with the Earth*". This area of biology deals with the relationships between the environments and organisms in a ternary vision. The triad "subject/interpretation/object" introduces the contingency of history and milieu; different culture could deploy different ways of living in the same natural environment. The main point while developing this project was never to limit it simply to my intervention nor to the material itself but to their emerging *co-arousal*[6] in a given context. Mesology forges this path where the object doesn't exist in itself but according to its relationship with the subject.

Augustin Berque states; "*Mesology is the study of the "milieus", in line with the meaning given to it by Uexküll and what he called* Umweltlehre, *ie. mesology. I approached the subject via a Japanese philosopher called Watsuji, and one of his books* Fudo *(which I translate as "milieu"). Mesology (...) is a form of dynamic coupling between a given being -that can be either an individual or a collective being such as living species for instance- and what surrounds it. These surroundings are the "environment" but they only become "milieu": the surroundings start out as an environment, and precisely only then become "milieu" in that dynamic and creative coupling, which will in fact produce both the "milieu" and the being in question. This happens gradually through evolution of course, over the long term, but also takes place over the course of human history*".[7]

Discovering mesology was like receiving "*a vast air necklace*" (Aragon, 1956) and if I am instinctively quoting poetry here, it is that this discipline,

5 Geographer, orientalist, price winner of the 2018 Cosmos Prize in recognition of his outstanding achievements in establishing a unique environmental anthropology.

6 *Co-arousal: "(...) conditioned coproduction, interdependence,* synonym of concrescence (...)". Berque, A. (2018). *Glossaire de mésologie, definition of co-suscitation* (p. 14). Bastia: Eoliennes.

7 URL: pca-stream.com/en/player/from-mediance-to-places-augustin-berque-stream-talk-63.

mesology, has the latitude to host a wide range of languages, ecological to ontological, an infinity generated by the consideration of a third part: the dynamic connecting a person with its milieu. Getting away from a binary system allows multiplicity and singularity, and overcomes the "or" for the "and". Pablo Servigne, collapsologue asks; Why shall we oppose "*nature to culture, reason to emotion, innate to acquired, body to spirit, individual to collective, local to global*" when it provokes ruptures and extinctions?

From the beginning of my design carrier I laboriously tried to find a way where I could embrace the multiplicity of who I was, and of the trans-disciplinary fields, which were arising all around me. I refused to confine myself to a static and specialized course, since my vital energy was bound to the articulation of multiple budding forces, collectively forming an order.

Warp (linen) and weft (Saccharina lattissima) forming an order ©Violaine Buet, 2019. *Photo credit: Violaine Buet.*

1.2 Cosmos

Terrestrial life could be an intense journey where we travel to find where our personal orderliness aligns with a more general and fundamental order. It is not about an already existing place, neither utopian nor a random one, but a contingent venue; a favorable adequacy of who we are within our milieu. Only we, can be there and yet it can happen in many different places …

Cosmos is originally a Greek word, κόσμος, meaning well-ordered whole, in opposition to chaos. It can concern the scale of the universe to the scale of the human. They meet in the space of cosmicity "*projecting the order of the skies (kosmos) here on earth, to create an ordered human world*" (Berque, 2019d, p. 19). "*The splintering of bonds between human-beings and bonds with all living beings, has left a vast void*" (Servigne & Chapelle, 2019, p. 23) and left our anthropocenic era unbalanced in an *acosmy*.[8] "*We do not leave pyramids*" says the architect Rem Koolhaas. Traditional societies were founded on the two intricate meanings embraced by the word cosmos; the general order, fundamental, macrocosmic, and a personal ornament, superficial. "*The cosmetic was referring to the cosmic, the cosmic was revealed through the cosmetic*".[5] Without cosmic rooting, the cosmetic has become errant.

Woven seaweed, unwoven franges, Cooper Hewitt design triennial, NY. 200cmx90cm ©Violaine Buet, 2019. *Photo credit: Laurent Desmaret.*

[8] *Acosmie; "Lack of cosmicity, resulting from a decosmisation (…)"*. Berque, A. (2018). *Glossaire de mésologie, definition of acosmie* (p. 9). Bastia: Eoliennes.

I thank seaweed for its complex nature, combined with my directive line of using soft and biodegradable products, and a quest for meaning, I was aware of these two directions. I am as honored as challenged to listen to the laws of nature. With seaweed I have to take into account seasons, species, tides, weight, smell, particularities, color-fade, hydrophilicity, decomposition ... etc. The advancement of my aesthetic exploration is moving forward more rapidly than its "scientific order", because of the nature of my creative skills. If I love and vibrate with this energy empowering Beauty, I flee mirages and speculation. The material itself always brings me back to its reality, its fundamental order, with the help of the experts I am conferring with; researchers in biopolymer, natural dye, biologists, seaweed-farmers ... The question begs; How is it still possible to conceive non multi-disciplinary teams in the complexity of our eco-techno- symbolic world? When are we going to blend right at its outset creative minds to scientist and technical ones?

The profound joy and well-being I am embracing while spending time on the rocks, observing and harvesting are signs of this fundamental order too. During these moments I flow with the elements, I re-invite my atoms to the sea. I tune my internal rhythm to the one of the waves breathing with seaweed. It is a resourcing time yet very down to earth; there is a precise tide window, short, a certain quantity of what can be harvested, kelp is heavy and protected. This step of sensing the material and its pace at the eve of a manufacturing process is essential to bind realities together; To experience the time macro-algae take to grow, the harvestable quantity at a low tide timeframe, the manageable volume without any waste ... All these movements are keys to further project a potential fabrication process, in another manner than the regular modern exploitation pattern. A regenerative system has to, amongst other factors, adopt the rhythm of its living partner. I believe in slow and local stories, brainstormed by international insights, far from a hubris leading us "*to the age of limits*" (Latouche, 2012). Standing before new materiality or envelopes, we need to choose our consistencies. I would like this seaweed adventure to be "*a slow arrow entering the heart*"[9] and the archer to be aimed by the heart.

[9] Céline Sciamma, film maker, talking about her cinema "*I would like my cinema to always be that; a slow arrow entering the heart*".

Knitted Saccharina Lattissima, Cooper Hewitt design triennial, NY. 40cmx50cm ©Violaine Buet, 2019. *Photo credit: Laurent Desmaret.*

1.3 Moving to substance

To me, luxury is the precious born out of nothing. For instance, when we think our fridge is empty and we finally make a succulent dinner. When we totally compose with what we are and what we are surrounded by, free from any speculation. When lack triggers our engagement and our inventiveness. It is frequently this situation or moment that will imprint on our memories.

When in doubt, I love to look back at my seaweed adventure and even if it is fragile, fresh and incomplete, I can now read a story. What started with in-substance -seaweed on the sea shore-has now opened the path to substance, with a tangibility of dialog, fabrication, process, network, consultation, orders, research, communication: the deployment of an eco-techno-symbolic system.

Seaweed palet exploration ©Violaine Buet, 2017. *Photo credit: Pierre-Yves Dinasquet.*

Seaweed (as a marine plant, from a certain ocean, a lineage, a species, with properties, seasonality, aesthetic ... etc) seen by an interpreter (human, brought up on the sea-shore, creative, designer ...) as a textile and artistic material. Here is the mesologic *as*; "*the operation that brings Being out of the wrap of its self-identity into a certain form of existence*" (Berque, 2018b, p. 15). The deployment of the story "seaweed *as* a textile and surface material" that would, once again, convey seaweed in forming the ecumene -the humanely inhabited earth (Berque, 2018c, p. 16) Once again because seaweed already has a large palette of existence, depending on who is looking at it from the living world; it is either oxygen, energy, shelter, shade, food, insulation, textile, sculpture, fertilizer, fuel, nutriment, bio-stimulant, polymer, gellant, anti-oxidant, iodine ... etc.

These different presences vary from macroscopic to microscopic roles: algae as a part of a large ecosystem, for example providing oxygen through photosynthesis, or the element taken as a whole body, for instance as a textile I use it integrally, or the use of a part of it as in any chemical extraction or benefit. They are all the result of a trajective process; "*The transformation of the universal object virtuality, into a singular reality through the sense of action (for all living beings), thought (for superior animals), speech (for human-beings only)*" (Berque, 2019b, p. 44).

It is impossible to compile a complete list of their multiple existences; we are here in the heart of dynamic, gestational movement, and our view is far too partial to embrace one of the whole living world. In my research I am benefiting from knowledge established from these different existences connecting seaweed to the ecumene, they will partly contribute to form what I am developing, and they, themselves, lay on previous or surrounding trajection. The more we move forward, the more we generate complexity and rise toward a superior ontological level. "*Summarily, the trajectory chain corresponds to the elevation of matter to life, then from life toward spirit; or from the emergence of the biosphere from the planet and from there to the ecumene.*" (Berque, 2017, p. 98).

I was wondering and asked Augustin Berque how one can know in this hypostasis process, when in-substance actually becomes substance. I could feel it required a slow process, observable with a certain distance and a progeny, but may be we could read it too in a "rapid current"? "*One can only know retrospectively (...) and therefore only when studying history; otherwise we are caught up within the present. But as you noted, there are also 'rapid currents', catastrophic events which change our vision of things*". I soon understood the importance of time and its maturation process; my seaweed adventure became substantial in my eyes when I could perceive the transformations in myself too.

In my studio, 2018. *Photo credit: Mirjam Bleeker*

2. Spinning the yarn

Seaweed is multifaceted in its composition, successive state and aesthetics. Its complexity challenge my designer's skills on a day to day basis, it roots me in the present, on the eve of a new-born craftmanship.

Putting my pictures from my archive project in this particular order, I tried to convey the primeval spring (Berque, 2018d, p. 43) of the narrative chain that I could gleen at the early stage of my seaweed development. The material samples reveal the evolution, from training a vegetal to the elaboration of a tree-like growing system. From a raw material to one of its progeny, shaped by belief, passion, doubts, audacity, engagement, discouragement, curiosity, awe, impatience, enchantment.

Narrative chain © Archive Violaine Buet, 2019.

2.1 Getting on board (first ring, from the center)

Seven years in India oriented me to complete my industrial design background with a recent textile training: a year in Les Arts décoratifs where I chose to work with a natural material, which changes over time. I first started looking at micro-algae, spirulina, with a particular interest at the scale of yarn - and this idea of combining algae and yarns through natural and self-growing coating. I realized I would lose the craft approach if working on micro-algae and on a biotechnological scale, although "making with my hands" was for me the key.

In these six centered pictures one can see a kind of distant view with the material. In design methodology, the first approach is naturally to turn around the subject, with a kind of fumbling allowing us to find the spark to enter, the space that motivates us; where we engage ourselves in the germination of a singular point of view, and eventually a process. I remember, in these very first intentions, after experimenting coloration and prints with sea-grown spirulina, I even wove some *nori* sheets, the one usually destined to make sushis.

In my years abroad, I think the sea is the element I missed the most. "My" sea, the one I was from; wild, sheltering light, and cosy corners to nest and observe "the end of the earth", *Finistère*! It is most probable that having spent time far from my native environment allowed me to look at it with a new "tonality", I imagine that depending on our major journey, we nuance our own inner geographies and draw the possibilities of looking at things differently. I love to observe how another relationship to a milieu we build somewhere, can generate a change in the relation we have with a primal place. It shows an incorporative arrangement, with infinite bounds, and finally the importance of dynamic spaces generating "*the reciprocal adequation of the Living and its milieu*" (Berque, 2018e, p. 24) which is always unique, non-reversible, structural, breeding ground of the equation to be, and that has been named Mediance in mesological terms.

When they bump into each other, some realities make the interlocutors gain ground: I have this feeling in front of Asian people discovering my work. They are my most surprised public, and in their reaction -elaborating the jump from "seaweed-as-food" to "seaweed-as-textile"- I can literally get a glimpse of where they come from, and where I take them. In a furtive second I travel to their place, they reach mine: milestone to open new worlds.

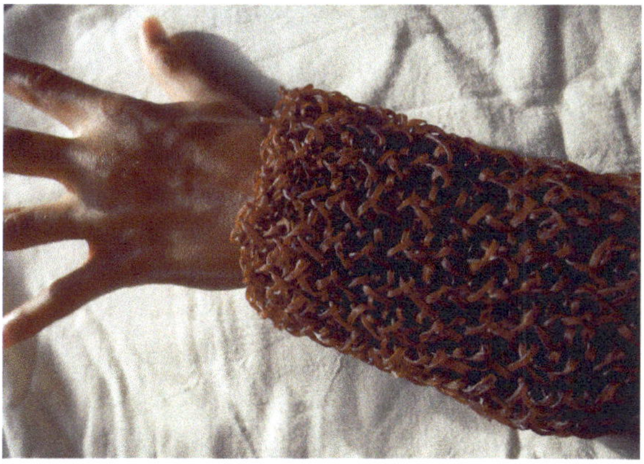

Knitted laminaria © Violaine Buet, 2017. *Photo credit: Violaine Buet.*

2.2 Diving (second ring)

When listing the various existences of seaweed above, we see that there are finally not many usages of seaweed in its integrality; it is mainly used as a component, an extract, and the user finally loses touch with the initial material and its emotional impact. As soon as I started experimenting, I was amazed by the nobility of this marine vegetal, and decided to adopt this angle of the material as a whole. I "dived" into brown seaweed with my creative palette. Initially with simple tools and process.

I collected some, observed them, dried and wet them, sun bathed, embossed, cut, indigo dyed, printed, painted. I was focusing on their reaction, testing the limits, and scanning their aesthetic spectrum. I soon could see they embraced multiple facets, they were challenging and playful once I knew how to stabilize them and keep their flexibility. They have a strong emotional impact; wet seaweed has an amazing resemblance to the tenderness of human skin, the vibrancy of the animal world, the benefit of the vegetal lineage ...

Curious about the imaginary of others about seaweed, I interviewed randomly people with this question; what does the word seaweed evoke for you? They definitively awake senses:

My childhood, walking along the beach, popping the bubble of seaweed.
It's a living thing, life coming out of the sea.
Seaweed in the sea really reminds me of women's hair. It cannot be only mysterious, also frightening.
Slippery, sticky, it scares me.
The smell!
Reassuring permanence, rooted fluidity, refreshing iodine. The bond with the primal.
Without them, it would be quite desert, they are the vegetation of the sea.
it inspires awe in me.
Japanese food. Wild creeks.
Attraction and repulsion, deep and mysterious. it espouses the movement of the water
Agar-agar, Strawberry tart.
Cosmetics.
Disgust.
The colour, dark green, the texture, softness, flexibility.
it's quite culinary at the moment for me ...
Seaweed is all sorts of things: diversity, form, poise, habitat, food, material, ... a key to that world beneath the waves.
It is scary, when you swim something just touches you and you wonder what it is.
Delicacy, freedom, sensitivity!;
Wild submarine forest, moving and colorful.
Soft, slimy, minuscule or giant, they sometimes bother us.
Hiding places for a surprising animal life. A wet aromatic garden.

When entering a new project, it is fruitful for a short while to leave our frame of references behind, in order to provoke innovative angles and allow for serendipity and improbable connections to emerge. For that we need to quiet the cerebral and engage our senses. At this step of my research the fascination for seaweed had already operated, my vision was enlarged. I soon could read the connivances of textile ennoblement techniques with this improbable partner. I had to be there, with no desire to conquer, to exploit, but to welcome what could eventually happen. In accepting that the meaning would get drawn in the process, I was making space for a third part, away from a duality conditioning our subject-centered economical reality. I wanted to take the journey further than the regular legitimacy: a wonderful resource is a wonderful market.

Golding on Saccharina lattissima© Violaine Buet, 2018 *creditphoto violaine buet.*

2.3 Blending (third ring)

Once this instinctive exploration had triggered complicity and a tangible palette of samples, I enlarged the set of tools and techniques using other's expertise too. I could fine-tune and validate some processes, distinguishing chancy results from controllable effects.

I saw seaweed could be woven, knitted, dyed, sewn, printed, embossed, tufted, engraved, braided, embroidered, pressed. I began building up my network, I got a small grant to work on natural coloring from a passionate French expert, we patiently tried to understand the affinities and dead-ends. The challenge finally not being the coloration itself but the preparation and decoloration of the material without damages. Grafting these two lineages, marine and terrestrial through colors carries something very strong and opens exciting unvisited paths. On seaweed we applied indigo, cochenille, campeche, garance, camomilla, gardenia, scutellaire, rhubarb, sophora, cachou, tara. sumac, iron sulfate, titanium ... etc.

Vegetal dyed Laminarias© Violaine Buet, 2016. *Photo credit: Pierre-Yves Dinasquet.*

The main issue being about the stabilization, the conservation of a soft and flexible surface. Very soon I distinguished two potential paths, which are finally complementary and could correspond to two different perspectives or phases;

- A light treatment enabling the soften seaweed to be used like a biodegradable skin, keeping a certain fragility, still being subject to its reversibility under water and potentiality to dry and harden. I naturally investigated this path, because of the simplicity of means required, and the fascinating question of impermanence behind: are we ready to possess disappearing objects and materials?

Exploring it further inevitably leads to question another angle:

- A deeper chemical bonding, overcoming the reaction to water and opening the field of seaweed as a predictable surface and new bio-material: the main challenge being the control of the biodegradability factor too. To cross this barrier one needs a proper framework, a multidisciplinary team, funds and dedicated global thinking; we all know the human ability to squander natural resources (let seaweed be our guide toward regenerative models!).

These two options reveal two rhythms: the production of unique pieces to potential series. At the beginning I felt torn between my artistic talents and my industrial design background, between my responsibility toward the resource, and the desire to convey seaweed in our daily life. People discovering my work sometime worry about the final applications of my research. I cannot design objects just for reassurance. I am first deploying the milieu of a story: seaweed as textile. I manufacture unique pieces, mainly for exhibition and performances. I blend an ancestral vegetal with an ancestral aptitude -weaving- and their combination brings novelty. I associate myself in drawing the identity of this "primitive innovation" as I believe it could open unvisited paths and models for the future. This maturating anchoring, slow and at the closest to the material, is the best legitimacy to move forward to other rhythms of development, for which I am developing international collaborations.

Material exploration© Violaine Buet, 2017. *Photo credit: violaine Buet.*

Juggling between craft and industrial environment (I work on prototypes with a few companies from the textile world) orients the possibilities of mixing tools to adapt to the singularities of this raw material. If process gives birth to new materials, new materials generate new processes too. "*We note the emergence of a form of inter-technical hybridization which has its source in traditional artisanal methods, with production stages defined by the logic of industry and process innovations both from precision industries, in particular the chemical industry, or from the more classic ones*" (Trapp, 2014). Algae lend themselves wonderfully to these hybrid approaches, as they call for emotion as much as performance. I am dreaming about new alliances, between the complexness from nature and the high tech from the human. Novelty arises from this; "*our capacity to inhabit several worlds*"[10].

2.4 Naming (Fourth ring)

Three years have now passed since I decided to invest in my studio of creative research and development around seaweed in Brittany. From pieces of laminaria japonica pieces bought at Frère Tang, when I was a student in Paris, to a budding, innovative expertise, patiently witnessing a singular encounter with the Living. I am moving forward in a fragile but persistent equilibrium, trying to harmonize a system, in the same way I require a balanced attention toward the heart, the body, the mind. While writing this chapter, within the same timeframe, I am weaving a one meter piece ordered by a museum (picture 14), I see both of them getting "written" like siblings, their manual and cerebral solicitations supporting each other.

Against the discomfort of such a professional bet (" Mama, why did you choose this job?") I constantly keep in mind the gratification of handling a complete journey, from harvests in any weather, to informal exploration with researchers, from improbable tests with yarn-makers, to dialogs with encouraging clients, from studying weaving to delivering custom-made pieces.

[10] Horvilleur, D. Les chemins de la philosophie. France culture, July 12th, 2017. 55 min.

Finalized piece, woven while writing this chapter. Collection of the Philadelphia Museum of Art, 100 cm × 70 cm ©Violaine Buet, 2019. *Photo credit: Morgane Le Guen.*

Over the three years I had followed interactions centered on my activity with: seaweed farmers, experts in natural dye, biologists, biopolymer and micro-algae researchers, seaweed harvesters, seaweed trainers, textile designers, food designers, textile engineers, fashion designers, sustainable fashion experts, design students, graphic designers, illustrators, weavers, basket weavers, glass-blowers, leather-workers, artistic directors, architects, food-engineers, decorators, naval architects, jewelers, tailors, poets, dancers, actors, costume-makers, fab-lab directors, international development advisors, marketing directors, journalists, film makers, editorialists, trend-makers, curators, artists, business innovators, business incubators, biotechnologists, innovation and prospective departments, circular economy consultants, geographers, landscapers, philosophers, hat-makers, watch-fabricants, yarn-makers, pleats-makers, carpenters, photographers, skippers, apneists …

The word *seaweed design* embraces a wide angle and my colleague Julia Lohmann, German artist and designer is its esteemed pioneer. The French name of my activity has not bloomed yet, but its living milieu is maturing at the confluence of an ancestral novelty in its aesthetic and direction. "*We are the new ancient*" sings the rapper Aloïse Sauvage, and so is the identity of seaweed. Coming from afar, taking us way ahead; partnering with a biodegradable material confronts us with the question of decomposition and impermanence. We inherently would like to conserve, possess, secure. I feel myself very tempted to see the development of a lasting seaweed skin become a reality but I believe the other path of a disappearing material could finally take us much further than a predictable option. It pushes us to think differently, to envisage impermanence like fluidity breathing regeneration into our exhausted production models.

Hat prototype, kelp and linen textile samples© Violaine Buet, 2016. *Photo credit: Pierre-Yves Dinasquet.*

3. Opening new worlds

Seaweed is gathering more and more interest and attention but is still unfamiliar, our bond with the marine vegetal is getting written, what kind of amplitude are we aiming at? Could we seize the opportunity of this brilliant living partner, which has the power to guide us toward new models, emerging from a global attention: what we learn from the material itself, from us, from the milieu they compose together.

3.1 The department of seaweed

In June 2017, after a year of email exchanges about our respective approaches and practices, I finally met Julia Lohmann. Under her impulse we founded the Department of seaweed[11]: a collaborative platform for the exploration of seaweed as a sustainable resource. We are 9, from different disciplines, maker, thinker, technician, researcher, spread around the globe. We joined last year the mentorship program of the Submariner Network in Denmark. We are connected by this idea that working together with the constraints and asset of seaweed in their milieu could provoke the emergence of regenerative proposals.

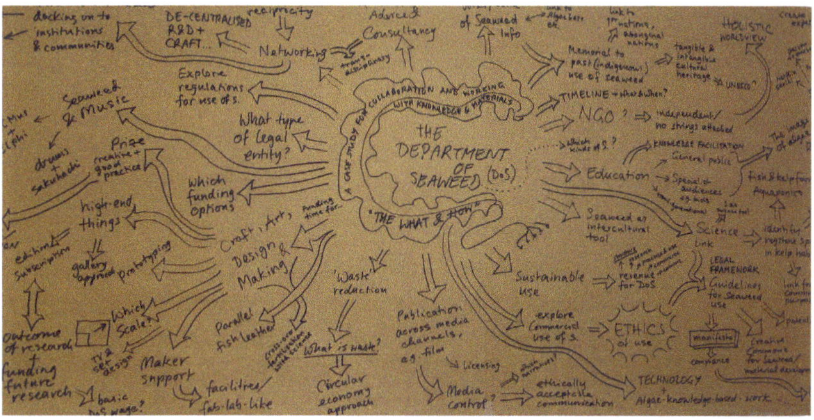

Mapping the Department of Seaweed, Hamburg, 2017. *photocredit department of seaweed.*

3.2 Singularities

Whilst harvesting seaweed, we are constantly astonished by the diversity of species present. It takes intense training to patiently discover where to find certain types and how to identify them. Throughout the divers geography of Brittany, there are particularities in the algal landscape. To us, the Department of Seaweed, this biodiversity is a pointer: each seaweed, brown, red or green, giant or minuscule can be the start to the development of a potential material. Depending on their attributes, they call for different kinds of directions; a workshop, a space, a methodology, a service, a product (in

[11] www.department-of-seaweed.com.

the vast categories of food, material, construction, textile, furniture, clothing, stationery, lighting, accessories ...). Instead of imagining a mass-market seaweed product that could be sold anywhere in the world, we aim to work on singular systems joining together to develop a resource (a chosen algae), with local know-how or industries, and local needs and their particular issues. Using this approach, taking into account the vegetal-human partnership in its eco-techno-symbolic environment, we reinforce the value of the process and the guidance of a context in the elaboration of a solution, imprinted by its milieu. As a trans-disciplinary, international, consultative group, we wish to create well-balanced teams, based on individual specifications both on and off field.

3.3 From Julia Lohmann

In 2007 I was an artist in residence in Hokkaido, Japan. Hokkaido is the northernmost island of Japan and is famed for its 'kombu', large brown kelp that has traditionally been eaten and is one of the most important flavors in the Japanese cuisine. When I discovered the large sheets of seaweed on the fish market, I immediately asked my host what they are making from it – fully expecting him to list numerous things, from book covers to shoes, furniture, clothing etc. It turned out that the seaweed is only eaten, and I realized that the things I imagined were either fantasy or future – if I took on the task of developing seaweed as a material for making. I took some seaweed back to the UK, where I lived in 2007 and got to work. I quickly realized that the potential of seaweed as a material for making was immense, and that it required time and dedication. I became a Stanley Picker Research Fellow in Kingston and developed the material further.

When the fellowship ended, I realized my work was far from done and concluded two things: I applied for a practice based PhD at the Royal College of Art and the Victoria and Albert Museum to further develop my work with seaweed in a research context. I understood that the scope of potential afforded and called for a new way of working and sharing knowledge, a kind of creative common of material development that required me to not only work on a material and object level but also to enable a Community of Practice to collectively develop this new field of knowledge and link it with existing crafts and technologies.

Imagining the Department of Seaweed, V&A's residency, London, 2009. *Photo credit: Petr Krejci.*

3.3.1 Seaweed is both our material and our method of engagement

When people initially encounter the Department of Seaweed (DoS), they see that we are developing methods for working with seaweed. We work with fronds of macro algae like kelp, maintaining the integrity of the kelp blades instead of processing them into powders, liquids, etc. However, the scope of the DoS, the community of practice we founded to explore the use of algae as a sustainable resource for making, is much wider than that. We want to engage in situations in which algae is a problem or can be part of the solution toward achieving balance in marine ecological-socio-technic systems.

Design Theoretician Richard Buchanan describes an expanded field of design that goes beyond 'posters and toasters' — the original domain of design that dealt with signs, symbols and objects — and expands to the design of services, interactions and, finally systems. In our view, this expansion of the field of design did not only affect the macro-scale toward systems, but also happened in the micro scale, with the design of materials, micro-structures, molecule chains and even DNA. The work of the Department of Seaweed requires engagement on all of the levels, and this in turn can only be achieved with a trans-disciplinary community. In it, we combine the knowledge of the life sciences and social sciences with the methods of design to engage a wide range of people who might not yet know that the topic matters to them.

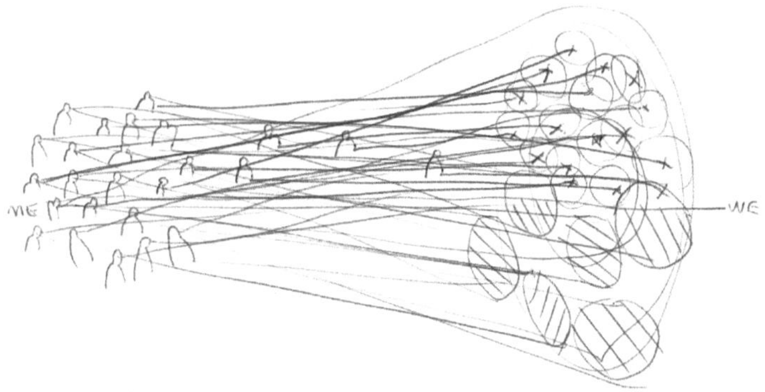

Co-speculative cone, sketch Julia Lohmann.

3.3.2 *Sharing knowledge*

There is so much to discover around the uses of kelp – and we have so little time to find sustainable ways of living on this planet – that it is futile if each and every one of us starts to look by themselves, starts from step one. We have to share with those that share our value, that share our aspiration toward sustainability and respect for the eco-systems we are impacting with our actions. In the Department of Seaweed we advocate an eco-centric (as opposed to human-centered), trans-disciplinary, empathic approach to design that understands the use of a resource as a chance to rebalance some of the imbalances we inflicted upon the oceans.

Design is the way of building bridges between knowledge from different disciplines and the wider society. It is a method of making scientific research actionable and enacted. Designers can help build communities of practice around issues, become the do-tanks that test the knowledge created in the think-tank researching sustainable ways into the future.

Seaweed teaches me to share, to go beyond the individual and strive for a collective endeavor. It is showing me how it is possible to be both plentiful and benign – harmless and helpful to the organisms that seaweed co-habits with. Brown algae is an eco-system builder, it is a mother. (...) Seaweed is like a partner in this research. It 'affords' – encourages, enables – certain actions, shapes, forms, applications and codes of conduct. As a material it is highly reactive, changing rapidly with the moisture content in the air and making visible atmospheric changes. In Victorian time Britain, seaweed was used as a natural barometer; hung outside seaside dwellings it would inform inhabitants about changes in the weather by the way it hung and swung. Its responsiveness makes seaweed an interesting agent in the process. This notion of 'listening' to the material, being attentive to its' responses, being 'in correspondence with' (Ingold, 2013) the material is the essence of craft practices.

Resilient, strong and rebalancing – kelp is not just our material but our metaphor and method to find ways toward a more harmonious co-existence with other inhabitants on the planet we call ours.

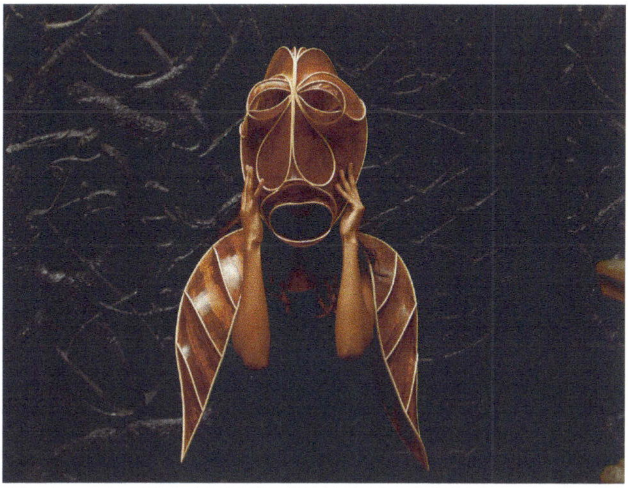

Mask, kelp and rattan© Julia Lohmann 2009. *Photo credit: Petr Krejci.*

Conclusion

The fast-growing international exposure of seaweed emphasizes the importance of local grounding -for us here, in Brittany, a unique seaweed spot in Europe. We are a group of complementary local actors in the field of algae but disciplines are profoundly partitioned in our system, we will have to shake mindsets to provoke porous, recognized frameworks. In the interstitial interaction offered by informal collaboration with researchers, we clearly see the power of blending the creative and scientific ways of thinking. Trans-disciplinarity is a potential catalyst for reconnecting with mutual narrative. Could the attention to singularities re-enchant our collective story. "*The answer beckons from the worlds when open; one of a researcher, the other of an artist, the finest outcome for them both: to become one world, to also welcome others, in the great trajective chain of history*".[12]

[12] Augustin Berque, Email from the 12th of July 2019.

Annexes

Glossary of mesological terms

ACOSMY (ACOSMIE) n Lack of cosmicity*, resulting from the decosmization* provoked by the MCWP*: *in architecture and city planning, acosmy manifests itself as junkspace and landscape killing* (shafengjing 殺風景)

CO-AROUSAL (CO-SUSCITATION) n (transl. of the Chinese *yuanqi* 縁起, itself a transl. of the sanskrit *pratitya samutpāda*, conditioned coproduction, interdependence) Syn. of concrescence*: *implying ternarity*, the co-arousal of things* is more complex than the mechanical cause of objects**

DECOSMIZATION (DÉCOSMISATION) n Loss of the landmarks which made the beings* and things* of a milieu * go together. Loss of cosmicity*. Akin to Heidegger's *Entweltlichung* (deworldization)

ECUMENE (ÉCOUMÈNE) n (from the Greek ἡ οἰκουμένη, the inhabited one) The human abode*, the whole of human milieus*. Eco-techno-symbolic relationship of Humankind with the Earth. NB: this mesological definition is distinct from the usual geographical one, i.e. that part of the Earth which is inhabited, *renjing* 人境 in classical Chinese (as opposed to *xianjing* 仙境, the wilderness*). The ecumene, in the present sense, is not limited to the *topos* of planet Sol III: it is the *chora** comprising all that which exists* for Humankind

EXCLUDED MIDDLE (TIERS EXCLU) n Theoretical impossibility, in the MCWP*, to be between A and non-A, i.e. neither A nor non-A (third lemma*), or to be both A and non-A (fourth lemma*); hence the practical impossibility to take milieus* into account. The law of excluded middle is no more than a logosic* abstraction*, since in a concrete* milieu, the reality*of things* necessarily belongs to the tetralemma*: *in the* Timaeus, *Plato's rationalism excluded thinking the « third and other gender » (τρίτον ἄλλο γένος) of the* chora* *as a matrix-imprint*, left therefore in the realm of dream*

MEDIANCE (MÉDIANCE) n Transl. of the Japanese *fūdosei* 風土性, which Watsuji defined as "the structural moment of human existence" (*ningen sonzai no kôzô keiki* 人間存在の構造契機), that is the dynamic coupling of Being* and its milieu*; corresponds to what Uexküll called *Gegengefüge* (counter-fitting), i.e. the reciprocal adequacy of the animal (or more generally the living) and its milieu

TRAJECTION (TRAJECTION) n 1. To-and-fro motion of reality* between the two theoretical poles of the subjective and the objective: *reality belongs neither to the sole realm of the object*, nor to the sole realm of the subject*; belonging to the trajection of both terms, it is trajective.* 2. Assumption of S* as* P*, syn. of ek-sistence*: *in the IVth century, in China, occurred the trajection of mountain waters (*shan shui 山水*) as landscape* (*shanshui 山水*)*

TRAJECTIVE CHAIN (CHAÎNE TRAJECTIVE) n Sequence of trajections*, progressively hypostatizing (substantializing) S/P* (thus hypostatizing P at the same time) into S′, S′/P′ into S″, S″/P″ into S‴, and so on. Represented with the formula (((S/P)/S′)/S″)/S‴… etc.: *trajective chains are analogous to Barthes' semiologic chains and Peirce's semiosis*

***URSPRUNG* (German n.) v** Primal spurt of a reality, owing to the work of art: *the invention of perspective symbolized the primal spurt of the modern subject, objectifying the world under its gaze and giving birth to dualism*

Acknowledgments

Thanks to Augustin Berque for its english translation from; Berque (2018a, p. 15, 2018b, p. 15, 2018c, p. 16, 2018d, p. 43, 2018e, p. 24) Glossaire de mésologie. Bastia: Eolienne. My warmest thanks also to Louise Assad, Bertrand Douhard and Julia Lohmann

References

Aragon, L. (1956). *Il n'aurait fallu, Le roman inachevé, poèmes*. Paris: Gallimard.

Berque, A. (2013). *Les conference publiques: Humaniser la nature, naturaliser l'humain, aujourd'hui*. Institut d'études avancées de Nantes.

Berque, A. (2017). *Sur les bords de l'Yvette, dialogues mésologiques*. Bastia: Eoliennes.

Berque, A. (2018a). *Glossaire de mésologie, definition of decosmization*. Bastia: Eoliennes.

Berque, A. (2018b). *Glossaire de mésologie, definition of as*. Bastia: Eoliennes.

Berque, A. (2018c). *Glossaire de mésologie, definition of ecumene*. Bastia: Eoliennse.

Berque, A. (2018d). *Glossaire de mésologie, definition of ursprung*. Bastia: Eoliennes.

Berque, A. (2018e). *Glossaire de mésologie, definitions of mediance*. Bastia: Eoliennes.

Berque, A. (2019a). *Descendre des étoiles, monter de la terre, La trajection de l'architecture*. Bastia: Eoliennes.

Berque, A. (2019b). *Descendre des étoiles, monter de la terre, La trajection de l'architecture*. Bastia: Eoliennes.

Berque, A. (2019c). *Descendre des étoiles, monter de la terre, La trajection de l'architecture*. Bastia: Eoliennes.

Berque, A. (2019d). *Descendre des étoiles, monter de la terre, La trajection en architecture*. Bastia: Eoliennes.

Chapman, J. (2005). Emotionally durable design: Objects. In *Experiences & empathy*. London: Earthscan.

Koolhaas, R. (2002). Junkspace. October. In *Obsolescence* (Vol. 100).

Latouche, S. (2012). *L'âge des limites*. Paris: Mille et une nuits/Fayard.

Reeves, H. (2017). *Le banc du temps qui passe*. Paris; Seuil: Méditations cosmiques.

Servigne, P., & Chapelle, G. (2019). *L'entraide, l'autre loi de la jungle*. Paris: Les liens qui libèrent.

Trapp, J. P. (2014). *How can the manufacturing process of luxury companies integrate a new form of sustainable craftsmanship?*. Master's thesis.

Kolibri 360 GmbH
Ferdinand-Jühlke-Straße 7
99095 Erfurt, Deutschland
produktsicherheit@kolibri360.de